THE ROSE BOOK

Phaidon Press Limited
2 Cooperage Yard
London E15 2QR

Phaidon Press Inc.
111 Broadway
New York, NY 10006

phaidon.com

First published 2025

ISBN 978 1 83866 880 8

A CIP catalogue record for this book is available from the British Library and the Library of Congress.

Commissioning Editor: Victoria Clarke
Project Editor: Olivia Clark
Production Controller: Gif Jittiwutikarn
Cover Design: Julia Hasting
Layout Design: João Mota
Typesetting: Cantina

Printed in China

INTRODUCTION 4
ROSES IN FASHION 6
THE SCENT OF THE ROSE 8

ROSE: THE WORKS 10

THE LANGUAGE OF THE ROSE 238
GLOSSARY OF ROSE TYPES 239
INDEX 241
ROSE GARDENS AND FESTIVALS 246
FURTHER READING 247

"OF ALL FLOWERS, METHINKS THE ROSE IS BEST": OUR ENDURING LOVE FOR ROSES

Kristine Paulus

Familiar as they may be, roses are perhaps more spellbinding than any other flower. Their everlasting allure is evident in the way they have permeated nearly every aspect of human life, from medicine to botany, politics, and fashion. Roses are so adored that they represent love itself—is it any coincidence that the word "rose" is an anagram of Eros? Steeped in mythology, they are the subject of fairy tales and folklore. There are so many idioms, adages, and proverbs referring to roses and their long association with romance and glamour that they border on the cliché, yet poets, painters, and philosophers cannot get enough of them. With incredible diversity of form and color, there is a rose for every person and every purpose.

Originating more than 30 million years ago, the genus *Rosa* contains about 150 accepted species, but—thanks to its endless ability to hybridize—there are tens of thousands of cultivars and varieties now in existence. Many of these have been bred since the mid-nineteenth century, although rose cultivation can be traced back to the fourth century BCE in China. The blossom that Greek poet Sappho crowned "the queen of flowers," and of which Shakespeare wrote "of all flowers, methinks the rose is best," remains one of the best-selling cut flowers among florists, with more than 250 million stems sold on Valentine's Day each year alone. The quintessential garden plant, the rose is grown by untold numbers of gardeners around the world, and it is a rare garden that is without them. What is this enchanting plant, then, and where did it come from? Botanical classification of the varied members of the genus *Rosa* is complicated, and even botanists have not reached a consensus on the exact number of species.

Roses are the namesake and most important genus of the family Rosaceae, and their relatives include many edible fruit, such as apples, pears, peaches, cherries, and strawberries. The familiar domesticated garden roses we know today have been bred to display a diverse array of colors, forms, and fragrances. Now incredibly varied, these cultivated types derived from just a few wild species, but their basic anatomy shares some common traits, such as five-petaled flowers (although their contemporary descendants are often bred to have numerous extra petals, sometimes to the disappointment of their pollinators, the bees); rose hips for fruit; oval, serrated leaves; and sharp thorns (the proper name for which is actually prickles).

The rose's anatomy is so archetypal that artist Christopher Dresser's simple yet powerful cross-section diagram of a rose flower from the late nineteenth century still teaches the plant parts to students today (see p.184). Beyond horticulture, roses are present in nearly every part of culture. We love them so much that we eat and drink them (see p.142); we perfume our bodies with them (with a splash of La Rose de Rosine, perhaps? see p.129) and ink them permanently onto our flesh; and we decorate our homes with them and sleep on beds of them (at least, some Romans may have done so).

The annals of the rose are complex, yet much has been lost to time or mired in myth. Roses christened before 1735, when Carl Linnaeus began to standardize plant-naming conventions, also cause confusion, because a rose by any other name . . . might just be a different plant. The rose's colloquial names, however, are nearly identical in dozens of languages. It is impossible to talk about the history of the rose without also considering its many uses, its symbolism, and its representation in art, literature, and fashion. A particularly important icon in the art of floriography—the early nineteenth-century practice of assigning meanings to flowers (see p.112)—the rose has been used to communicate everything from expressions of affection in Victorian calling cards (see p.121) to Cuban liberation (see p.92).

The ancient Greeks loved roses, transforming them into golden wreaths worn at ceremonial events (see p.209). Some even say they named their historical capital Rhodes for these blooms. In classical mythology, the rose sprang from the ground to greet Venus, Roman goddess of love (see p.55), as she emerged from the sea, with the foam turning into white roses and her blood to red roses. But archaeological evidence shows that the rose is far older than antiquity. At least 35 million years ago, a primitive rose left an imprint of its flower, leaf, and stem in mud during the late Eocene epoch (between 56 million and 33.9 million years ago), to be immortalized in fossil rocks at Florissant Fossil Beds National Monument in Colorado. Around the same time, wild rose species appeared all over temperate regions of the northern hemisphere. Their exact origins are debated, but they probably originated in northeastern Asia and have since evolved in different ways, although they are not found growing naturally south of the equator.

About half of all naturally occurring rose species are native to China, and they probably were first cultivated there and in Persia about 3000 BCE. Celebrated Chinese teacher and philosopher Confucius described roses growing in the imperial gardens of the Zhou dynasty about 2,500 years ago—in one of the earliest rosariums, or rose gardens—and Chinese art from all periods is filled with roses (see pp.35 and 97). Some of the oldest pictorial evidence of roses, which also constitutes some of the earliest records of the flower's cultivation in Europe, is found in frescoes from the first century BCE depicting a garden at the Villa of Livia in Rome (see p.208). Roman naturalist Pliny the Elder frequently wrote about roses, particularly their medicinal uses, in the first century CE as did his contemporary Pedanius Dioscorides in his treatise *De Materia Medica* (see p.122).

Romans even created the first rose festival, Rosalia, and petals were often used as fragrant, decorative, and sometimes even deadly (see p.16) flourishes at ceremonies and banquets. In ancient Egypt, roses were associated with the afterlife. Tombs excavated at Hawara, dating from the second century CE, revealed a sarcophagus with remnants from a funerary wreath composed of rose petals. Apart from fossils, this is the earliest known rose in existence.

Although the rose is considered a symbol of the prophet Muhammad in Islamic mysticism—often acting as an important visual motif throughout the Koran (see pp.44 and 172)—they were not notably popular among early Christians until the early medieval period in Europe, when the rose began to symbolize Christ, martyrdom, and especially the Virgin Mary.

An icon of the purity of faith, Mary is often depicted emerging from the white rose of Heaven or within a garden, surrounded by roses (see pp.91 and 130). Roses were even incorporated into the architecture of cathedrals, such as the famous stained-glass rose windows at the cathedral of Notre-Dame in Paris (see p.131). The plant was not commonly cultivated in England until the Middle Ages, when roses—especially *Rosa gallica*, which was known as the apothecary's rose—were grown as medicinal plants in the gardens of religious institutions. The religious rose garden was also transformed into secular romantic allegory, as evidenced by the illustrations for the thirteenth-century French poem *Roman de la Rose*, which Geoffrey Chaucer, purported inventor of Valentine's Day, translated into English (see p.206).

Interest in horticulture in general, and roses in particular, thrived during the Renaissance and the Enlightenment. This was also the era of great herbals, such as John Parkinson's *Theatrum botanicum* (1640) and John Gerard's *The Herball* (1597), which claimed that "the rose doth deserve the chiefest and most principall place among all floures whatsoever" (see pp.76 and 180). The rose even became a political icon during the fifteenth-century dynastic struggle for the British crown known as the Wars of the Roses, which symbolically ended when the red rose of Lancaster and the white rose of York were combined to create the Tudor rose. That symbol was used in portraits of Queen Elizabeth I and well beyond the Renaissance to signify one of England's greatest dynasties (see pp.33 and 164). France has long been a leader in rose cultivation, but interest in the flower did not begin there until the eighteenth century, when such figures as King Louis XV, his mistress Madame de Pompadour, and Queen Marie Antoinette championed roses not only for use in gardens, but also as elements in elegant fashions and fragrances (see pp.36 and 124). Fields of highly fragranced Centifolia roses were established at Grasse in southern France for the specific purpose of manufacturing rose-based scents, and Grasse remains the epicenter of the fragrance industry today (see pp.47 and 143).

Likewise, Empress Joséphine, wife of Napoleon Bonaparte, is credited with the rose's surge in popularity during the early nineteenth century. Forever changing the course of its history, she was responsible for creating what many consider to be the first garden devoted entirely to the plant at Château de Malmaison, introducing hundreds of new varieties to France. In the early nineteenth century, she hired Pierre-Joseph Redouté to produce a florilegium, or book of flower portraits, of roses—some of which came from the 170 different varieties growing in her garden—called *Les Roses* (see p.13). Redouté would be known ever after as the painter of roses (although he had competition from other notable artists, among them Henri Fantin Latour, Pierre-Auguste Renoir, and Claude Monet; see pp.75, 155, and 169).

Up to this point, cultivated roses had not strayed too far from the origin of their wild ancestors. Native, or species, roses are still those that are found growing in nature and usually bloom just once a year, in early summer. They include such varieties as *Rosa canina* (dog rose, the most common European wild rose species for much of history), *R. rugosa* (beach rose), *R. moschata* (Musk rose), *R. palustris* (swamp rose), and more.

The native roses of Europe and North America generally have a limited color palette, whereas wild Asian roses include the only naturally yellow roses and most have trailing or climbing habits. In the eighteenth century, botanical expeditions to China brought back the China rose, *R. chinensis* (see p.35), which is noteworthy for its ability to flower repeatedly over a growing season. The influence of Chinese roses revolutionized Europe's gardens and changed the trajectory of the garden flower that is known today (see p.84). Although its true lineage remains a mystery, the cultivar 'La France', a Hybrid Tea rose with Chinese rose parentage, was introduced in 1867 by plant breeders who wanted to create European blooms that could repeatedly flower.

From one rose, a new era dawned, and now roses are categorized as either Old Garden (also called antique or heritage) roses, which are those that existed before 1867, or modern roses, those bred after that watershed year. Heritage roses—the Gallica, Alba, Damask, Centifolia, and Moss roses—are known for their deep fragrance and large blooms, while modern rose varieties—Hybrid Tea, Polyantha, Floribunda, Grandiflora, and Hybrid Persicas—are often bred for disease resistance or hardiness, or to exhibit a rainbow of hues. According to the American Rose Society, there are more than forty different contemporary classes of rose (including Bourbons, Portlands, Noisettes, and Hybrid Perpetuals), and tens of thousands of varieties available.

The rose's tendency to mutate is one cause of so much variety, holding an assortment of habits, such as climbing, rambling, and miniature, and adapting to a range of climates. Small mutations, or sports (genetic modifications that may be passed on to the next generation), occur regularly within the genus *Rosa*, bringing a multitude of breeding possibilities. These changes might result in new colors or habit, or a change in the number of petals (the green rose, *R. odorata* 'Viridiflora', is one such interesting mutation; see p.93).

The Plant Patent Act of 1930 in the United States marked the beginning of a new period of horticultural production. Large commercial nurseries began to operate extensive mass-hybridization programs and account for most of the registered plants, while several significant gardens that are still in existence today were opened to the public and modern rose festivals were established. The ebullient rose parades in Pasadena, California, and Tyler, Texas, continue to be sources of joy and revelry centered on the rose (see pp.187 and 211).

Roses remain esteemed as a muse for modern times. While it is no surprise that they are celebrated in the flower-centric artworks of Robert Mapplethorpe and Georgia O'Keeffe (see pp.146 and 147), roses have also been realized in surprising ways in contemporary art: in yellow neon by Welsh sculptor Cerith Wyn Evans (see p.18) or vibrant collages by American artists Romare Bearden and Shinique Smith (see pp.205 and 226). The rose has also played a role in popular culture, from the films of Alfred Hitchcock (see p.101) to *American Beauty* (1999), which takes its name from the cultivar that was long the most popular rose in the United States. The film's so-called rose-petal scene remains a famous moment in cinematic history. Perhaps one of the most evocative uses of the rose in music is as the symbol for the ultimate jam band, the Grateful Dead. Their psychedelic album artwork and posters featuring a skeleton with a garland of roses continue to represent the peace, love, and ephemerality that the band's music evokes (see p.34).

The new wave of rose infatuation probably percolated in the mid-twentieth century, with new attitudes toward rose breeding and cultivation. Inspired by Old Garden roses, famous English rose breeder David Austin (see p.105) introduced his first commercially available rose in 1961: 'Constance Spry', named after the innovative floral designer (see p.113). His aim was to create a robust modern rose with the charm and fragrance of antique roses. With their wide color ranges, abundant petals, fragrance, and ability to blossom from early spring to late fall, these English roses (also known as David Austin Roses) represent the future of cultivation as rosarians continue their attempts to tame an ancient wild species (see pp.14 and 77). Today, almost infinite new varieties are being developed by hybridizers in the constant quest for the perfect rose. At the same time, gardeners and consumers around the world seek plants that support sustainability and a harmonious coexistence. It is clear that the demand for roses is here to stay, whether it be for roses in garden beds and as cut flowers for home bouquets or the not-so-secret ingredient in Chanel No. 5 and inspirational fashions, patterns, and jewelry designed by the likes of Christian Dior, Victoire de Castellane, Alexander McQueen, Thom Browne, and many more (see pp.37, 189, 219, and 227).

The collection of images presented in *The Rose Book* constitutes an insightful survey illustrating the symbiotic relationship between humans and this beloved flower in art, botany, popular culture, and history. Our collective worship at the altar of the rose is evident in the temples we have erected to it: some of the most beautiful rose gardens in the world, including the Parc de Bagatelle (see p.41), Roseraie de L'Haÿ (see p.196), Peggy Rockefeller Rose Garden at the New York Botanical Garden (see p.152), and Vita Sackville-West's White Garden at Sissinghurst Castle (see p.110). The world's first rose museum, housed in a stunning building designed by NEXT Architects in Beijing, may be the most ambitious shrine built yet (see p.198). We still confer heroes and winners with orders and awards signified with roses (be it a military honor or a Kentucky Derby winner; see p.204), and we proclaim our rosy exaltation. Our abiding admiration for this favored flower continues to be expressed by artists, authors, gardeners, songwriters, photographers, perfumers, filmmakers, and seemingly anyone who has met a rose. Their work is wonderfully represented in the pages that follow.

ROSES IN FASHION

Amy de la Haye

Roses, like fashion, are decadent and evanescent. Appealing to all the senses, both can be obsessively collected and displayed—in gardens and on the runway—and the tension between the exquisite delicacy of the flower's petals and its sharp, prickly stems has long inspired designers, writers, and artists. For fashion creatives, the rose is compelling for its sheer beauty and transience. But it is significant also for the breadth of stories it can tell that intersect with trade, travel, colonialism, political ideologies, sexuality, gender identities, shifting notions of beauty, and the cycle of life.

Like the genus itself, the influence of the rose upon dressed appearances is enduring and pervasive. Since at least ancient Egyptian times, humans have decorated their bodies with fresh roses and applied the blooms' charred petals to fashionably darken their brows. The flower has been used to beautify and adorn over the centuries, including inspiring tattoo designs that are permanently drawn, using colored inks, onto human flesh. Although gendered feminine for much of history, particularly in the nineteenth century, roses were not always associated with women, and in antiquity it was men who wore rose garlands and scented themselves with rose perfume (see pp.46 and 209). A few themes become apparent when we attempt to take in the centuries-long presence of the rose in fashion—the rose's natural shape, its ephemerality, its potential to be gendered, feminized, and abstracted—and a brief study of these themes highlights the flower's impact on the ways in which people look and dress.

While it is omnipresent within textile design, there are certain periods and occasions when naturalistic rose expression was especially bountiful. From the 1730s, the colonial trade in roses from China, and the new profusion of botanically correct illustrations of them, inspired European and North American fashion designers to create designs for woven, hand-painted, and embroidered textiles that incorporated the rose as it appears in nature. In Paris, the capital of international fashion, Madame de Pompadour (the mistress of King Louis XV; see p.36) and Queen Marie Antoinette—both rose lovers and style icons—championed the flower beyond the garden, commissioning portraits of themselves wearing rose-ornamented gowns and nosegays, flower clusters worn both for decoration and for perfume. A proliferation of accessories, such as shawls, fans, and the small drawstring bags known as reticules (see p.192), became sites for rosy flourishes. The grandest occasions necessitated engaging skilled stylists to weave the fresh flower and its bud into spectacular, towering pouf headdresses.

Throughout the nineteenth century, fashion's depiction of the rose remained mostly true to life, relying on the flower's inherent soft romance. By the 1850s, the Parisian haute-couture industry was well established, supported by a range of craft workshops, including artificial rose makers who created silk flowers for dress objects (see p.176). The most heavily rose-laden fashions were special-occasion garments for balls and ceremonial events. There was an insatiable international demand for designs by Paris-based haute couturier Charles Frederick Worth, who was instrumental in establishing the modern fashion industry. His adoration of flowers is legendary, and swags of silk roses cascading down evening gowns were his signature (see p.104). At the turn of the twentieth century, Art Nouveau, more than any other design movement, glorified flowers in the decorative arts. French artist-jeweler René Lalique often worked directly from nature, surrounded by flowers, and this is apparent in such pieces as his pendant crafted to look like a bouquet of cascading roses (see p.167).

In the twentieth century, Anglo-American designer Charles James captured perfectly the romantic neo-Victorian vogue with his *La Sylphide* dress, a corseted gown with a décolletage bedecked with silk roses (see p.49). Christian Dior, a skilled gardener whose landscape at his home in Granville was filled with roses, conceived his "flower-woman" silhouette in 1947, which would soon become known as the—now iconic—"New Look." Dior would become the mid-twentieth century's fashion rosarian; a white silk evening gown from 1951, decorated with a garland of pink roses, exemplifies the refined elegance of his floral designs (see p.37). Following his premature death, a rose (a double red Hybrid Tea) was named after him, an honor later also bestowed upon fashion designers Valentino and Vivienne Westwood. American couturier Ann Lowe, the first African American designer to achieve international stature, is also revered for her luxurious, romantic gowns that heavily feature roses, inspired by the rural gardens of her youth in the American South (see p.154).

The rose as an event in nature has continued to inspire creatives, many of whom combine the organic aesthetic with modern technology or artisanal techniques. Zac Posen collaborated with General Electric Additive and Protolabs to design a high-tech 3D-printed red rose garment for fashion's most significant event, the Met Gala, inspired by the roses growing in his own garden (see p.144). Chopard's organic design for a rosebud necklace is intended to drape around the neck in a naturalistic manner (see p.39). Alexander McQueen, modern fashion's most creative, provocative, political, and romantic rosarian, incorporated silk flowers into the decadent, decaying grandeur of his *Sarabande* collection (see p.219), revealing continuities in the use of roses across centuries of elite fashion.

Beyond sartorial expression inspired by the appearance of the flower itself in nature, roses are central to the ceremonial dress worn for rites of passage from birth to death, and in their ephemerality they have stood as symbols for the cycle of life since antiquity. It is here that the various symbolic meanings attributed to the color of the petals were—and, to a lesser extent, still are—most pronounced. Within some cultures, white and pale pink are associated with debutantes and brides; and black (although there are no truly black roses in nature, just deepest purple) with death and mourning.

The genus *Rosa* has such a broad geographic sweep, and is so beloved, that communities around the world stage festivals and parades to celebrate its flowering (see p.187). For many, the ceremonial crowning of a folk-costumed Rose Queen is central (see pp.175 and 211). In everyday life, holding a bouquet or bunch of roses—whether at the altar or on the street—comprises part of a dressed appearance.

Even wearing a simple bud or bloom tucked into a buttonhole or pinned onto a garment or hat marks a day as special. Within many nations, especially those known for growing roses—China, Iran, Turkey, and India—the rose is a popular decorative and symbolic motif for dress textiles. In North America, Indigenous cultures have used the symbol of the rose for generations, connected to its cultural significance and importance as a plant resource. A bold red rose motif set within a white circle, for example, is often present in the intricate beadwork of the Eastern Shoshone tribe from central-west Wyoming applied to garments and ceremonial regalia (see p.203).

The power of transformation is central to roses and fashion, not only to ceremonial dress but also to the politics of gender. Anthropomorphically, the rose has been personified through the ages in the female form. The Hindu deity Vishnu is said to have fashioned his wife Lakshmi from 108 large and 1,008 small red rose petals, and the rose is the floral symbol of both the pagan goddess Venus (see p.55) and the Christian icon of the Virgin Mary (see pp.91 and 130). Women to this day have often been portrayed as *femmes-fleurs,* flower–human hybrids with their heads emerging from open blooms, in fashion advertising or editorial photography (see p.61 and 237).

From the 1840s, in particular, the association of roses with women was solidified through the development of floriography, or the language of flowers (see p.112), which not only ascribed specific meanings to different kinds of flowers, but also cemented the rose as a symbol of gentle femininity and romantic love. This is apparent in French illustrator J. J. Grandville's exquisitely costumed yet satirically depicted female roses in *Les fleurs animées* (*The Flowers Personified,* 1830; see p.151). Exceptionally for the period, Walter Crane, the British illustrator and radical dress reform supporter, depicted a wild rose couple clad in near-matching rose minidresses as male or gender-neutral in his book for children *Flora's Feast* (1889; see p.109). These provide interesting examples of the ways in which the rose, dress, and gender were inextricably entwined during the nineteenth century. A century later, in 1999, Mattel introduced the Rose Barbie (see p.229). Costumed in a gown of pink and red, with roses encircling her sleeves, the doll introduced the flower to a new generation of girls for the next millennium.

Although Gertrude Stein famously pronounced "a rose is a rose is a rose" (see p.222), the rose has not always appeared as itself in the history of fashion. Since the beginning of the twentieth century, it has been stylized, abstracted, and deconstructed for clothes and textiles expressing a series of design movements. From Charles Rennie Mackintosh's Art Nouveau roses—bulbous in form, with sinuous lines—to a modernistic Cubist design by the Viennese Wiener Werkstätte (see pp.38 and 218), textile designs have played with the shape, form, and color of the flower out of the garden. In high fashion, Elsa Schiaparelli famously combined rounded Art Deco roses with an illusory Surrealist graphic (see p.71), a motif revived by the House of Schiaparelli's current creative director, Daniel Roseberry, in the 2020s. Milliners have also played with the shape of the rose, from the Pop-infused headpiece designed by Sally Victor that turns the wearer's head into an upside-down rose in full bloom (see p.150) to a stylized red satin spiral rose perched on top of a miniature black top hat designed by Stephen Jones (see p.158).

Fashion and dress are among the most visible, immediate, expressive, and accessible forms of human creativity and experience. The rose—above all other flowers—has been twinned with dressed appearances for centuries to convey beauty, longing, and ephemerality. Whether connected to its presence in nature or abstracted out of its shape, personified as a *femme-fleur* or worn by a queen, the rose can be found in every era of fashion history. Contemporary designers—including Thom Browne (see p.227), Simone Rocha, Erdem, and Sarah Burton—have returned to the naturalistic shape of the rose in their diverse fashions for all kinds of sartorial, gendered, or ceremonial expressions, crafting leather bags in the shape of cut flowers or filling sheer dresses with rosettes until they are nearly bursting. Now more than ever, the rose in all its forms provides fashion creatives with fresh inspiration to adorn. It truly seems as though the rose will never go out of fashion.

THE SCENT OF THE ROSE

Victoria Gaiger

The scent of few flowers fascinates us so completely as that of the rose. Since antiquity, roses have been used medicinally and nutritionally, as inspiration for fashion and dress, and as a source of perfume. When highly fragranced, a rose's scented petals captivate, while its thorns repel and attract, demanding that gardeners and perfumers alike take their chances and bury their noses in the flower's pungent promise. The fragrance of an aromatic rose rarely disappoints, which a dawn visitor to the peppered pink fields that surround Grasse in southern France soon discovers. As the early mist disappears, a dewy scent lies on the thermals; tantalizingly familiar, yet fleeting, it evaporates as quickly as it arrives. For a few short weeks in early summer, the roses that grow on small farms between the Mediterranean coast and the western end of the Alps are harvested by hand. To watch the flower pickers at work is to witness a ritual that is as old as fragrance itself.

There is something almost sacred about the rose in perfumery. It has crisscrossed the centuries, rambling its way into the heart of human rituals, anointing, scenting, and cleansing as it goes. The Greeks wove it into their myths and scented themselves with it; the Romans bathed in it and used it to fragrance everything from their houses to their banquets (see p.16). Cleopatra was said to have scented the sails of her ships with roses, and the floors of Babylonian palaces were strewn with rose petals.

Some of the earliest extracts of rose date back to Persia in about 1000 BCE; they were called *attar,* the Persian word for "scent." These early extracts were made by steeping the fragrant petals in oil or fat, a method called *enfleurage,* but it was the discovery in the ninth century CE of distillation by steam that transformed perfumery. Distillation allowed for the use of alcohol, instead of oil, as the perfume carrier, and had the added benefit of producing the prized rose water that enabled the Persians to dominate the early perfume industry. Roses and their scented products traveled from the East through the trade routes and the Crusades. In medieval Europe, rose water was used to ward off plague, scent clothes, and sanitize the body. The early eighteenth-century custom of scenting gloves to take away the smell of animal hide in France turned the southern village of Grasse into a new hub for fragrance and brought about the rise of France, in general, as the home of modern perfumery (see p.143).

Later in the century, two important women in France—Madame de Pompadour (see p.36), the mistress of Louis XV, and Queen Marie Antoinette—were credited with the increased popularity of both the aesthetic and the fragrance of the rose. The latter's favorite fragrance was concocted by royal perfumer Pierre-François Lubin and inspired by the gardens of the palace of Versailles, including its roses. Later, Christian Dior (see p.37) was so inspired by Grasse that he planted his own roses to harvest at his country home nearby. These fields are still used for the heart notes of his Miss Dior perfume, which debuted in 1947 and has been in production ever since. When Dior told his perfumer what he wanted for his initial foray into fragrance, he reportedly declared, "Make me a perfume that smells like love."

Among the thousands of varieties of roses that are grown today, contemporary perfumers tend to use only two: *Rosa × centifolia* (literally "hundred leaved," on account of its full petals), a hybrid rose that is known variously as the cabbage rose, Holland rose, *Rose de Mai,* and Provence rose (see pp.24, 180, 212); and *R. × damascena,* the Damask rose (see pp.60 and 76), another hybrid with an intense fragrance. *R. × centifolia* is thought to have been developed by Dutch flower breeders in the late sixteenth century, and genetic analysis has shown that it is probably the offspring of *R. × damascena,* which was first cultivated in the Middle East and is thought to have been brought to Europe by the Crusaders; *R. gallica,* the Gallic or French rose, and a favorite of the ancient Romans (see p.122); and an unknown rose, possibly the Chinese Tea rose. The Damask rose is itself a hybrid, derived also from *R. gallica* and *R. moschata* (the Musk rose). Considered one of the world's oldest and most precious roses, the Damask is characterized by its large, fragrant, delicately fringed petals in a wide range of colors, from white to pale pink, bright red, and purple.

With their complex and long-lasting scents, both varieties are used by the greatest "noses" to create many of the renowned rose fragrances. The aroma of a Centifolia rose is fresher, brighter, and more transparent than that of a Damask, which has variations of a deeper, sticky, spicy fragrance. The Damask rose is the main variety cultivated for perfumery today and is grown commercially in Bulgaria, Morocco, and Turkey. As is the case in the wine industry, different conditions and soil is a question of "terroir," where specific growing conditions can give the perfume extracts different scents, facets, and even uses. The rich alluvial soil, mild temperatures, humidity, and sheltered valleys of the region around Grasse produce the perfect terroir for growing Centifolias, while the Kings' Valley plains around the town of Kazanlak, between the Balkan and Sredna Gora mountains of Bulgaria, give the shelter and conditions that the Damask rose needs.

A rose's scent comprises a mix of hundreds of molecules released from various parts of the plant. A trained nose can detect many different accords, from powdery vanilla honey to sparkly citrus; raspberry and plum to green tea; nutty, amber, spicy, and peppery to musk; or even mint and artichoke. The power of the rose's classic floral scent is in its petals (see p.116), but other parts of the plant give a range of tonalities. The sepals, which protect the flower in bud, and ovules, which form the seed, add further green notes to the fragrance, as well as a unique pepperiness that becomes part of the final extract.

The rose is unique in its ability to lend top, middle, and base notes to a fragrance, and this rich, multifaceted chemical composition means it partners easily with other floral, woody, and citrus notes, giving perfumers the flexibility of a large palette from which to work. Rose oil contains hundreds of compounds, and the quantity of each molecule depends on the rose variety and the terroir in which it is grown. The main fragrant components responsible for the faceted notes are citronellol (clean-citrus-floral), phenethyl alcohol (soft-honey), geraniol (sweet-geranium), beta-damascenone (jammy-spicy), nerol (citrus-fresh-floral), and rose oxide (green-metallic). All these qualities are manipulated to form some of the most famous rose fragrances. These include important perfumes of the early twentieth century, which combined modern techniques for synthesizing molecules and industrial methods of bottling fragrance in large quantities with the new world of advertising. Significant examples are Paul Poiret's La Rose de Rosine (see p.129), created in 1912; Chanel No. 5 (see p.47), launched in 1921; and Lanvin's Arpège (see p.61), which debuted in 1927.

The Centifolia is nicknamed *Rose de Mai*, a reference to the month in which it blooms and is harvested in Grasse. The roses are gathered at dawn, before the heat of the sun can evaporate the flowers' precious oils. Their scent changes according to the time of day at which they are picked, and the flowers are richest in their fragrant components at dawn. The open flower heads must be picked on the morning they bloom—as soon as the stamens are visible—and the process of extracting the perfume starts immediately, so that the petals retain their quality and fragrance. The color of the open rose changes from fuchsia to pale pink by the afternoon, and the fragrance changes with it, so time is of the essence.

Up to twenty buds sit in a cluster at the end of the stems, and it takes a quick, deft, and experienced hand to pick the open roses without damaging the unopened buds. The rose is picked just below the receptacle with a two-fingered grasp, a quick twist, and a pull. Each open bloom is then placed in the picker's apron, which is worn to avoid damage to the tender rosebuds. Pairs of harvesters move along the rows of bushes, which are planted with enough room on each side for a picker to work. The full aprons of harvested flowers are transferred into burlap sacks that allow the petals to breathe and stay cool. The roses are then transported in their sacks for processing within two hours in factories that are normally close to the fields. The bags are weighed at the facility's entrance, and pickers are usually paid by weight. This is the first stage in the process that takes the roses from flower to bottle.

The extracts of *R.* × *damascena* and *R.* × *centifolia* are today obtained using one of two processes—steam distillation and solvent extraction—to create either the essential oil or what is called the absolute oil, the deeper and sweeter variant of the rose's scent. The process of solvent extraction results in a "concrete" (a brownish semisolid substance), which goes through several solvent washes to produce a rose absolute (as opposed to the lighter, less fragranced essential oil). In steam distillation, steam penetrates the freshly picked petals, forcing them to release their oil into the vapor. The steam is then condensed, and the weightier, denser oil is separated and collected from the vapor by condensing through a coil.

As fragrance technology evolves, innovative, more sustainable, and more efficient processes offer new possibilities for the industry by refining the stages of extraction to accentuate or keep only the best facets of the process. In the 1930s, when the demand for natural perfume ingredients was at its peak, some 2,000 tons of Centifolia rose petals were harvested from Grasse. As fashions in fragrance changed, the flower farmers turned to more lucrative crops and perfumers to new, often synthetic ingredients manufactured in labs, so the number of rose farms fell. However, the resurgence of interest in all things natural in the twenty-first century has meant the production of rose absolute has doubled, and some of the most in-demand fragrances today are rose compositions, including Byredo's Rose of No Man's Land and Maison Francis Kurkdjian's À la rose. The need for highly skilled pickers, the sheer quantity of petals required to produce the final product (it takes somewhere between 200,000 and 250,000 flowers to obtain just a little over 2 pounds/1 kg of rose absolute oil), and the risk of bad weather or disease affecting the quality explain why rose absolute is one of the most expensive ingredients in perfumery.

Watching the harvest of these small pink roses offers a reminder of perfumery's close link to the soil in which it is grown, from the people involved in the picking to the creation of the final fragrance. The rose sustains countless industries worldwide with its recognizable scent. The size of these famous scented roses is inversely proportionate to the fragrant punch they pack. Both types of rose are much smaller than one would imagine—a potent reminder that something tiny and fragile can be powerful.

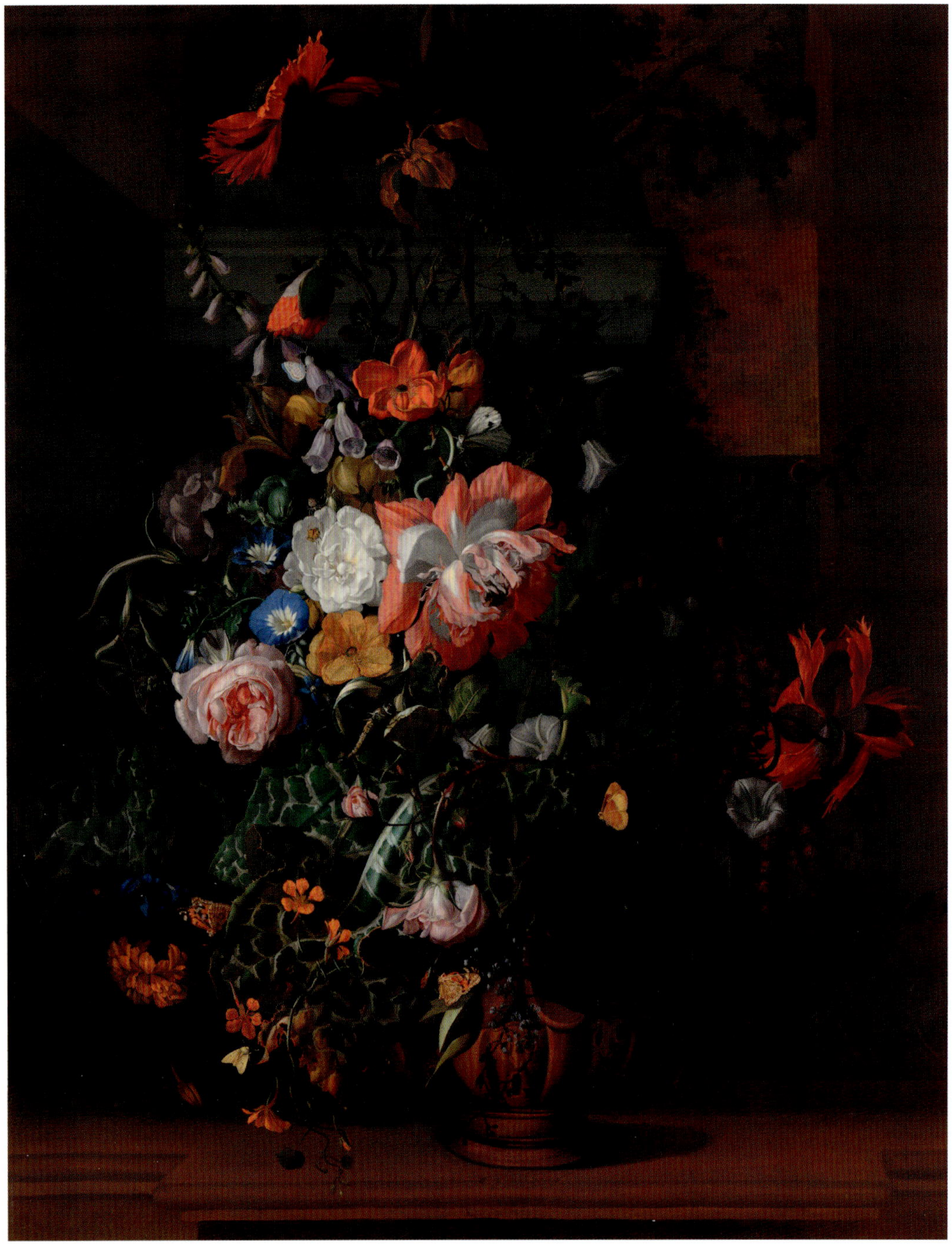

RACHEL RUYSCH

Roses, Convolvulus, Poppies, and Other Flowers in an Urn on a Stone Ledge, c. late 1680s
Oil on canvas, 41½ × 33 in. / 105.4 × 83.8 cm
National Museum of Women in the Arts, Washington, DC

The evolution of flower still lifes over the course of the seventeenth century shows them displayed at first like jewels, each precious blossom upright on its stalk, glowing with enameled color against a dark background. The later generation of artists, to which Dutch painter Rachel Ruysch (1664–1750) belonged, allowed the flowers to twist and droop, their petals fluttering and their colors nearly clashing. The roses shown here prove the point. Usually, Dutch artists chose the newly bred *Rosa* × *centifolia*, now commonly known as the cabbage rose. Centifolia roses were so named because they appeared to have a hundred tightly folded petals (see pp.24 and 180). Although Ruysch's unnamed roses remain center stage, they also compete with poppies and orange nasturtiums that outshine the roses' delicate pink, as well as wildflowers and a gigantic thistle leaf. The composition is dramatic and finely arranged, its one seemingly dying rose fitting with the *vanitas* tradition, whereby specific objects were used as symbols of transience. In the period during which she created this work, Ruysch, who was the daughter of a scientist, also painted "forest floors," where roses and lilies burst from stumps of rotten wood encircled by brambles and toadstools, reflecting the Dutch fascination with blending the macabre and beautiful, as well as science and art. Ruysch grew famous as a flower painter, becoming the appointed court artist of the Elector Palatine in Düsseldorf in 1708 and commanding high prices for her still lifes, which she painted for nearly seventy years, well into her eighties.

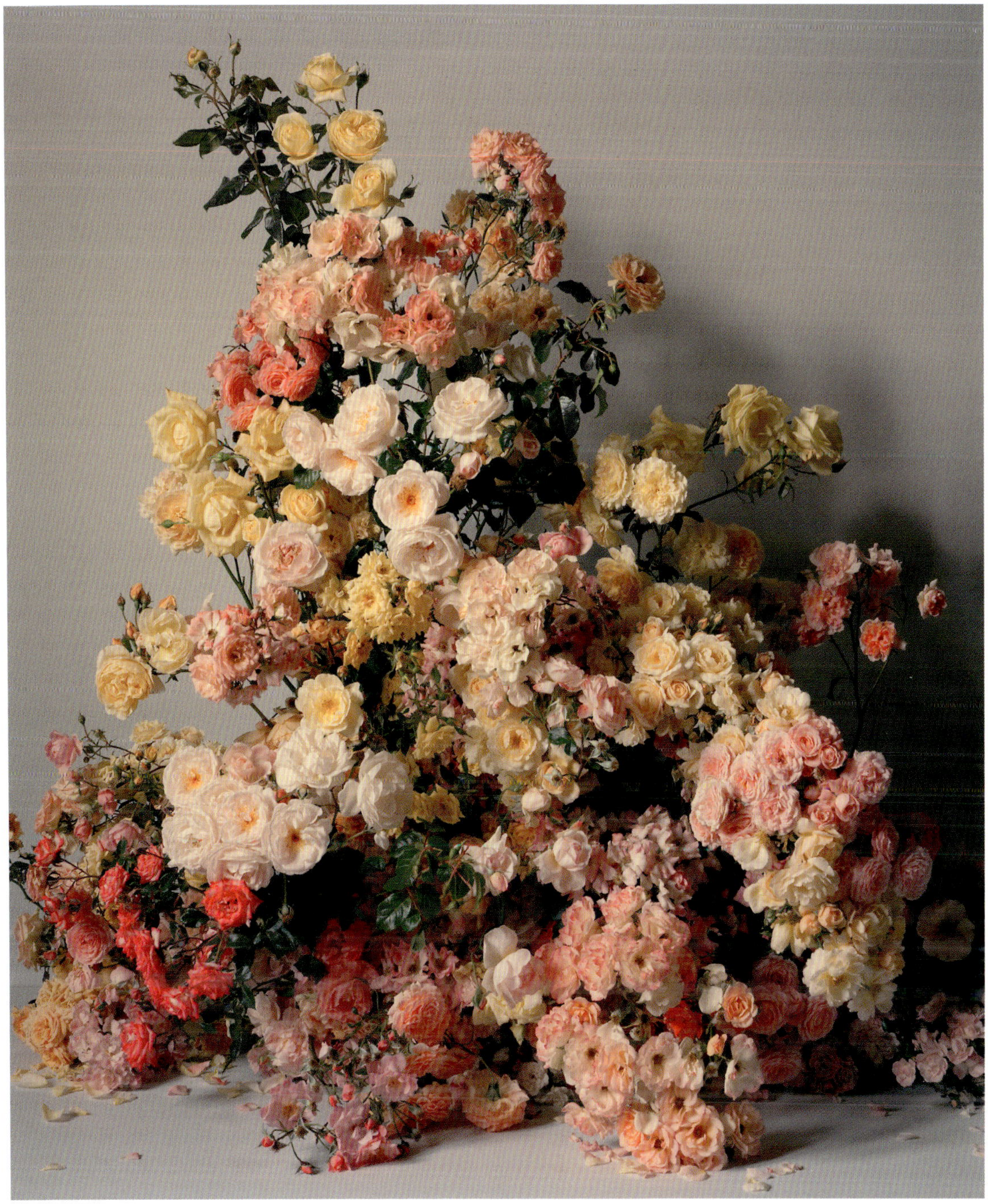

SIMONE GOOCH AND DEREK HENDERSON

Rosa #12, from *Rosa*, 2020
Inkjet print, 47¼ × 37¾ in. / 120 × 96 cm

A voluptuous display redolent of seventeenth-century Dutch still-life paintings, this image captures the ancient spirit of the rose in quiet elegance, fusing classical and contemporary aesthetics as well as photography and floristry. This ethereal portrait of roses arranged sculpturally in a palette of sunset hues is the work of Australian floral artist Simone Gooch (b. 1975), who operates under the artistic name Fjura (Maltese for "flower"), and New Zealand photographer Derek Henderson (b. 1963). Henderson was originally commissioned by *Pleasure Garden* magazine to photograph the renowned rose breeders Matthews Nurseries Ltd. at their farm in Whanganui, New Zealand, where they have been cultivating the flowers since 1947. Inspired by the experience, he enlisted Gooch to develop their own collaborative publication, *Rosa*, which celebrates the roses of Matthews Nurseries in luxurious, hyperrealistic portraits of floral arrangements that, like paintings, play with light and form. Closer scrutiny of such works as *Rosa #12* reveals that Gooch challenges the idea of a floral "arrangement" itself by using unexpected, overscaled, and layered elements, giving the roses the space to reveal their beauty as the viewer's eyes trace their shape, a style influenced by Japanese *ikebana*. Using film instead of digital imagery, Henderson's large-format field camera provides greater detail and tonal range, imbuing the images with a particularly painterly quality that further blurs the lines between photograph and arrangement, roses and art.

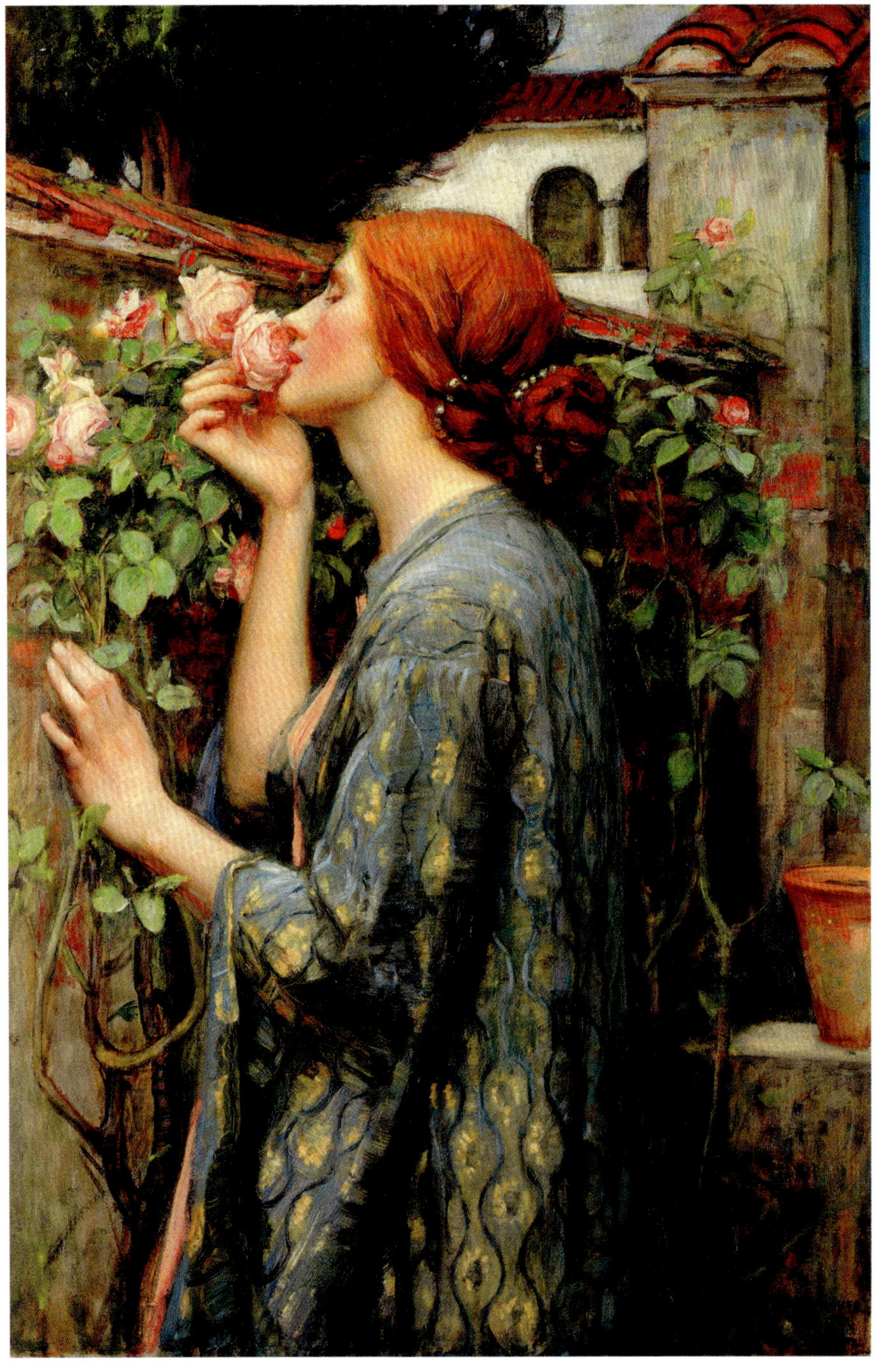

JOHN WILLIAM WATERHOUSE

The Soul of the Rose, 1908
Oil on canvas, 34¾ × 23¼ in. / 88.3 × 59.1 cm
Private collection

In this painting, a beautiful young woman is depicted in a walled Tuscan garden, lost in her thoughts, perhaps imagining a romantic encounter inspired by the fragrant aroma of the rose that she holds sensually to her nose. Famed for his large-scale canvases of women from ancient mythology, Shakespearean drama, and Arthurian legend, John William Waterhouse (1849–1917) painted *The Soul of the Rose* toward the end of his career. Unlike many of the British artist's works, it is not a scene taken from an ancient tale of love but is loosely inspired by Alfred Lord Tennyson's poem "Maud" (1855), which tells of the narrator's love of the eponymous heroine and features the line "And the soul of the rose went into my blood." Strongly influenced by the Pre-Raphaelite Brotherhood's medieval revivalist style, as well as paintings by such fourteenth-century artists as Fra Angelico, Waterhouse's composition balances exquisite detail with passages of looser brush-work, while its palette of bright pinks and earthy oranges and browns evoke a warm summer's day. The artist places the viewer in a voyeuristic position, watching the red-headed figure as she faces the wall; her eyes half closed, she is absorbed in the moment. Waterhouse often linked women to nature in his paintings, and here he associates the protagonist with roses, which probably symbolize a lost or absent lover. He produced several similar works featuring women and roses, but this is the most well known and, indeed, one of his most celebrated paintings.

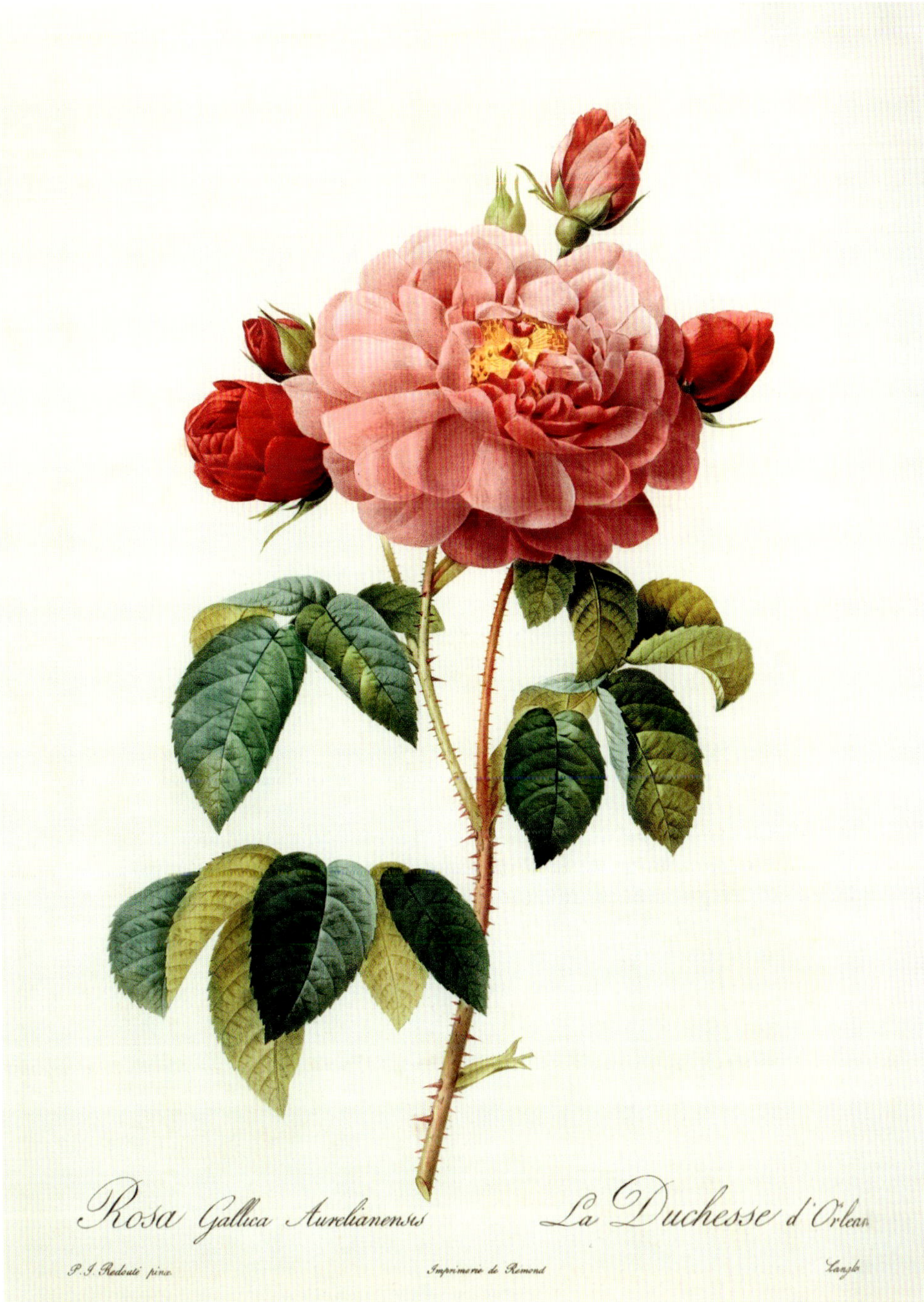

PIERRE-JOSEPH REDOUTÉ

Rosa Gallica Aurelianensis, from *Les Roses*, 1817–24
Stipple-engraved plate printed in color and finished by hand, 14⅛ × 10¼ in. / 36 × 26 cm
Private collection

Pierre-Joseph Redouté (1759–1840) was the most celebrated botanical artist of his day. By the 1810s, after becoming French empress Joséphine Bonaparte's official Painter of Flowers, he was known to such an extent that his works appeared under his own name instead of the botanist who wrote the text (in this case Claude-Antoine Thory). Published in parts between 1817 and 1824, *Les Roses* features masterpieces of color printing, but Redouté portrays only the flowers, without dissections, making them more flower paintings than botanical art. Each rose is named twice, in Latin and in French. Here, *Aurelianensis* is a Latinized version of "Orléans," the city named in honor of its founder the Roman emperor Aurelian. Redouté dedicated the accompanying text to the Duchesse d'Orléans, dating it carefully September 30, 1821. The previous duchess had died in June, so the dedicatee was probably the current duchesse Maria Amelia, who, as wife of Louis Philippe, would become the last queen of France. There are still some mysteries about the actual flower; who bred the rose, and when? Was it named after Maria Amelia or her predecessor? A different rose with a similar name, 'Duchesse d'Orléans', a Hybrid Perpetual bred by Pierre-Victor Quétier in 1851, became popular in the 1850s. Quétier's rose was presumably named after Duchesse Hélène, the daughter of Redouté's duchesse. He acknowledged that the rose he drew was not found in many collections. Presumably, it had died out before 1850.

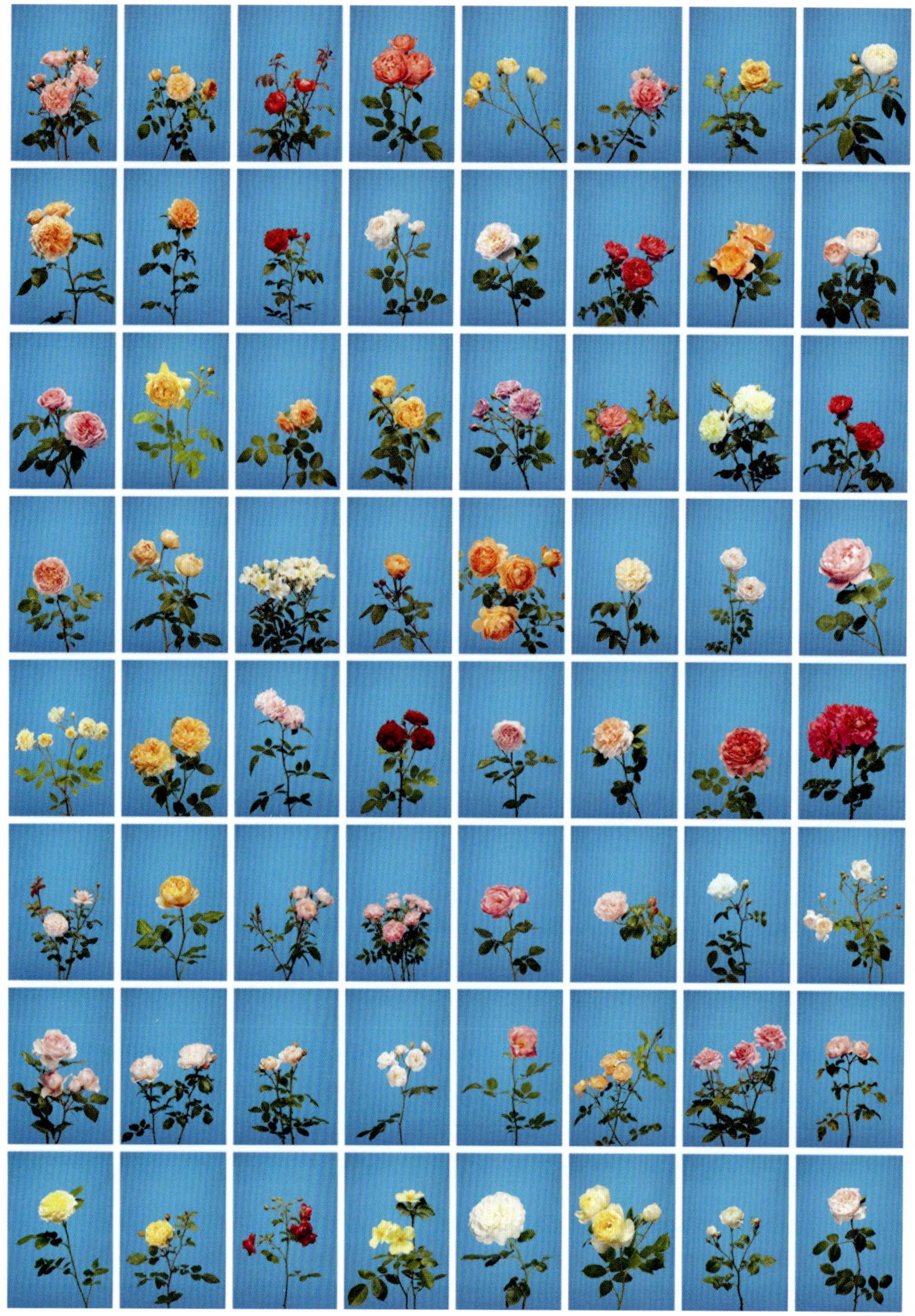

LUKE STEPHENSON

The English Rose, 2019
Printed poster, 16½ × 23⅜ in. / 42 × 59.4 cm

Showcasing the rich diversity of the rose, the national flower of England, British photographer Luke Stephenson (b. 1983) assembles sixty-four modern varieties in this vibrantly colorful grid. Traditionally associated with royalty, religion, and beauty, the symbolic flower has for centuries dominated the nation's poetry, art, and literature. However, the modern English rose was popularized in the 1960s by the celebrated British breeder David Austin (see pp.77 and 105), who crossed antique varieties (those introduced before 1867) with modern roses (hybrids and Floribundas) to create plants that have the character of Old Garden roses but flower for longer. Stephenson spent two summers at the headquarters of David Austin Roses in Shropshire, England, photographing the flowers in a small, makeshift studio. He shows each stem in isolation against a rich blue backdrop with a retro quality reminiscent of 1970s flower books. The color gives the blooms a rich luminosity, intensifying their lush pinks, deep crimsons, lemon yellows, and milky whites. Each rose is named after a famous English person, such as 'Roald Dahl', which has peach-colored blooms that evoke the eponymous fruit from the celebrated writer's children's tale *James and the Giant Peach* (1961). Another, 'Maid Marion', is named after the companion of legendary English hero Robin Hood and has clear mid-pink, saucer-shaped rosettes. Others are named after writer Emily Brontë, actor Judi Dench, Tudor queen Anne Boleyn, and Austin's own daughter Claire.

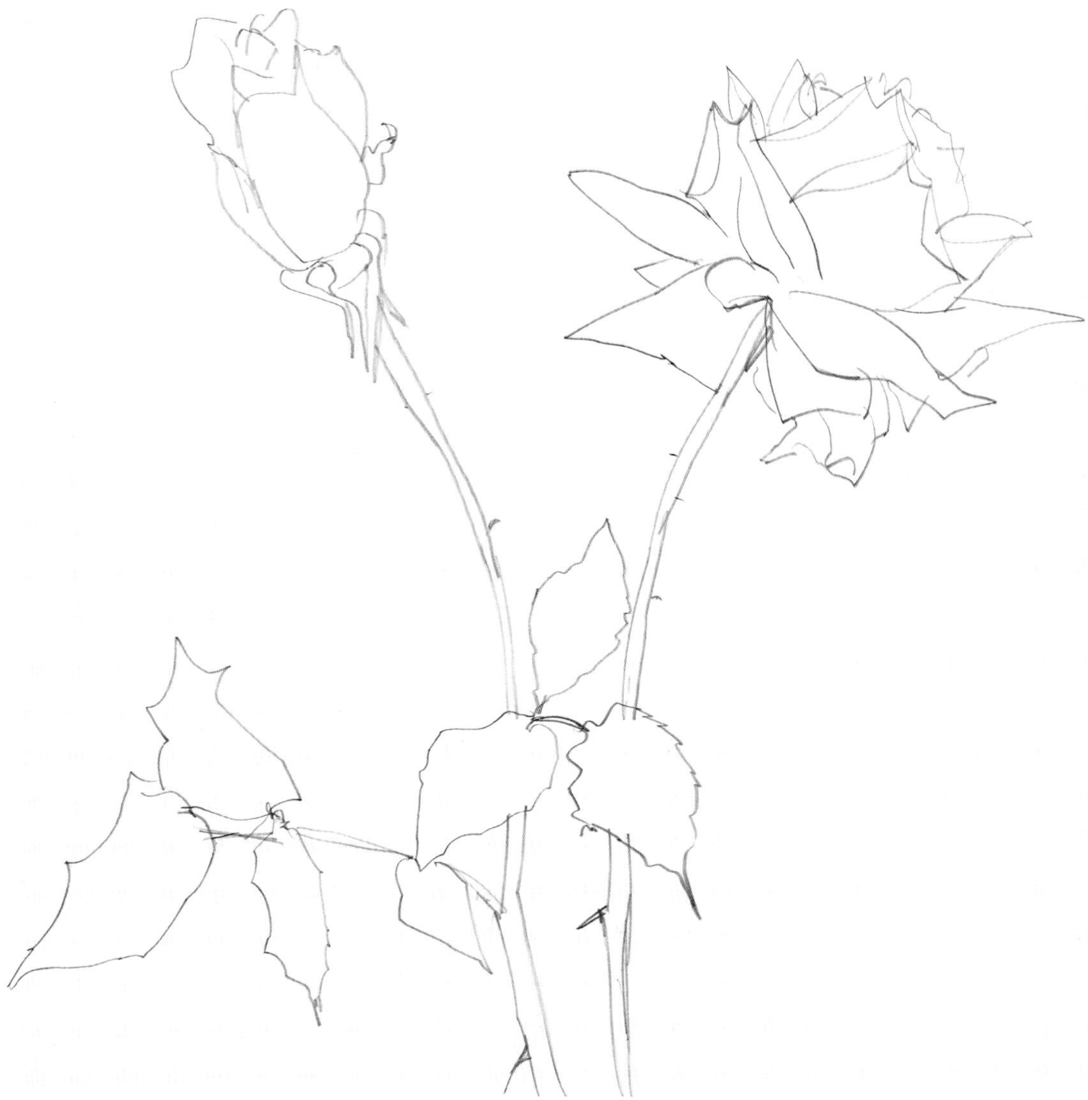

ELLSWORTH KELLY

Rose, Palm Springs, 1984
Graphite pencil on paper, 20⅛ × 24 in. / 51.5 × 61 cm
Whitney Museum of American Art, New York

Sketched lightly in pencil, this sparse outline of a rose was drawn from direct observation and records a tightly furled bud, a fully opened bloom, a cluster of leaves, and thin stems with fine, sharp thorns. Dashed off in five minutes, it captures the plant's character and form with swift precision and economy. It is just one of the thousands of studies of flowers and plants that American artist Ellsworth Kelly (1923–2015) made throughout his career. He insisted that they are "portraits, not anonymous," each found by chance and corresponding to a particular place and time. "They are exact observations of the form of the leaf or flower or fruit seen," he said. "Nothing is changed or added." Such naturalistic drawing would appear to have little relation to the bold abstract canvases for which Kelly is best known, but he believed his plant life drawings, which he began in 1949, to be an extension of the practice of seeing that underlies all his colorful, hard-edge abstractions. Kelly was never interested in creating illusions of spatial depth, and his work is characterized by a singular focus on line, form, and flat, uninflected shapes. The same is true of his deceptively simple plant drawings, which are always two-dimensional, without shading, and lacking details such as veins. The interest, for Kelly, lay in the formal relationship of one part to another, how they overlapped or abutted, creating new shapes in the spaces between.

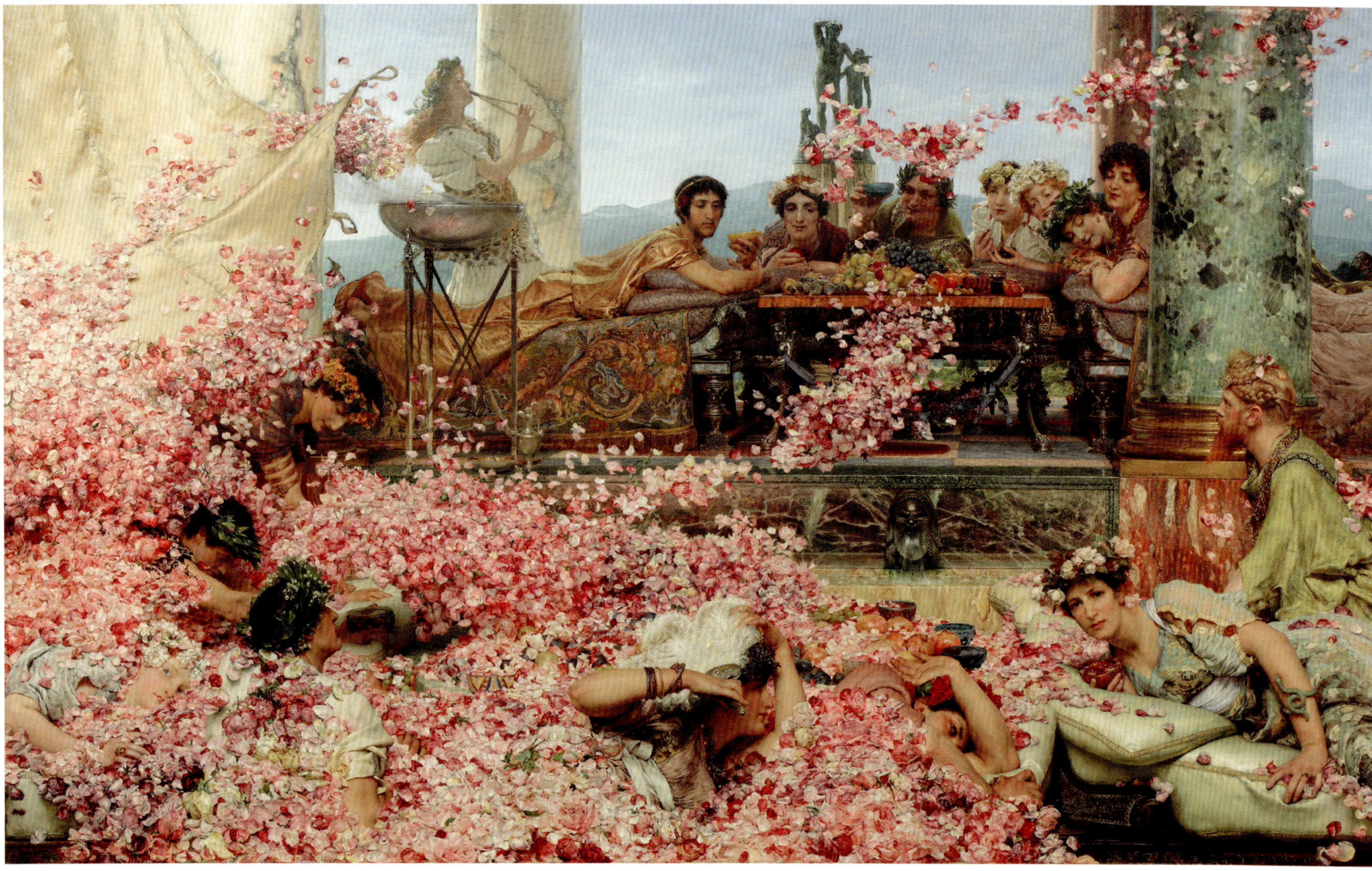

LAWRENCE ALMA-TADEMA

The Roses of Heliogabalus, 1888
Oil on canvas, 4 ft. 4 in. × 7 ft. / 1.3 × 2.1 m
Private collection

According to the *Historia Augusta*, a fourth-century compilation of anecdotes about the Roman rulers, Emperor Heliogabalus had a banqueting room with a reversible ceiling that could be turned to shower flowers on the guests. The story goes that he once rained down so many petals, "that some were actually smothered to death, being unable to crawl out to the top." (Nero had a similar reversible ceiling a century and a half earlier, but there is no report of anyone dying as a result; perhaps he was just more careful. Heliogabalus was a teenager at the time, after all.) Dutch-English painter Lawrence Alma-Tadema (1836–1912), who was famed for his depictions of antiquity, caused a sensation when he exhibited *The Roses of Heliogabalus* in 1888. Its colors, its unorthodox composition, and the blank, affectless faces of the guests puzzled viewers, with the British newspaper *The Observer* reporting, "the girls smothering under the rose leaves do not seem to mind." Many questioned the artist's motive. Did the painting portray an amusing prank or a sadistic exercise in murder? The second interpretation became standard, and some were outraged by the lack of morals the artwork displayed. Collectors had no such qualms, and it sold for £20,000 (about $1.6 million today), which probably helped to offset the massive costs Alma-Tadema incurred while creating the work. Painting in the depths of the English winter, the artist had a fresh harvest of roses shipped weekly from the south of France to ensure that he painted the petals perfectly.

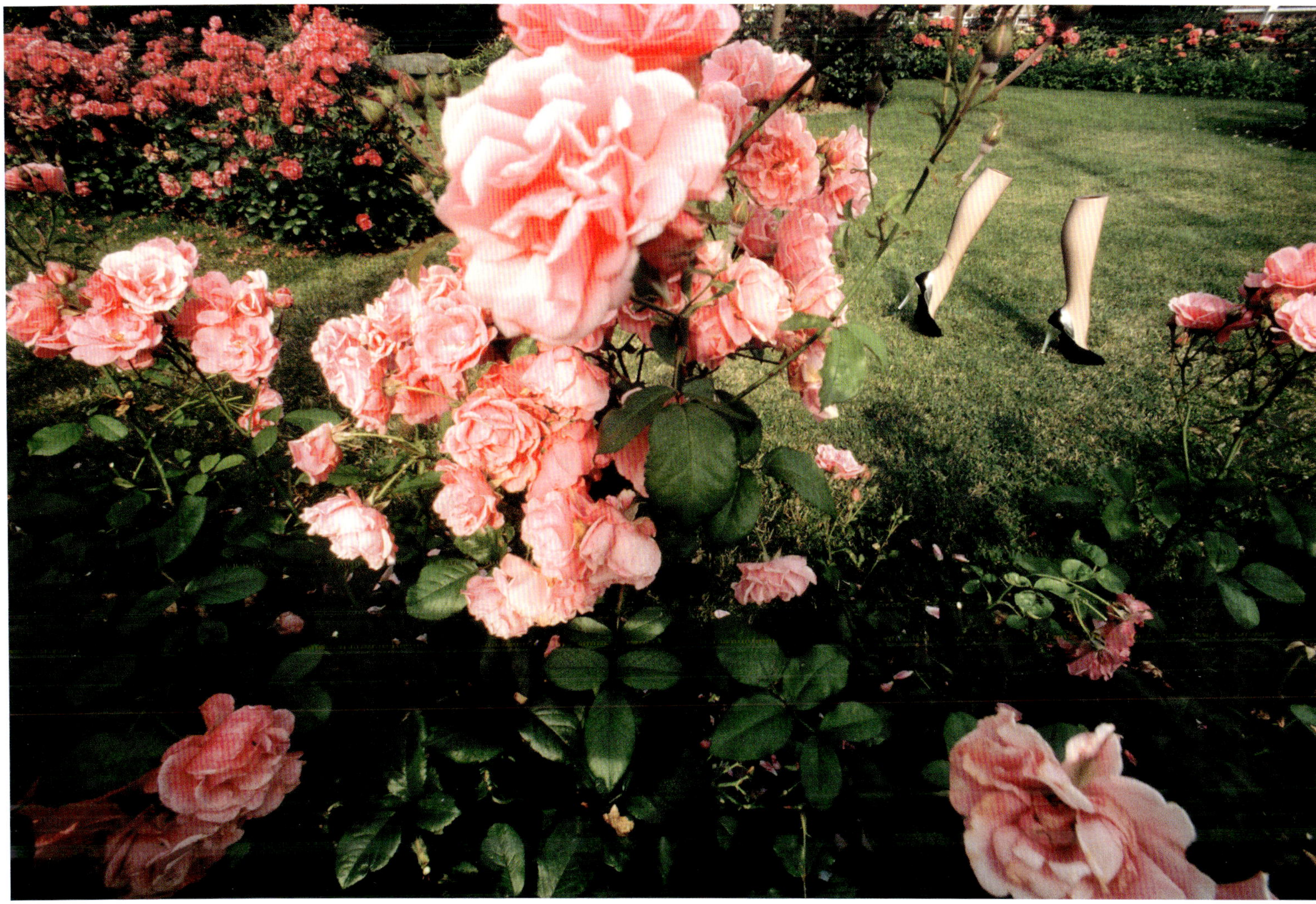

GUY BOURDIN

Charles Jourdan, 1979
Fujiflex crystal archive print, 36⅝ × 47⅞ in. / 93 × 121.5 cm
Private collection

A profusion of blousy pink roses dominates a photograph that, at first glance, appears to capture a green lawn surrounded by beds of roses on a summer's day. The pink blooms are wide open, and many show signs of decay, suggesting that the garden is slightly past its best. Look more closely, however, and the image takes on a surreal quality. A pair of female legs in towering black-and-gold stilettos strides across the grass—but without any accompanying body. The image is the work of French artist and fashion photographer Guy Bourdin (1928–1991), a longtime contributor to *Vogue* who, along with Helmut Newton, revolutionized fashion photography in the mid-twentieth century. Bourdin's unsettling juxtapositions, theatrical compositions, and bizarre hyperreal mise-en-scène owe much to one of his major inspirations, Surrealist artist Man Ray. So determined was Bourdin to work with his hero that he is said to have turned up at Man Ray's Paris studio many times, only to be repeatedly turned away. It was only on Bourdin's eighth visit that the older artist let him in. Bourdin worked closely with Man Ray throughout the 1950s and absorbed his influence, combining it with his own unique take on capturing a fantastical realm through photography over a successful career, as reflected in this 1979 advertisement for French shoemaker Charles Jourdan. Known for stiletto shoes with their spiked heels, Jourdan found in Bourdin the perfect photographer: his surreal images became inextricably linked to Jourdan's iconic footwear.

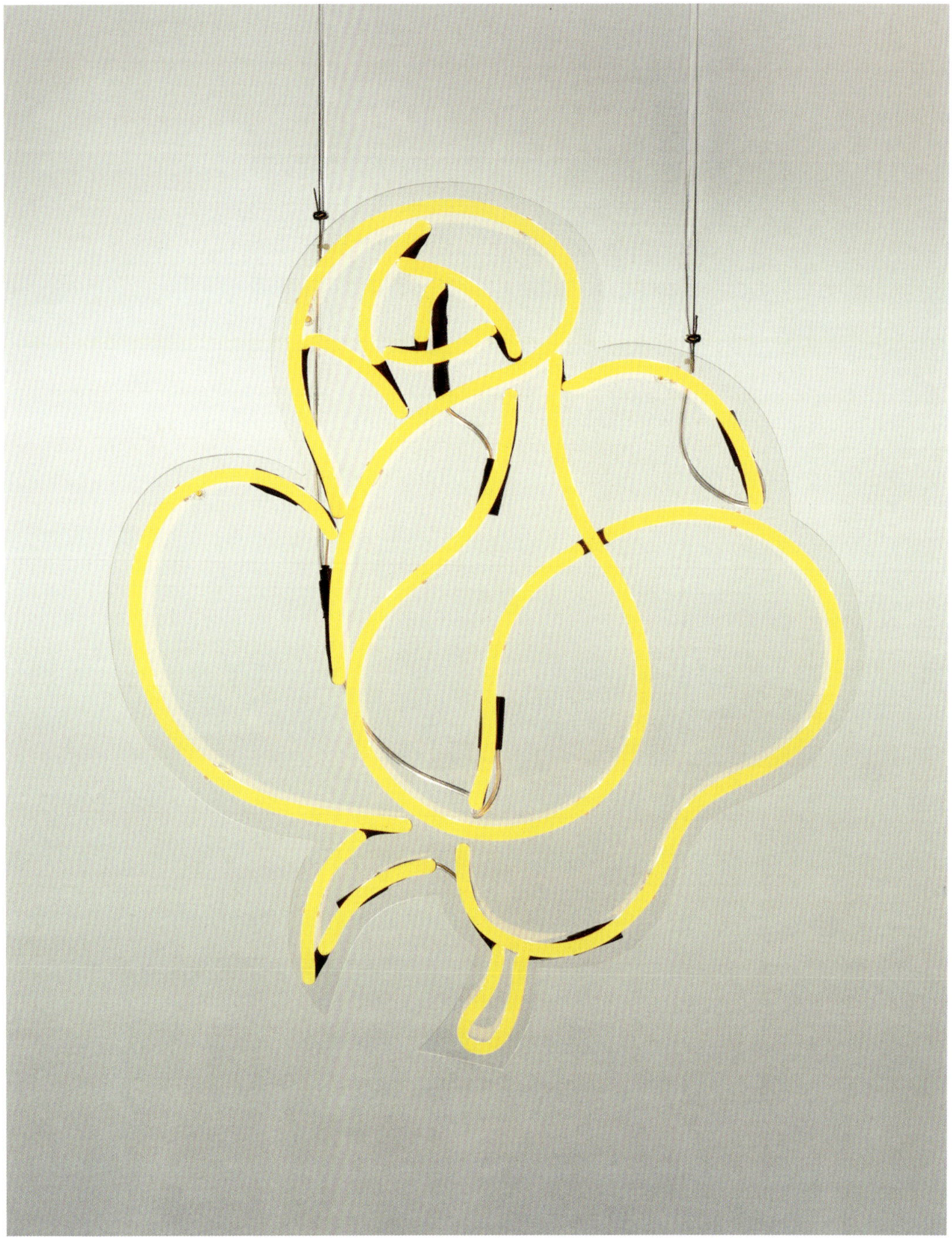

CERITH WYN EVANS

Untitled (yellow rose), 2007
Yellow neon and Plexiglas, 36⅝ × 32⅝ in. / 93 × 83 cm
Private collection

The delicate and complex form of a rose in bloom is reduced in this sculpture to a few simple lines rendered in yellow neon. Welsh sculptor Cerith Wyn Evans (b. 1958) based its design on the logo of the famous Japanese Takashimaya department store. Although the glowing artwork recalls retail signage, it points not to shopping but to the long history of the rose as a motif and symbol in art and literature. As with other works by the artist, the sculpture is both an object in its own right and a signpost to a multiplicity of references. Here, the rose might evoke for viewers Eikoh Hosoe's dramatic photographic series *Ordeal by Roses* (1961), in which controversial Japanese author Yukio Mishima is depicted in surreal and erotic poses; or perhaps the strange and elusive female alter ego of radical French artist Marcel Duchamp, who went by the name Rrose Sélavy (a pun on the French phrase *Eros, c'est la vie*, meaning "Love, such is life"). Others might be reminded of Gertrude Stein's attempt to unpack the symbolic complexity of the flower in the phrase "rose is a rose is a rose is a rose" (see p.222). For Stein, the sentence expressed that, simply by using the name of an object, one invoked the imagery, meanings, and emotions associated with it—a concept that lies at the heart of Evans's practice, which explores how ideas can be communicated through form.

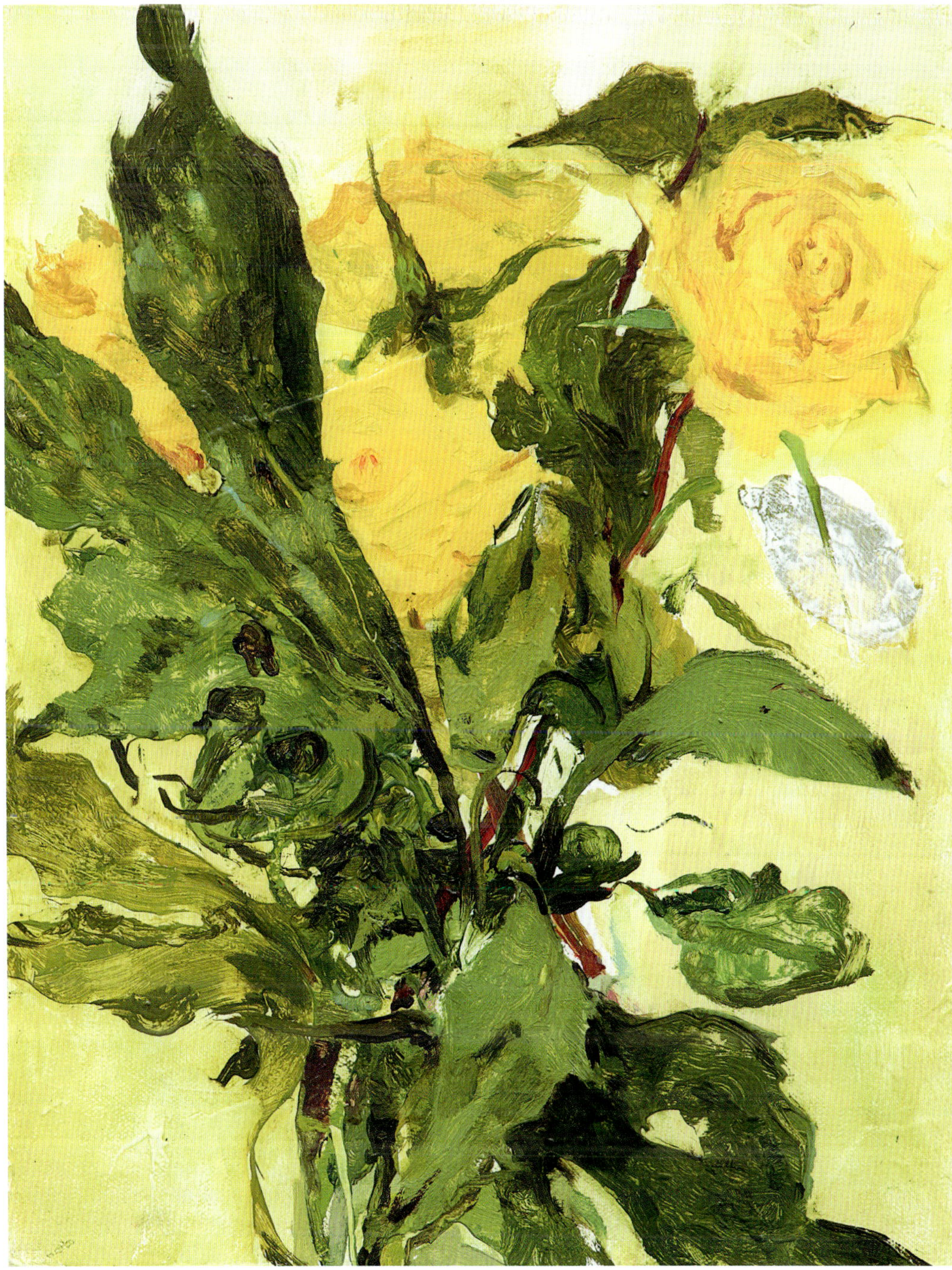

JENNIFER PACKER

Yellow Roses, 2014–15
Oil on canvas on board, 12 × 9 in. / 30.5 × 22.9 cm
Private collection

Painted with loose, spontaneous brushstrokes and a limited palette, this sketchlike depiction of cut yellow roses typifies the flower paintings of American artist Jennifer Packer (b. 1984). Expressionistic green foliage dominates the composition, while the blooms themselves seem to recede, almost merging with the thin wash of lemon yellow behind. Alongside flowers, Packer is known for her intimate portraits of friends and family, which similarly slide between abstraction and representation, combining observation, improvisation, and memory. She started painting bouquets of flowers in 2012, initially as a respite from a complicated portrait painting. Over time, as the Black Lives Matter movement brought increased awareness of the violence suffered by innocent Black men and women in the United States and elsewhere, her floral studies became inspired by specific individuals and she began dedicating her flower compositions to named and unnamed victims. For Packer, painting is a political act, which she regards as a way of bearing witness and sharing testimony as she responds to the tragic killings of Black Americans. Her works have been compared to seventeenth-century Dutch *vanitas* still-life paintings, in which the ephemerality of flowers reminded viewers of the brevity of life (see pp.10 and 140). Reinterpreting this theme for contemporary audiences, Packer likens her flowers to funerary bouquets that speak to loss and bereavement. Such pictures as *Yellow Roses* are, therefore, less reminders of death than meditations on grief and mourning.

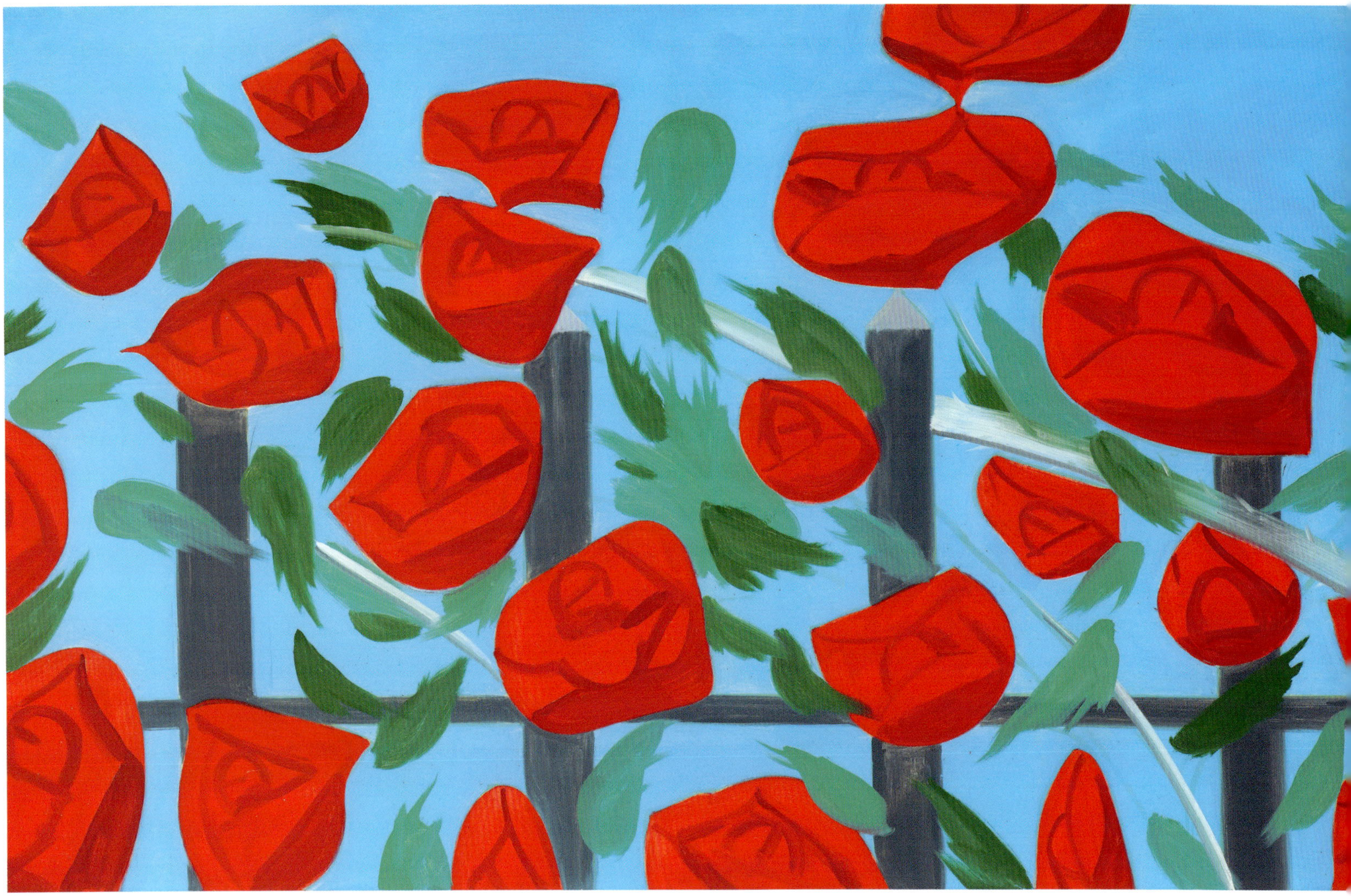

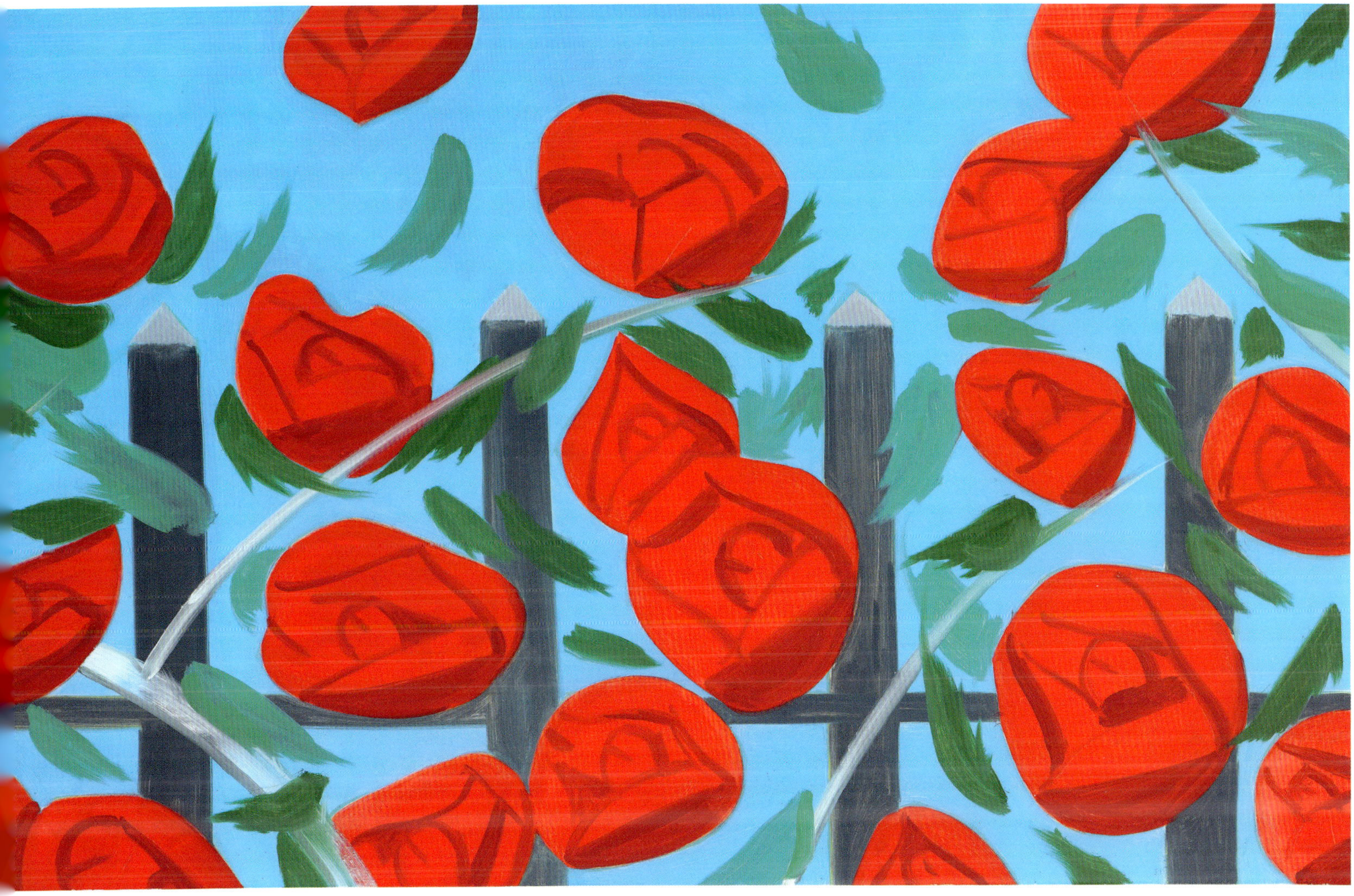

ALEX KATZ

Roses on Blue, 2002
Oil on canvas, 4 ft. × 12 ft. 6 in. / 1.2 × 3.8 m
Private collection

Dominated by bold colors and geometric shapes, this close-cropped detail of a fence is immediately suggestive of a garden as a whole. Scarlet roses riot over the posts, their stems shown sparingly and, in some cases, completely absent, leaving flowers and leaves hanging untethered against the flat blue ground that suggests the sky. Here are abundance and profusion, painted in colors so vibrant that they have an almost psychedelic intensity. American artist Alex Katz (b. 1927) has described flowers as being some of the most difficult forms to paint, so, to give his flower pictures a sense of immediacy and animation, he adopted a wet-on-wet technique in which each stroke of paint is applied before the previous one is allowed to dry. The leaves are dashed in, so that they appear to flutter, and the roses themselves have a dynamic vivacity in the bright light that saturates the scene. This is one of many flower paintings that Katz has made over many summers spent in Maine, and it is typical of his elegantly reductive approach. The roses are pared back to their essentials, painted as flat, stylized motifs; their shapes are sharp-edged and sketched with the minimum of modeling, characteristic of the way in which Katz blends abstraction and realism. The panoramic composition, with no sense of depth or recession, is so tightly cropped that the viewer is immersed in an intimate close-up with these dazzling blooms and can only imagine the lush garden of which they are a part.

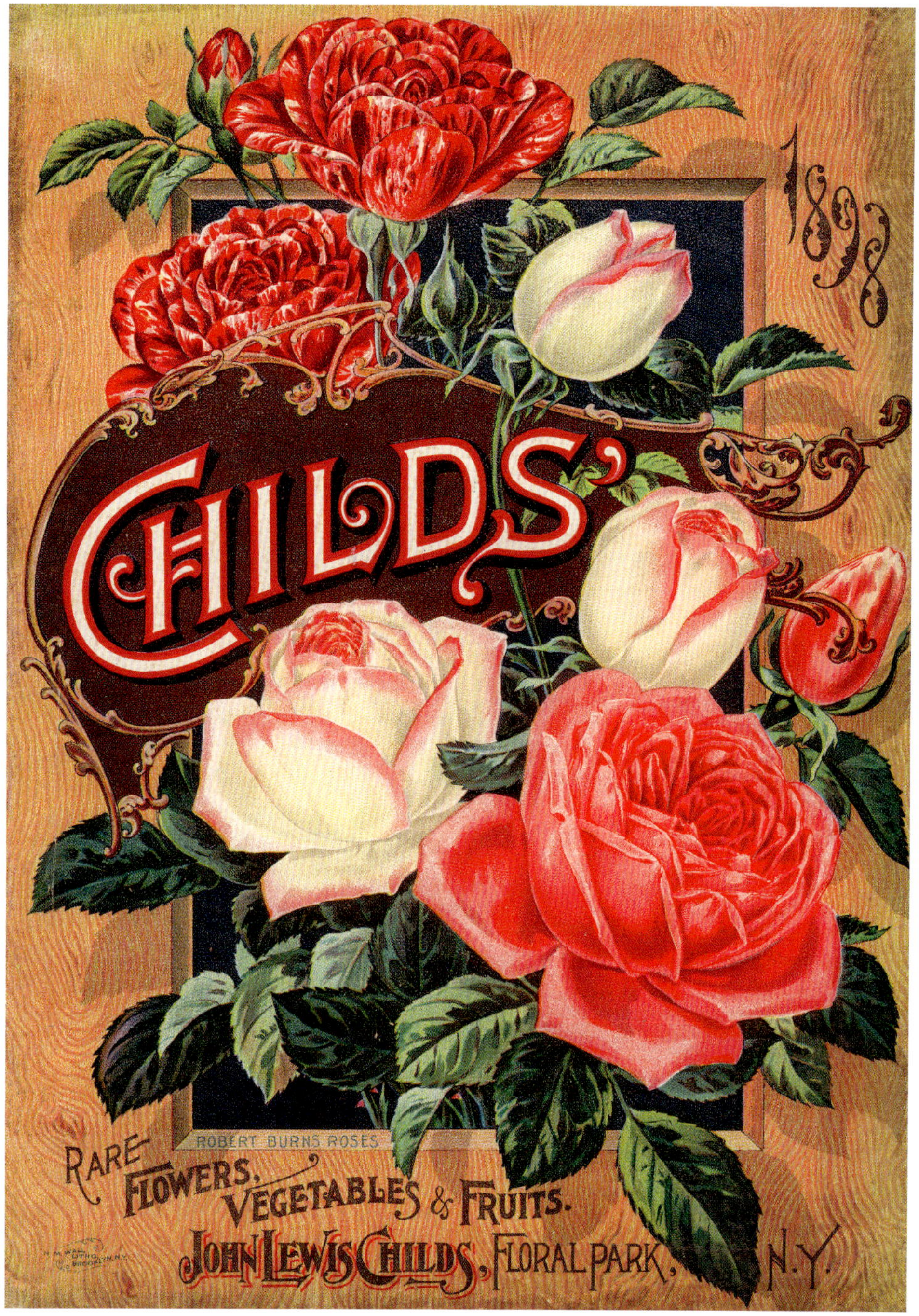

JOHN LEWIS CHILDS

Childs' Rare Flowers, Vegetables and Fruits, 1898
Chromolithograph, H. 9½ in. / 24 cm
Smithsonian Libraries, Washington, DC

It was a marketing idea, capitalizing on the admired reputation of a Scottish poet known for his ode "A Red, Red Rose." John Lewis Childs (1856–1921), a floral entrepreneur from Long Island, New York, was credited with starting the first mail-order seed catalog business in the United States. In 1898 he decided to tie his rose introductions to Robert Burns and feature them on the cover of his spring catalog. Blooms and buds enclose the company name in a vibrant surround. Using steam-driven printing and the chromolithography process, Childs could reproduce this eye-catching design at a reasonable price. Although the Scottish bard had been long dead when Childs brainstormed this marketing move, his fame had grown in the late nineteenth century. Composers used his verses as lyrics for popular songs performed in parlors and concert venues across the English-speaking world, and "Burns Night," celebrated on the anniversary of his birth in January, had quickly become a global phenomenon. In 1889 the *Boston Globe* covered the event, which included songs, such as Burns's famous ballad "Auld Lang Syne," and speakers, among them Ralph Waldo Emerson. If "Rabbie" Burns offered the name recognition, Childs provided the roses. His 1898 catalog offered three different kinds, christened after titles of Burns poems: the striped 'Halloween', the white 'Highland Mary', with its "rosy lips," and the pink, fragrant 'Sweet Afton'. Of the three, only 'Sweet Afton' remains in the nursery trade today.

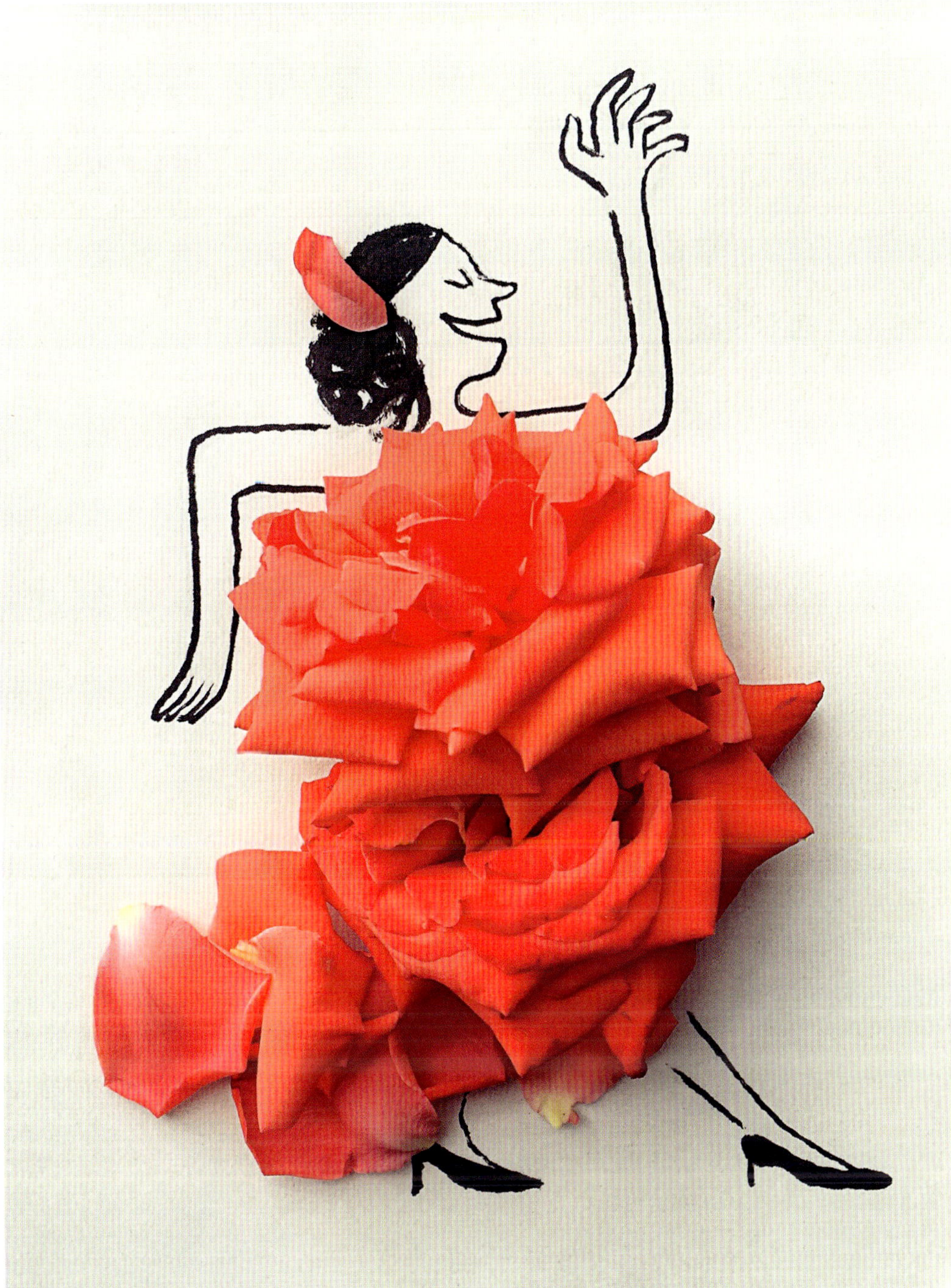

JEAN JULLIEN

Untitled, c. 2023
Ink on paper and rose petals

Real rose petals form the elaborate ruffles of a flamenco dancer's dress in this quirky drawing by prolific French graphic artist Jean Jullien (b. 1983), who is known for his lighthearted illustrations of everyday life. The elaborate outfits worn in Spanish flamenco are designed to accentuate the performer's elegant movements; here, however, the carefree figure, drawn with the bold and expressive lines typical of Jullien's distinctive style, is virtually engulfed by her enormous dress of vibrant red roses. As alluded to by the single petal in her hair, which has been swept into a bun, flamenco dancers traditionally decorate their hairstyles with flowers. Carnations and peonies are popular choices and are often combined, but for those seeking a seductive and sensual look that complements the fiery, passionate art of flamenco, a single large red rose can create a stunning effect when combined with a classic Manila shawl, red or black ruffled dress, and flamenco heels. According to tradition, the practice of dancers wearing flowers in their hair dates to the early days of the *Feria de Abril*, the annual spring festival held in the Andalusian capital of Seville. Originally created as a livestock fair in 1846 or 1847, the event was known for its unpleasant aromas, which attendees attempted to mask with the scent of flowers. It was not long before Andalusian women began placing blooms in their hair and a beautiful custom was born, one that has become a staple of the flamenco costume.

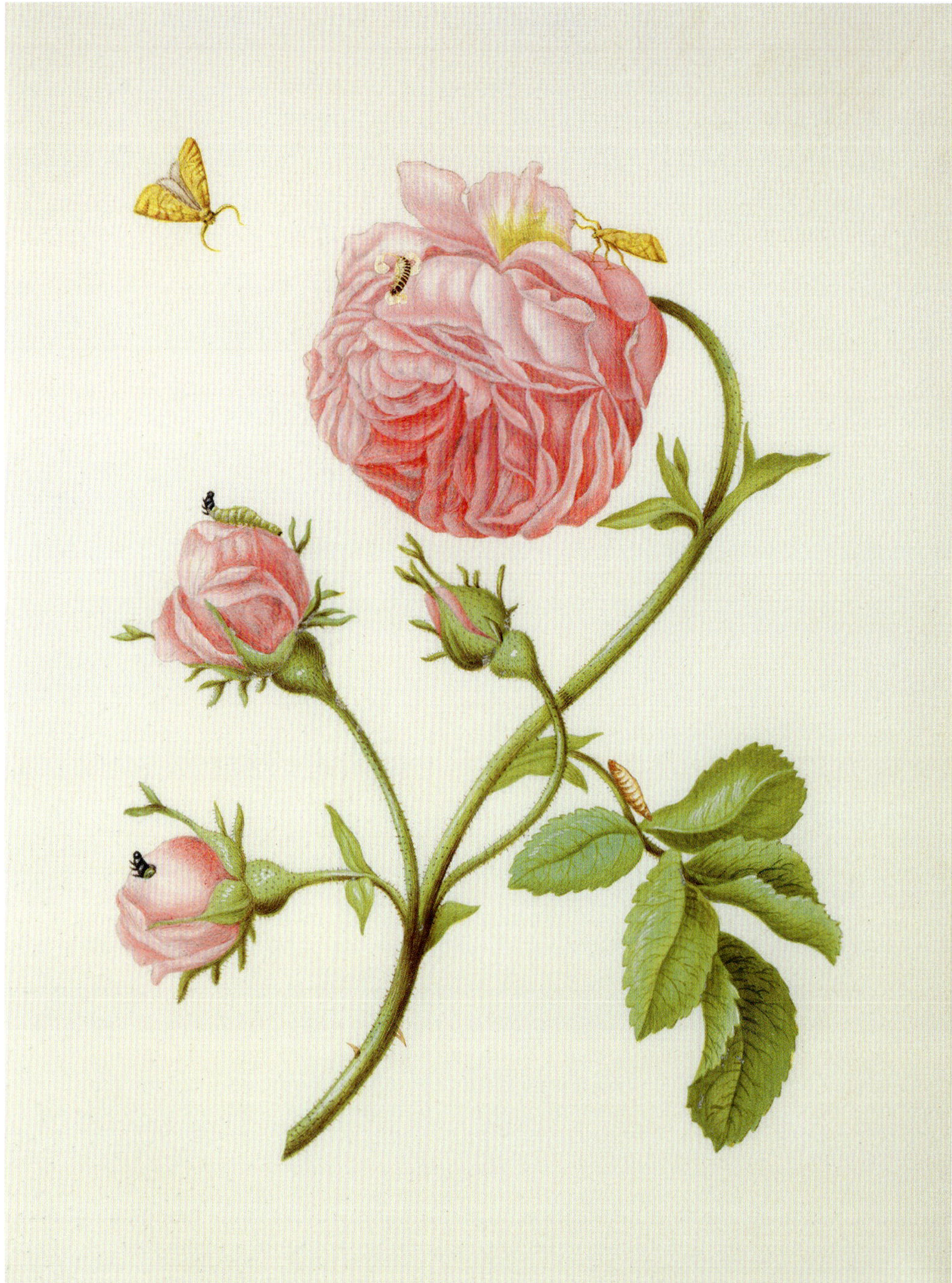

MARIA SIBYLLA MERIAN

Rose with Metamorphosis of Leaf Roller and a Glued Beetle Larva, c. 1679
Watercolor and gouache over transfer print on vellum, 7⅜ × 5¾ in. / 18.8 × 14.6 cm
Städel Museum, Frankfurt

Fascinated by insects since childhood, with a special interest in the butterflies and moths of the order Lepidoptera, German naturalist and scientific illustrator Maria Sibylla Merian (1647–1717) made immeasurable contributions to the study of natural history. While she was not the first person to paint butterflies and moths, nor to document ecological interactions, her direct observations from life were rendered with precise scientific accuracy and aesthetically pleasing compositions. She was one of the first naturalists to explore the process of metamorphosis—a controversial concept, then considered an act of magic. In her multivolume book *Der Raupen Wunderbare Verwandelung und Sonderbare Blumen-nahrung* (*The Wondrous Transformation and Particular Food Plants of Caterpillars*), she produced a series of beautiful illustrations, accompanied by her own detailed texts. Plate 24 depicts *Rosa* × *centifolia*, a newly hybridized garden rose developed by Dutch breeders (see pp.180 and 212). Merian was innovative in pairing insects with their preferred host plants, as seen here with the black-headed, green caterpillar-like larvae of a rose leaf-rolling sawfly feeding on buds. She describes the flower damage they inflict by consuming the blossom from the inside out, as shown in the lowermost rosebud, and their transformation into "pretty moth-birds." Merian made numerous images of this cultivated rose, but this print has an unusual detail—a preserved specimen glued to the large open blossom.

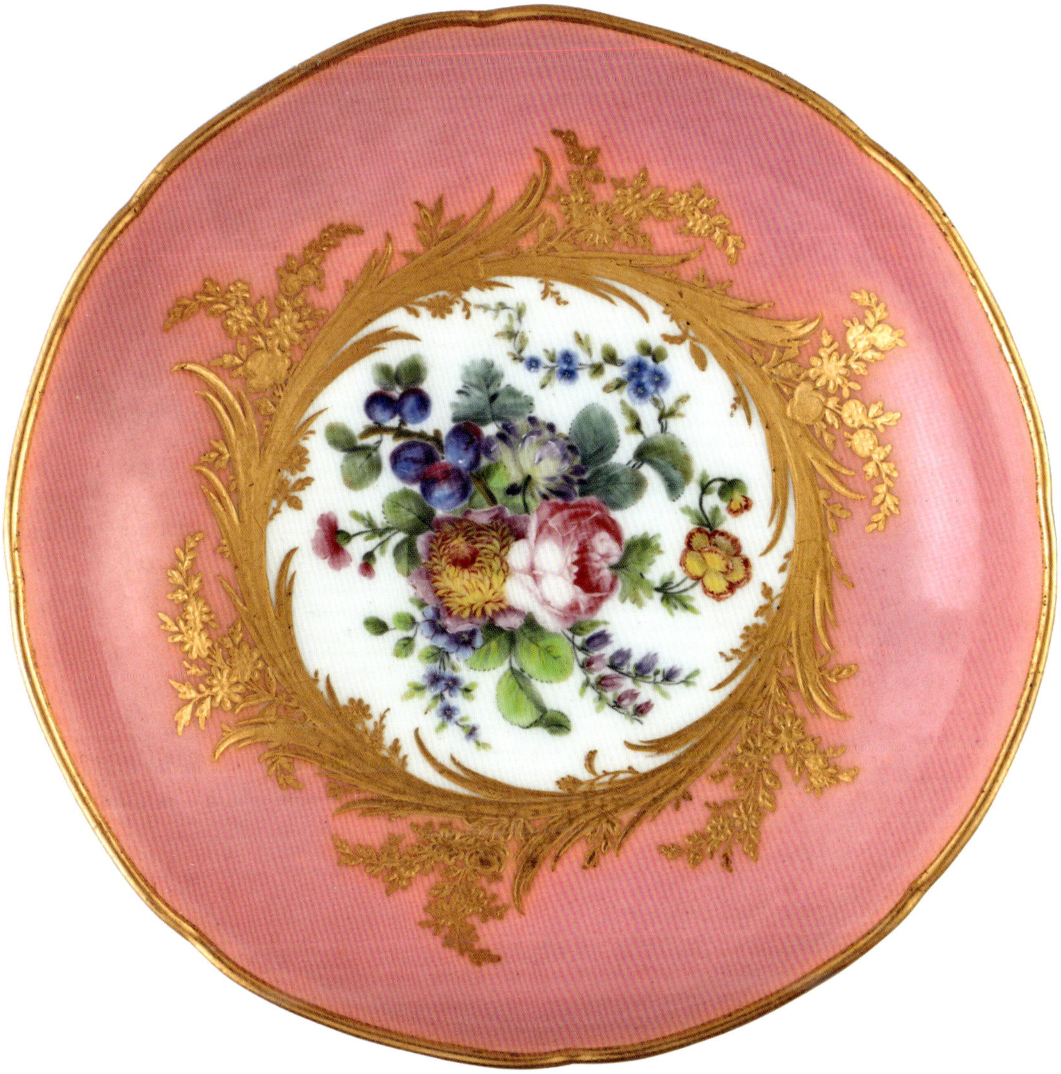

SÈVRES PORCELAIN FACTORY

Deep plate with Pompadour pink base, 1775
Soft-paste porcelain and enamel, Diam. 8½ in. / 21.5 cm
Petit Palais, Musée des Beaux-Arts de la Ville de Paris

This deep plate edged in gold with a delicate centerpiece of flowers—including a pink Centifolia rose in full bloom—and a variety of fruit painted directly onto the white porcelain, was made in the Sèvres Porcelain Factory in the suburbs of Paris. Almost overwhelming the intricate central floral decoration and the surrounding gold festoons is the dramatic color of the rest of the plate: a shade named Pompadour pink after Madame de Pompadour, mistress of Louis XV of France (see p.36). The couple fueled the popularity of the delicate china, which had been shipped at great cost from East Asia to Europe until German alchemist Johann Friedrich Böttger discovered how to make hard-paste porcelain in 1707. Soft-paste porcelain, which allows paint to be absorbed into the china, was made in France from 1738 at Vincennes. However, the king ordered a new factory to be built at Sèvres in 1756 to meet the growing demand for china pieces, particularly from Madame de Pompadour herself. The following year, the factory introduced a new color to honor its chief patron. Formulated by chemist Jean Hellot, who created five new hues at Sèvres, Pompadour pink is notable for the luxurious depth of its tone. The floral centerpiece of the plate also pays homage to Madame de Pompadour, whose love of flowers inspired Sèvres not only to include floral decorations on many of their pieces but also to create three-dimensional porcelain flower arrangements, which were so lifelike that the king mistook them for real flowers.

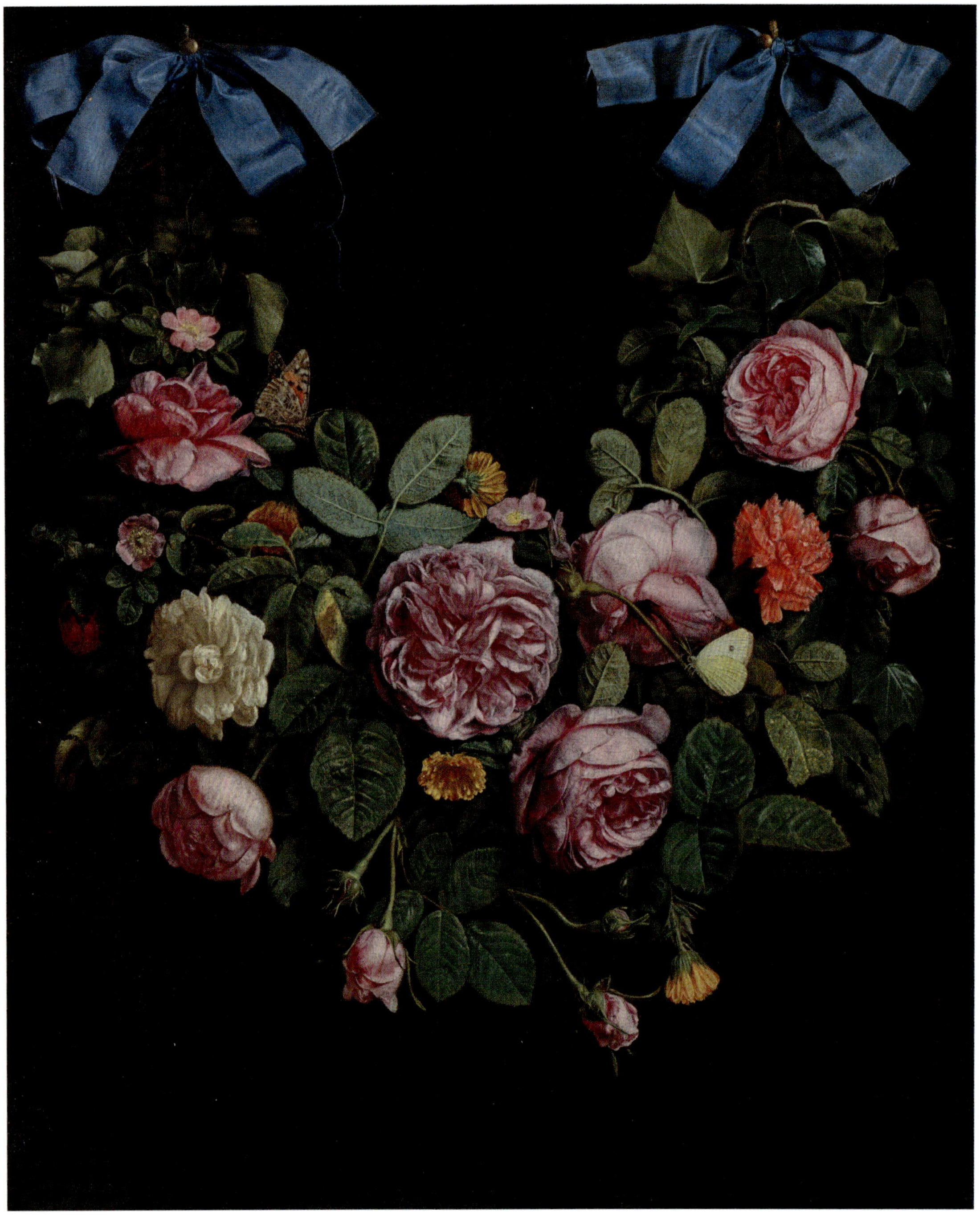

JAN VAN KESSEL THE ELDER

A Swag of Roses Suspended from Blue Ribbons, mid-17th century
Oil on canvas, 23⅝ × 19½ in. / 60 × 49.5 cm
Private collection

Positioned against a black background, a garland of roses hangs loosely from its ties of fraying ribbons of cornflower-blue silk. The gorgeous mixture of flowers includes wild and cultivated roses ranging in hues from dusky pink to white, interspersed with ivy, daisies, and carnations. Red admiral and cabbage white butterflies nestle on the foliage, and visible dewdrops glisten on the petals. Flemish artist Jan van Kessel the Elder (1626–1679) painted this garland in minute detail, and his attention to the intricacies of this sumptuous bouquet reveals his artistic lineage, both literal and inspirational, as the grandson of renowned painter and draftsman Jan Brueghel the Elder, and a follower of self-taught artist and polymath Joris Hoefnagel, a pioneer of still-life painting. Van Kessel's central pink rose is probably a Centifolia, or "hundred-petaled," rose (see p.180), a favorite floral subject of Dutch still lifes and symbol of fragile imperfection as well as undeniable beauty. On close inspection, some of the roses' leaves have been eaten by insects, while others are browning. Many of the rose petals even look as if they are ready to fall, giving the painting a vague air of decay. Highly prolific, Van Kessel painted more than three hundred panels during the 1650s and 1660s, often featuring garlands and flowers against dark backgrounds, as here. His images were commercially highly successful, but Van Kessel is better known for his small, minutely detailed paintings of insects and shells, typically painted against a light background.

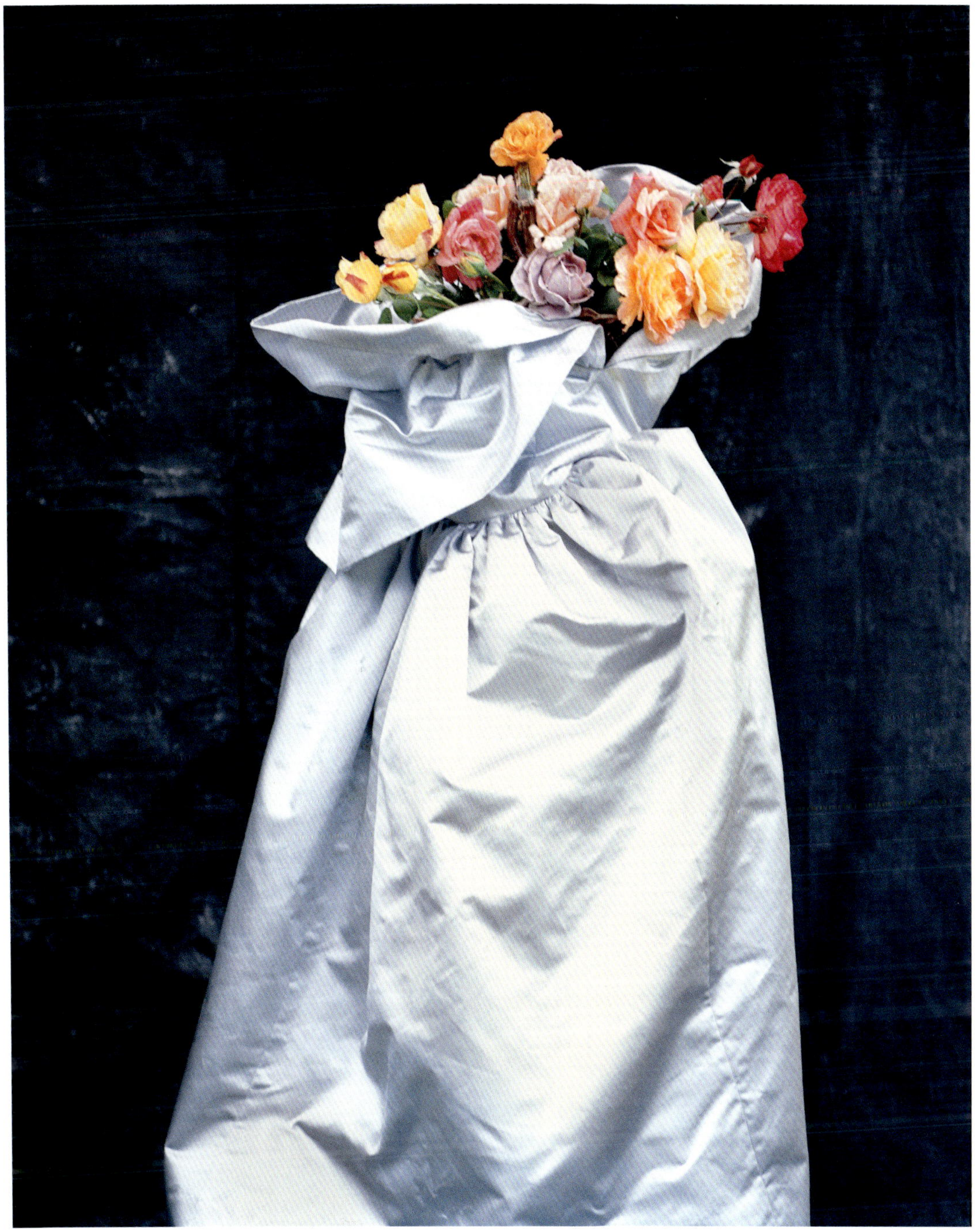

BRUCE WEBER

Charles James Dress and Roses, Kent, 1982
Iris print, 20 × 15⅞ in. / 50.7 × 40.2 cm
Private collection

At first glance, this 1940s Charles James (see p.49) dress looks as though it is being worn, gracefully filled out and standing on its own. On closer inspection, it becomes clear that the gown's bodice is stuffed with a colorful arrangement of roses, making for a bold still life by American fashion photographer Bruce Weber (b. 1946). Known for his seductive fashion images and his celebration of the male body, Weber instead captures here the feminine sculptural form of this vintage dress with freshly picked English roses while on a shoot in Kent. One of the most revered couturiers in fashion history, James was known for his exquisitely constructed gowns, which were worn and photographed by the celebrities of the early twentieth century. Without formal training, James created technical garments that folded around the female form, visualizing his complex patterns in three dimensions before wrapping or draping them directly on the body. The manipulation of material was one of his strengths, along with his extraordinary use of color. His signature silhouette—a cinched waist, firmly supported décolletage, and an excess of luxurious materials in a voluminous skirt—stands out as the ideal in femininity and contained eroticism. James's magic was to combine bodily geometry and dressmaking techniques with the sensual act of wearing a perfectly fitted piece of clothing. Weber plays with this idea by extracting the body, juxtaposing the empty dress with a bouquet of the most seductive of flowers—the rose.

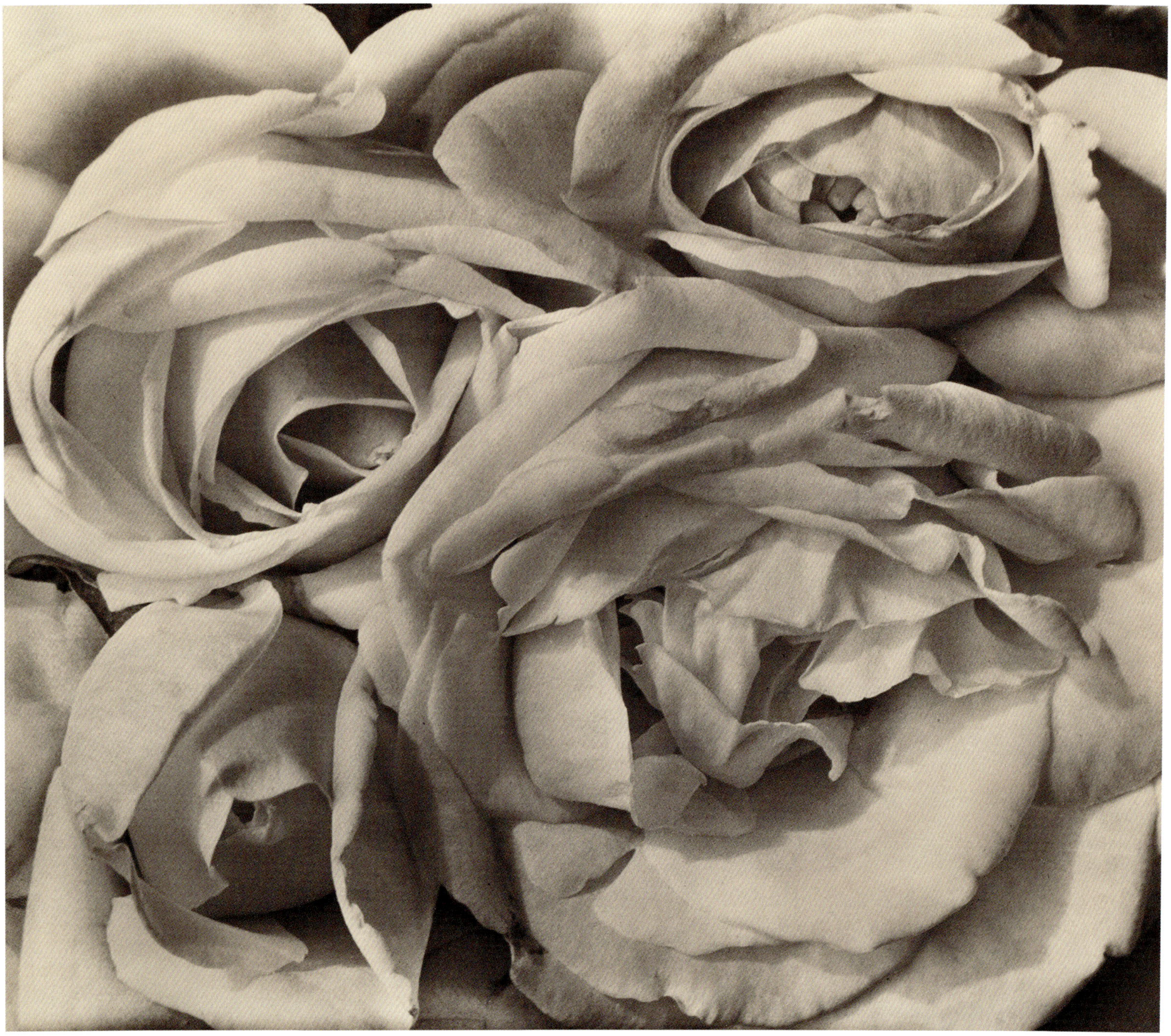

TINA MODOTTI

Roses, Mexico, 1924
Platinum-palladium print, 7⅜ × 8½ in. / 18.9 × 21.6 cm
Museum of Modern Art, New York

The softest of blossoms piled up in a pillowy mound fill the picture plane, a gentle light cast upon their fading petals, in a still life by Italian artist Tina Modotti (1896–1942). After immigrating at age sixteen to the United States, where she was an actor in silent films and an artist's model, Modotti moved to Mexico City in 1923 with her lover and mentor, photographer Edward Weston. Together they ran a photography portrait studio and became part of an international circle of political exiles and cultural figures, including Sergei Eisenstein, Leon Trotsky, and Mexican artists Frida Kahlo and Diego Rivera. Motivated by her interest in urban life and her admiration for the Mexican working class, and inspired by Mexican folk art, landscapes, and plants, Modotti worked with a large-format camera to capture the city and its inhabitants. Her studio work—especially still-life photography, such as this arrangement of individual roses presented as one subject—helped Modotti understand and fine-tune the formal considerations of her work. While these roses exemplify her aesthetic concerns, they also recall the socialist slogan "Bread and Roses" from the early 1910s, an appeal for fair wages and working conditions that would have been familiar to Modotti, a committed Communist. Her political leanings caused her exile from Mexico in 1930, whereupon she settled in Moscow, abandoning photography for political work as part of the Soviet Communist Party, before dying unexpectedly at the age of forty-five.

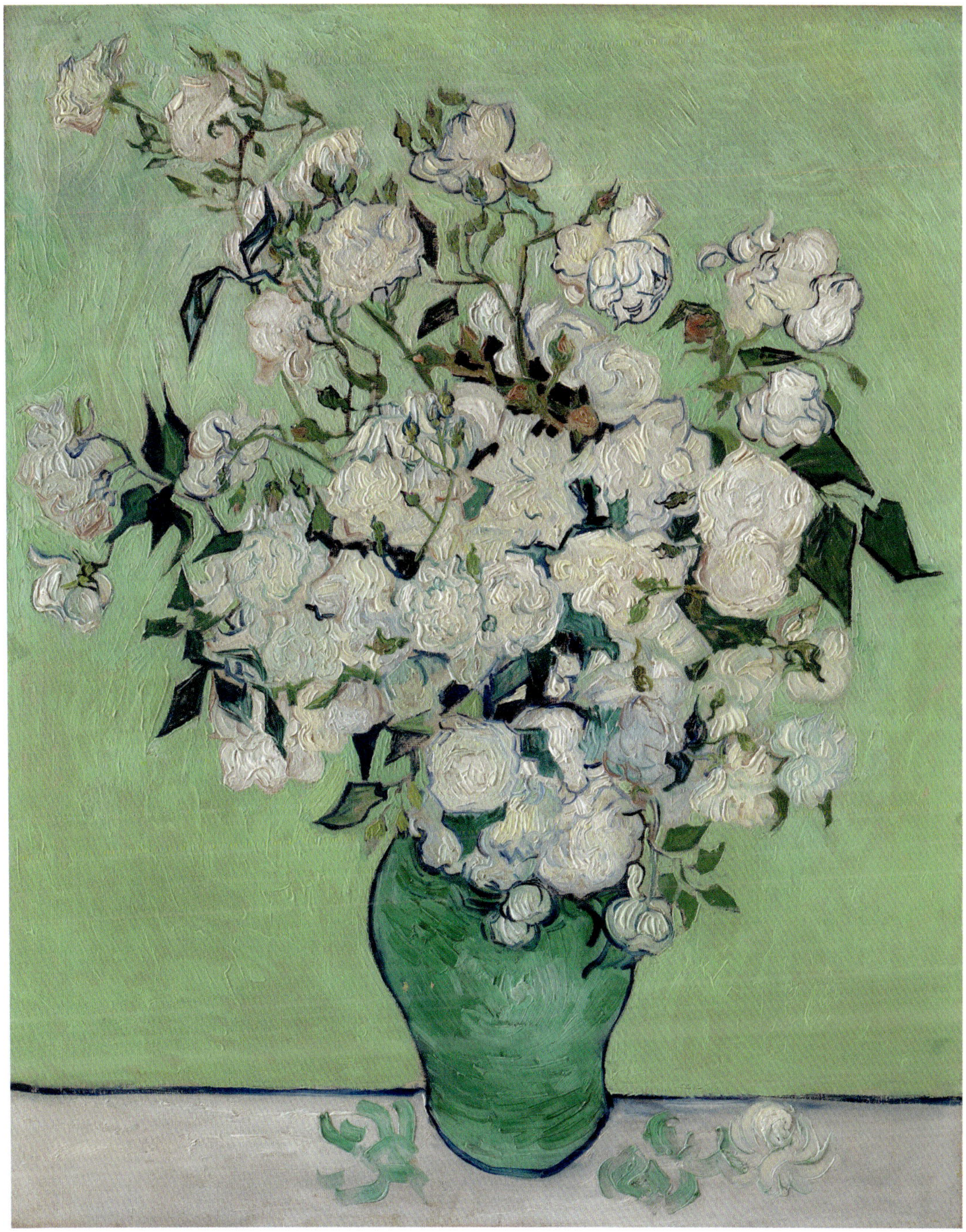

VINCENT VAN GOGH

Roses, 1890
Oil on canvas, 36⅝ × 29⅛ in. / 93 × 74 cm
Metropolitan Museum of Art, New York

While the name of Vincent van Gogh (1853–1890) may be synonymous with sunflowers, the famous Dutch Post-Impressionist artist loved painting other flowers as well. Thanks to the copious correspondence he left behind, we know much about his garden musings. In the many letters to his brother Theo, Vincent wrote passionately about gardening and flowers, and in a letter to his sister, he declared that his thoughts on gardens could fill an entire book. Although success eluded Van Gogh in his lifetime, preventing him from cultivating a garden of his own, he had friends with lush and beautiful gardens, and some regularly delivered bouquets for him to paint. Theo, in particular, encouraged his brother to paint flowers, seeing them as marketable and the ideal subject for working in a cramped studio. This floral still life of pale pink roses in a green vase is one of a series that Van Gogh produced just before the end of his yearlong stay at the Saint-Rémy asylum. In his distinctive style and impasto technique, bold blue lines define sections of the composition, contrasting with the softness of the flowers and adding vibrancy to their muted hues. The influence of Japanese prints, as well as of the seventeenth-century Dutch tradition of flower painting, can be seen. Spent petals and dried leaves rest on the tabletop, a reminder of the bouquet's ephemerality. In returning to his beloved floral arrangements late in his career, Van Gogh shows in *Roses* the optimism he then felt, however fleetingly.

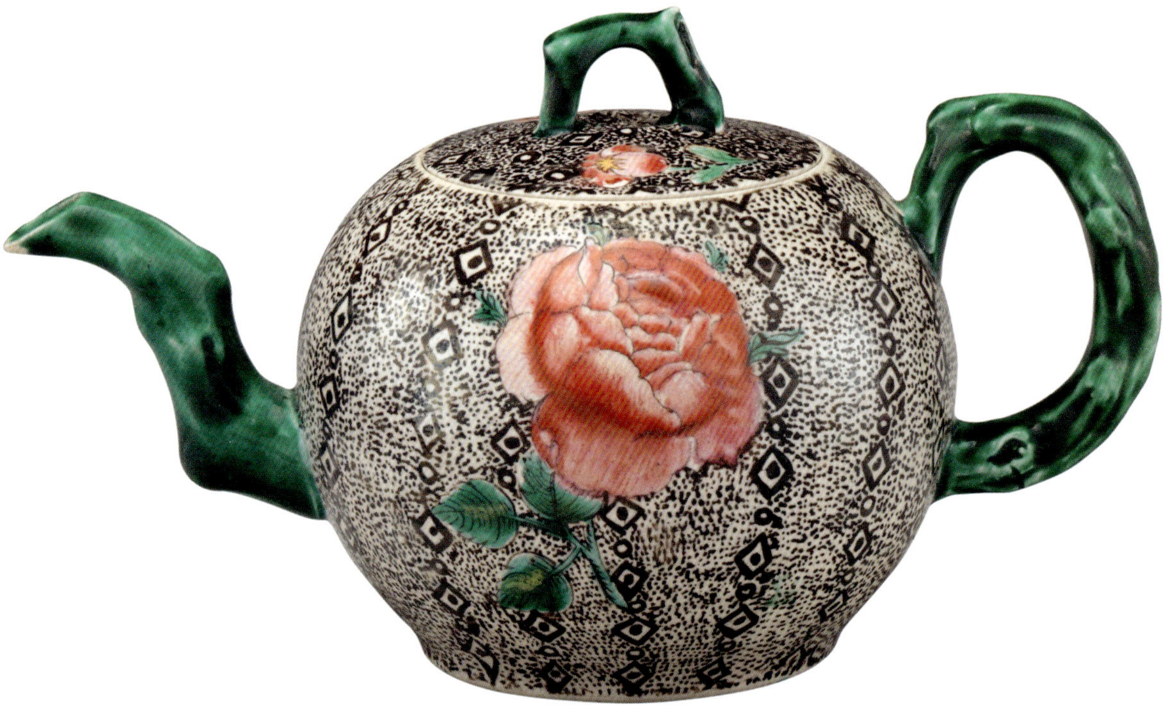

ANONYMOUS

Teapot, c. 1765
Salt-glazed stoneware with enamel decoration,
3⅝ × 6½ in. / 9.2 × 16.5 cm
Metropolitan Museum of Art, New York

Once more valuable than gold, and said to be the most popular drink in the world after water, tea has been drunk since the tea plant (*Camellia sinensis*) was discovered in China more than three thousand years ago. This striking enameled ceramic teapot from the mid-eighteenth century comes from a period in British history when the drinking of tea was still restricted to the wealthy; its refinement shows that it is a luxury object. Against an abstract black-and-white background, a single red rose appears on its lid and a large, flat rose decorates the two sides of the pot; the green twisted spout and handles resemble the rose's stem. British manufacturers had discovered how to make stoneware—which had until then been the preserve of the Chinese—in the mid-eighteenth century. This teapot was manufactured in North Staffordshire, the largest producer of ceramics in Great Britain and a global exporter of teapots. The object's rounded shape echoes that of the earliest Chinese pots, which were designed to allow the tea leaves to steep in hot water; the earliest British teapots, in contrast, more closely resembled conical coffee pots. The choice of a rose for decoration speaks to the enduring popularity of the flower, but teapots have over the centuries been graced with designs from the sophisticated to the decorative to the crudely humorous. In 1784 the British government reduced the punitive duties levied on tea, making the drink more affordable—and initiating a national obsession.

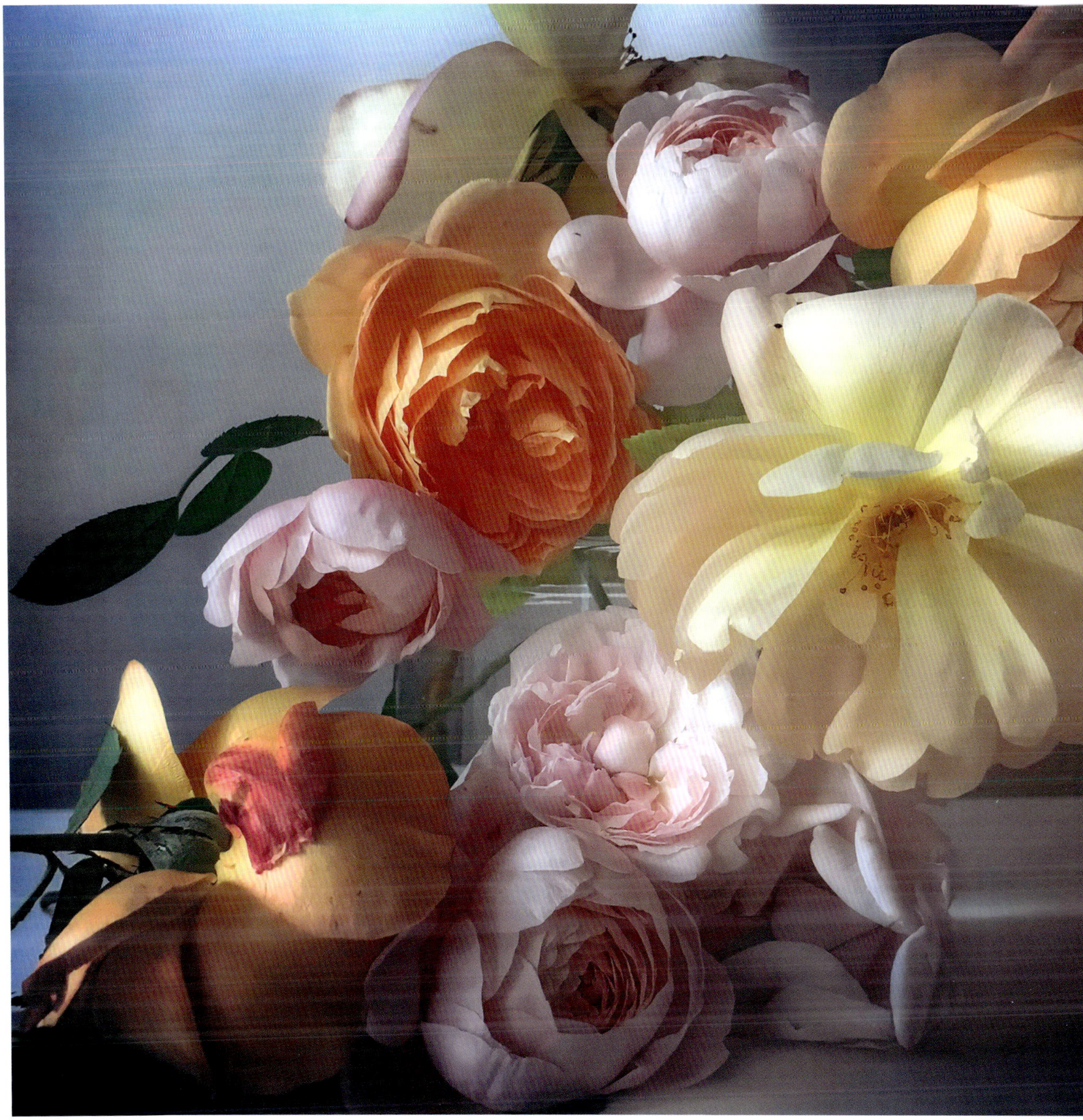

NICK KNIGHT

Saturday 22nd October, 2016, 2019
Hand-coated pigment print, 20 × 20 in. / 50.8 × 50.8 cm

Roses in shades of pale pink, lemon, and apricot cascade from a glass jar, while others lie in front and fill the frame of this image taken by British photographer Nick Knight (b. 1958) on Saturday, October 22, 2016. Better known for his fashion photography—Knight has worked with the designers Jil Sander, Alexander McQueen (see p.219), Gareth Pugh, and Yohji Yamamoto, among others—he also made a series of painstaking studies of roses from his garden in southwest London, each titled with the date of its creation. Inspired by sixteenth- and seventeenth-century still-life painters, such as Jan Brueghel the Elder and Jan van Huysum (see p.140), Knight picks his roses from the garden and sets out to capture their fragile beauty using just his iPhone and daylight. For a photographer more used to the busy intensity of a packed studio, part of the joy of photographing his favorite flower is the solitary nature of the project. Each photograph takes four or five hours, during which time the rose changes; it might unfurl or start to shed its petals, but it is never static. Roses have long featured in Knight's life. His mother's middle name was Rose and he sports a single tattoo of a rose, but his real passion for the plant began in the early 1990s, after he spent three and a half years photographing pressed flowers in the collections of London's Natural History Museum. He sees the rose as possessing an intrinsic poetic tragedy: its doomed beauty captures the transience of life with every fallen petal.

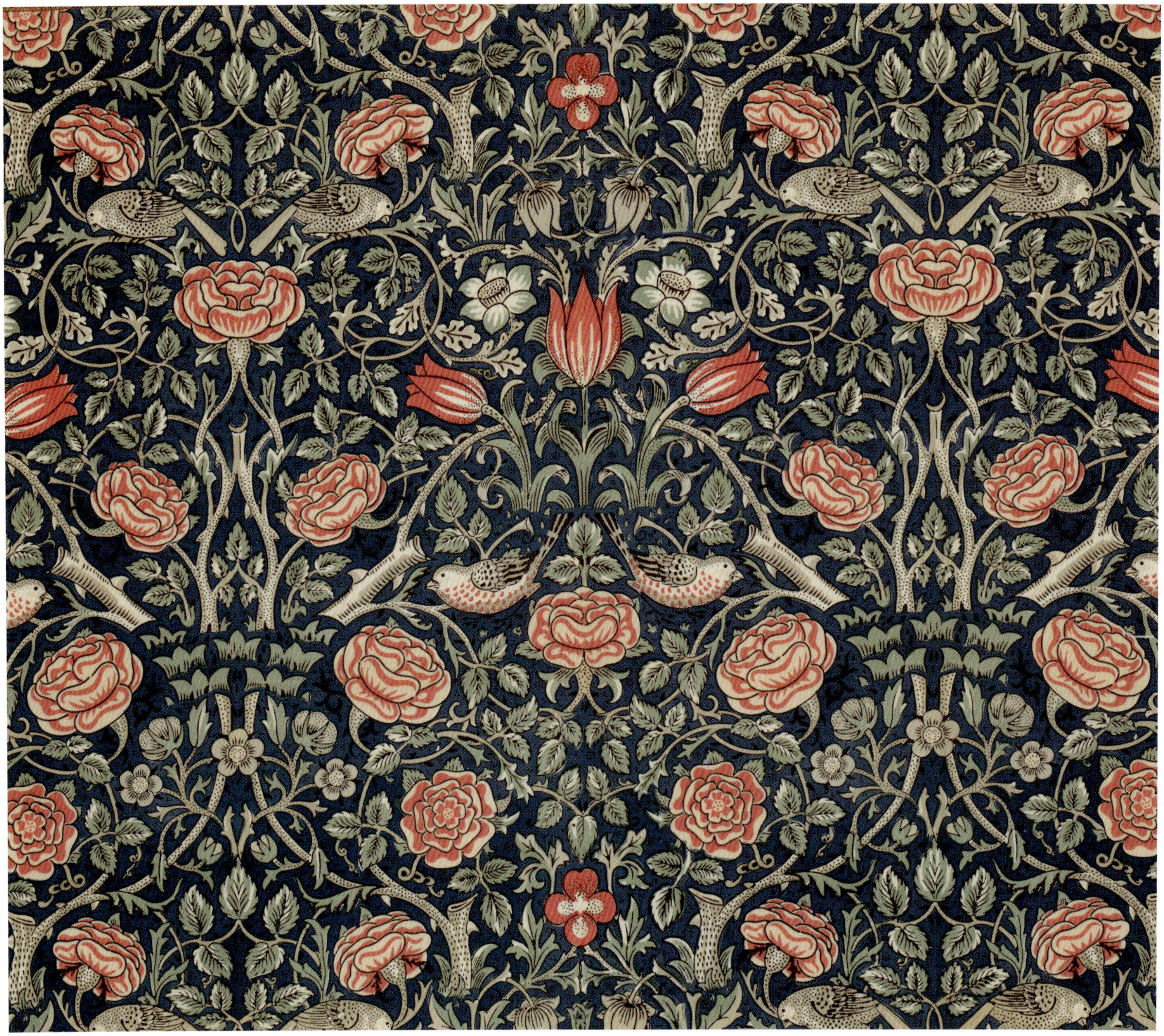

WILLIAM MORRIS

Tudor Rose Pattern Printed Fabric, c. 1883
Cotton, 36 × 35 in. / 91.4 × 88.9 cm
Museum of Modern Art, New York

Red-and-white bicolor roses on vigorous canes dominate this textile by English designer William Morris (1834–1896). Morris—writer, artist, entrepreneur, and social activist—was a key proponent of the British Arts and Crafts movement, a creative response to the rapid industrialization of the nineteenth century. Favoring handcrafting for home and garden, Morris & Co. produced a range of decorative furnishings including tiles, wallpaper, furniture, and floor coverings, but Morris's first and lasting love was textiles. The designer and his fellow artist-members of the Pre-Raphaelite Brotherhood drew inspiration from English history, particularly the medieval and early modern periods. He even called roses the queen of the garden, the "flower of flowers." For this fabric design, Morris chose the name "Tudor Rose," an allusion to Henry VII, the first Tudor king of England, who adopted the emblem of the two-tone rose to symbolically unify the white rose of the House of York with the red rose of the House of Lancaster after the Wars of the Roses, a drawn-out dynastic conflict during the fifteenth century (see pp.33 and 164). No actual plant species was associated with the heraldic symbol, but the honorific was later conferred on *Rosa gallica* 'Versicolor', a bicolor variation of *R. g.* 'Officinalis' from the sixteenth century (see p.80). With its heritage, fragrance, and simple form, this cultivar fit Morris's preference for Old Garden roses. This was one of Morris's most successful patterns and is still produced by fabric manufacturers today.

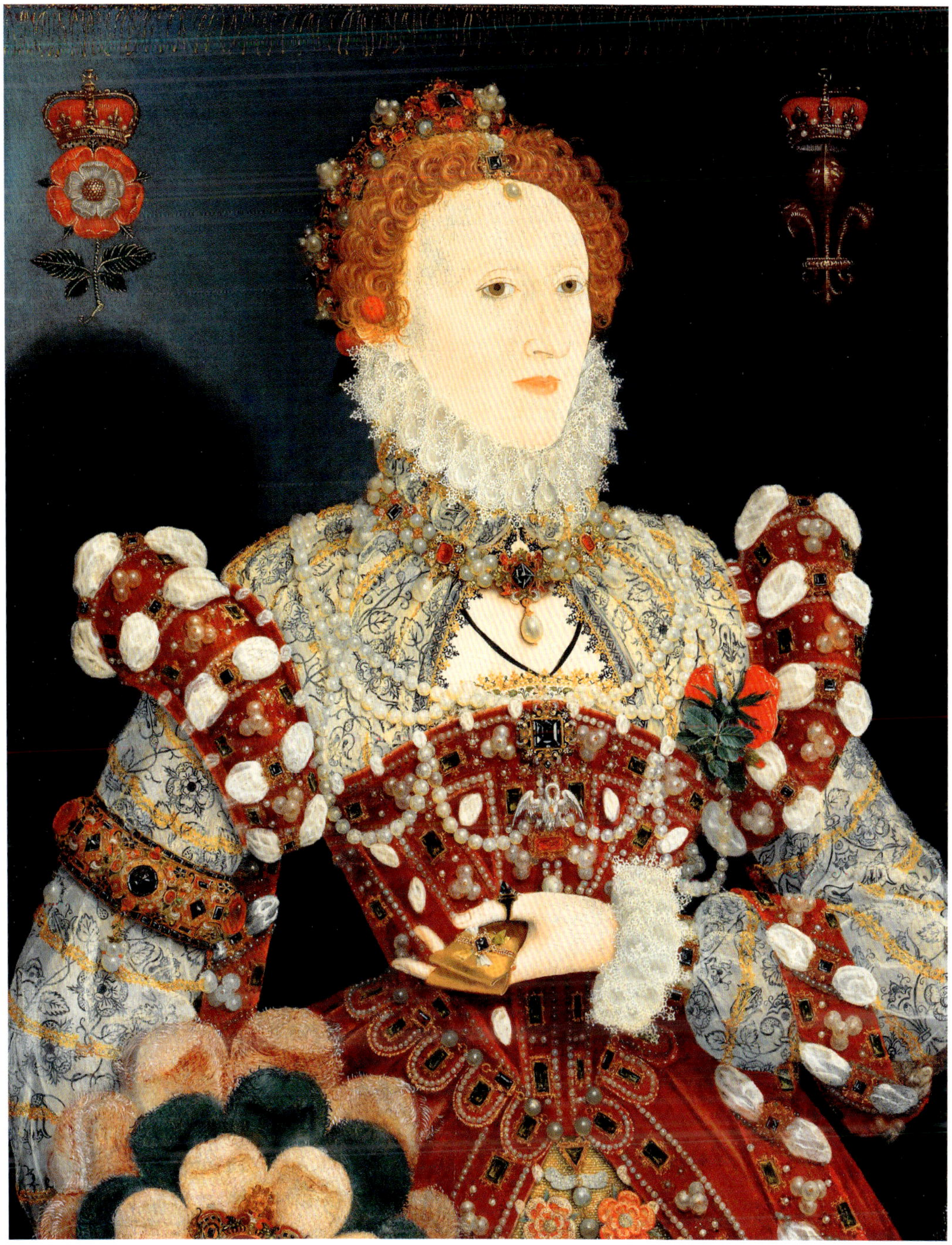

NICHOLAS HILLIARD

Portrait of Queen Elizabeth I, c. 1575
Oil on panel, 31 × 24⅛ in. / 78.7 × 61 cm
Walker Art Gallery, Liverpool, UK

Queen Elizabeth I loved emblems, and a startling variety from ears to rainbows appear in her portraits. She especially favored roses, whether red, white, or pink, and was often painted holding or wearing them. This portrait, attributed to English artist Nicholas Hilliard (c. 1547–1619), significantly includes a Tudor Rose floating at top left, surmounted by the crown. This famous flower, which combines the white rose of York with the red of Lancaster, symbolized the peaceful union achieved by her grandparents after the conflict known as the Wars of the Roses ended in the 1480s (see p.164). For good measure, there are also Tudor roses embroidered in blackwork on the Queen's chemise. Elizabeth had her own dynastic problems, being unmarried with no direct heir, but to her subjects she was a glorious, feminine figure. She created a cult as the Virgin Queen, and in this context roses added the mystique of a flower associated with the Virgin Mary herself. Hilliard's painting is also known as the "Pelican portrait," after the pendant Elizabeth wears in it. Another appropriation of Christian iconography, the symbol of a pelican feeding its young from the blood in its breast reflects Elizabeth's role as the mother of the nation. Hilliard's delicate portrait miniatures initiated a distinctive genre in England, and Elizabeth gave them as gifts. His half-length courtly portraits of her were painted in the 1570s, when, after the age of forty, the Queen welcomed his exquisite touch to maintain her image of masklike youth.

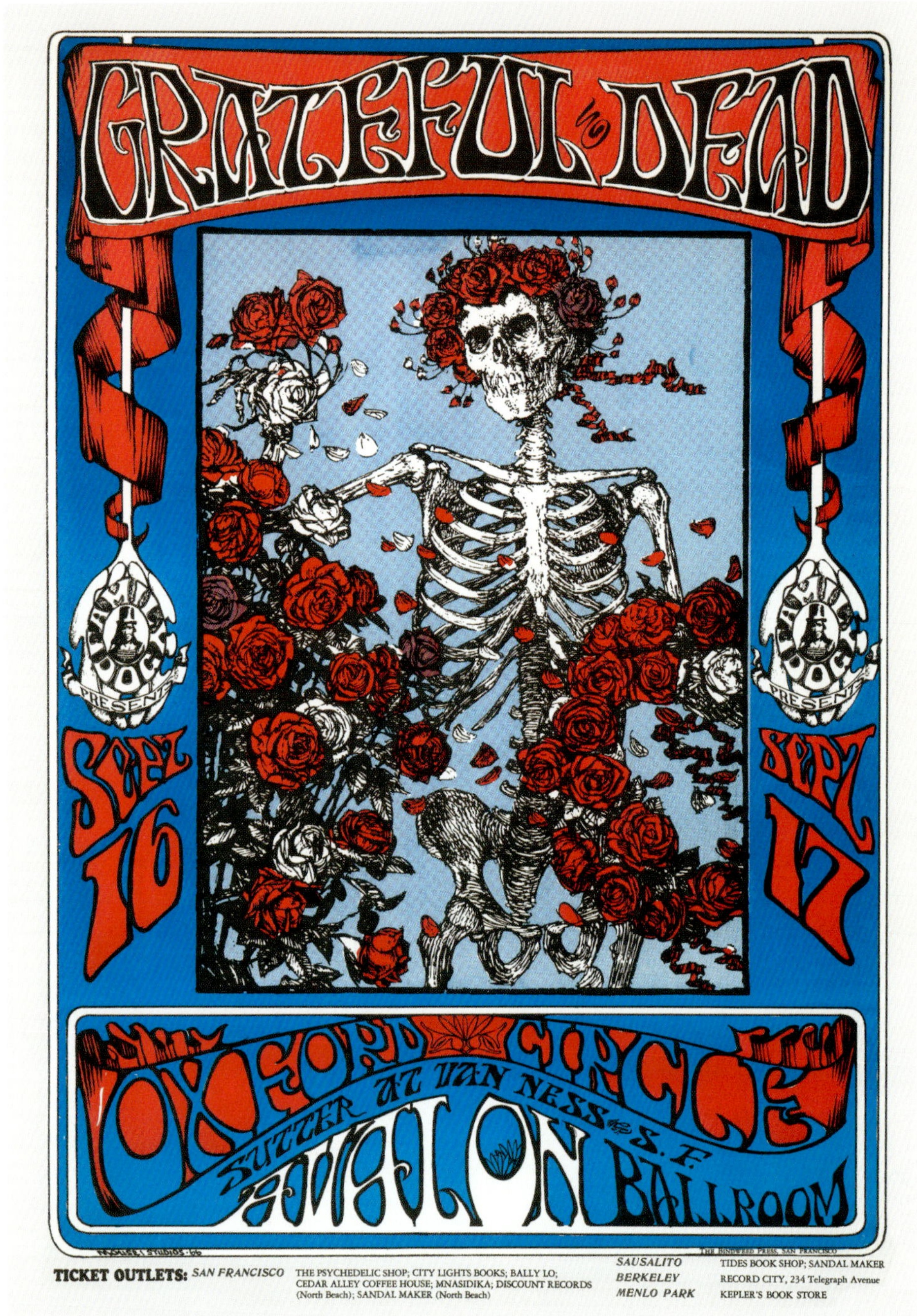

ALTON KELLEY AND STANLEY MOUSE

Skull and Roses; Grateful Dead and Oxford Circle at Avalon Ballroom, 1966
Offset lithograph poster, 20¼ × 14¼ in. / 51.4 × 36.2 cm

Recognized by die-hard Deadheads—the aptly named followers of famed jam band the Grateful Dead—and nonfans alike, the band's iconic symbol of a skull wearing a crown of roses has been endlessly reproduced around the world. The artwork was created by American artists Stanley Mouse (b. 1940) and Alton Kelley (1940–2008), known for their psychedelic poster designs during the 1960s. Heavily influenced by Art Nouveau, particularly the works of Alphonse Mucha and Edmund Joseph Sullivan, Mouse and Kelley appropriated this particular design from one of Sullivan's own drawings to use in a poster advertising a concert at San Francisco's Avalon Ballroom in 1966. They found their inspiration at the public library in an edition of Edward FitzGerald's *Rubáiyát of Omar Khayyám* (1859), a translation of twelfth-century Persian poetry. Sullivan's drawing "A Skeleton Amid the Roses" illustrates the poem's twenty-sixth quatrain, which contains the verse, "One thing is certain, that Life flies . . . / The Flower that once has blown forever dies." Apparently, the rose's transience and the ideas of Persian mysticism intrigued late nineteenth-century English bohemians and twentieth-century American hippies alike. A far cry from a collection of lifeless bones, Mouse and Kelley's skeleton wears a rose crown while holding a flower wreath in one hand and a bouquet in the other, as petals fall. Kelley and Mouse added color by imbuing the roses with a vivid, psychedelic red, thus creating one of the most famous concert posters in history.

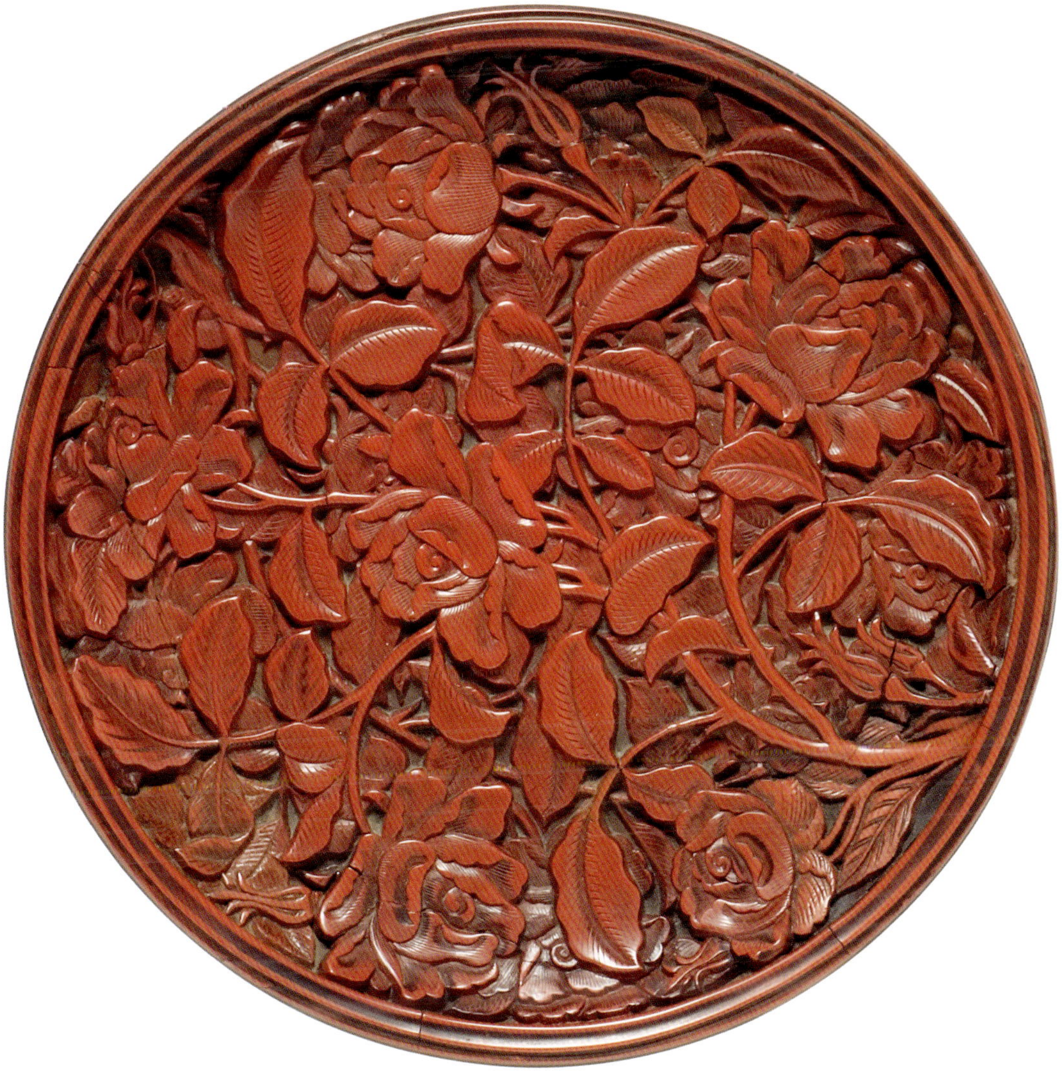

ANONYMOUS

Dish with roses, 1279–1368
Carved red lacquer on wood, Diam. 6⅜ in. / 16.2 cm
Los Angeles County Museum of Art, California

This striking lacquer serving dish was made during the Yuan dynasty (1279–1368), and its floral design probably depicts a rose discovered in China nearly 1,300 years ago, *Rosa chinensis*. With examples dating back as far as 450 BCE, lacquerware as an art form has a long history in China that evolved along two distinct paths: carved lacquer (*diaoqi*) and pictorial (or surface) decoration. The former technique, which was first developed in the Yuan period and continued through the Ming (1368–1644) and Qing (1644–1911) dynasties, involves the carving of built-up layers of thin coats of lacquer into a three-dimensional design. A natural plastic, lacquer is a resin made from the viscous sap of the highly poisonous, aptly named Chinese lacquer tree. The raw lacquer is a capricious material, resistant to water, acidity, abrasion, and heat. Once purified, it can be applied to the surface of any object in a thin layer, and the object can then be placed in a warm cabinet to dry. Because lacquer may require thirty or more coats to become completely protective, its production is laborious and expensive. *R. chinensis* was notable for the fact that it did not bloom just once a year, but repeatedly—hence its local name *yue yue hong* (every month red). Chinese plantsmen crossed it with *R. gigantea* to produce the oft-blossoming roses from spring to the first frosts. These cultivars were imported into Europe in the nineteenth century as part of the tea trade, which is why they are referred to as "tea" roses.

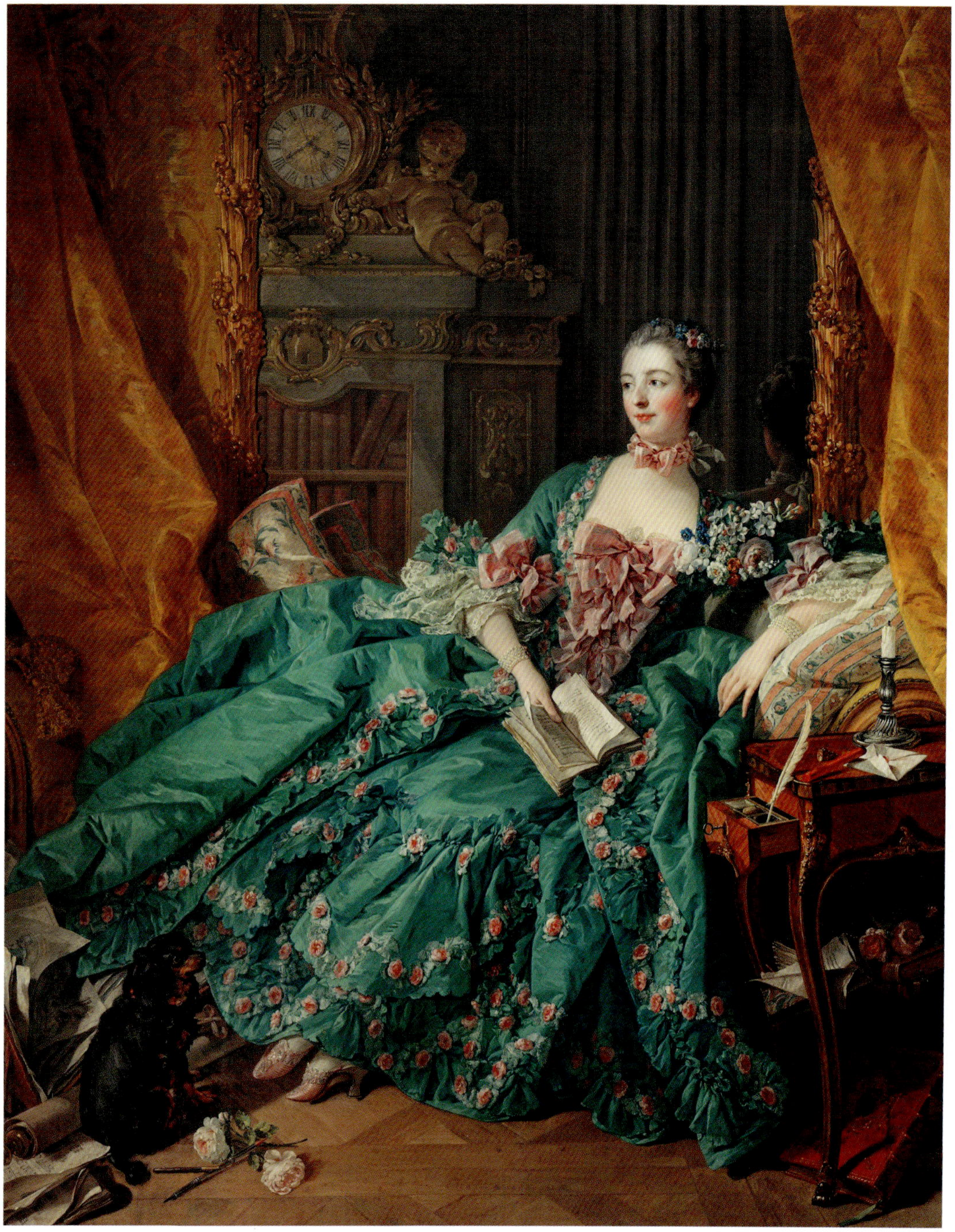

FRANÇOIS BOUCHER

Madame de Pompadour, 1756
Oil on canvas, 80¾ × 63⅜ in. / 205 × 161 cm
Alte Pinakothek, Munich

Madame de Pompadour, born Jean-Antoinette Poisson, became a woman of considerable influence in the court of Louis XV. Her position as the official mistress of the French king between 1745 and 1751 was, of course, a determining factor in her powerful courtly status, and allowed her to transcend her bourgeois family background. However, Pompadour's intelligence and discerning eye for decorative arts and fashion allowed her to adroitly realize her role as bellwether for modish concerns at the court of Versailles. She has been described as the personification of the Rococo style that was the dominant aesthetic in this period, characterized by pastel colors, curvilinear design, and floral adornment. Roses were a recurring feature in portraits of Madame de Pompadour, and this painting by French artist François Boucher (1703–1770) is an exemplar of her predilection for rosy adornment. She rests in a languid position, clothed in a green silk-taffeta gown, with garlands of small pink roses embellishing the trim of the bodice and skirt. Roses decorate her hair and feature in the nosegay attached to her bodice, and in the pink silk ribbons that jostle along the front of the dress. Creating a choker fancily fastened at her throat, the silk recalls the color and form of blossoming roses. Next to the faithful spaniel that sits at her feet are two discarded rose stems, as if to underscore the abundance of roses in her accoutrements.

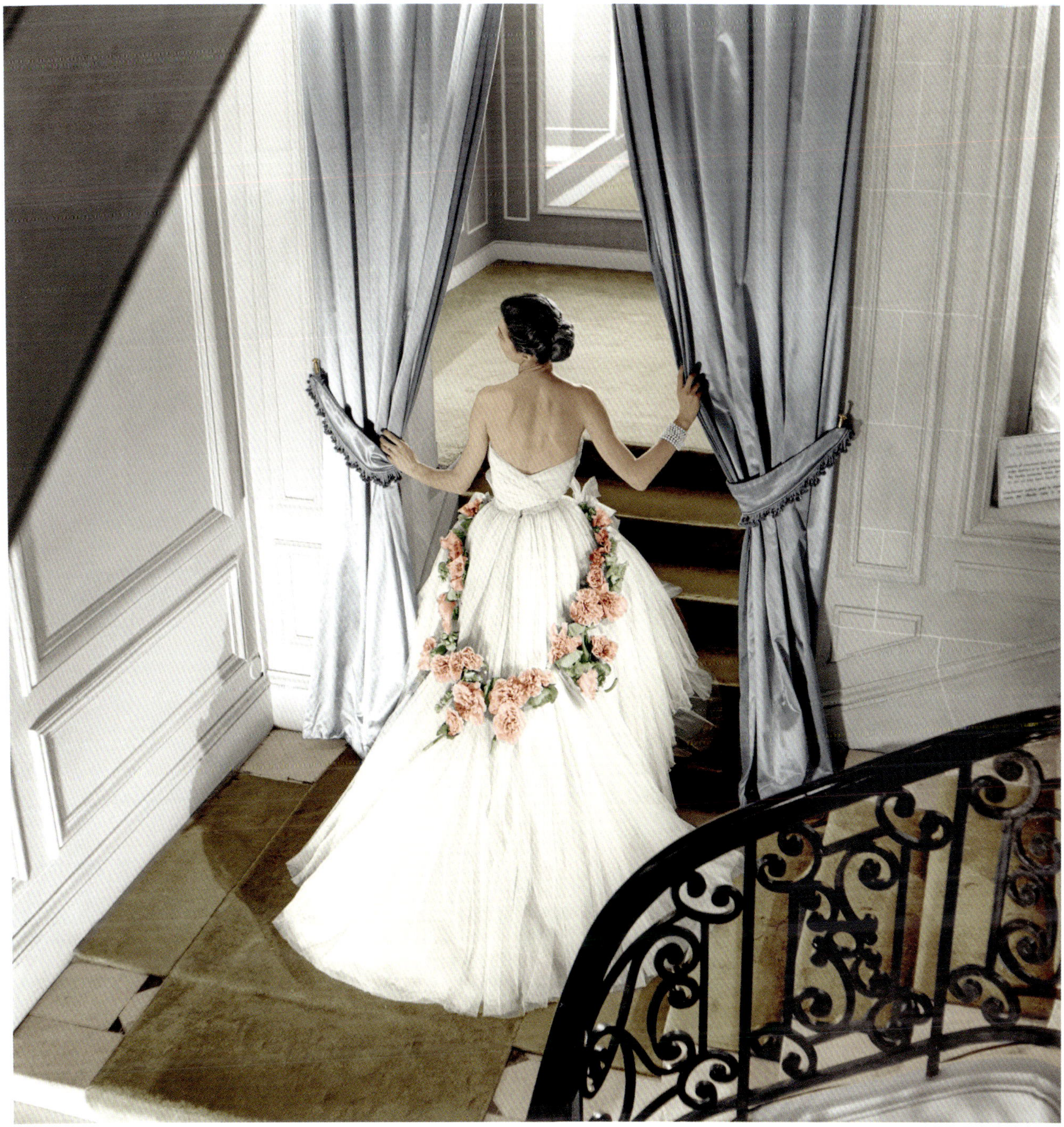

CHRISTIAN DIOR

Tableau Final formal gown, from the haute-couture collection *Naturelle Line*, spring–summer 1951
Photograph by Willy Maywald

A loose garland of full-bodied pink roses laced with ivy sits far below the cinched waist of a white tulle gala gown worn elegantly by a model for the Spring/Summer 1951 haute-couture collection of French designer Christian Dior (1905–1957), the twentieth century's most influential fashion designer. Unseen in this image by photographer Willy Maywald, the roses continue on the front of the strapless dress in a pink double-rose corsage. In 1947, just two years after the end of World War II, Dior had stunned the world with his "New Look," which debuted an innovative silhouette for women's clothing with rounded shoulders, a small waist, and a full skirt that used a remarkable 16 yards (14.6 m) of fabric. Dior claimed later that the look was inspired by the rose, and, indeed, flowers were his lifelong passion. He once said, "After woman, flowers are the most lovely thing God has given the world." His love of flowers was inspired by his mother's rose garden close to Granville in northern France, as well as by the career of his beloved sister Catherine, who was a professional flower farmer. Roses appeared in many of his garments—as rose-printed fabrics, accessories, or additions to his dresses—and in his profitable fragrance line. His first perfume, Miss Dior, which he launched in 1947, was even inspired by the rose. At his final home, the Château de La Colle Noire near Grasse in southern France (a town itself renowned for its own fragrance industry; see p.143), Dior planted rose fields that are still used for the perfumes that bear his name.

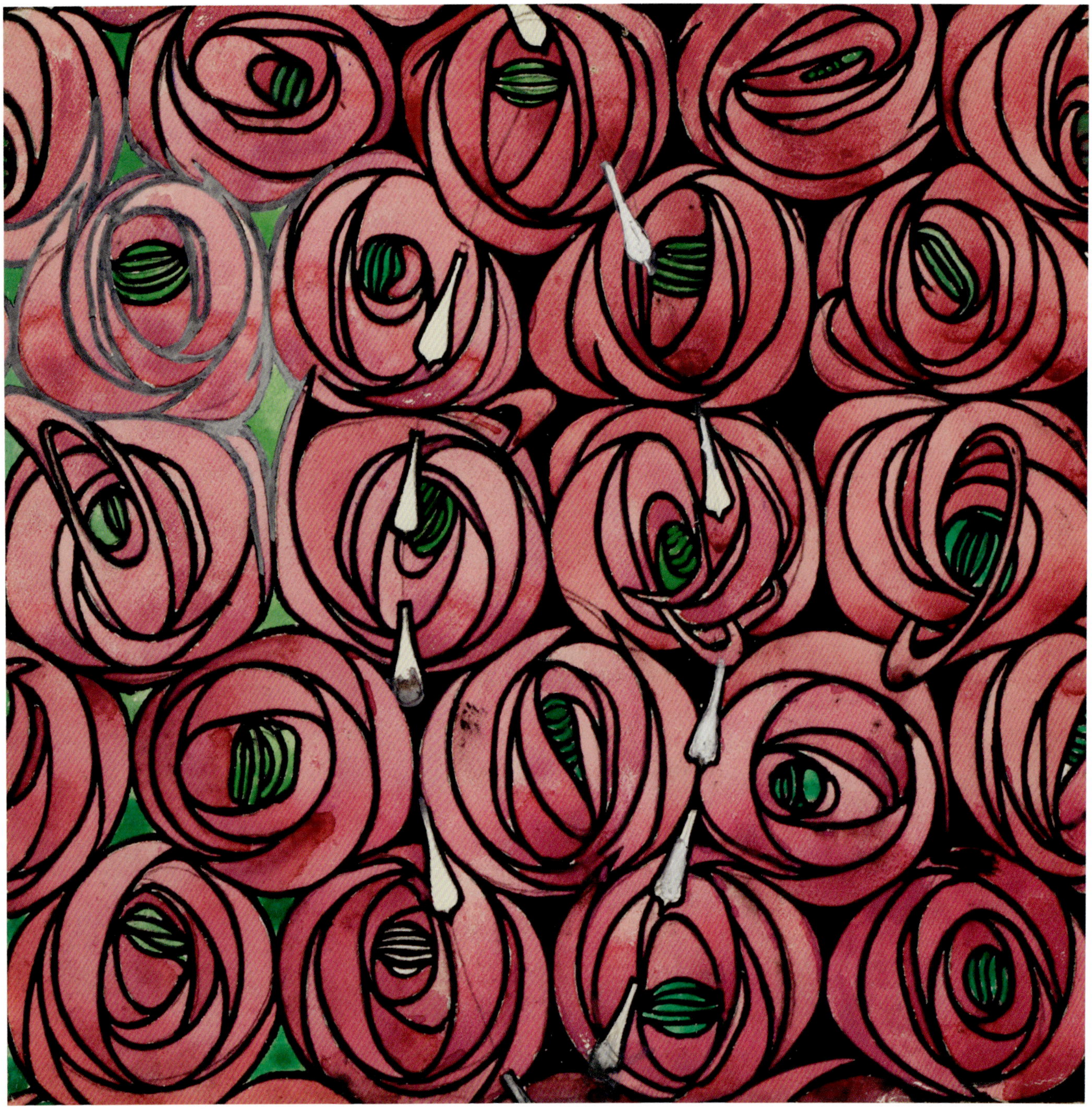

CHARLES RENNIE MACKINTOSH

Rose and teardrop textile design, 1915–23
Pencil, watercolor, and gouache on paper, 10 × 10 in. / 25.5 × 25.4 cm
Hunterian Museum and Art Gallery, University of Glasgow

Looked at closely, each of these highly stylized roses is unique, although all are made up of the same elements: a series of curved black lines shaded in pink, with green centers and white teardrops scattered across them. These rows of roses are the signature of Scottish architect and designer Charles Rennie Mackintosh (1868–1928), who is better known as an architect than as a painter, having helped to create a whole new architectural idiom, the Glasgow Style. Working alongside his wife, Margaret Macdonald Mackintosh, he originated an innovative aesthetic that emphasized function over beauty and was heavily influenced by the Art Nouveau movement that flourished in Europe between 1890 and 1910, in particular its concentration on organic patterns and flowers, along with Celtic and Japanese art. In this pattern, Mackintosh incorporated the roses to soften the boundaries between the blooms; the curved lines of each rose represented his vision of reality. These roses became his signature, and they still appear today on a wide range of textiles and products. After leaving his Glasgow architectural practice, he retired in 1914 to the English village of Walberswick, Suffolk, where he spent much of his time painting flowers. So strong was the link between Mackintosh and the rose that iconic British rose grower David Austin (see pp.14, 77, and 105) later bred an English shrub rose to closely resemble Mackintosh's stylized ones, naming it after the Scottish architect.

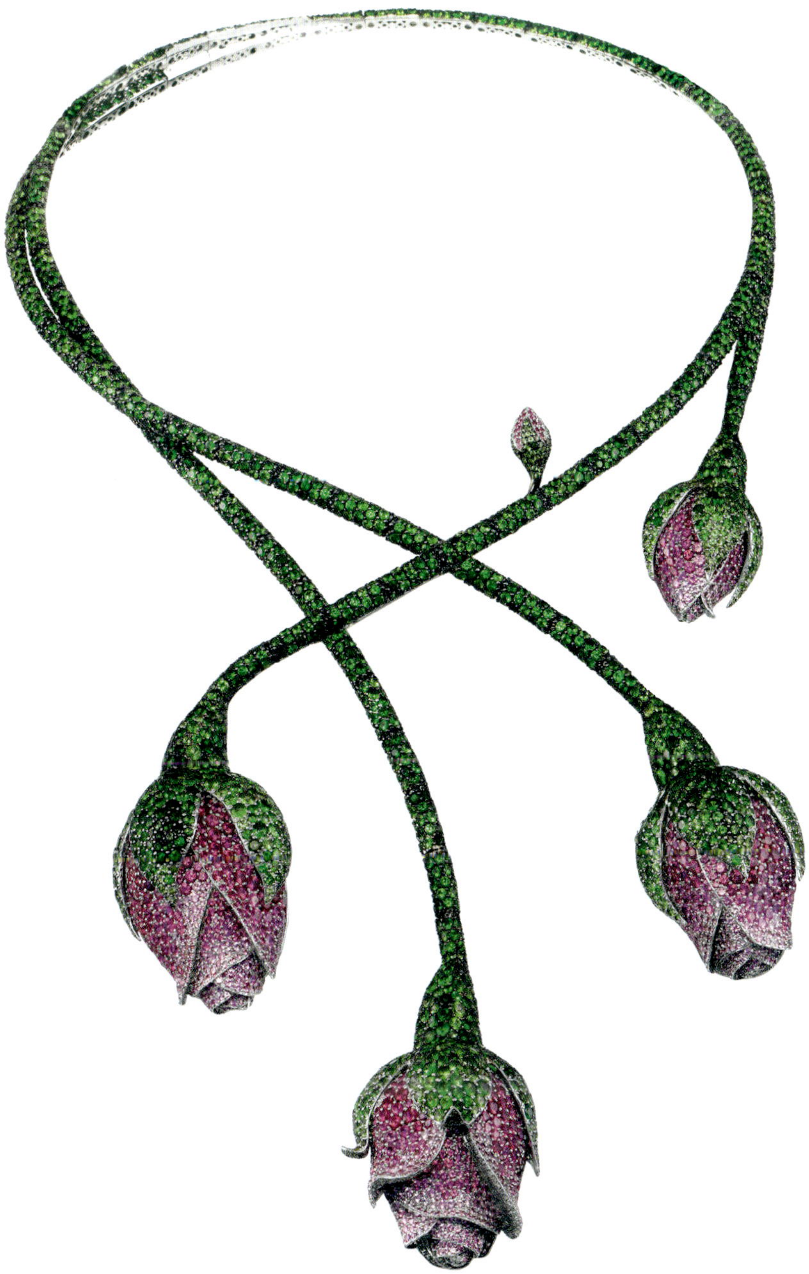

CHOPARD

Floral necklace, from *Red Carpet* collection, 2014
Tsavorites, pink sapphires, diamonds, and rubies

This glittering necklace takes the form of a gorgeous grouping of wild pink roses in tight bud, which drape about the wearer's neck as if climbing and growing around the body. Designed by renowned Swiss jewelry house Chopard in 2014, this piece was part of the *Red Carpet* collection, inspired by the glamour of actresses from the Golden Age of Hollywood in the 1950s and 1960s, as well as by the natural world. Worn by Australian actor Cate Blanchett on the actual red carpet of the Trophée Chopard after-party at the Cannes Film Festival, the piece twists on itself like a rambling rose, the flexible stems made from thousands of tsavorites, a green variety of garnet known for its intense color and brilliant luster. The green sepals reveal petals intricately created with diamonds, pink sapphires, and rubies of a slightly darker hue, which have been set to fashion the tight buds. More than a century and a half old, Chopard is a family-owned business, overseen by artistic director Caroline Scheufele. Known for her innovative approach to fine jewelry, Scheufele takes inspiration not only from nature and art, but also from the gems themselves, often designing pieces based on the look, feel, and cut of the stones. In the case of this highly crafted necklace, the verdant green and romantic pinks of the gemstones must have immediately conjured the image of budding roses in a lush garden.

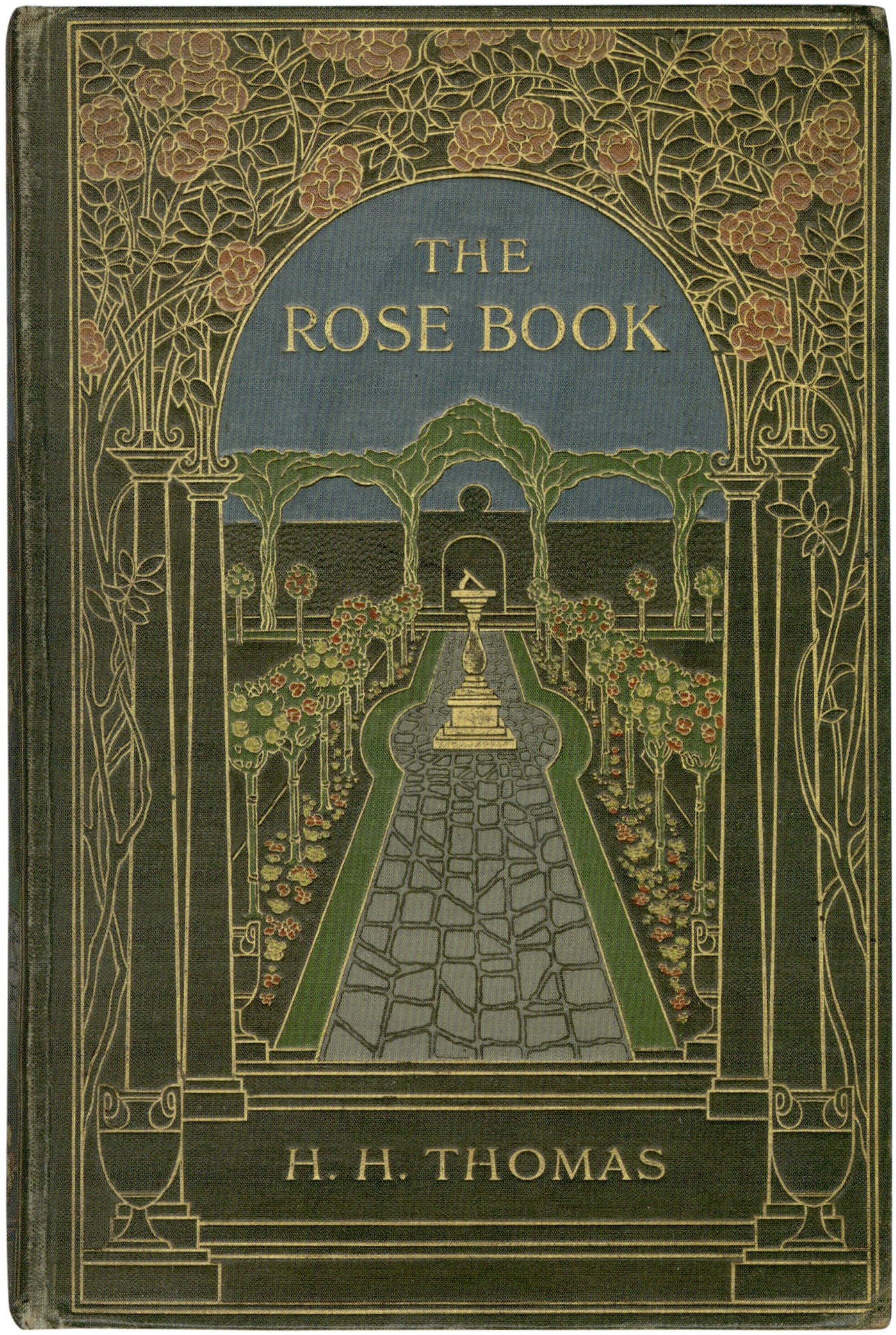

H. H. THOMAS

The Rose Book: A Complete Guide for Amateur Rose Growers, 1913
Printed book, 8⅛ × 5½ in. / 20.8 × 14 cm

Harry Higgott Thomas (1876–1956) was the son of Owen Thomas, royal gardener at Windsor Castle in England, where young Harry trained, emerging with a sound knowledge that ran the gamut of horticulture. In the early twentieth century, the name of H. H. Thomas was one of the most famous in English gardening; he was editor of the weekly magazine *Popular Gardening* for forty years, as well as editor of more than one multivolume encyclopedia of gardening, a prolific author of books, and gardening correspondent for the *Daily Telegraph* newspaper. Some of his early books, published before World War I, were sumptuously produced with ornate front and back covers. *The Rose Book* (1913) is one such, illustrated with color photographs by Henry Essenhigh Corke (one of the most prominent plant photographers of the period) and monochrome photographs by Thomas himself. The gilded cover design shows a path with intricate paving, standard roses, and a pergola in the distance. Thomas begins this book, meant to be a guide for amateur rose gardeners, with the statement: "Everyone who writes gardening books is morally bound, I think, to help the beginner . . . There is also a debt to the roses, for some are so much more accommodating than others . . . So 'twere shame not to let the beginner win his way to the finer rose glories through those that are more easily attained." Even with such elaborate rhetoric, Thomas's books fostered a love for and understanding of roses for an entire generation of British gardeners.

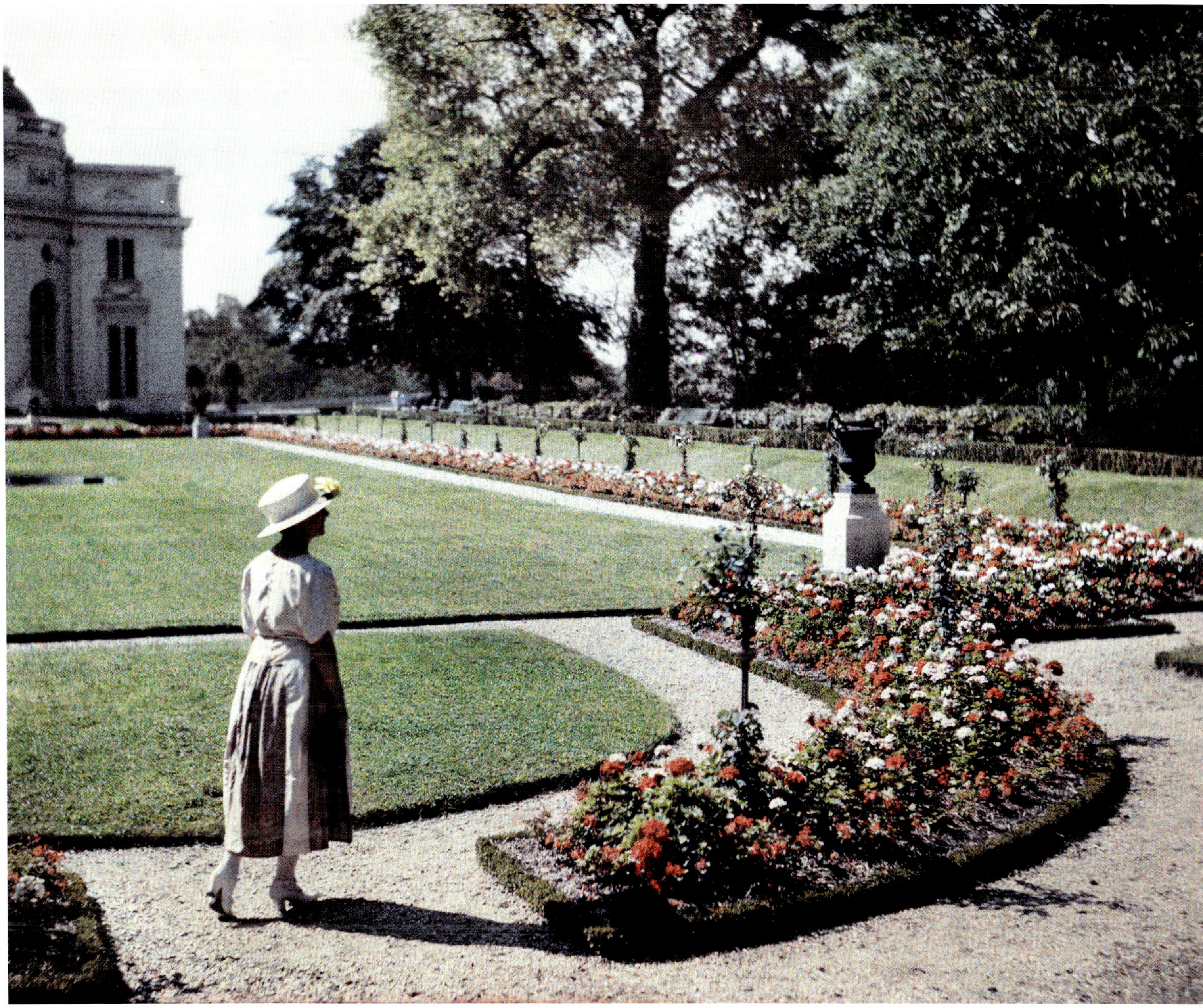

ANONYMOUS

Untitled, c. 1920
Photograph, 14½ × 18⅛ in. / 36.7 × 46.1 cm

The stylish Parisian lady in this photograph is enjoying a summer stroll through the charming Parc de Bagatelle, home to two of the oldest, most important rose gardens in France. Located on the western outskirts of Paris over an area of 62 acres (25 ha), the park is one of the four main botanic gardens in the city (the others being the Parc Floral de Paris, the Arboretum de l'ecole Du Breuil, and the Jardin des Serres d'Auteuil). Its neoclassical chateau was created in 1777, being built in just sixty-four days following an audacious wager between Marie Antoinette and her brother-in-law, the Count of Artois. The formal gardens that initially surrounded the chateau were redeveloped in the nineteenth century in an Anglo-Chinese style. They were expanded in 1905 by French landscape architect Jean-Claude Nicolas Forestier with the introduction of collections of horticultural plants and two magnificent rose gardens—the landscape rose garden and the classic rose garden—which today contain more than nine thousand rose plants, representing more than one thousand cultivars that comprise France's national collection of modern roses. In the summer, the turfed flower beds bordered with boxwood burst with the vibrant blooms of rosebushes and shrub, climbing, and rambling roses. The gardens are also home to the Concours International de Roses Nouvelles, an international competition for new roses. Held every June since 1907, it is one of the world's most prestigious events for rosarians, awarding prizes for the most beautiful and fragrant rose hybrids.

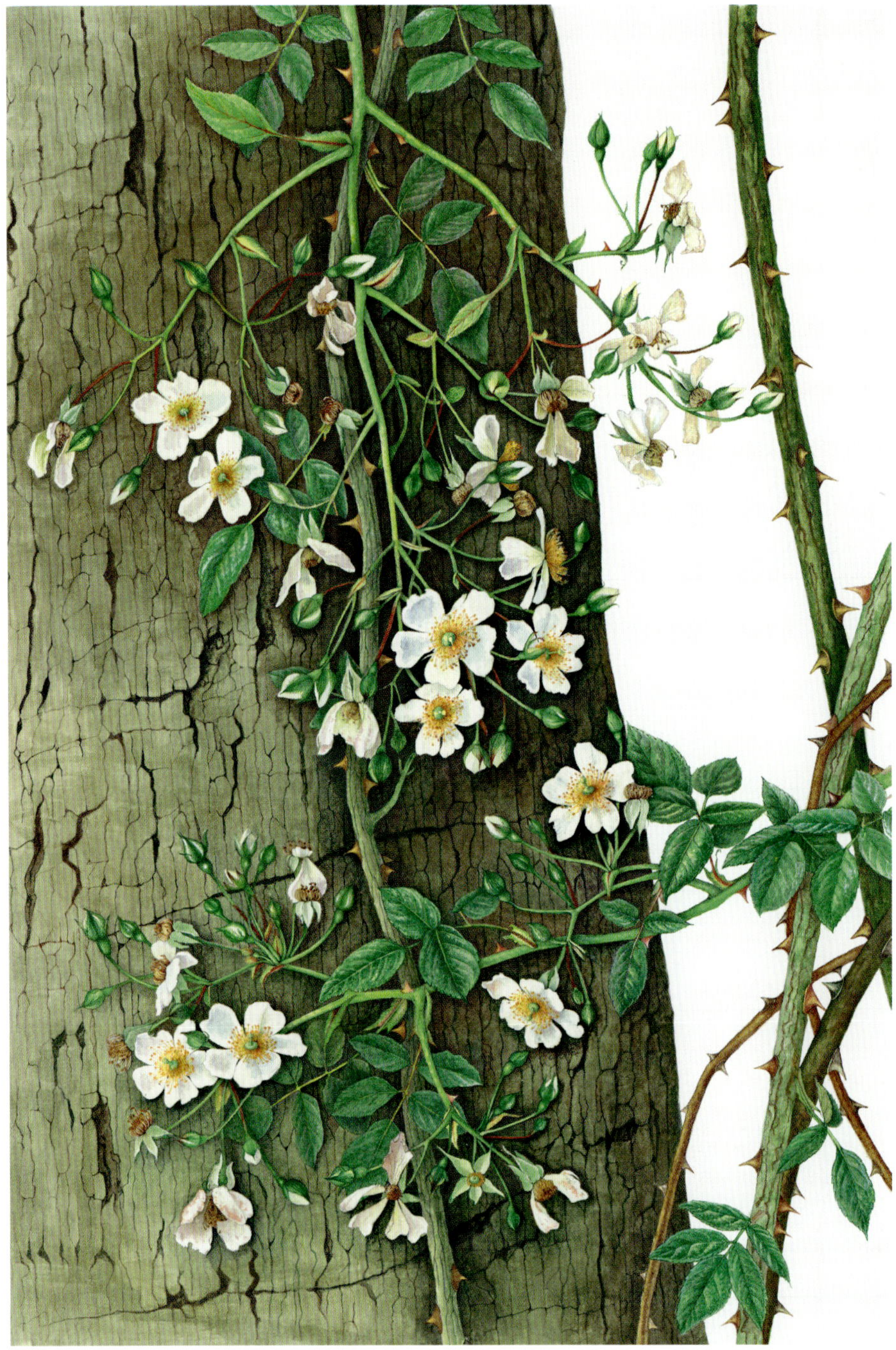

NORIKO WATANABE

Rosa filipes 'Kiftsgate' in May, 2011
Watercolor on paper, 26¾ × 20 in. / 68 × 51 cm
Shirley Sherwood Collection, Royal Botanic Gardens, Kew, London

Rosa filipes 'Kiftsgate' is renowned for its vigor and floriferous nature. The largest known rose specimen at Kiftsgate Court Gardens in Gloucestershire, England, the 'Kiftsgate' rose rambles over three mature trees and is covered from midsummer by a cloud of small, white flowers with a musky fragrance, followed by eye-catching coral-red rose hips in fall. Japanese botanical artist Noriko Watanabe (b. 1954) captured the detail of this rose just as its flowering begins, with a sprinkling of buds and early blooms with their golden stamens. She was inspired by a specimen entangled in an old Japanese stone oak (*Lithocarpus edulis*) at Ofuna Botanical Garden in Kanagawa Prefecture, Japan. After making rough sketches of the flowers and taking photographs, she returned to her home studio to create the paintings and bought similar blooms from florists for reference. Observing the rose throughout the year, Watanabe visited the gardens multiple times in each season to sketch the changes before selecting the best compositions. She completed a series of six paintings as bud became bloom and then faded, and depicted the color transitions of the berries before they ripened then withered away. Watanabe is renowned for her ability to capture natural transformation in her works. She rose to prominence as a botanical artist in 2006 after she won a Royal Horticultural Society Gold Medal for a series of watercolor paintings demonstrating seasonal change in oakleaf hydrangeas (*Hydrangea quercifolia*).

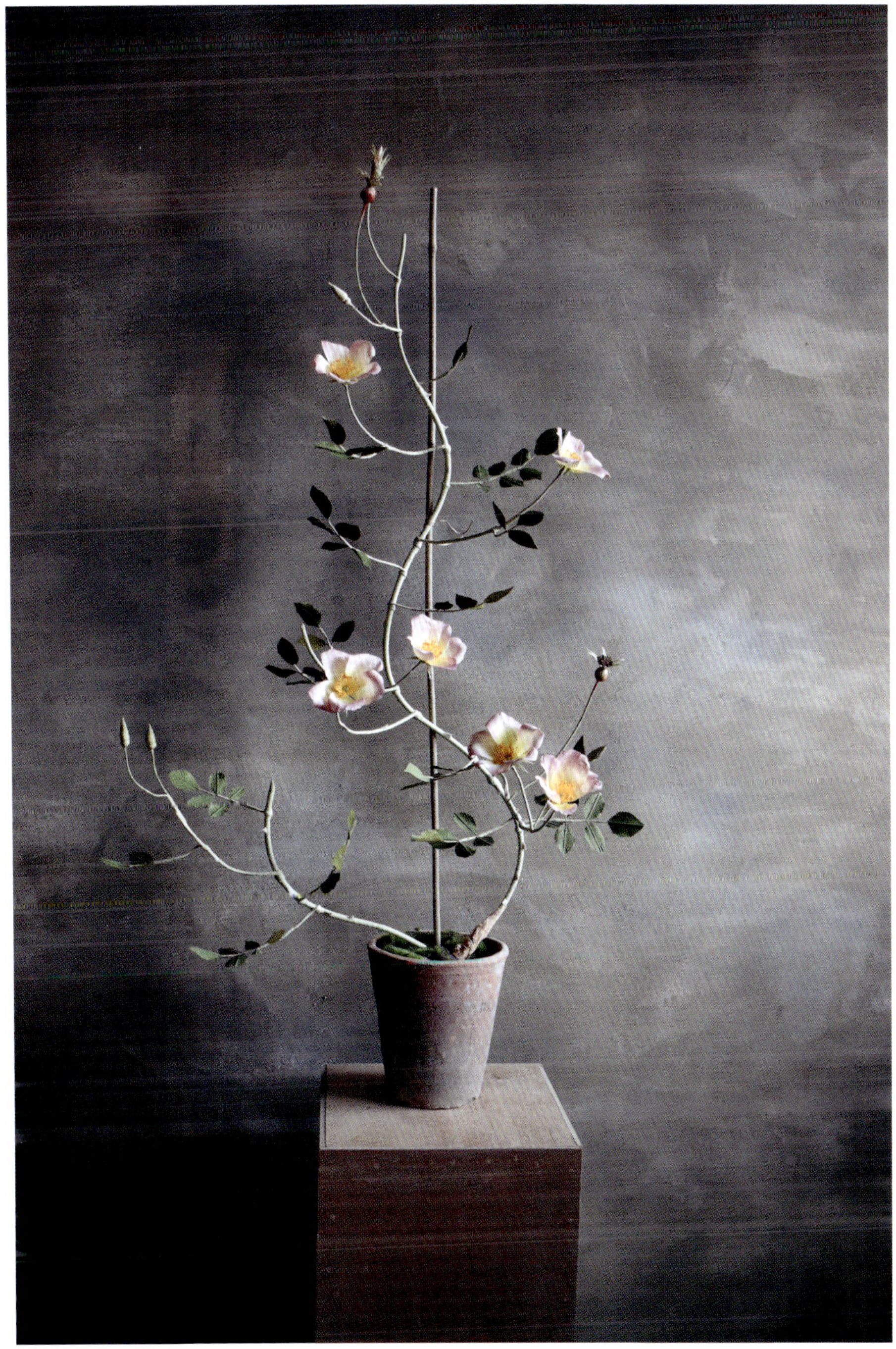

SOURABH GUPTA

Always Romantic, 2022
Handcrafted paper roses and terracotta pot, dimensions variable

The elegant, sculptural forms of this climbing dog rose (*Rosa canina*; see pp.98, 103, and 185) make for a balanced and harmonious display. Yet, despite appearing to grow from a pot, the staked plant requires no maintenance, for each flower, bud, leaf, and stem has been painstakingly crafted from paper tinted with food coloring. The handmade blooms of Indian-born, New York–based artist Sourabh Gupta (b. 1990) are tied to a long-standing fascination with nature, which began when he was a schoolboy in the small town of Hiranagar in the territory of Jammu and Kashmir. After training and working briefly as an architect, he won a scholarship to Parsons School of Design in New York in 2018. There, living in a tiny space with little money for supplies, Gupta turned to working with paper out of necessity, using simple and accessible materials to make his botanical sculptures. Just like real flowers, each of his creations is unique in its design, created in his Harlem studio using tweezers and a magnifying glass. From afar, these pink-tinged rose petals are lifelike, but up close, it is clear that they are fabricated. Indeed, realism is not Gupta's aim; instead, he seeks to create works that capture the essence of a plant. Likening his process to that of a painter depicting flowers in a landscape, he strives with each sculpture to create an impression, conjured through memory, observation, and improvisation.

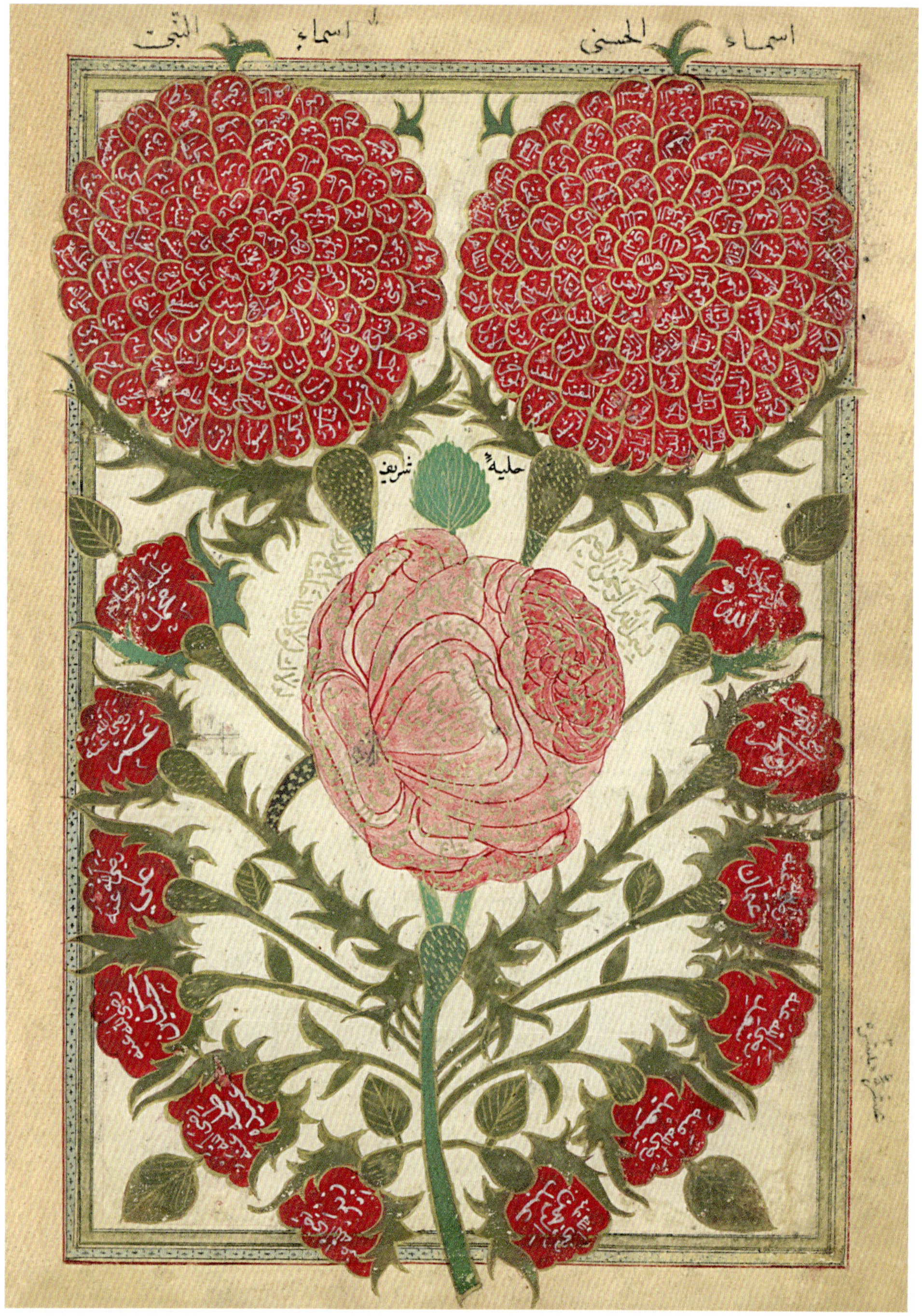

ANONYMOUS

Roses and a Carnation, 1708
Illuminated manuscript, dimensions variable
Staatsbibliothek, Berlin

This page from an early eighteenth-century manuscript, probably a copy from a Koran, describes, among other things, the names of God and the names of the Prophet Muhammad. In Islamic mysticism, it is believed that roses are the symbol of the Prophet—a rose whose fragrance scents all of creation. The manuscript comes from the Ottoman Empire, which ruled much of the Middle East from Turkey between 1299 and 1922. In Islamic culture, the rose also symbolizes the temporality of the world and the brevity of life. Referred to as *verd* in Arabic, the term "rose" is used differently in the contemporary Turkish language. The word "rose," which means "flower" in Persian, has become common in Turkish, and other flowers are called *çiçek*. According to some Sufis (believers in a sect of Islamic mysticism), the letters of the Arabic word *verd*—*waw, ra,* and *dal*—have meaning themselves. The first, *waw,* signifies the Prophet Muhammad being *wali,* a Muslim term for an Islamic saint. The second, *ra,* reveals the Prophet being *rauf* and *rahim,* which means merciful. The last letter, *dal,* refers to the word *davet,* which suggests the Prophet calling all to the right religion. In *divan* poetry—a rich Ottoman poetic tradition that lasted for nearly seven hundred years, and its influence can still be felt in the modern Turkish poetic tradition—the rose is an essential flower. According to legend, the Ottomans developed their passion for the rose when they conquered the city Edirne, which, when they arrived, supposedly contained vast gardens covered in roses.

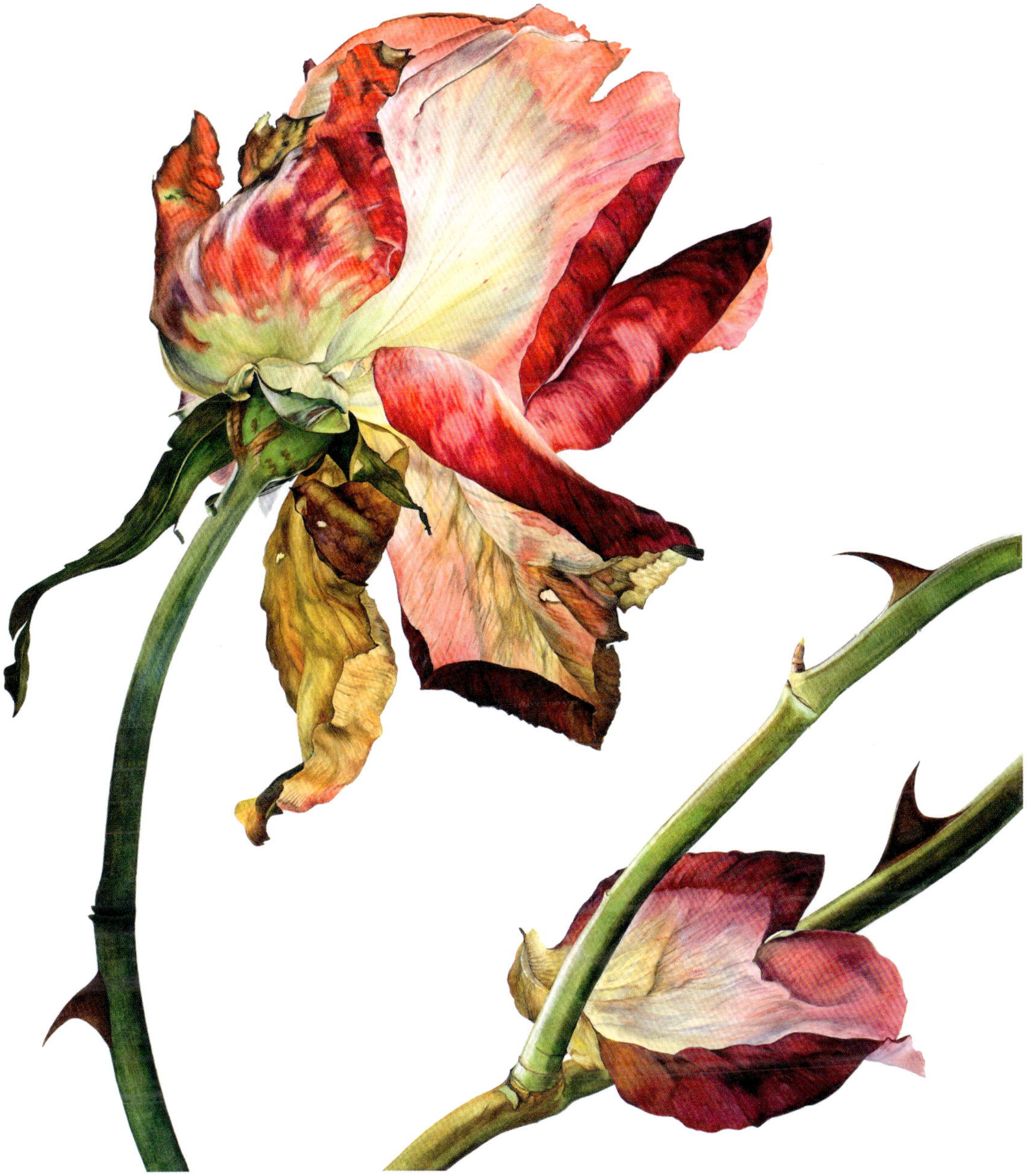

ROSIE SANDERS

A Rose in Winter, 2012
Archival digital print, 28 × 24⅛ in. / 71 × 61 cm

Against a bare background, a Hybrid Tea rose that is clearly past its prime is the subject of this bold print by award-winning English botanical artist Rosie Sanders (b. 1944). Painted in the winter, her image celebrates the sheer bravado of the rose, even if some of its outer petals are browned, its leaves withered, and its large thorns appear slightly menacing. Sanders has described how she spotted the vibrant red flower on a dull January day, hidden behind shrubs, inspiring her not simply to paint the rose but to capture its defiant blooming amid the cold to, as she said, "convey the strength and tenacity of the flower." For Sanders, the rose is a special flower that she paints frequently. She even has a whole book dedicated to the plant, *Rosie Sanders' Roses* (2019), which includes more than eighty works focused on roses. Instead of being merely photographic, Sanders's paintings transcend pure reproduction; her depictions are instantly recognizable as their distinct species but also appear otherworldly thanks, in large part, to her fascination with light and color. Working in her studio in Devon, England, she lights her flowers from behind, as here, to highlight the translucent quality and delicacy of the flowers' petals. A key participant in the inaugural exhibition of the Shirley Sherwood Gallery of Botanical Art at London's Royal Botanic Gardens, Kew, in 2008, Sanders began her painting career as a miniaturist, working with the close observation that still characterizes such images as this.

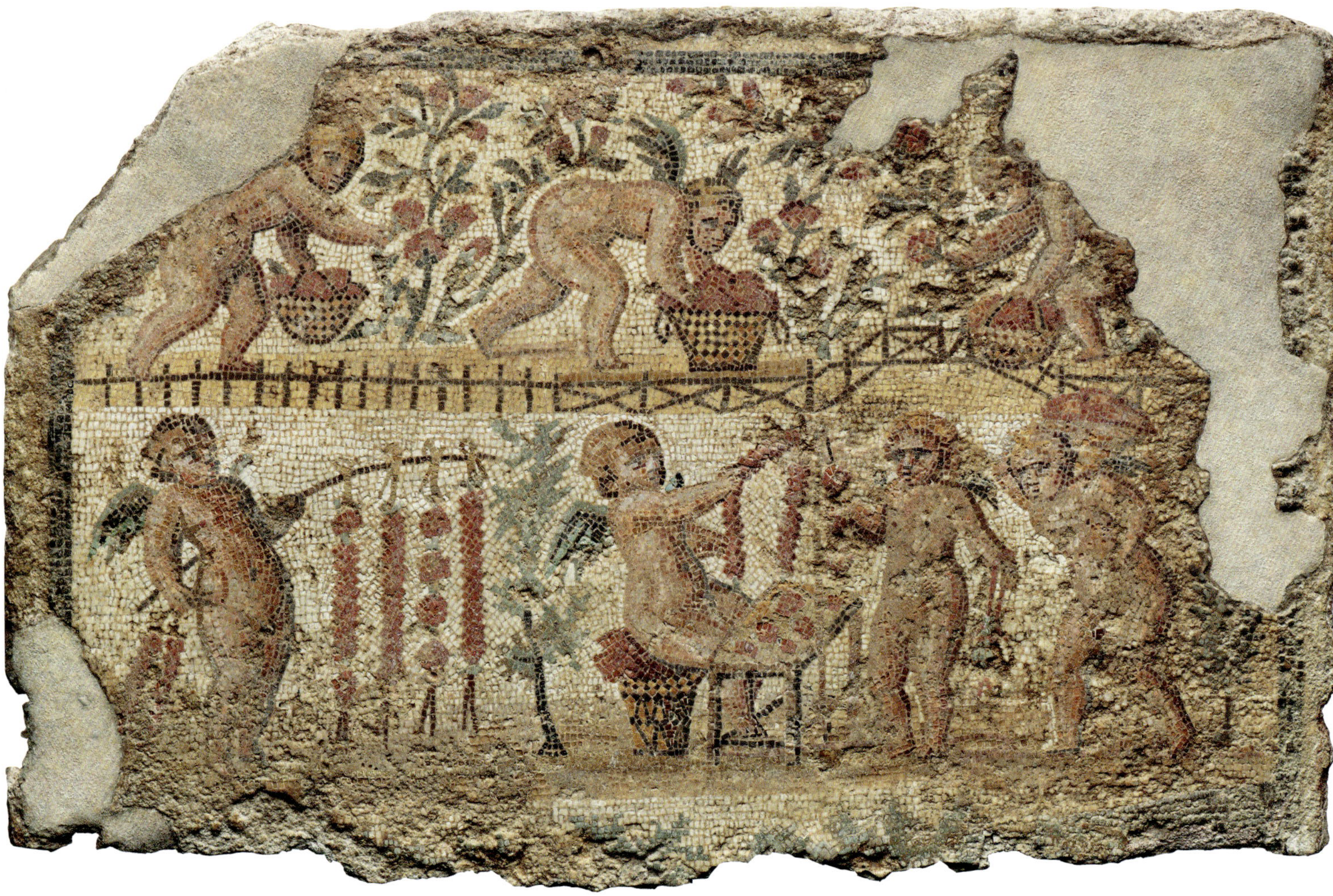

ANONYMOUS

Mosaic *emblema* with cupids gathering roses in a garden, late 2nd–mid-3rd century CE
Fine stone and glass tesserae on terracotta panel, 19⅛ × 24 × 4⅞ in. / 48.5 × 61.1 × 12.3 cm
Museum of Fine Arts, Boston, Massachusetts

This delightful *emblema* (a portable panel mosaic that would have been made in a workshop and installed at the center of a mosaic floor) consists of three registers, although the uppermost is truncated; it illustrates the rose harvest and the garland business during the Roman Empire. At the top, three cupids gather roses in a garden, while in the lower register a cupid carries a basket of roses on his back. Before him stands another cupid, who holds up a string of roses with a pomander around his wrist. At the center, a third cupid—who may be the garland maker—is sitting on a basket of roses, facing a table with bunches of the flowers tidily organized. A final cupid at left holds a pole from which four garlands hang with two more garlands in his hand, recognizable by their four- or five-petal designs. Typically made from such flowers as roses, violets, or oleander, garlands are found represented in Roman frescoes and mosaics, often draped over panels to brighten the surface with the appearance of fresh flowers. Roman art is full of symbols and motifs associated with religious practice, and there are signs that in Roman houses, latrines, and streets, physical garlands were actually attached to the walls as votive offerings. An inscription found near the Sacred Way in Rome, close to monuments that received habitual religious worship, indicates that flowers and garlands could be purchased there, although this *emblema* comes from as far afield in the empire as Tunisia.

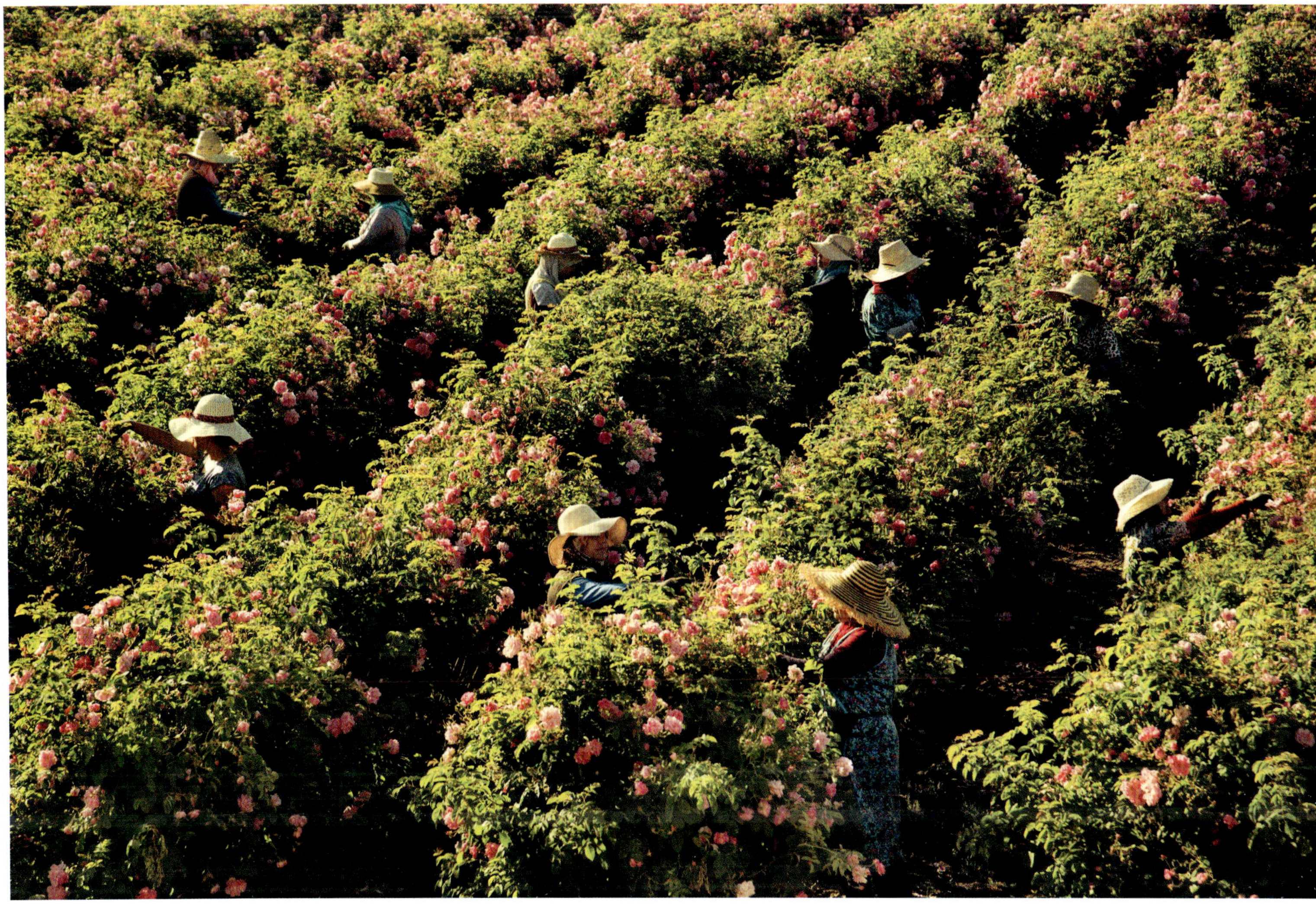

PARI DUKOVIC

Fragrant Harvest: How Many Roses Go Into a Bottle of Chanel No. 5?, 2018
Photograph, dimensions variable

There's a rhythm to the pickers shown here as they move in pairs down the carefully cultivated rows of roses, collecting their fragrant harvest. Pinching the bud just below the calyx, each opened rose is placed into an apron and then transferred into the distinctive burlap sacks. Time is really of the essence; as the day warms up, so do the roses, and when the picked flowers start to oxidize, their scent changes, becoming more spiced in the heat. The fragrance of roses is at its best at dawn. This image, taken by Turkish photographer Pari Dukovic (b. 1984) in 2018 for the *New Yorker*, captures a process that has not changed much since the French started to cultivate roses for perfume in the seventeenth century. The commission took Dukovic to Pégomas, slightly north of the French Riviera and close to Grasse, the spiritual home of French fragrance (see p.143), the region of the world where *Roses de Mai*, or Centifolia roses (see pp.24 and 180), are mainly grown. The rich, fertile soil and ideal microclimate, sheltered from the wind, are central to the powdery floral notes of Chanel No. 5 and many other rose-based fragrances. When the fields bloom over three weeks in May, about 37 tons (33,565 kg) of rose petals must be harvested by hand to keep the delicate petals intact. It takes tens of thousands of roses to yield a little more than 1 pound (450 g) of intense rose absolute for perfume. Sweet but with a watery green facet, it is a scent so highly prized that its value, ounce for ounce, is higher than the price of gold.

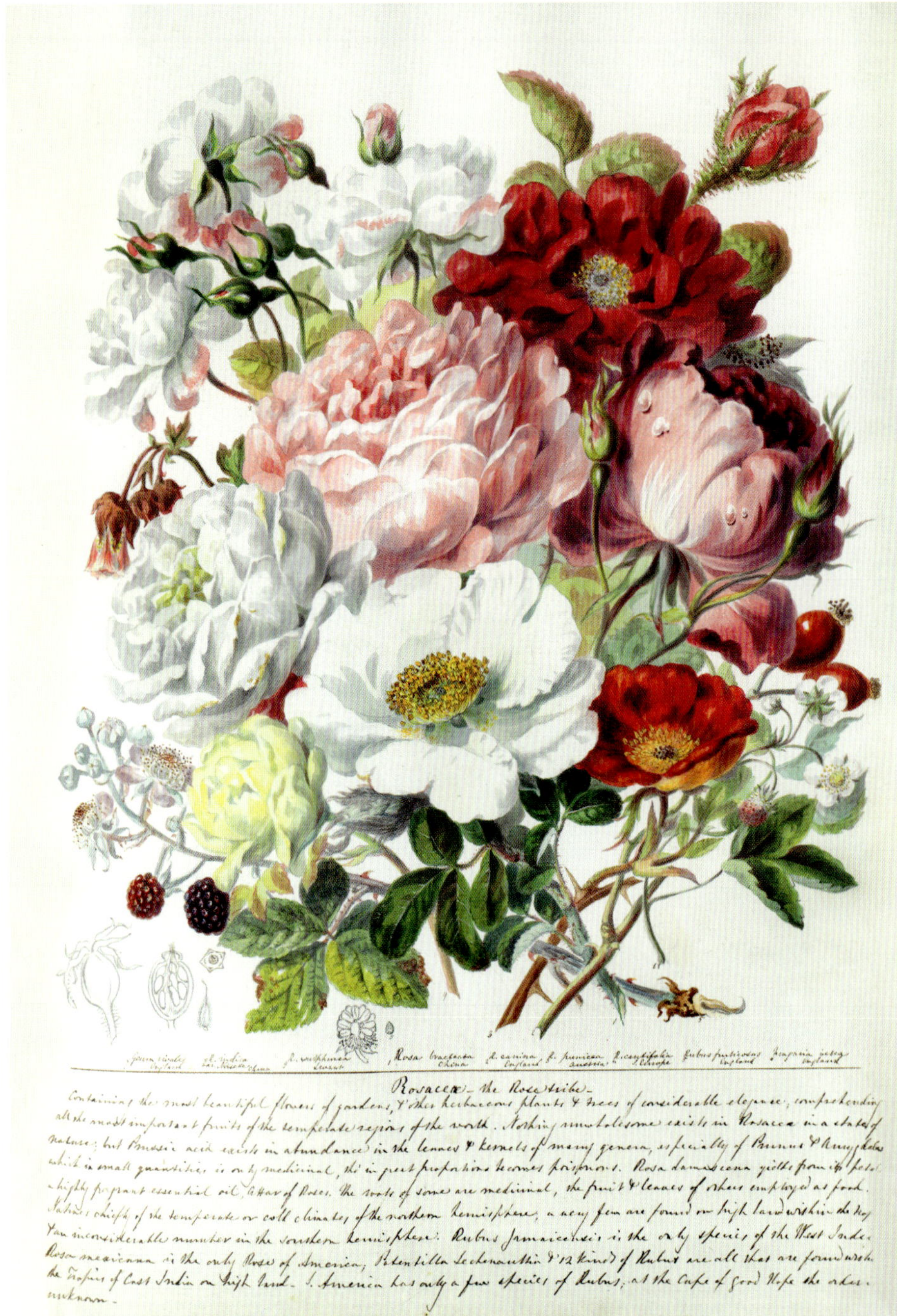

ELIZABETH TWINING

Various Species of Rosa, from *Illustrations of the Natural Order of Plants*, 1855
Watercolor, 19½ × 12⅝ in. / 49.5 × 32.1 cm
Natural History Museum, London

This lively watercolor depicts nine varieties from the rose family, grouping them as if in a decorative bouquet that may have helped to demystify botanical content for contemporary Victorian readers. This illustration comes from English artist Elizabeth Twining's (1805–1889) seminal *Illustrations of the Natural Order of Plants*, published in two volumes in 1849 and 1855. Born into the famous tea trading family, Twining took up botanical illustration after being inspired by the plants she saw at the Royal Botanic Gardens, Kew, London, and by the paintings of William Curtis. The *Natural Order* sought to illustrate the main groups of British plants alongside some notable species from other countries. This watercolor is an initial sketch for the Rosaceae plate; the final hand-colored lithograph has a slightly different set of plants. The central double, pink flower here is a Centifolia, commonly called a cabbage rose (see pp.24 and 180). Beneath it, the white flower is the Macartney rose (*Rosa bracteata*), a native of southeastern China. The shiny red hips at right belong to the dog rose (*R. canina*; see p.103), a common native of Great Britain, and to their left, a single bicolor bloom from *R. foetida bicolor* is visible, orange-red on top and yellow underneath. The dark red Gallica rose at the top (*R. punicea*) is probably now extinct, but it is very similar to the contemporary rose 'Tuscany Superb'. Also easily identifiable in the top right corner is the bud of a Moss rose, so called for the mossy growth around the receptacle and sepals, which is sticky and has a strong resinous smell.

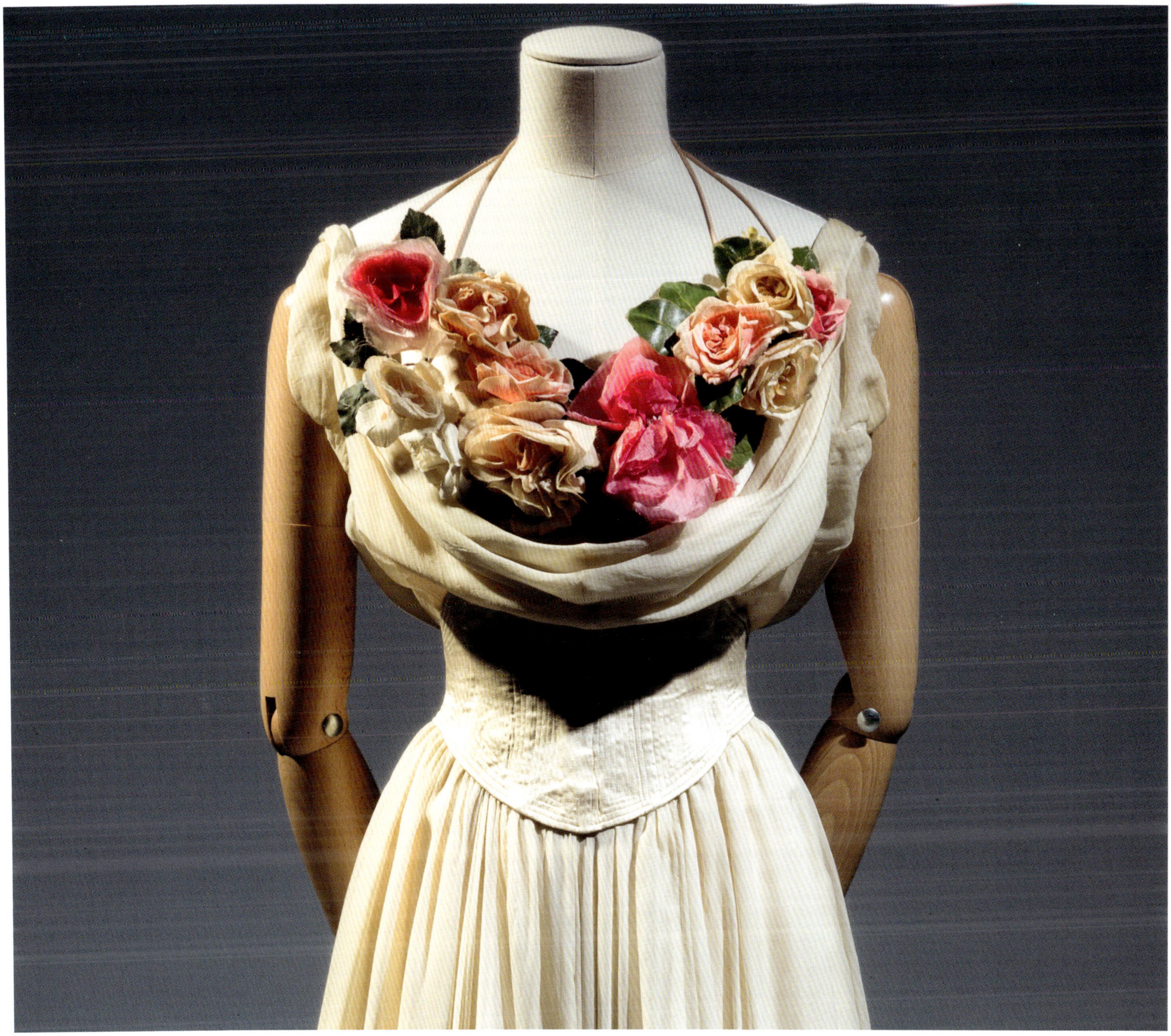

CHARLES JAMES

La Sylphide dress, 1937
Ivory silk organza, satin, pink silk, faille, and grosgrain ribbon
Museum at the Fashion Institute of Technology, New York

A semicircle of blooming silk roses, mixing shades of deep pink and pale gold, demurely covers the décolletage of this off-white silk organza and satin dress. Worn by the debutante socialite Esme O'Brien in New York in 1937, *La Sylphide* is the work of the Anglo-American fashion designer Charles James (1906–1978), who was celebrated for his ability to sculpt shapes from fabric. Since antiquity, roses have been used to mark significant life events, from birth to marriage and death. White and pink roses, in particular, are often associated with young women, so they featured in the gowns of many debutantes. Although he was known for creating floral shapes in the evening gowns he made for noteworthy clients, including Mrs. Randolph Hearst and Millicent Rogers, James rarely used artificial flowers. When he did, he used them sparingly, understanding the power both of roses and of fashion itself. As he expressed it, "Fashion is a little like radium . . . or like essence of roses, which if undiluted would asphyxiate." Considered a genius by his peers but little known to the public, James produced fewer than a thousand garments during his forty-year career. *La Sylphide* was an early design, inspired by the ballet of the same name—one of the world's oldest, having first been performed in 1832—and particularly by the romantic costume worn by Marie Taglioni dancing in the title role, with its floral corsage above her tiny waist.

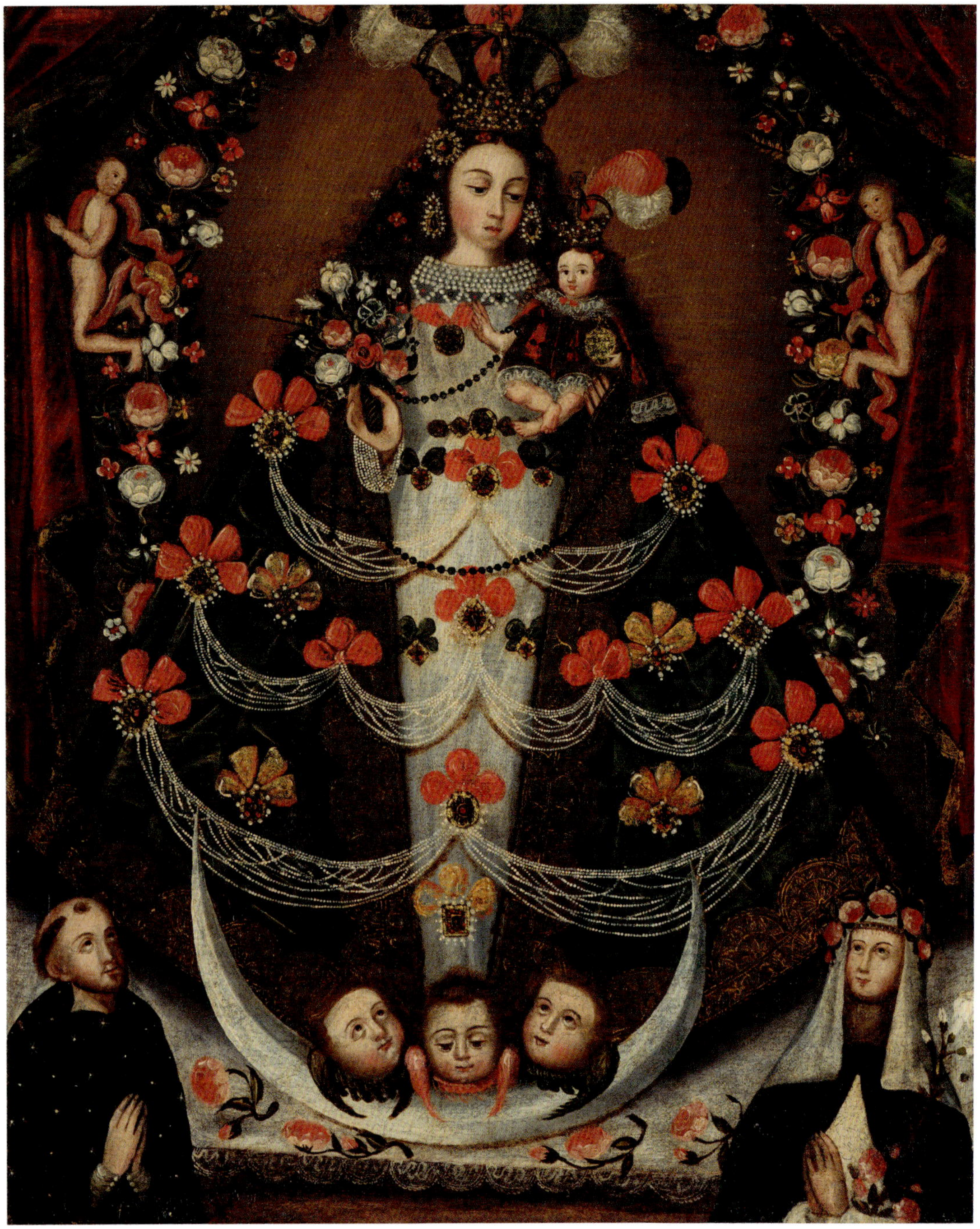

ANONYMOUS

Virgin of Pomata with St. Nicholas Tolentino and St. Rose of Lima, c. 1700–50
Oil on canvas, 26 × 21 in. / 66 × 53.3 cm
Brooklyn Museum

In eighteenth-century colonial Latin America, the town of Pomata in the highlands of Peru, above Lake Titicaca, became a popular Christian pilgrimage site, thanks to the miracles associated with the statue of Mary, mother of Jesus, that stood on the side altar of the church. Known as Our Lady of Pomata, in this painting the statue appears dressed in ornate robes and garlanded with roses and other flowers; she wears a crown of feathers and is bedecked in precious metals. This iconic image was re-created throughout Latin America by unknown artists in the service of the Roman Catholic Church. It departs from other images of the Virgin of Pomata by including two saints looking up at Mary: Saint Nicholas Tolentino, who was an advocate for the souls of those trapped in Purgatory, particularly on All Souls' Day (November 2), and Saint Rose of Lima. Saint Rose was born into a wealthy Peruvian family in the late sixteenth century and spent her life caring for the city's poor. The patron saint of gardening and flowers, she raised funds to help others by selling the roses and other blooms she grew in her garden. In recognition of her importance within Peru, the painting is full of roses, not just in the garlands but also at Mary's head and feet. According to legend, when Saint Rose died, rose petals began to fall from the sky and the whole of Lima was filled with the scent of the flowers.

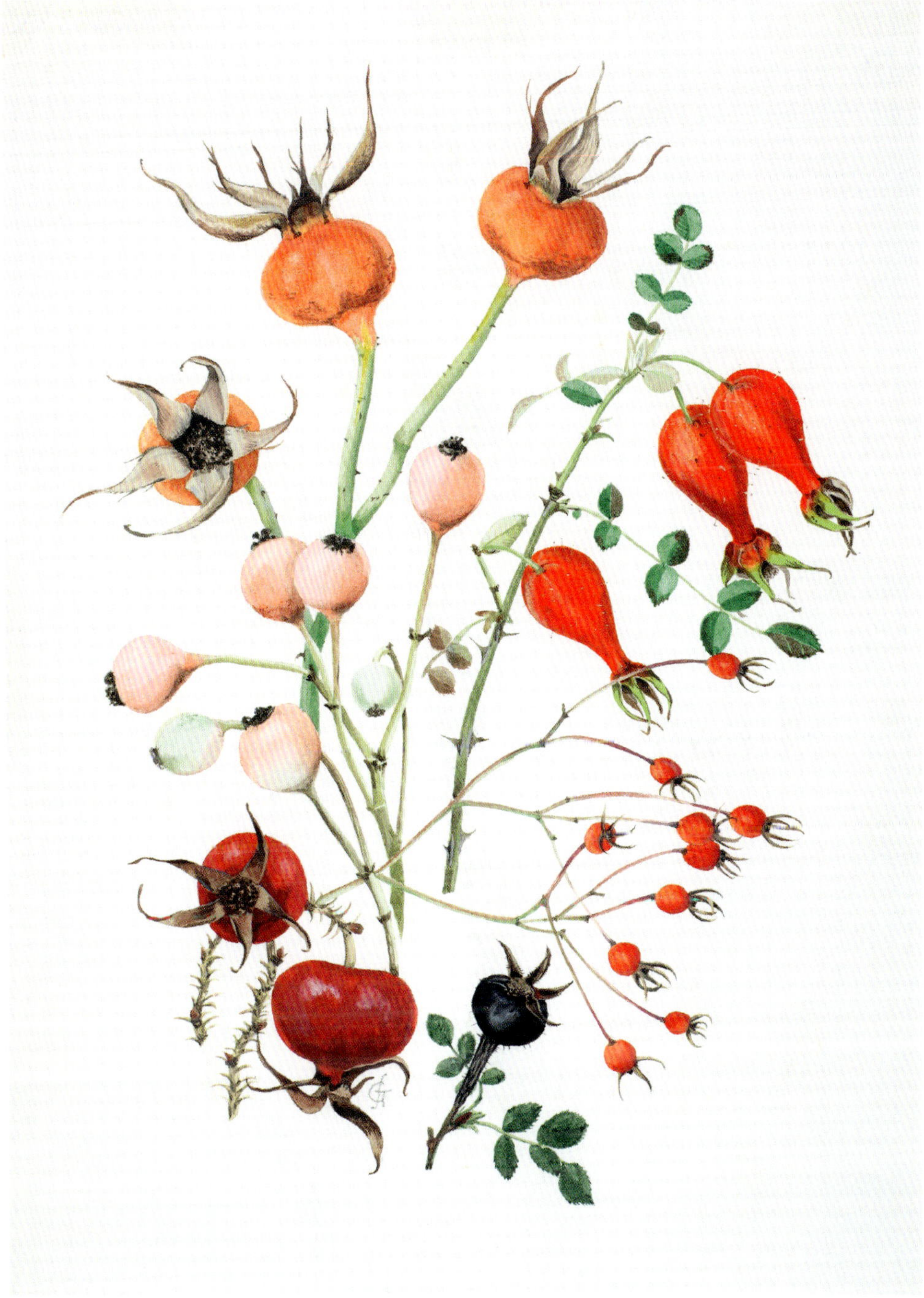

GRAHAM STUART THOMAS

Rose hips of 'Nymphenburg', R. moyesii, 'Penelope', R. filipes 'Kiftsgate', R. rugosa, and 'Ormiston Roy', c. mid-20th century
Watercolor, 13 × 9½ in. / 33 × 24.2 cm
Royal Horticultural Society, Lindley Library, London

Given the word "rose," one immediately thinks of the flower in bloom, but, in fact, rose hips—the fruit produced by the rose plant—can themselves be attractive and last right through the winter. In this exquisite watercolor, English horticulturist Graham Stuart Thomas (1909–2003) has chosen some of the best rose varieties to produce hips, demonstrating the wide range of colors and shapes that one can find in the rose world. Their colors vary from pale yellow to orange, pink and red through mahogany and black. In size, they range from about ¼ inch (5 mm) to the circumference of a cherry tomato, as in the hips of *Rosa rugosa* in the bottom left corner here. The black hip at bottom right belongs to 'Ormiston Roy', a hybrid of *R. spinosissima* (known as the Scots or Burnet rose) and notable for its dark hue. The smallest hips above belong to the monstrous rambler *R. filipes* 'Kiftsgate' (see p.42), and the paler pink ones at center left to the Hybrid Musk 'Penelope'. The flagon-shaped hips at center right are immediately recognizable as those of *R. moyesii*, while 'Nymphenburg', rounding out the top, sports hips that resemble turnips. Thomas was highly regarded as a gardener and particularly well known for his extensive knowledge of roses. He was responsible for finding and identifying many Old Garden roses introduced in the nineteenth century and before and was instrumental in creating the National Collection of Old Roses at Mottisfont Abbey in Hampshire.

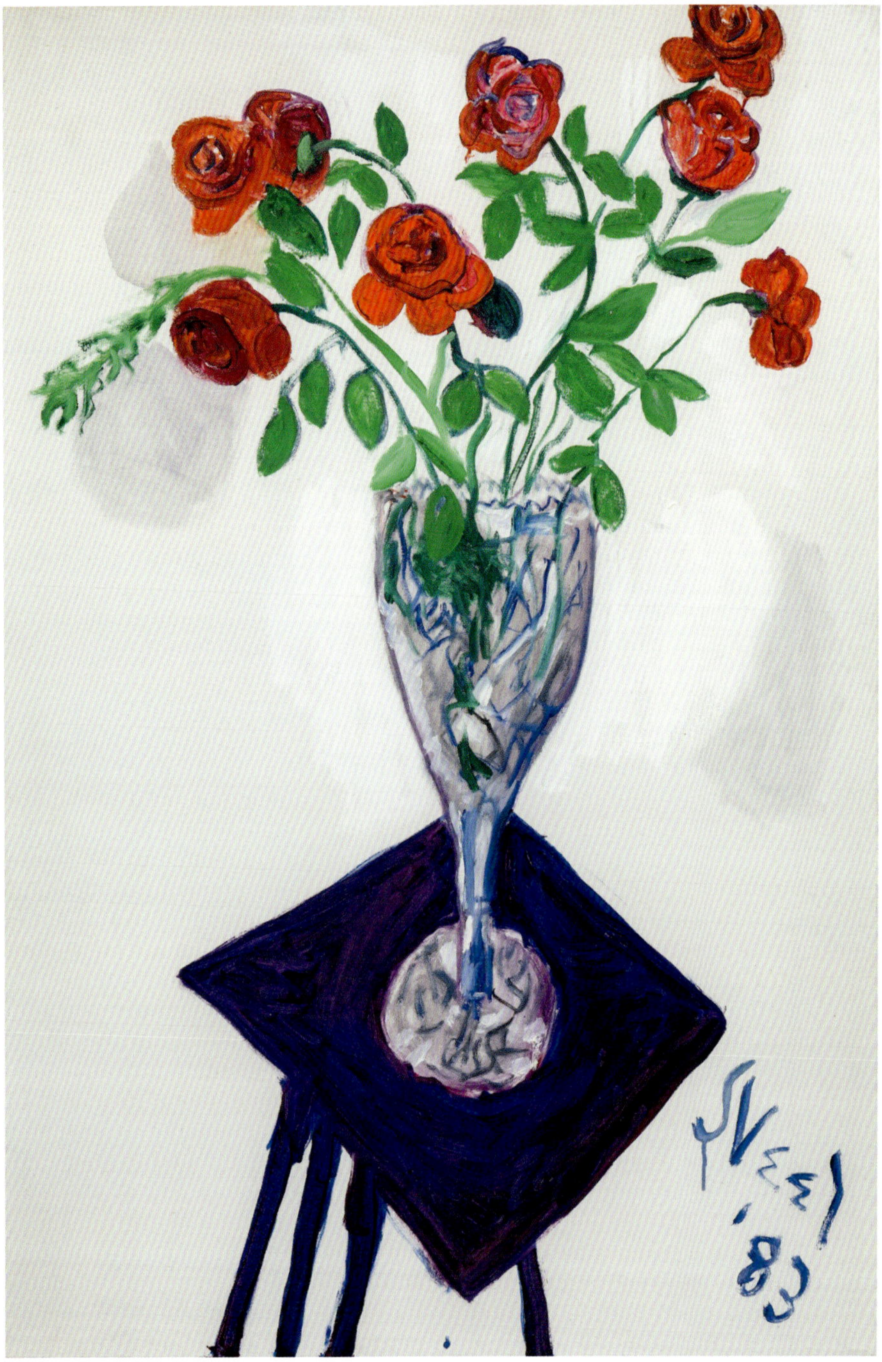

ALICE NEEL

Roses, 1983
Oil on canvas, 36 × 24 in. / 91.4 × 61 cm
Private collection

Painted toward the end of her prolific career, *Roses* epitomizes the unique style of American painter Alice Neel (1900–1984), with its idiosyncratic perspective, expressive brushwork, and use of bold color. Known for her figurative portraits of the ordinary people she knew—a cast of creatives and characters from the streets of New York, where she lived, as well as friends and family—Neel spent her entire creative life painting the subjects she wanted to paint, spurning the strict abstraction that was popular in twentieth-century art. Her eclectic portraits were of people and the objects that surrounded them, and her still lifes were often of cut flowers and houseplants that were close at hand. Here, Neel paints a bunch of blood-red roses with bright green foliage—perhaps a gift or purchased from a local convenience store—arranged in a crystal vase and sitting atop a precariously angled table. The roses are not artfully arranged, nor are they particularly lively; she paints them as she sees them, slightly drooping and plonked into their container, maybe approaching their last days. A striking example of Neel's deliberate distortion of her subjects, the roses appear flattened against the wall, and they tower in scale compared to their hazardous support. Through her fluid and sketchlike strokes and simple yet dynamic colors, Neel's expressive painting of roses is more portrait than still life, imbuing the ephemeral bouquet with an anthropomorphic aura and bringing the near-dead flowers to life.

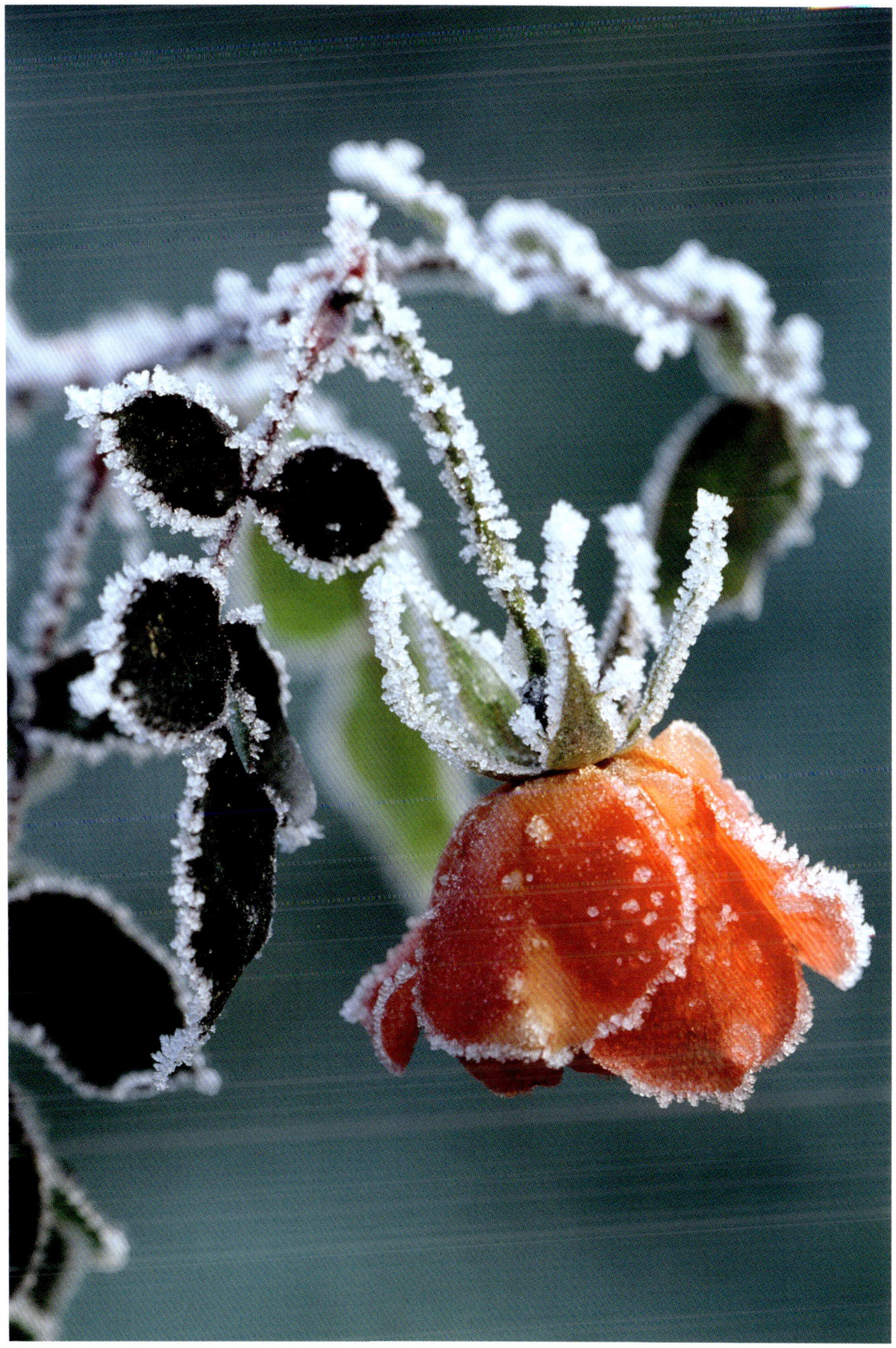

CLIVE NICHOLS

Frosty Rose 'Warm Welcome', 2010
Photograph, dimensions variable

The usual aim in photographing a rose is to show the plant at its very best, but British garden photographer Clive Nichols (b. 1962) has seen beauty in this bloom of the climbing rose 'Warm Welcome' in the depths of winter, looking rather forlorn and dusted by hoarfrost. As ever in his photography, the source of light is perfectly found; here, it highlights the intricate detail of the ice crystals but also allows for an appreciation of the rose's distinctive color. With its small, almost single flowers, *Rosa* 'Warm Welcome' is described as a miniature climber that can reach a height of nearly 8 feet (2.5 m), making it ideal for fences, walls, and arches, where easy management is important. The color is a vibrantly bright orange or vermilion, so careful thought must be given to its position and association with other flowers, which is perhaps why Nichols decided to photograph it on its own. This cultivar was bred in 1986 by British rosarian Chris Warner, who is well known for introducing a wide range of short climbers and ramblers as well as *R. persica* hybrids, all of which are noted for their excellent resistance to disease. Nichols has been photographing gardens and plants for more than thirty years. Although many in his profession prefer to shoot in the warmer months, he has said, "Winter gardens can be just as stunning as any summer or spring garden, particularly with a dusting of hoarfrost or a fresh coating of snow."

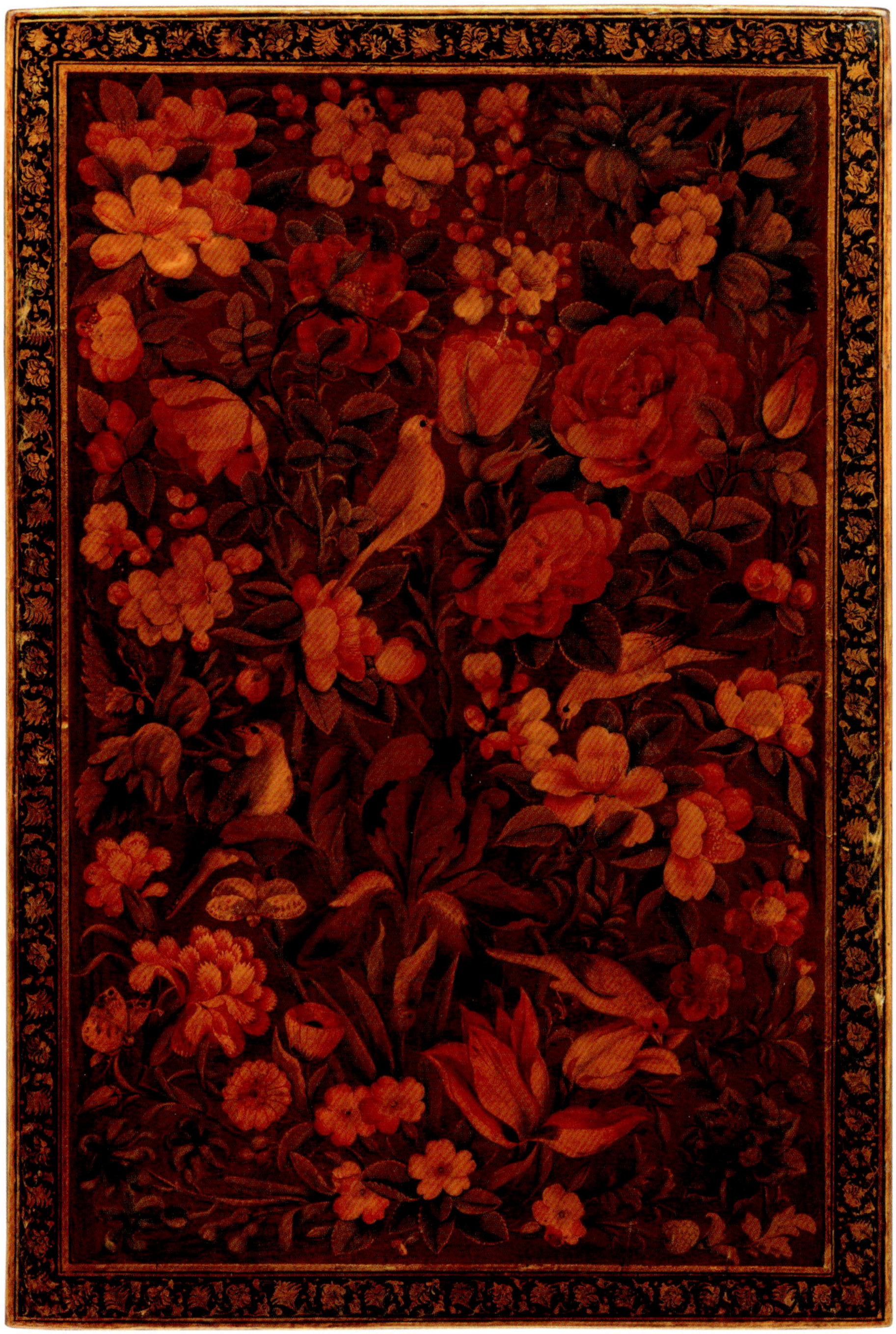

MUHAMMAD BAQIR

Vanity box, 1789
Painted and lacquered papier-mâché and pasteboard, 13¾ × 9⅜ × 1⅜ in. / 35.1 × 23.9 × 3.6 cm
Metropolitan Museum of Art, New York

The rose is one of the most important symbols in Islamic culture. The flower, particularly the Damask rose, is considered a symbol of the Prophet Muhammad, whose perspiration purportedly smelled of roses and whose grave is called the "rose garden." Another common motif in Islamic poetry and visual art from the seventeenth to the nineteenth centuries is the theme of flowers and nightingales (*gol o bolbol*). The rose indicates divine, feminine beauty, while the nightingale represents the male admirer. This beautifully painted vanity box by Persian artist Muhammad Baqir (fl. 1740s–1800) comprises a drawer with compartments for implements, such as scissors, tweezers, and a comb, and a removable lid covering a mirror. The painting on the case depicts nightingales and roses, among other flowers, while the inner surface of the lid features a feasting scene. On it, a prince and his lover sit on a veranda surrounded by courtiers and musicians, who serve them wine and provide music. This box exemplifies the emotion and beauty of painting from the Zand dynasty of southern and central Iran (1750–79)—which was known particularly for the decoration of bookbindings and lacquer boxes, such as this one—at its best. The disproportionate use of shading sometimes endows these works with a dark quality, but it does display an improved understanding of the play of light on three-dimensional forms. Supremely influential, Zand art became the foundation of later Iranian art made during the Qājār dynasty (1794–1925; see p.60).

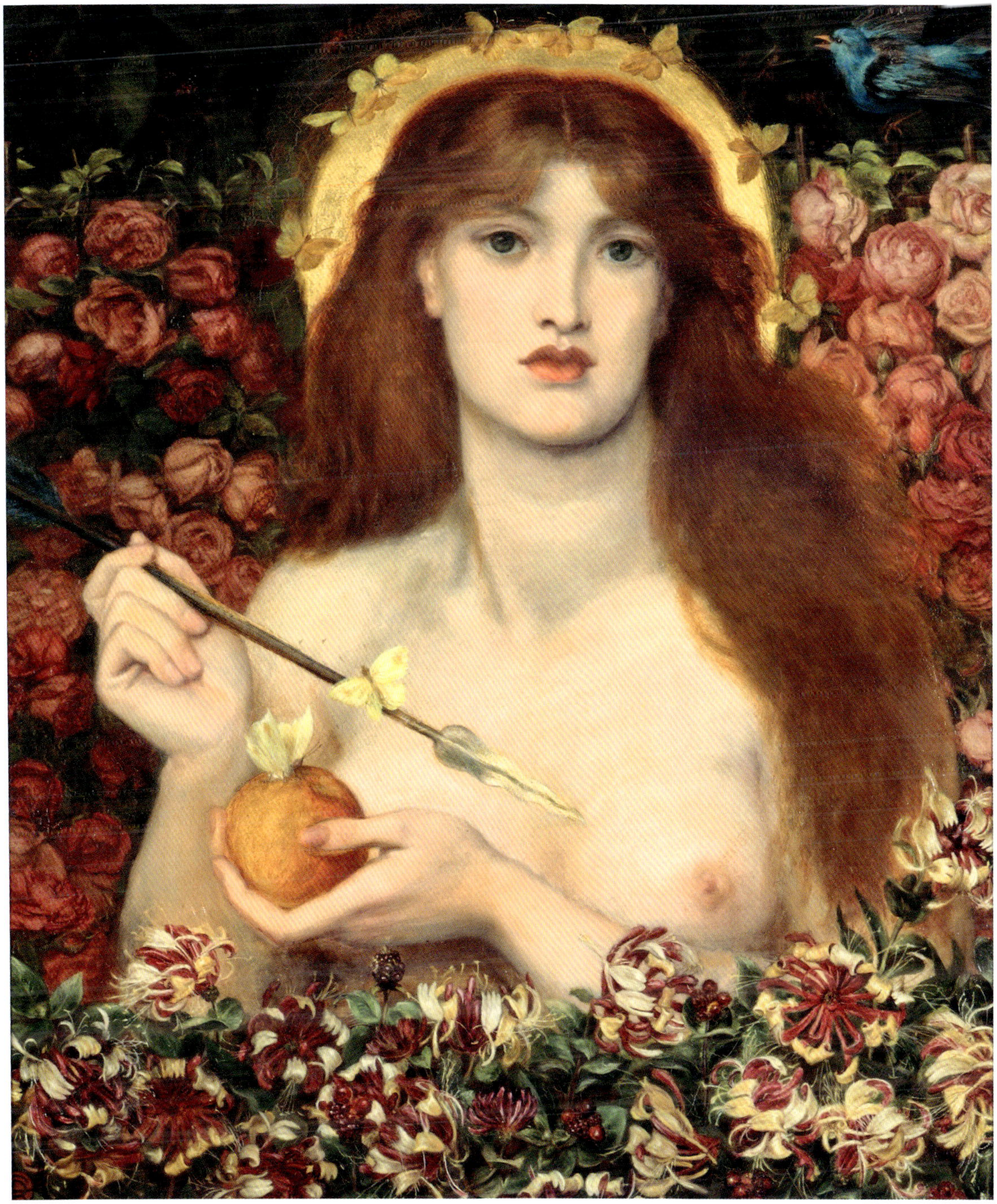

DANTE GABRIEL ROSSETTI

Venus Verticordia, 1864–68
Oil on canvas, 32 × 26¾ in. / 81.3 × 68 cm
Russell-Cotes Art Gallery & Museum, Bournemouth, UK

One of the most erotically charged paintings by Italian artist Dante Gabriel Rossetti (1828–1882), *Venus Verticordia* exudes an elegant and delicate charm that in large part rests on the lush representation of flowers. Rossetti employs a rich palette of colors to imbue the roses that frame Venus's bold and youthful portrayal with symbolic meaning. The vibrant reds and deep pinks are not merely decorative, but are strategically positioned to draw the viewer's eye to the paleness of Venus's skin. Rendered with meticulous detail, their petals lush and full, the roses appear to almost convey a tactile quality in their realism and evoke the intense emotions of passion and desire. Rossetti's attention to detail underscores natural beauty and truth, ideals traditionally associated with Venus's mythological character: the Roman goddess of love, who was known by the epithet *Verticordia*, "changer of hearts." The fullness of the roses also suggests a sense of ripeness and maturity, and Rossetti's choice to include roses in various stages of bloom—from buds to completely opened flowers—mirrors the transformative journey of love and beauty, from nascent attraction to full-blown passion. The blooms also reflect the cyclical nature of life and love, emphasizing that both are never static but ever-changing and evolving. Occupying the entirety of the painting's background, the flowers act as a barrier between the outside world and the purity that Venus represents, suggesting that true beauty and love are precious and require care and protection.

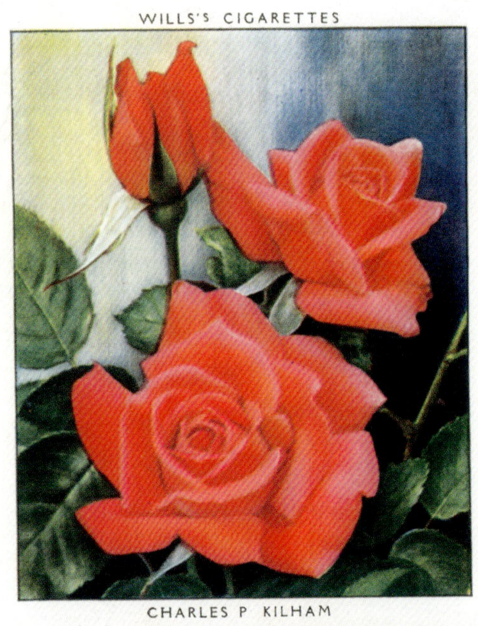
WILLS'S CIGARETTES
CHARLES P KILHAM

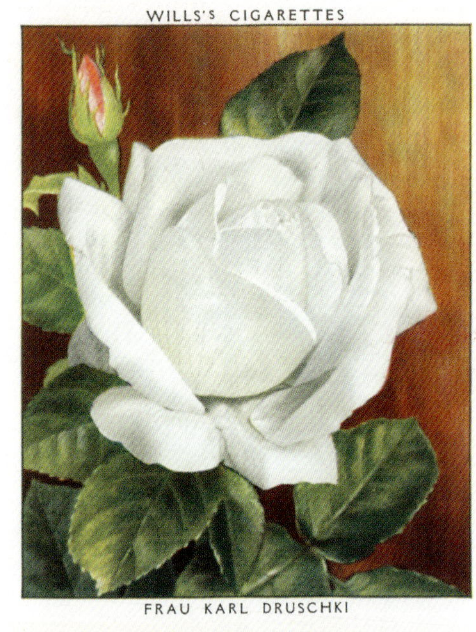
WILLS'S CIGARETTES
FRAU KARL DRUSCHKI

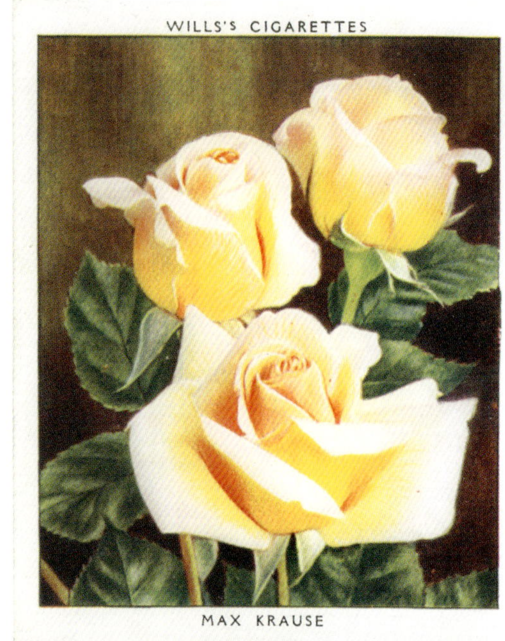
WILLS'S CIGARETTES
MAX KRAUSE

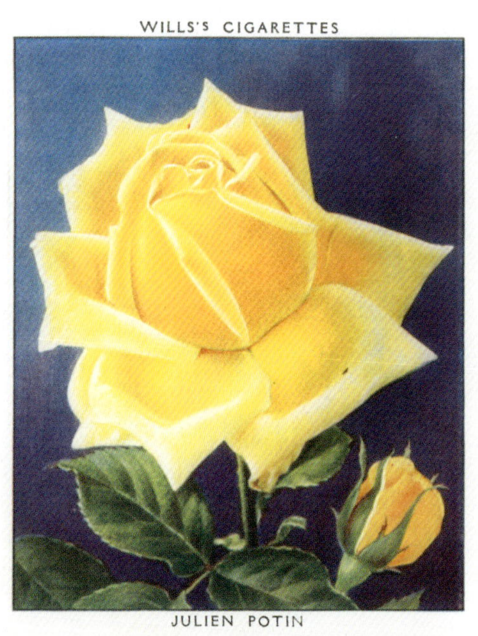
WILLS'S CIGARETTES
JULIEN POTIN

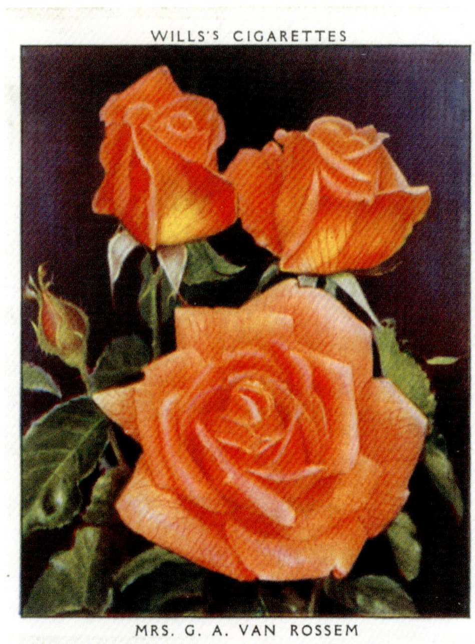
WILLS'S CIGARETTES
MRS. G. A. VAN ROSSEM

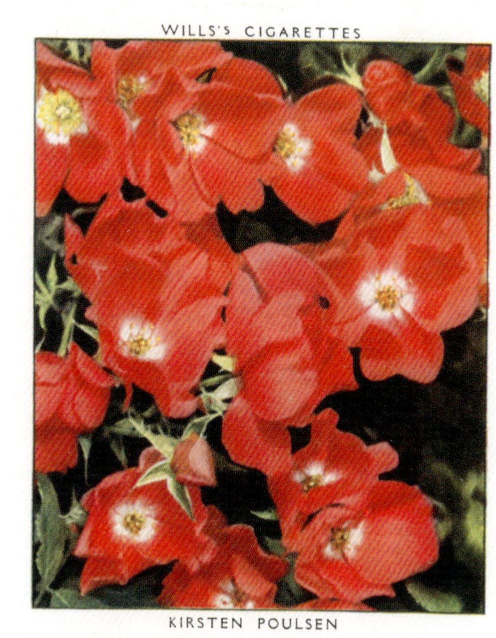
WILLS'S CIGARETTES
KIRSTEN POULSEN

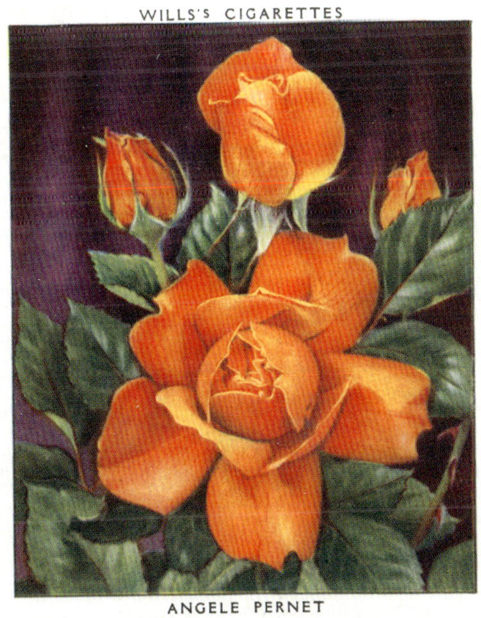

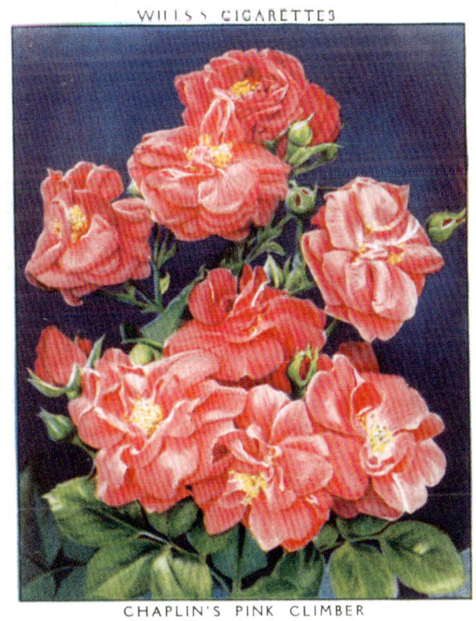

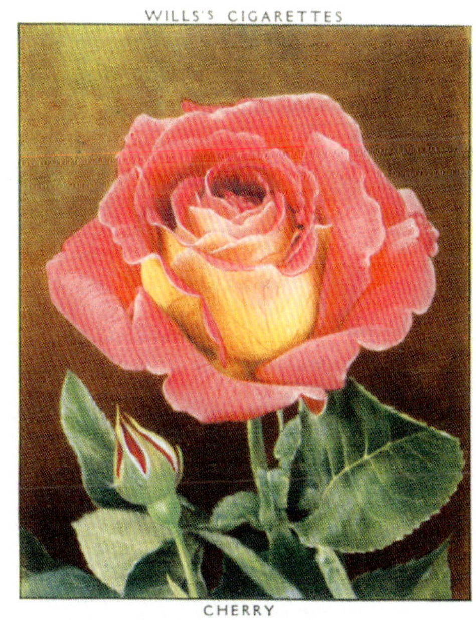

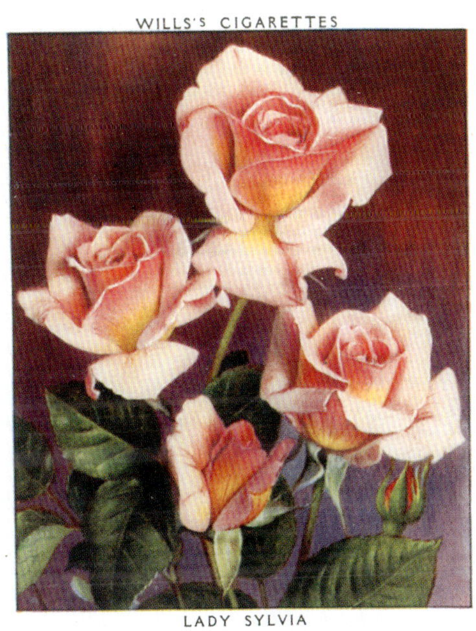

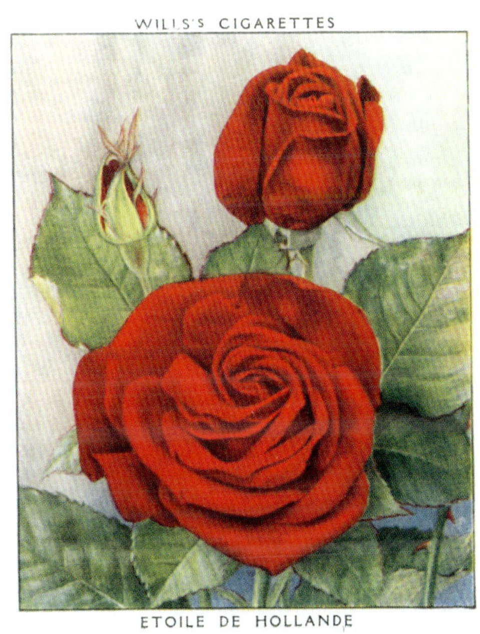

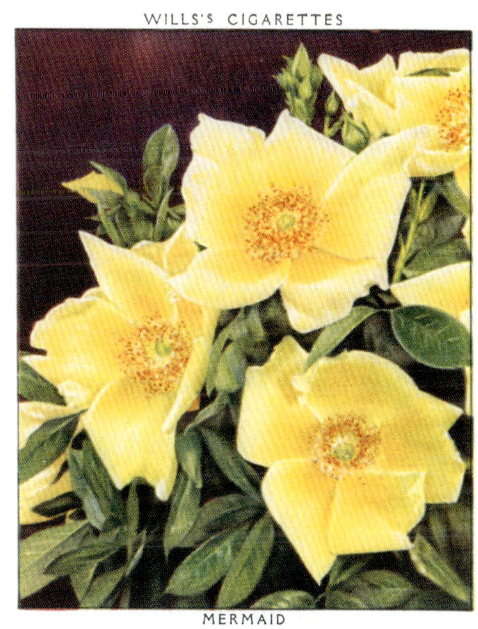

W. D. & H. O. WILLS

Roses cigarette cards, 1936
Color offset lithographic cards, each $3\frac{1}{8} \times 2\frac{3}{8}$ in. / 7.9 × 6.1 cm

Designed to educate green-fingered smokers on the different cultivars of roses, this series of richly illustrated cards was given away in packs of tobacco company W. D. & H. O. Wills's cigarettes at a time when gardening had become a popular pastime in Great Britain. The colorful series, which features forty botanically accurate illustrations and short, informative texts, includes such roses as 'Kirsten Poulsen' with its vivid reddish-pink flowers. Bred by Svend Poulsen, the Danish pioneer of cluster-flowered roses, and named after his niece, it was introduced in 1924. Another notable breeder represented in the series is the Frenchman Joseph Pernet-Ducher, an important figure in the development of the modern Hybrid Tea rose. Named after his second daughter in 1923, 'Angèle Pernet' has rich, coppery-orange blooms and glossy foliage, while his light yellow 'Julien Potin' has a high-centered bloom form that flowers in flushes; introduced in 1927, it is dedicated to the eponymous former vice president of the National Horticultural Society of France. An older rose is the stunning Hybrid Perpetual 'Frau Karl Druschki' from 1895. As its card illustrates, its bud displays a splash of dark pink, but its bloom is pure white. Bred by German rosarian Peter Lambert and dedicated to the wife of the then president of the German Rose Society, it is completely scentless. Wills pioneered the use of collectible picture cards in Great Britain and issued several horticultural themed series, each one forming an attractive miniature encyclopedia on its respective subject.

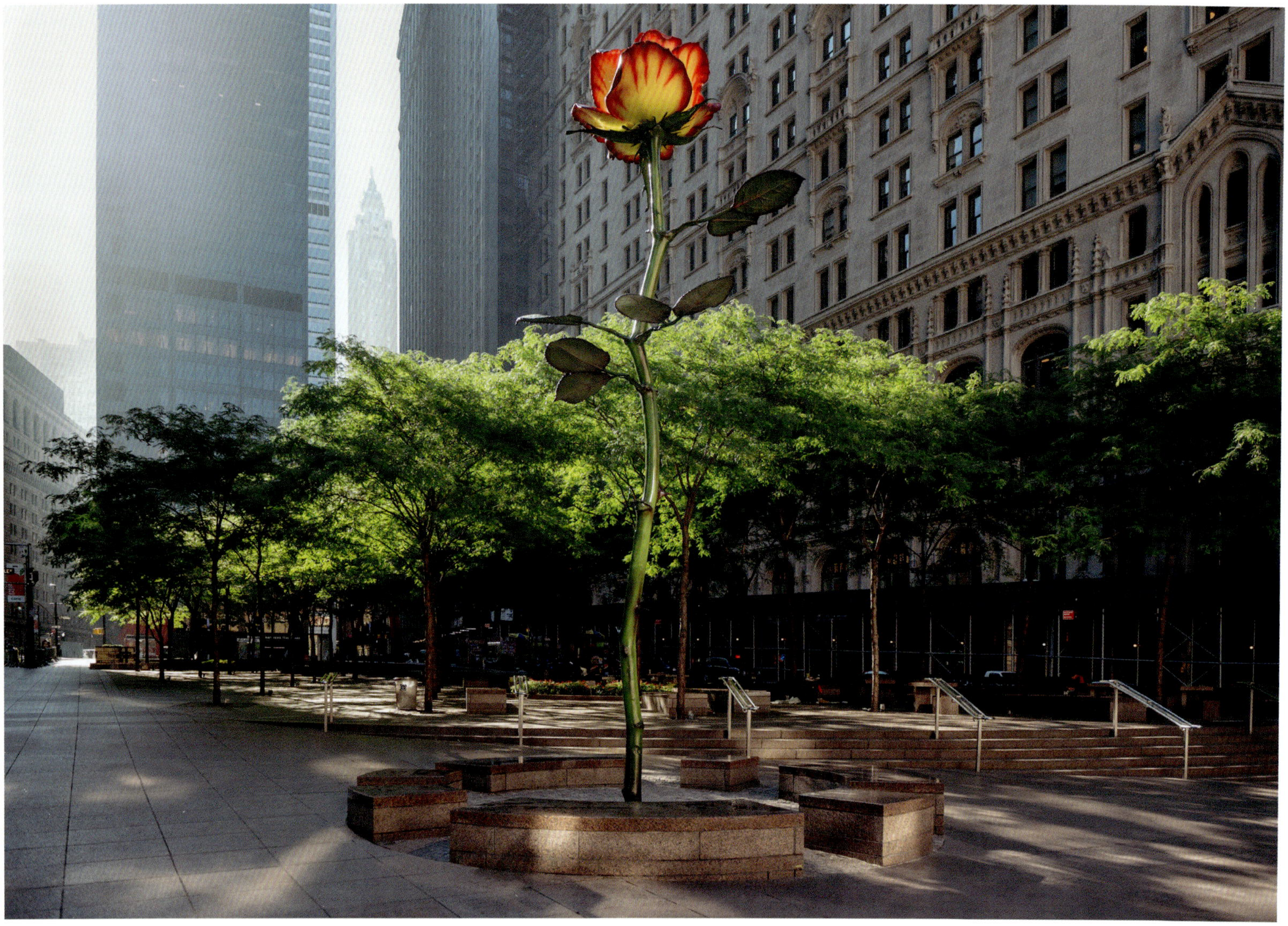

ISA GENZKEN

Rose III, 2016
Aluminum, galvanized steel, and lacquer, H. 26 ft. / 8 m
Installation view, Zuccotti Park, New York

This supersized rose injects a vibrant splash of color into the urban setting of Zuccotti Park in Lower Manhattan, New York. Standing in the shadow of the soaring skyscrapers that surround the area, the 26-foot (8-m) tall sculpture of a long-stemmed rose in cast aluminum and steel is modeled on a flower handpicked by renowned German artist Isa Genzken (b. 1948). The monumental red-tip, yellow rose, which towers over pedestrians, is encircled by a series of stone benches that offer an opportunity for quiet contemplation. The giving of a single rose is normally an intimate exchange between two people; by transforming this personal act into a public sculpture, the artist offered her rose as a gesture of love and appreciation for the inhabitants of New York. Genzken initially conceived of her rose sculptures in the early 1990s, when she was commissioned to make a new work for collector Frieder Burda, a resident of Baden-Baden, Germany, which is famed for its new roses competition. Every year, rose enthusiasts, growers, and an international jury gather in the German town to assess new rose varieties and present the Golden Rose of Baden-Baden, Germany's most prestigious award for roses. An edition of Genzken's first rose sculpture, *Rose* (1993), is on permanent display in the Roppongi Hills development in central Tokyo, while *Rose II* (2007), her first public sculpture in the United States, was installed on the facade of the New Museum in New York before being bought by the Museum of Modern Art.

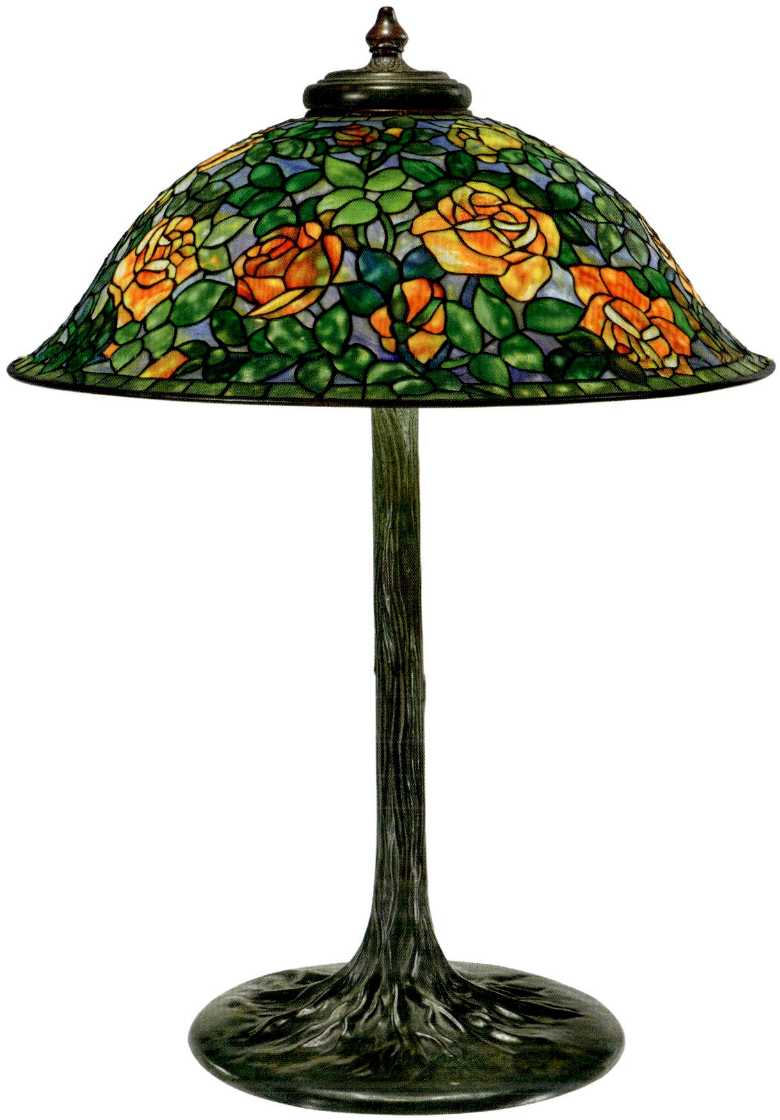

TIFFANY STUDIOS

Rosebush table lamp, c. 1915
Leaded glass and patinated bronze, 32¾ × 24¾ in. / 83.2 × 62.9 cm
Private collection

Yellow roses adorn this stained-glass Art Nouveau table lamp made early in the twentieth century by American decorative arts company Tiffany Studios, led by head designer Louis Comfort Tiffany. Against a blue glass sky, different shades of yellow and green suggest that dappled light is illuminating the rosebush's blooms and leaves. A talented and successful painter for many years before he joined the family business, Tiffany found inspiration for his art on his many trips to Europe and North Africa, and his love of flowers would find an expression in many of the company's most popular globe-shaped table lamps. To create the subtle shades and intensity of color in glass for which Tiffany Studios became known, the company's English glass-worker, Arthur Nash, worked with Tiffany to develop a technique of blending pigments while the glass was still molten. This lamp was made using a technique that Tiffany called "Favrile," which comes from the Old English word *fabrile*, meaning "handwrought." Pieces of colored glass were soldered together using lead and other alloys to create precise shapes that allowed for the lamp's curved shape. Tiffany Studios also produced stained-glass windows and expanded into other areas, including metalwork, jewelry, and enamels. Highly progressive in its outlook, the company employed women to work at the studio, including Clara Driscoll, who was head of the Women's Glass Cutting Department. It is thought that Driscoll designed many of the studio's floral designs, including *Daffodil* (her first) and the popular *Butterfly*.

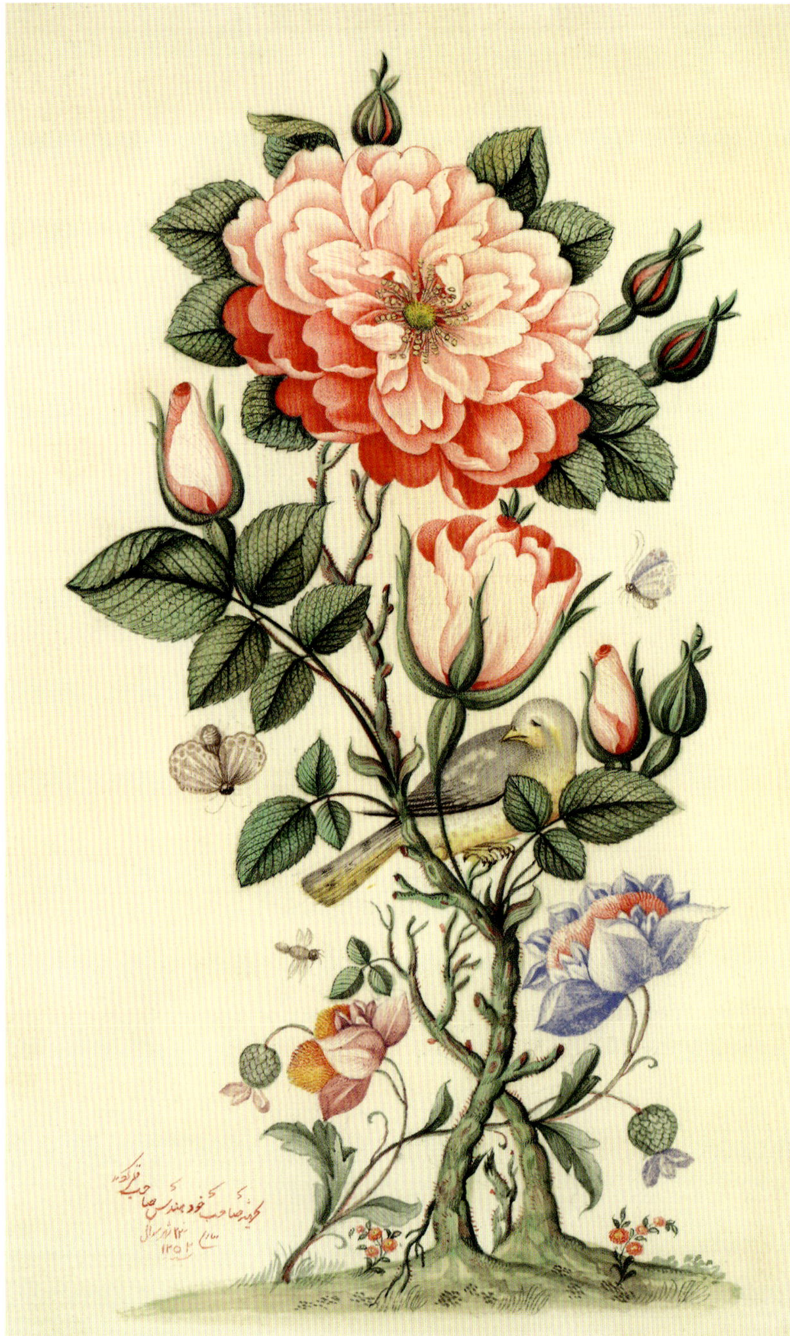

ANONYMOUS

Rose and nightingale, from *Portraits of Celebrities of the Persian Court*, 1837
Watercolor on paper, 8⅝ × 7⅛ in. / 22.4 × 18 cm
British Museum, London

This Persian watercolor of Damask roses was painted in 1837 and included in an album of flower studies known as the *Portraits of Celebrities of the Persian Court*. Compiled in 1852 during Iran's Qājār dynasty (1794–1925), the album was made when expansionist rivalries between Britain and Russia opened the area to European missions and influences. Here, Western botanical art has influenced the artist's accurate portrayal of the slightly recurved petals, rounded leaves, and elongated sepals typical of *Rosa* × *damascena*, from which the perfumed oils, known as *attar*, had been extracted for centuries in Asia. Roses are a venerated flower, and the less realistic flowers beneath this painted example seem to pay homage to it. In Islamic art, the rose is often connected to the adoration of the nightingale, seen here nestling against its stem. The Persian term for this motif is *gol o bolbol*, and it appears in many art forms (see p.54). Because roses are the flower of Muhammad, this could be a mystic representation of the yearning for union with God. Conversely, the flower could also symbolize a lover's passion for a beautiful yet cruelly unattainable woman. The rose-and-nightingale theme first appeared in Persian literature in the eleventh century and reached its height with the poems of Hafiz in the fourteenth century. His verses ranged from the ritual, "the nightingale with drops of blood nourished the rose," to the erotic, "the red rose is open and the nightingale is drunk," mirroring the diversity of rose symbolism itself.

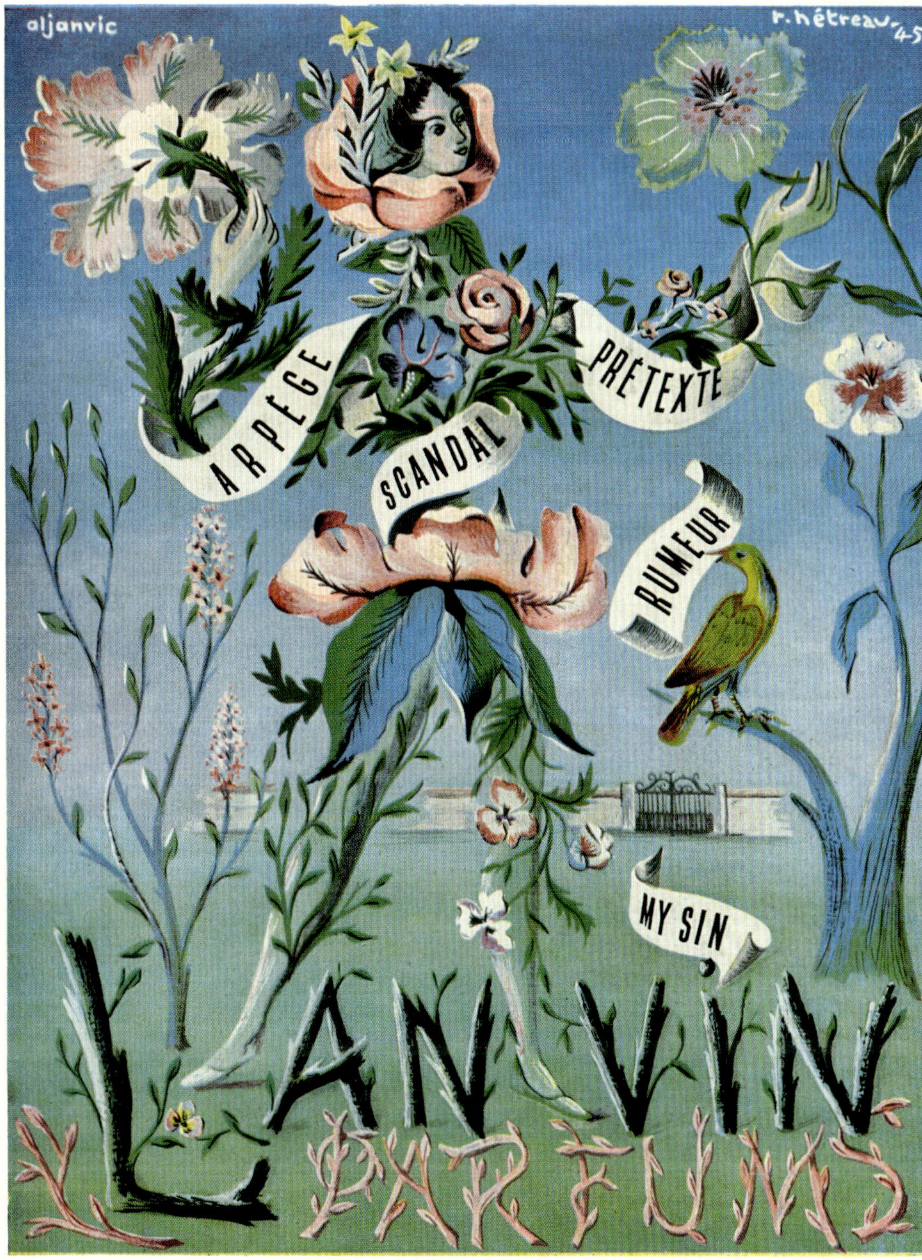

RÉMY HÉTREAU

Arpège, Scandal, Prétexte, Rumeur, My Sin, 1945
Lithograph, 11¾ × 8¾ in / 30 × 22 cm

In 1925, acclaimed Parisian haute couturier Jeanne Lanvin introduced perfumes to complement her romantic fashion designs, many of which featured rose ornamentation. The first fragrance was given the racy title My Sin (1925) and featured middle notes of rose petals. Two years later, Lanvin commissioned famous perfumer André Fraysse to create a scent for her musician daughter. The resulting perfume was named Arpège (a play on the term "arpeggio") and known as the "fragrance of 1,000 flowers," its core ingredients including Bulgarian rose, Grasse jasmine, honeysuckle, and lily of the valley. Further scents Scandal (1931), Rumeur (1934), and Prétexte (1937) followed. This advertisement for the family of Lanvin Parfums was designed by French artist Rémy Hétreau (1913–2001). The print portrays some of the flowers used in the fragrances, along with some fantasy blooms. Hétreau's fantastical and anthropomorphic *femme-fleur*, placed in a dreamy flowery landscape, has Surreal overtones. The image also exploits the trope, popular since the late nineteenth century, for portraying a female head within a rose bloom. From the mid-1930s onward, the more decorative aspects of Surrealism were appropriated for advertising, and its fascination with dreams, human desire, and transformation made it ripe for commercial translation. Lanvin would have placed this advertisement in such exclusive fashion magazines as *Vogue* and *Harper's Bazaar* to attract flower-hungry women to its lushly scented perfumes.

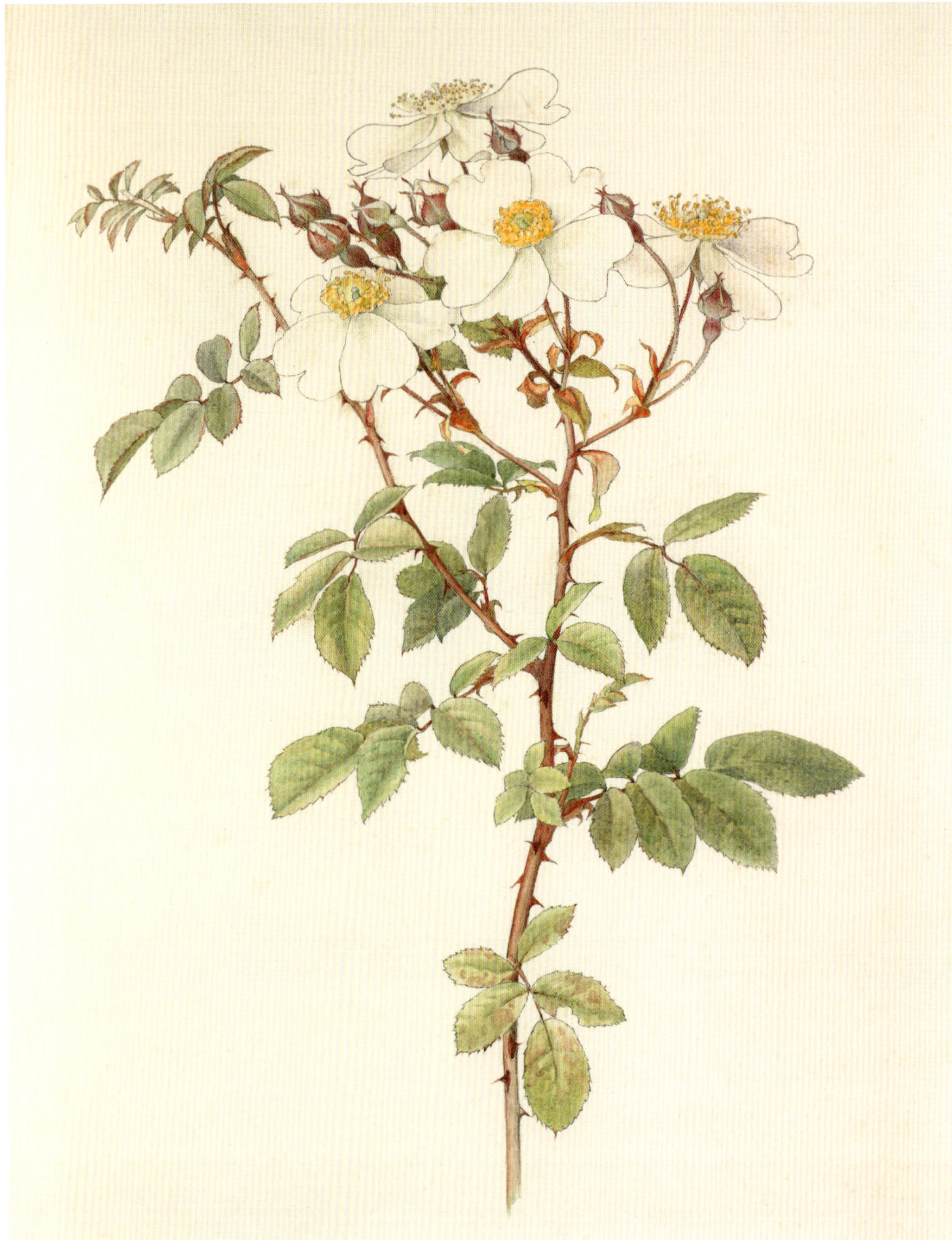

ALFRED PARSONS

Rosa arvensis Huds., from *The Genus Rosa*, 1910–14
Watercolor, 17¾ × 10¾ in. / 37.5 × 27.5 cm
Royal Horticultural Society, Lindley Library, London

Rosa arvensis, called variously the field rose, Ayrshire rose, and a variety of other ambiguous and confusable names, is the second most common native rose in Britain. A climbing and trailing rose with an unusual capacity to tolerate the shade of trees, it is widespread in western Europe and has been used in hybridization, especially with the Scotch roses fashionable in the early nineteenth century. Curiously, Carl Linnaeus never had a specimen in his herbarium, so it fell to English botanist William Hudson to coin the name *R. arvensis* for it in his *Flora Anglica* (1762). It had been known to previous botanists, such as John Ray, who treated it as a variety of *R. sylvestris*, but it was never described by the pioneering herbalists of the sixteenth century. This illustration comes from the massive *The Genus Rosa*, published in parts from 1910 to 1914 and compiled by horticulturist Ellen Willmott, whose garden at Warley Place in Essex was one of England's greatest at the beginning of the twentieth century. She hired Kew botanist John Gilbert Baker to write the text—who himself had called *R. arvensis* "the most beautiful of all our English wild roses"—and painter and garden designer Alfred Parsons (1847–1920) to illustrate the plants. Much to Parsons's disappointment, the three-color printing process made all the plates come out with a yellow tint. Willmott had his originals bound facing the published plates in her own copy of the work, which is now in the Royal Horticultural Society's Lindley Library in London.

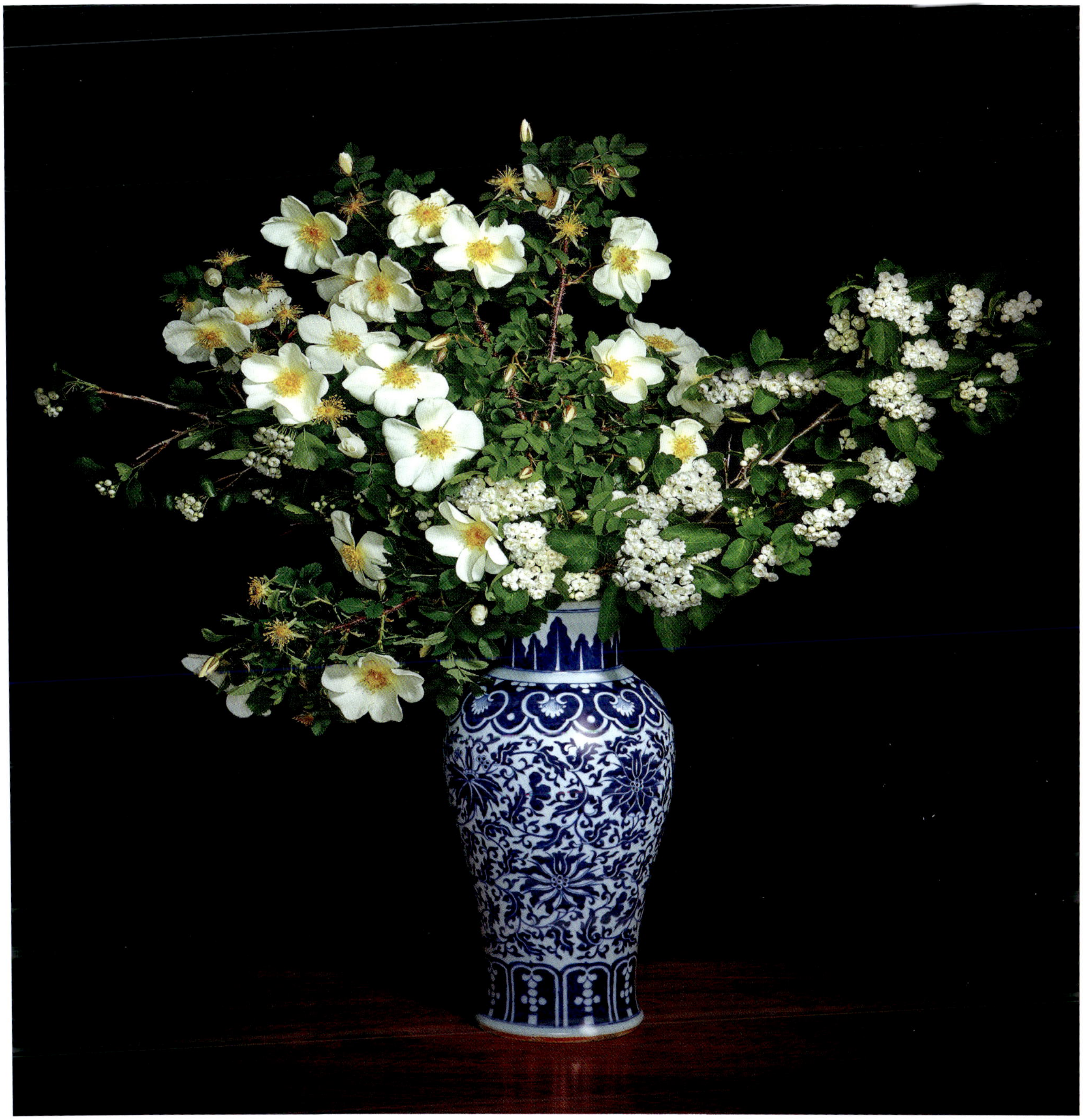

T. M. GLASS

White Hawthorne and White Shrub Rose in a Blue and White Chinese Vase, 2018
Archival pigment print on handmade Italian rag paper, dimensions variable

Every delicate petal, bud, leaf, and stem of this floral arrangement is captured in perfect focus and stunning definition. The showy white flowers of the shrub rose are complemented by white hawthorn blooms, and both contrast sharply with their dark green foliage. Displayed in an antique blue-and white Chinese vase, the whole bouquet is set against a black background, intensifying the colors and textures of the plants. The picture is one of many floral still lifes created by Canadian artist T. M. Glass, who employs cutting-edge photographic technology to digitally collage images. Using a digital painting technique, Glass spends weeks and sometimes months creating these botanical displays, painstakingly stitching together imagery and altering colors on the computer before printing the completed work at very high resolution in limited-edition runs. The resulting lens-based digital painting evokes Dutch still lifes (see pp.10 and 140) but provides a level of detail unimaginable to seventeenth-century painters or analog photographers. Fascinated by the mathematics of natural beauty, the artist strives for harmony by following age-old principles of proportion and balance. Glass began the original still-life flower series with cut blooms from the backyard, although the project has since expanded to include flowers from celebrated gardens around the world, paired with vessels from museum collections. This image was created in collaboration with the Jardins de Métis (Reford Gardens) in Quebec.

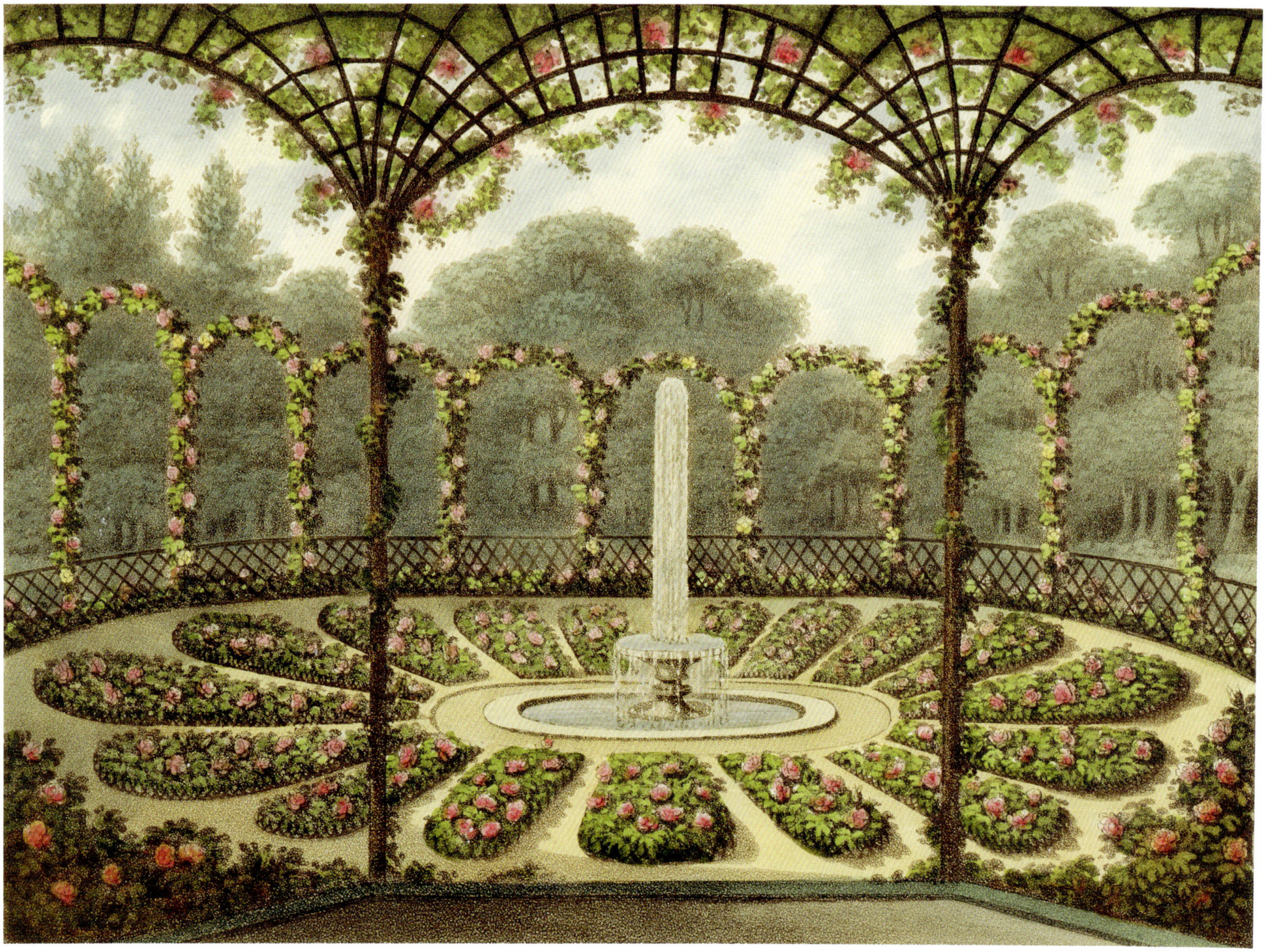

HUMPHRY REPTON

The Rosary at Ashridge, from *Fragments on the Theory and Practice of Landscape Gardening*, 1816
Color lithograph, 11 × 13¾ in. / 27.7 × 35 cm
Private collection

Humphry Repton (1752–1818) was the first English landscape designer to deliberately create new gardens in outmoded historical styles. Having spent his early career in the late eighteenth century making large informal parks in the manner of his predecessor Lancelot "Capability" Brown, in 1800 he turned to placing formal architectural gardens around the sides of the house and bringing flower gardens into the main views from the windows. In his final book, *Fragments on the Theory and Practice of Landscape Gardening* (1816), he included case studies of some of his recent commissions, reproducing text and illustrations from the "Red Books" he had made for his prospective clients, frequently showing before-and-after views of the gardens to depict his proposed changes. Ashridge, a country estate in Hertfordshire, was the largest in scale, comprising a series of fifteen gardens in different styles, including the Rosary (or Rosarium), which Repton depicted in a drawing that was then engraved in aquatint by publisher John Taylor in 1816. This engraving shows a circular garden with a fountain, surrounded by beds of dwarf roses, and a series of arches for climbing roses. Repton did not name the varieties he proposed to use, but it is difficult to think of climbing roses in that era that could have graced the trellis as shown. Architect Jeffry Wyatville completed the Rosary after Repton's death, and later developments in growing roses would have allowed climbing roses to be used as Repton originally proposed.

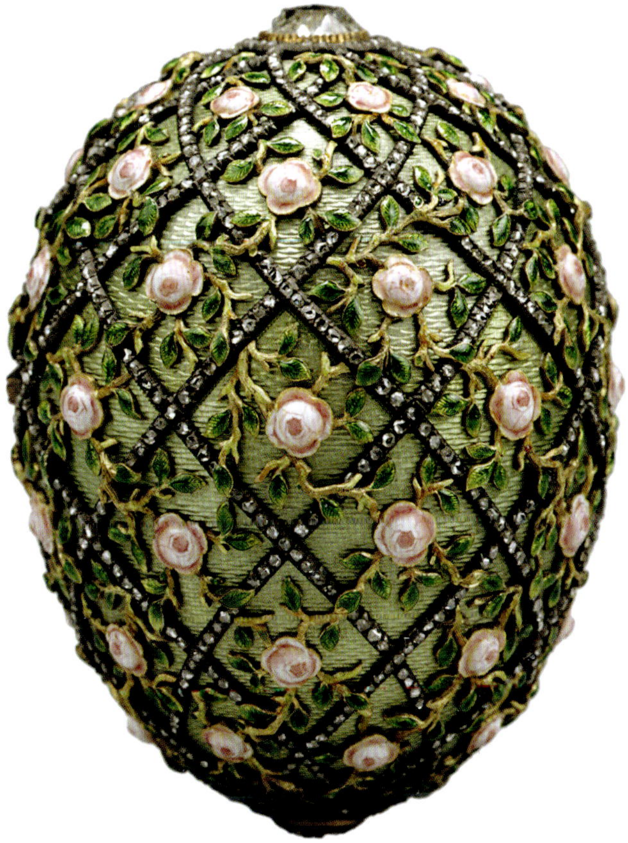

FABERGÉ

Rose Trellis Egg, 1907
Gold, enamel, and diamonds, 3 × 2⅜ in. / 7.7 × 5.9 cm
Walters Art Museum, Baltimore

Tiny enamel roses in shades of light and dark pink stud this diamond-encrusted green enamel egg, which was given by Tsar Nicholas II of Russia to his wife, Alexandra Feodorovna, for Easter 1907 to celebrate the birth of his heir, Alexei Nikolayevich. The boy had been born some three years earlier, but Russia's involvement in the Russo-Japanese War had halted the tradition of imperial eggs being produced every year by the House of Fabergé. The tradition had started in 1885, when Alexander III, Nicholas's father, commissioned Fabergé—the Russian jewelry manufacturer founded by Gustav Fabergé in 1842, and later made famous by his son Peter Carl Fabergé—to make an Easter egg for his wife, Maria Feodorovna. He stipulated only that it must be unique and must contain a surprise. So well received was the first egg that Fabergé, official goldsmith to the Russian imperial court, was subsequently commissioned to design an egg every Easter. Known for his unique style, his fascination with enameling, and his designs that showcased every stone—precious or not—Fabergé became a favorite of Nicholas's mother. Nicholas continued his father's tradition, presenting eggs each Easter to both his mother and his wife. The *Rose Trellis Egg*, made by chief craftsman Henrik Wigström (1862–1923), stands just 3 inches (7.7 cm) tall, with a portrait diamond set at top and bottom. The surprise inside was a diamond necklace for the tsarina and a miniature ivory portrait of the couple's baby son, both now lost.

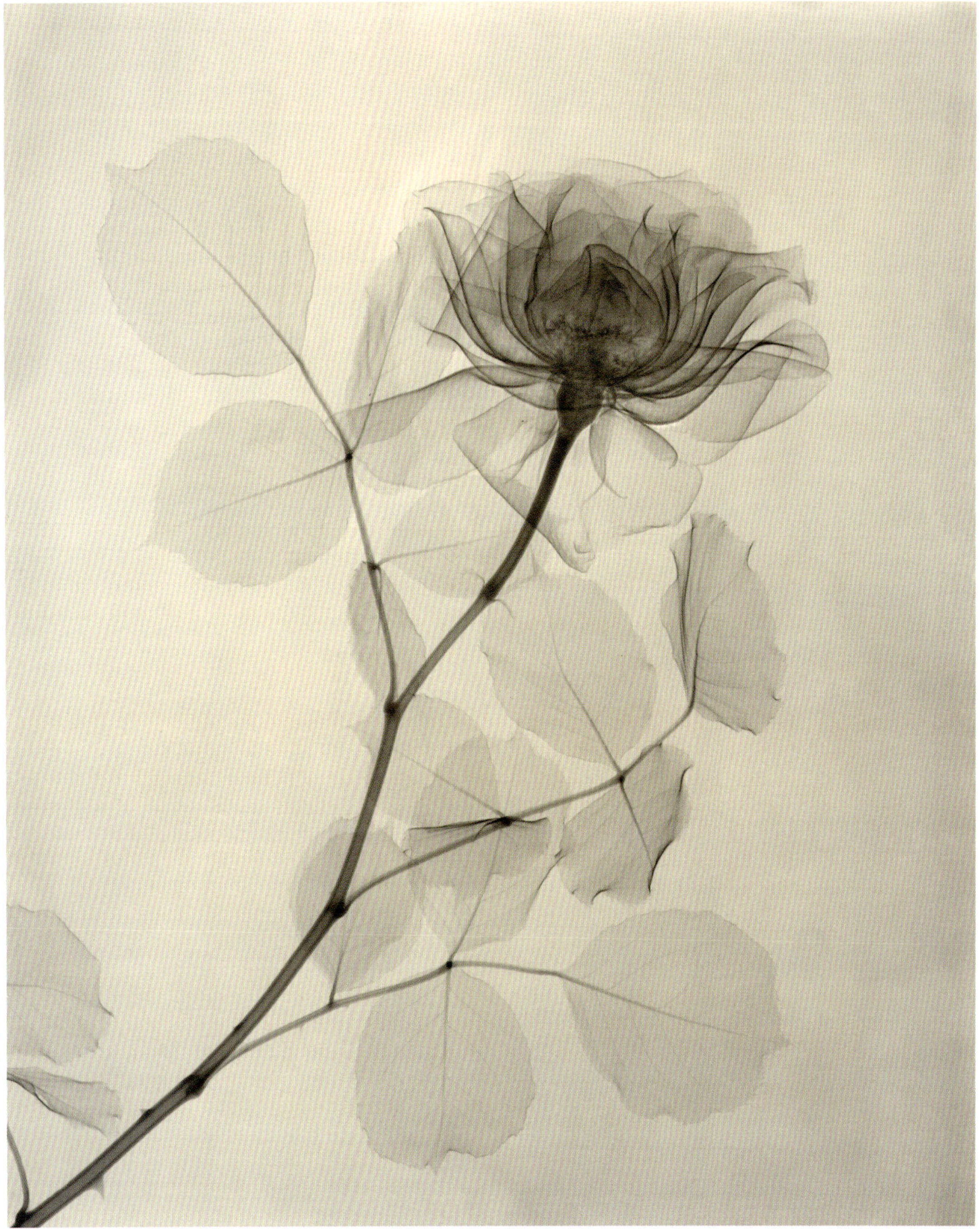

DAIN L. TASKER

A Rose, 1936
Gelatin silver print, 11¼ × 9¼ in. / 28.6 × 23.2 cm
Private collection

A generation before this photograph of a rose was made in 1936, the only way of portraying the internal anatomy of a flower was by cutting it in cross section, as was done by such notable illustrators working at the turn of the twentieth century as Arthur Harry Church. But the rapid development of X-ray photography made it possible to create images that were seemingly unmediated expressions of the planet's structure, in the same way that early photographs seemed to their initial audience to be unmediated presentations of reality. American doctor and radiologist Dain L. Tasker (1872–1964) played an important role in developing X-ray photography as a medium, experimenting with a wide range of subjects and having his work displayed in galleries and published in popular magazines. This image of a rose would be less helpful to a botanical student than Church's lithographs, which, with their deliberate stylization and use of distinct blocks of color to differentiate organs, made the floral structure intelligible. Here, it is difficult to assign the different layers of tissue to their correct planes. However, Tasker's image has its own value and beauty, conveying as it does the messiness and complexity of a floral dissection rather than the clarity of an artificially colored diagram.

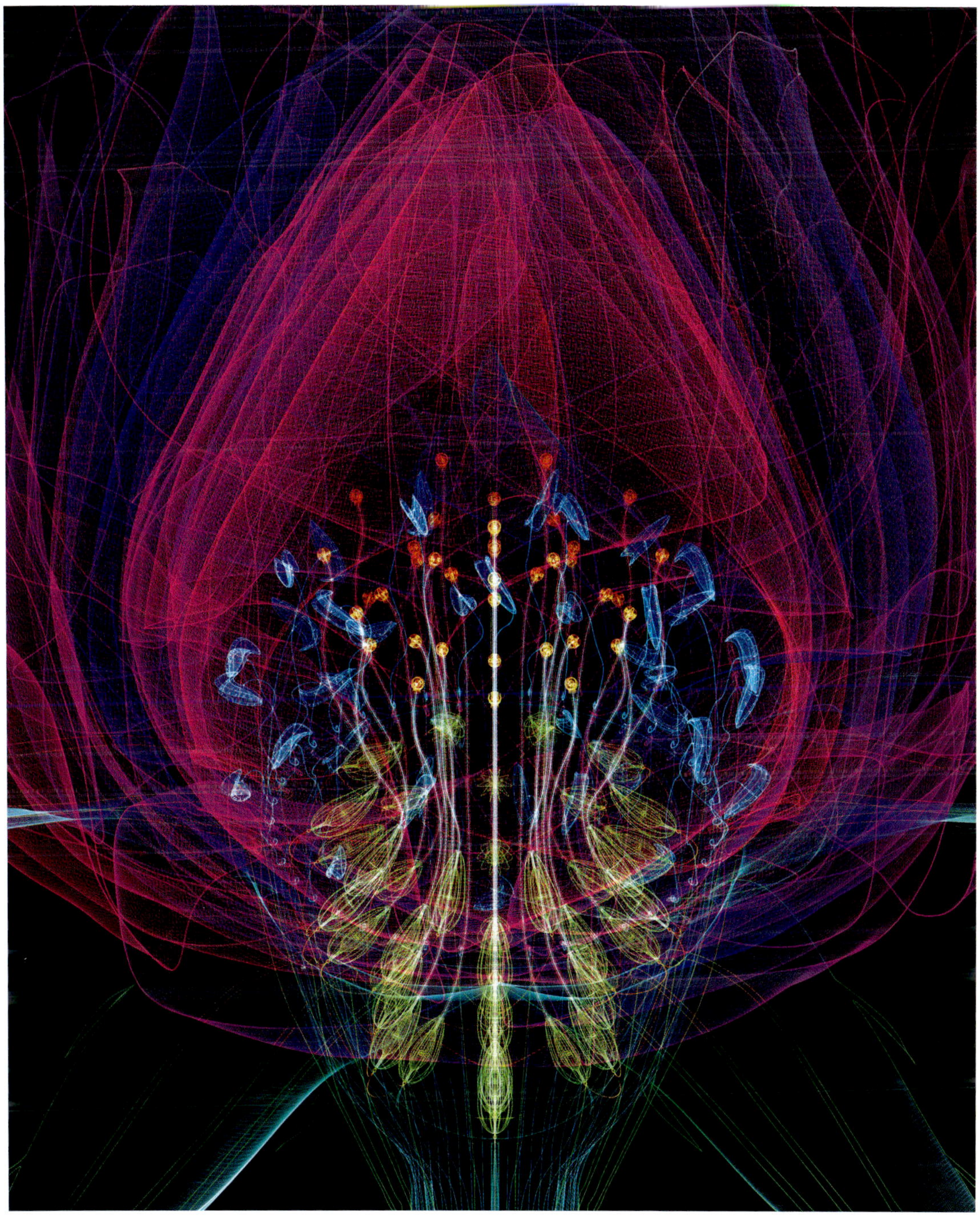

MACOTO MURAYAMA

Rose-iii-bc, 2008–10
Inkjet print, 22 × 18 in. / 56 × 45.7 cm
Private collection

This 3D rendering of a rose brings together botany and architecture. The two disciplines have long been associated—the capitals of ancient Greek columns were carved with acanthus leaves—but the contemporary Japanese architect-turned-artist Macoto Murayama (b. 1984) takes his fascination with plants into a luminous digital world. He uses software to render hand-drawn anatomical studies of dissected plants and flower organs with mathematical exactness, like engineering blueprints for an alien world. Extending the conventions of architectural drawing, Murayama fuses several elevations and plans in a single image, often with dimensional information and labeling. The diaphanous linear mesh of the rendering program describes unfolding clusters of petals wrapped around stamens and anthers, giving the viewer X-ray vision. The ability to ascribe color to individual parts enables the artist to combine technological analysis and romantic artistry. In *Rose-iii-bc*, the flower takes on a balletic swirl in which the tullelike petals enmesh the fragile organs within. The artist has exploited the potential for animation by making a film in which flowers are projected onto the gallery floor, with buds swelling and blooming as they intertwine to form a pastoral tapestry.

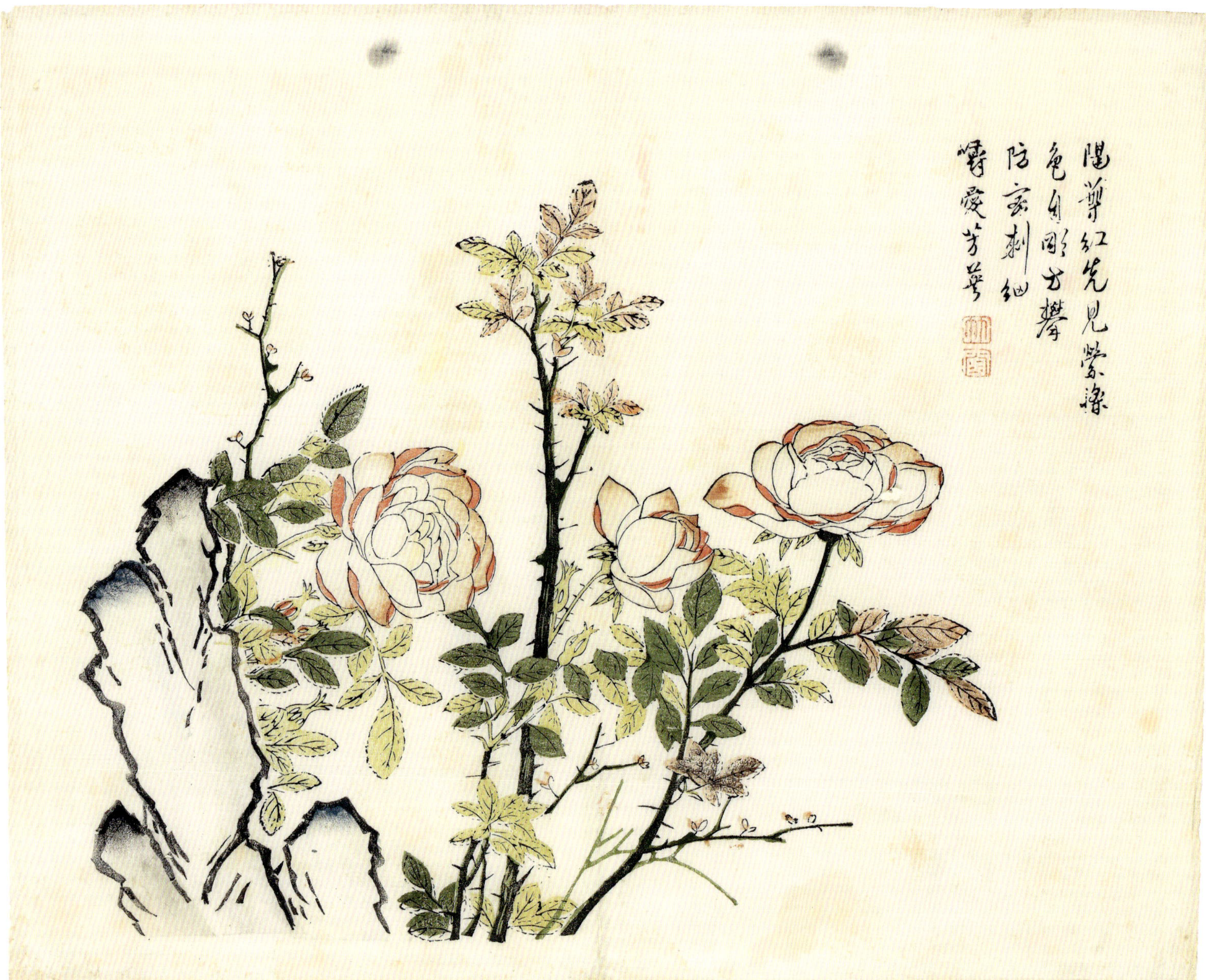

WANG GAI

Blooming Roses Near a Rock, from *Mustard Seed Garden Manual of Painting*, 18th century
Woodcut print, 10⅝ × 13 in. / 27 × 33 cm
Musée Guimet, Paris

In early Qing-dynasty China (1644–1911), painting was an extension of the calligrapher's skill, practiced by scholars and even emperors. Loading a brush with watery ink and twisting it deftly to form leaves or landscapes, the artist could curve or fade lines with absolute precision. Based on centuries of traditional copying exercises, calligraphy brings the natural world vividly to life, as in the white roses evoked here. Then, equal skill was needed to cut the woodblocks for printing the image, and to apply the delicate graded colors of leaves, petals, and rocks. An exceptional example of Qing-dynasty calligraphy and early color woodcut printing, the *Mustard Seed Garden Manual of Painting* is one of the first teaching resources to detail how to represent landscapes in paint. The manual was the idea of Li Yu and Shen Xinyu, publishers from Nanjing, whose property was known as the Mustard Seed Garden (*Jieziyuan*). They realized that, while there was a plethora of manuals for painting flowers and birds, there were none that taught landscapes, so they commissioned Wang Gai (1645–1710) as the principal artist of their own landscape painting resource. The first five volumes, concerning how to paint trees, rocks, and other landscape elements, were printed in 1679, while the final volumes, which expanded the teaching to plants and insects, appeared in 1701. An invaluable resource, the *Manual* was widely republished, and generations of aspiring Chinese painters used its pages as models.

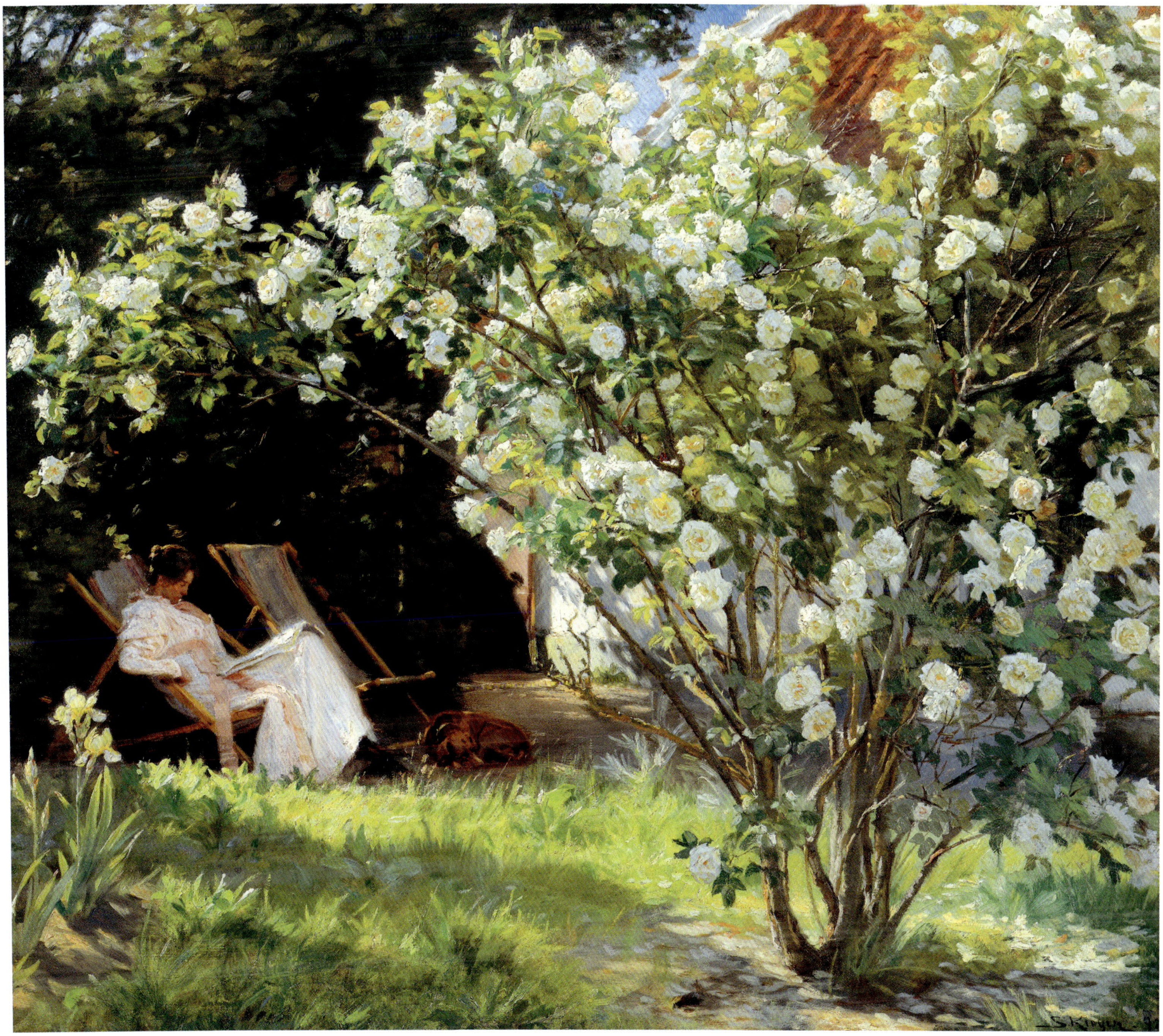

PEDER SEVERIN KRØYER

Roses, 1893
Oil on canvas, 26⅝ × 30⅛ in. / 67.5 × 76.5 cm
Skagens Museum, Denmark

There can be little doubt about the true subject of this oil painting by Norwegian-born painter Peder Severin Krøyer (1851–1909). The stunning 'Alba Maxima' rosebush (see p.72) dominates the image of an idyllic summer's day in the cottage garden where the artist and his wife, Marie Krøyer, stayed in the Danish seaside town of Skagen. The white rose-bush in the foreground, with its exuberant blooms, obscures the house almost completely. Behind the rosebush, but still in view, the artist's wife sits in a deck chair with their dog, Rap, by her side. Krøyer's seat on her left is empty because he is busy painting. Known for its fragrance, 'Alba Maxima' flowers for a brief period once in the summer, and here Krøyer captures it at its peak, laden with blooms and dappled with sunlight. A celebrated portrait painter, Krøyer was heavily influenced by the French Impressionists and set out in his painting to capture the special light of the Jutland coast of Denmark. Just as Claude Monet (see p.169) was inspired by the light of the French Riviera, Krøyer was drawn to the light of the northern Danish coast; he and his wife settled in Skagen permanently in 1889, having first visited in 1882. Krøyer often painted beach scenes, so this image marks something of a departure in its depiction of domestic bliss as symbolized by the rosebush, a variety of which survives to this day in the garden of the local Skagens Museum.

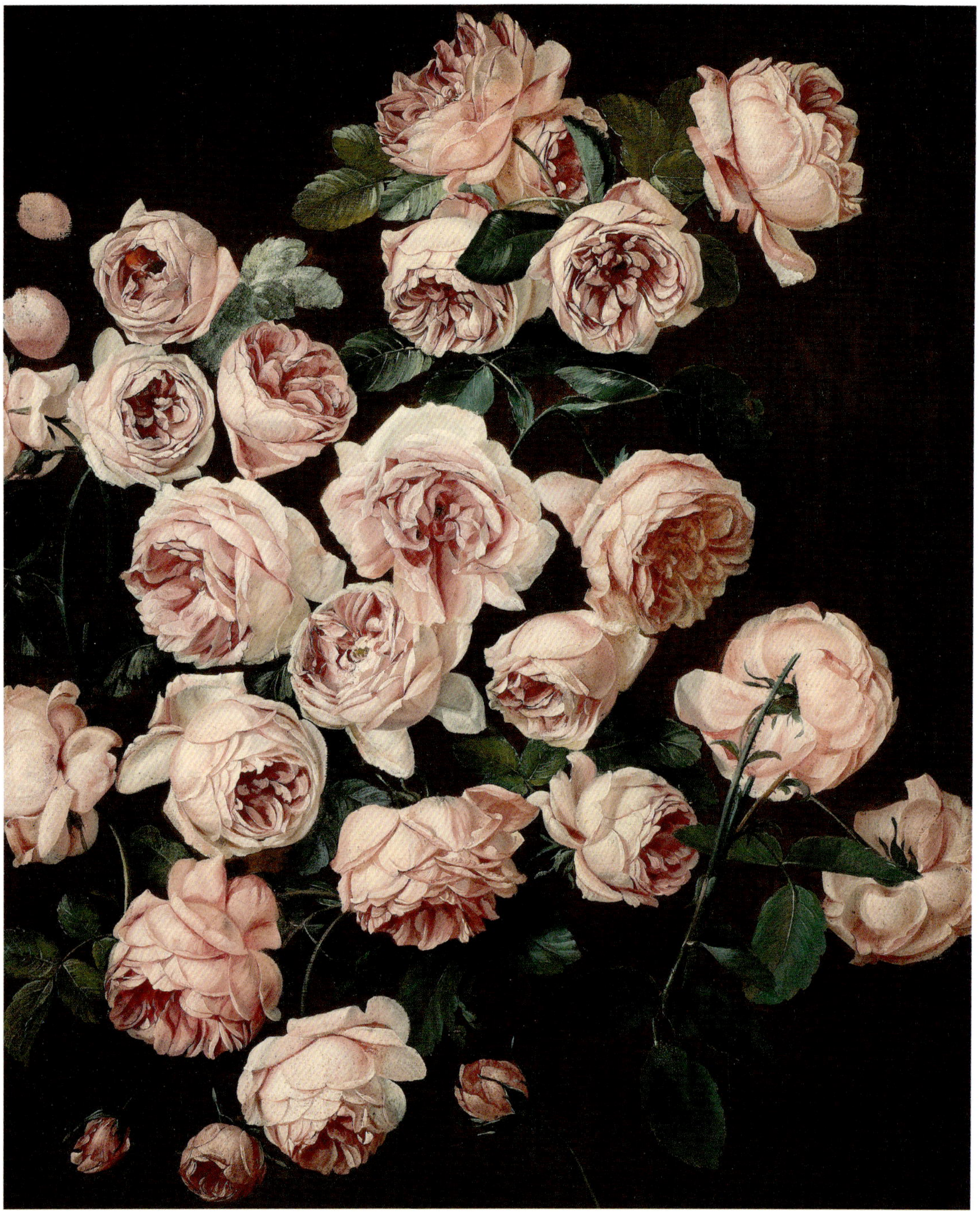

PANCRACE BESSA

A Rosebush, date unknown
Oil on canvas, 22 × 18¼ in. / 56 × 46.5 cm
Private collection

This exuberant, almost photorealistic painting of a rosebush in bloom by French artist Pancrace Bessa (1772–1846) was discovered unstretched in his studio after his death and remained in his family until the late 1980s. Bessa was a pupil of both master engraver Gerard van Spaendonck and Pierre-Joseph Redouté (see p.13) at the Museum of Natural History in Paris. Although he is now less well known than his teachers, in the early nineteenth century, Bessa was considered as equally skilled as Redouté. He was highly respected for his talents as both painter and engraver and enjoyed the patronage of the Duchess de Berry and Empress Joséphine Bonaparte. In 1823 Bessa succeeded Van Spaendonck as painter to the Museum of Natural History, where he was commissioned to create studies of flowers on vellum. Among Bessa's important works in this medium are 572 watercolors for the journal *Herbier general de l'amateur* (*General Herbarium for the Amateur*), commissioned by King Charles X, on which Bessa worked from 1816 to 1827. In 1830 Bessa brought his skill in portraying roses to a broader audience when he published *Album des Roses*, which contained twenty-three color engravings of the plant. More popular, however, was *Flore des Jardiniers* (*The Gardeners' Flora*; 1836), which contained nearly four hundred plates first published in *Herbier general*. Unusual for his oeuvre, this oil on canvas showcases Bessa's ability to depict plants even within their garden habitats, tending less toward the botanical and more in the direction of the dramatic and romantic.

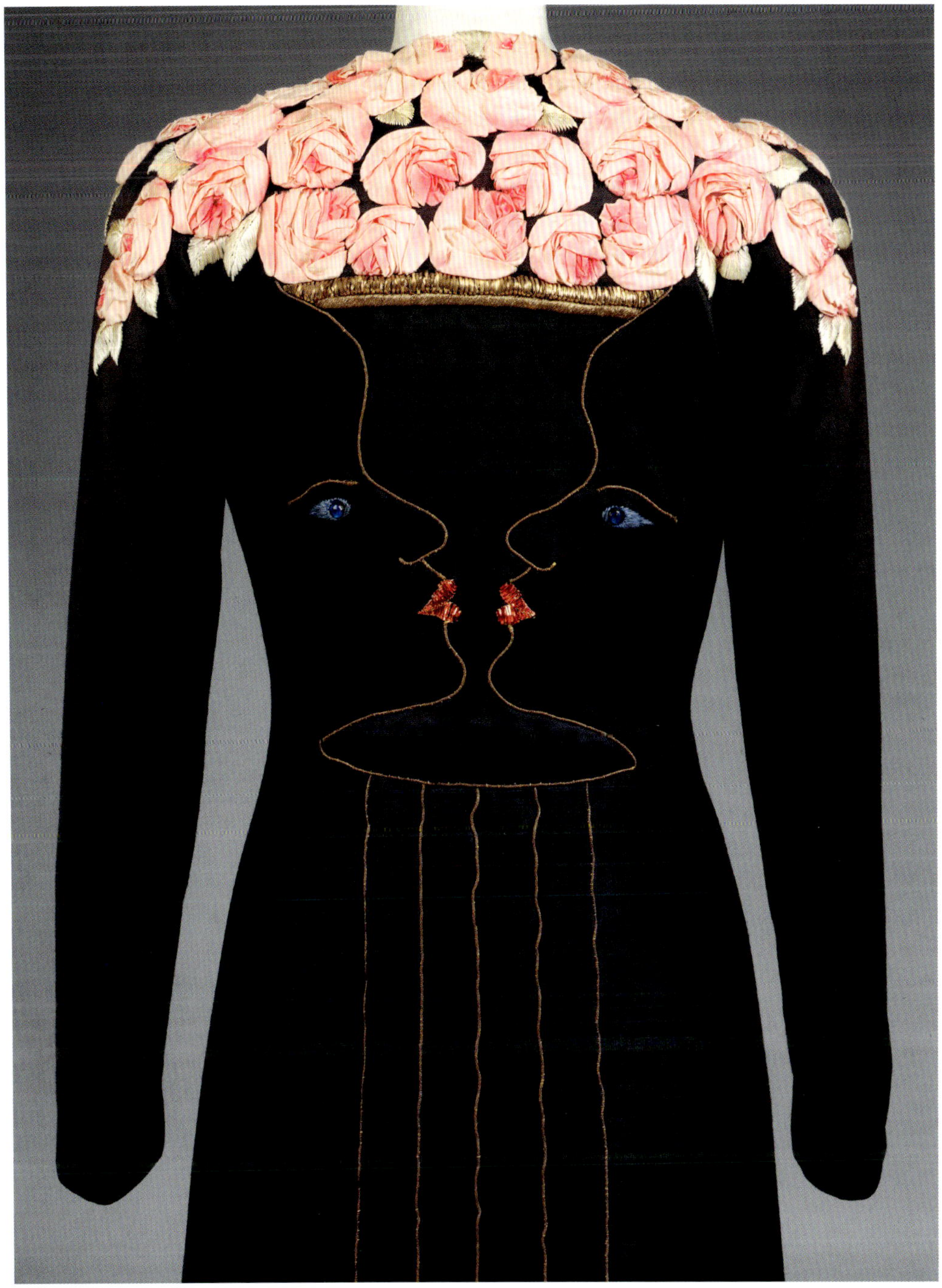

ELSA SCHIAPARELLI

Evening coat, 1937
Silk jersey, with gold thread and silk embroidery and applied decoration in silk
Victoria and Albert Museum, London

Les Deux Visages is an illusory double image depicting two face profiles that combine to create the silhouette of a classical urn, brimming with roses, placed on a column. Here, it appears on the back of an evening coat as a decoration in silk, designed by French artist, poet, and filmmaker Jean Cocteau for radical Italian-born, Paris-based haute couturier Elsa Schiaparelli (1890–1973). Highly skilled artisans working for Paris's premier embroidery house, Lesage, were responsible for stitching the outline design, the rosebud lips, and the Art Deco–style appliqué rose decoration. This full-length evening coat is cut in the fashionable style; it was originally blue and has faded to mauve. Unlike many fashion designers who subsequently exploited Surrealist devices, Schiaparelli's work was not derivative. From the mid-1930s she worked closely, and mixed socially, with the Surrealists, most notably Salvador Dalí (see p.87) and Cocteau. The collaboration with Cocteau was reciprocal, and she later made costumes for his plays. While this Surreal design is romantic in style, Schiaparelli also created hats that took the form of a shoe and a lamb chop, a handbag that resembled a telephone, and, in conjunction with Dalí, a trompe l'oeil printed silk design that depicted torn flesh. The House of Schiaparelli closed in 1954, but it was revived in 2012 with Daniel Roseberry as designer, whose surname could not be more appropriate in this context. He has reimagined many of Schiaparelli's original Surrealist designs, including her various depictions of roses.

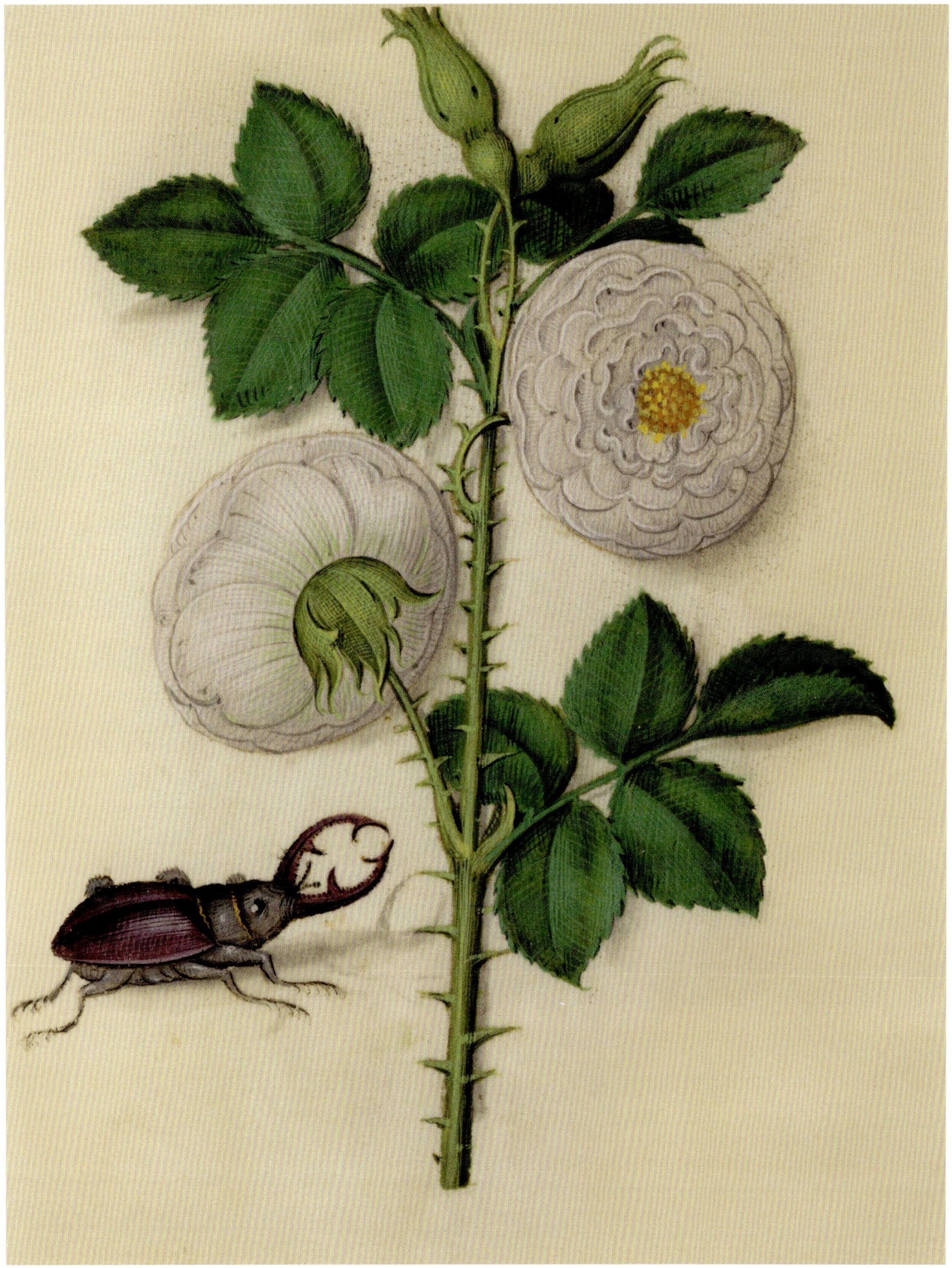

MASTER OF CLAUDE DE FRANCE

Plate 37, from *Book of Flower Studies*, c. 1510–15
Opaque watercolor, organic glazes, gold and silver paint, iron and carbon-based ink, and charcoal on parchment, 6 × 4 in. / 15.4 × 10.2 cm
Metropolitan Museum of Art, New York

This illuminated manuscript of the white rose of York (*Rosa alba*) with accompanying stag beetle is one of thirty-nine beautifully illustrated pages in the early sixteenth-century *Book of Flower Studies*, each specimen chosen not just for its symbolism and medicinal uses but also for its loveliness. The book's illustrations follow the late medieval style of northern European manuscript illumination, and some scholars deem this collection to be the last of its kind. The commissions were linked to Claude, Queen of France, and Antoine de La Barre, the Archbishop of Tours. It is probably a flower model book that depicts common plants in a fourteenth-century garden for other artists to copy. *Rosa alba* is the medieval name for *R.* 'Alba Maxima', a tall shrub rose with a strongly fragrant, full-petaled flower (see p.69). It is an old rose, dating back to the early fourteenth century, if not before. Its origin is actually uncertain, and some botanists say it resulted from a sport (a genetic mutation that results in differences in part of a plant) of the semidouble *R.* 'Alba Semi-plena', which may have been the rose grown by the Romans for both scent and beauty. There is a debate about whether *R. alba* or *R.* 'Alba Semi-plena', as depicted here, is the true white rose used as the symbol of the English House of York, which, when combined with the red rose of Lancaster, became the icon of the end to the fifteenth-century dynastic struggle known as the Wars of the Roses (see pp.33 and 164).

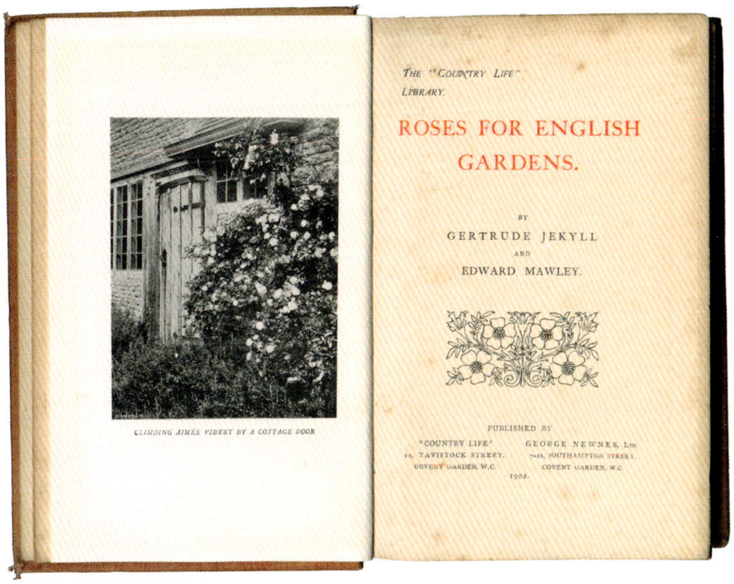

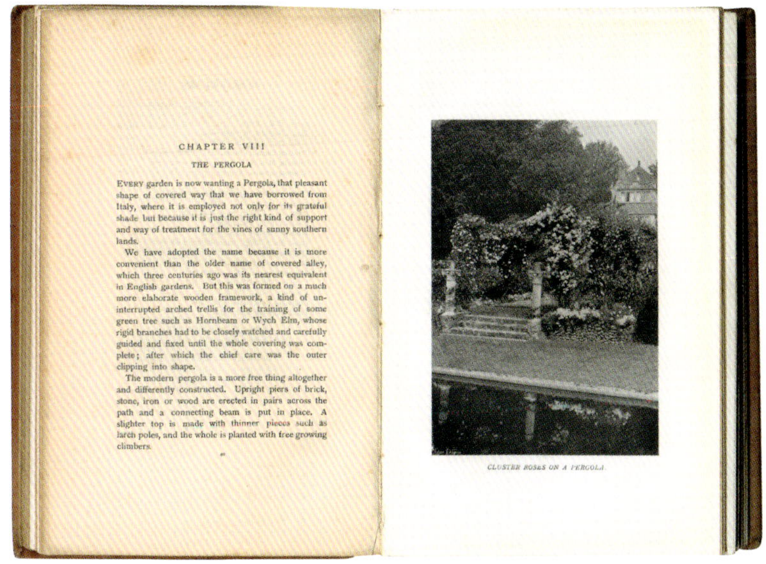

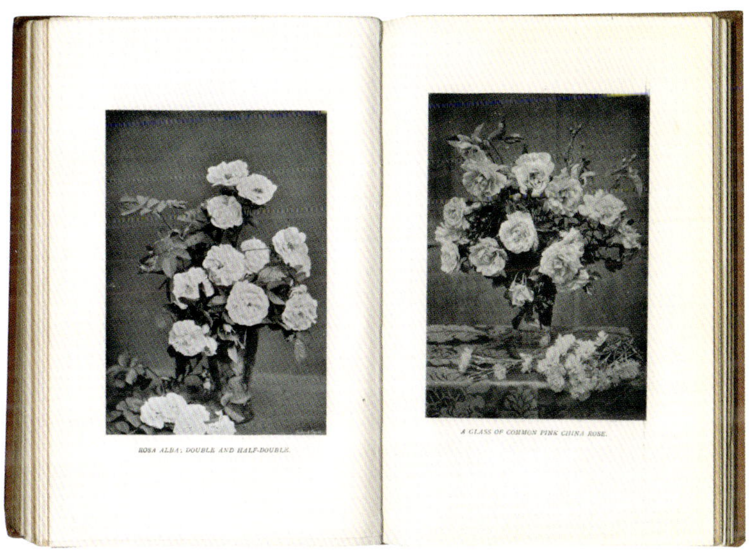

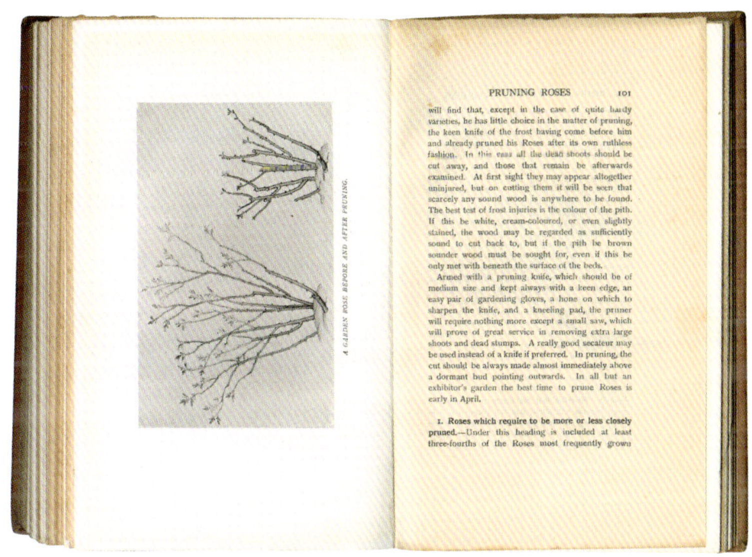

GERTRUDE JEKYLL AND EDWARD MAWLEY

Roses for English Gardens, 1902
Printed book, 9 × 6 in / 23 × 15 cm

Famed rose breeder David Austin (see pp.14 and 105) announced a new introduction to his English Rose line in 1986: a pink shrub rose with lush, fragrant blooms. He named it 'Gertrude Jekyll', a fitting honor since Jekyll (1843–1932) was one of the most celebrated designers in British gardening history. Members of the genus *Rosa*, with their range of hues, heights, and habits, suited Jekyll's aesthetic and her application of color theory to gardens. Roses were also the subject of one of her many books, writing projects that reinforced her design tenets and supplemented her income. In addition to books, she produced more than a thousand articles in periodicals including *Country Life*, which also published *Roses for English Gardens*. In its preface she wrote, "There is a distinct need for a book that shall not only show how Roses may best be grown, but how they may be most beautifully used." With her authoritative voice, she recommended specific cultivars and elaborated their best applications, whether grown on their own or trained over arches, pergolas, or walls. Jekyll handed over the horticultural half of the book, its second section, to British rosarian Edward Mawley (1842–1916), who covered the four "P's": planting, pruning, propagating, and pest management. The book is also rich in photography, although whether Jekyll was the photographer goes unsaid. *Country Life* had pioneered the use of high-quality photographs in its magazine, a technique it carried over into its successful books, including *Roses for English Gardens*.

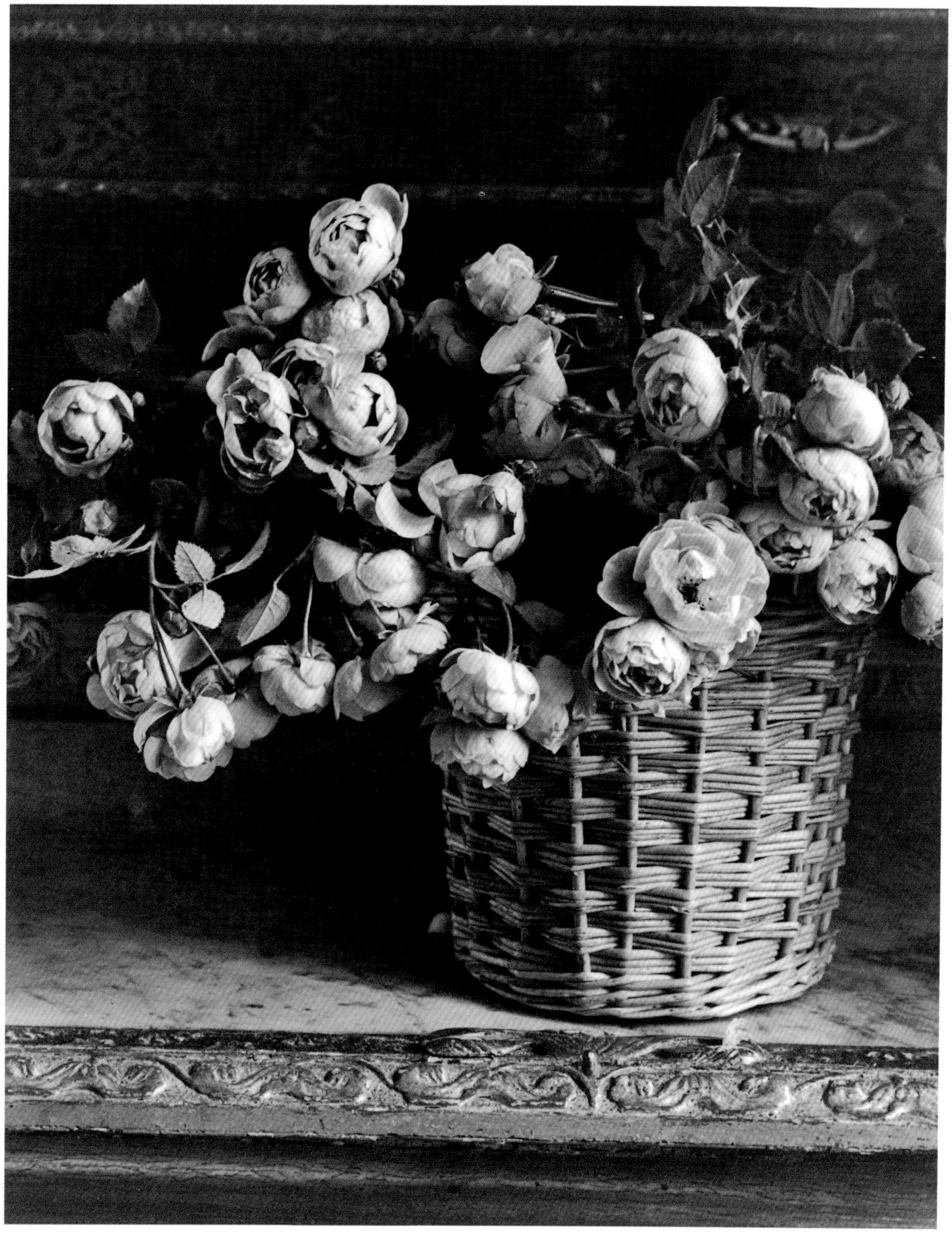

CECIL BEATON

Still Life of Roses, c. 1954
Photograph, dimensions variable
Private collection

British photographer Cecil Beaton (1904–1980) connected with flowers early in life, even playing the pivotal role of a buttercup in a school play. He discovered photography as a teenager and spent hours attempting to re-create glamorous society portraits with his sisters. His career blossomed from the mid-1920s, after his photographs and illustrations appeared in *Vogue* magazine, where he worked for the next thirty years, photographing famous faces from Hollywood, the theater, high society, and the British royal family. He also became a successful sets and costume designer for stage and film productions, and it was the theatrical composition of his images that made them so compelling. He photographed using as little equipment as possible, other than lights and lens.

Beaton's innate creativity is shown in the simplicity of this rustic basket of roses and the use of shadows to highlight the fine, intricate details of the blooms and their humble container, suffusing the image with the romance of roses. Beaton thought white flowers were incredibly chic, and his home studio was filled with white Easter lilies, as well as cherry blossom, chrysanthemums, and hydrangeas according to the season. An enthusiastic gardener, he often crowded his home with cut flowers, and many of his images for *Vogue* featured blooms as backdrops or cascades. These arrangements were sometimes produced by legendary floral artist Constance Spry (see p.113), but the majority were the fruits of his own artistic genius.

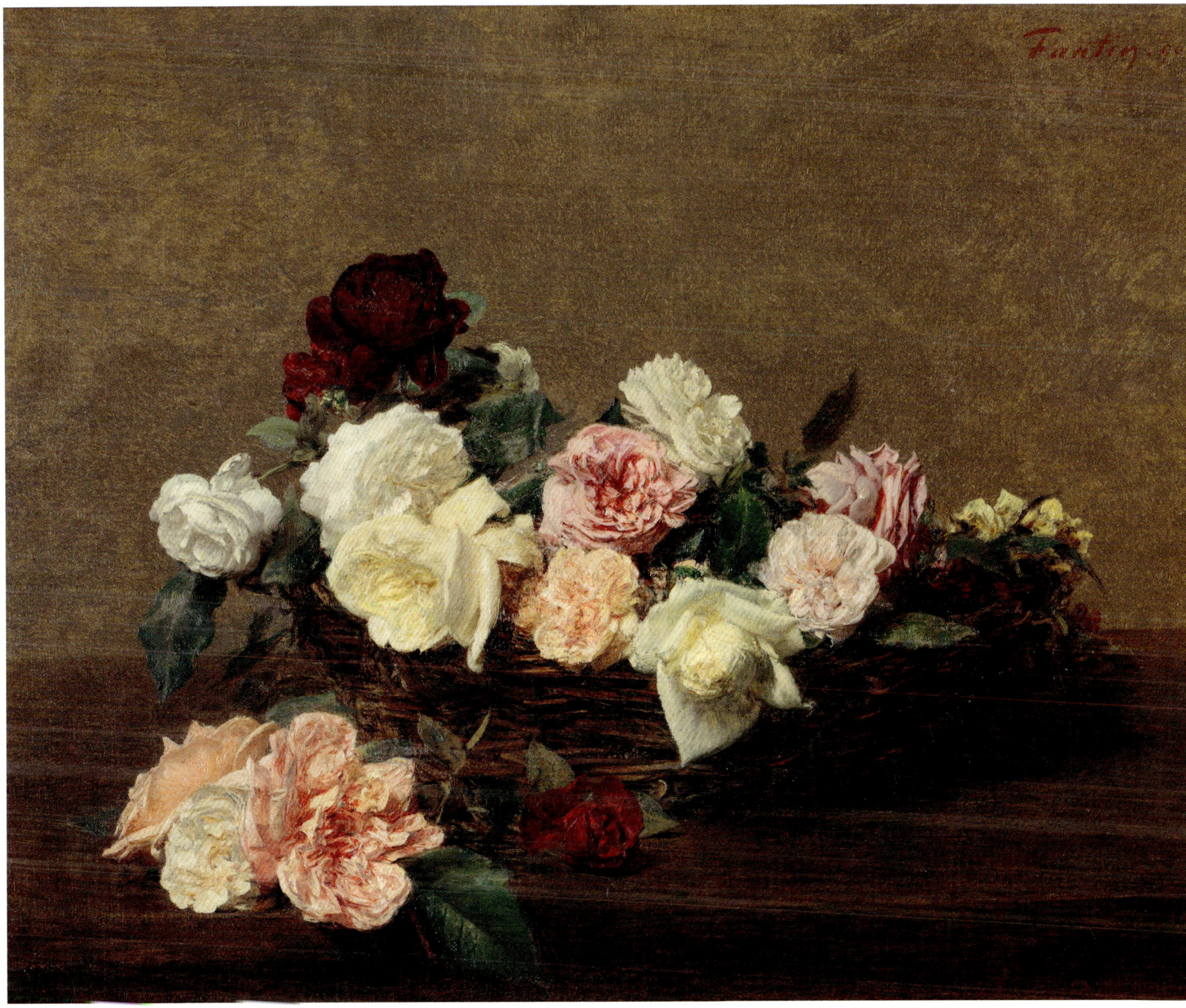

HENRI FANTIN-LATOUR

A Basket of Roses, 1890
Oil on canvas, 19¼ × 23¾ in. / 48.9 × 60.3 cm
National Gallery, London

French artist Ignace-Henri-Théodore Fantin-Latour's (1836–1904) portraits of roses were so critically acclaimed that a lush, highly fragrant, pink Centifolia rose was named in his honor. During the nineteenth century, the artist became France's foremost painter of flower still lifes, and roses were his forte. As a child, Fantin-Latour was taught to draw by his artist father, and he later studied at the École des Beaux-Arts in Paris before working in the studio of painter Gustave Courbet, who led the Realist movement. In Paris, Fantin-Latour exhibited alongside leading Impressionists, including Édouard Manet, Pierre-Auguste Renoir, and Claude Monet (see pp.155 and 169). However, it was in England—which by this time had usurped France as the leading rose-growing nation—that he found his greatest market. Fantin-Latour worked directly from nature and did not sketch, which means his paintings were seasonal. As he studied the look and feel of the fresh roses before him, he would have smelled their various exquisite fragrances, and perhaps this olfactory experience helped him to portray his floral subjects better. This informal composition of garden roses in a mix of white, cream, pale apricot, and pink blooms, overspilling from a wicker basket, appears effortless. On closer inspection, it becomes apparent how carefully each bloom was arranged to capture the light and create such harmony. Fantin-Latour's sensuous and botanically correct paintings of roses, from tightly furled bud through to *déshabillé* bloom, remain highly popular today.

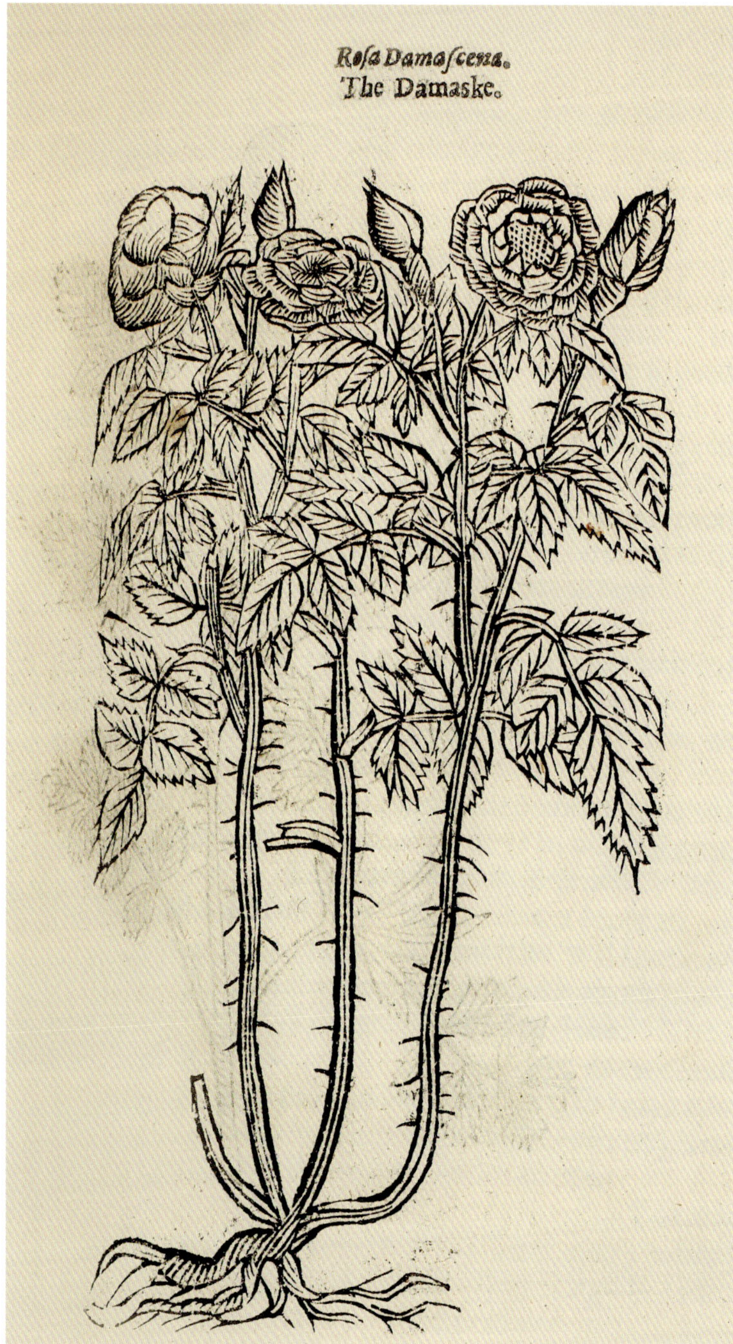

JOHN PARKINSON AND WILLIAM MARSHALL

Rosa damascena, from *Theatrum botanicum*, 1640
Woodcut print, 5⅛ × 2¾ in. / 13 × 7 cm
Royal Horticultural Society, Lindley Library, London

The Damask rose (*Rosa × damascena*) was probably first cultivated in Syria or Iran, where it has historically been grown for its perfumed essential oils (see pp.54 and 60). In those heartlands, it was considered Muhammad's own flower. The Greeks and Romans valued its scent in antiquity, and the plants themselves were reintroduced to medieval Europe, but remained rare, finally arriving in England about a century before the time of this description by English botanist John Parkinson (1566/7–1650) and engraving by English artist William Marshall (1617–1649). Published in 1640, when he was seventy-three years old, his *Theatrum Botanicum* incorporated a lifetime's expertise. Parkinson was a renowned plantsman, a founder of the Worshipful Company of Apothecaries, and royal botanist to King Charles I and Henrietta Maria, to whom he dedicated his earlier herbal *Paradisi in Sole Paradisus Terrestris* in 1629. True to his academic approach, Parkinson tried to establish the provenance of the Damask rose, puzzling over which names had been ascribed to it by earlier botanists. As a physician, he commented on its curative properties, which modern research confirms to be antibacterial and anti-inflammatory, and when preparing the syrup he recommended that "the nails of Damask roses should be retained as improving its medicinal quality." This woodcut illustration emphasizes not only these efficacious thorns but also another feature that defines Damask roses: the slight backward curl of the petals.

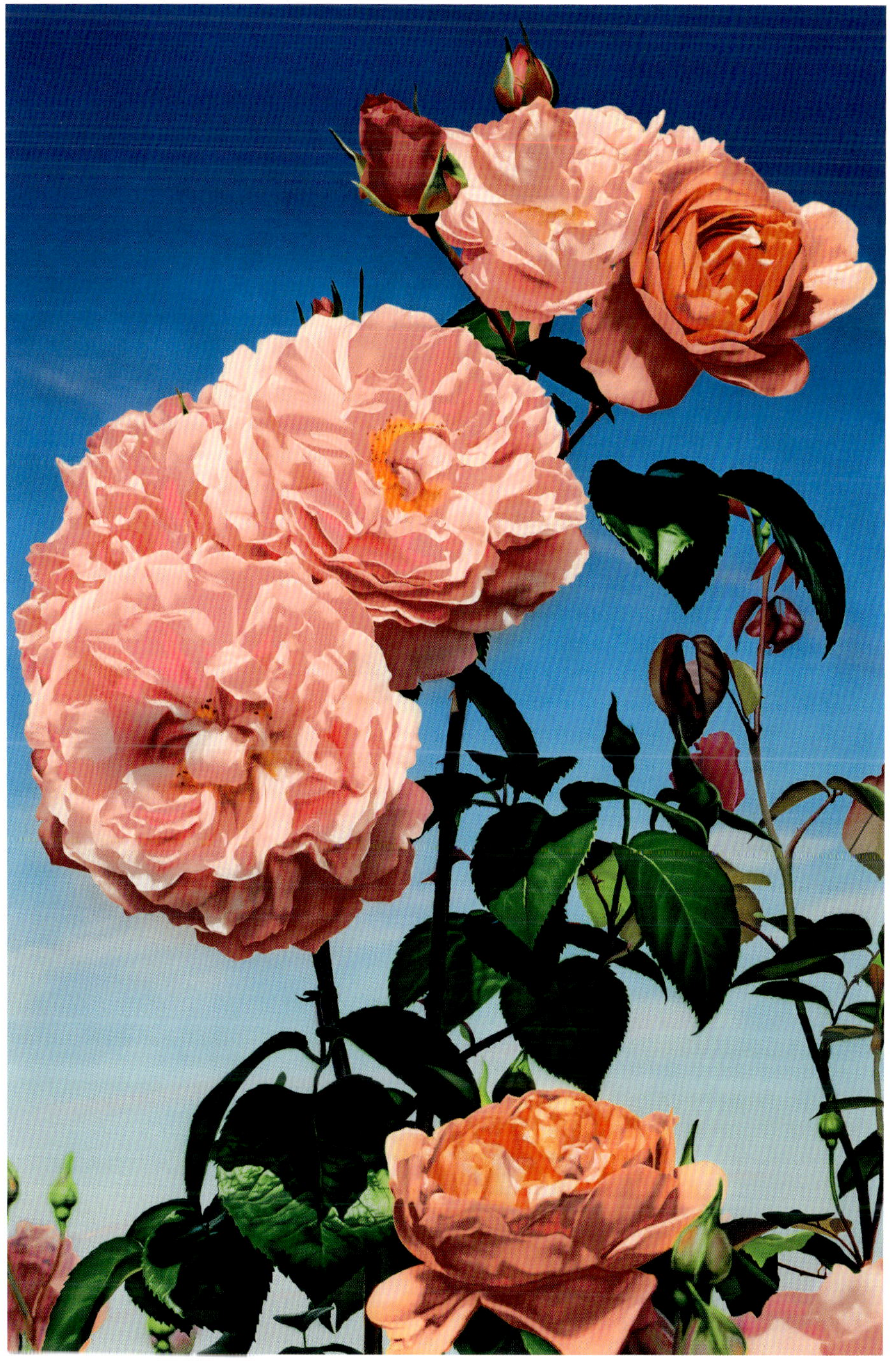

DES LAWRENCE

David Austin, 2021
Enamel on aluminum, 51⅛ × 33⅞ in. / 130 × 86 cm
Private collection

Pink flowers, green leaves, and a vibrant blue sky fill what might look like a photograph of a rosebush but is, in fact, a hyperrealistic painting by British artist Des Lawrence (b. 1970). Lawrence chooses the subjects of his artworks—which include paintings, drawings, and installations—from the obituary pages of British newspapers. The name he gives this 2021 painting is perhaps the most famous name in rose cultivation, that of English rose grower David Austin (see pp.14 and 105), who had passed a few years earlier. Lawrence created a fitting homage to Austin, whose passion for cultivating the national flower led him to release more than 200 varieties of the English rose over his lengthy career. After he released his first rose, 'Constance Spry' (see p.113), in 1961, each rose was named after a famous person from English culture and history, including both the contemporary and the long dead. Austin's passion was for highly fragranced Old Garden roses that flower once a year, but when he started out in the early 1950s, modern hybrid roses were all the rage. His genius was to realize that by combining the two he could create new roses that, while having the attributes he most loved in antique roses—fragrance and beauty—would also have the repeat-flowering tenacity of the hybrids. Austin's skill for cultivating roses led to a successful family business in David Austin Roses (where the flowers are still grown on the family farm in Shropshire), international recognition, and many awards.

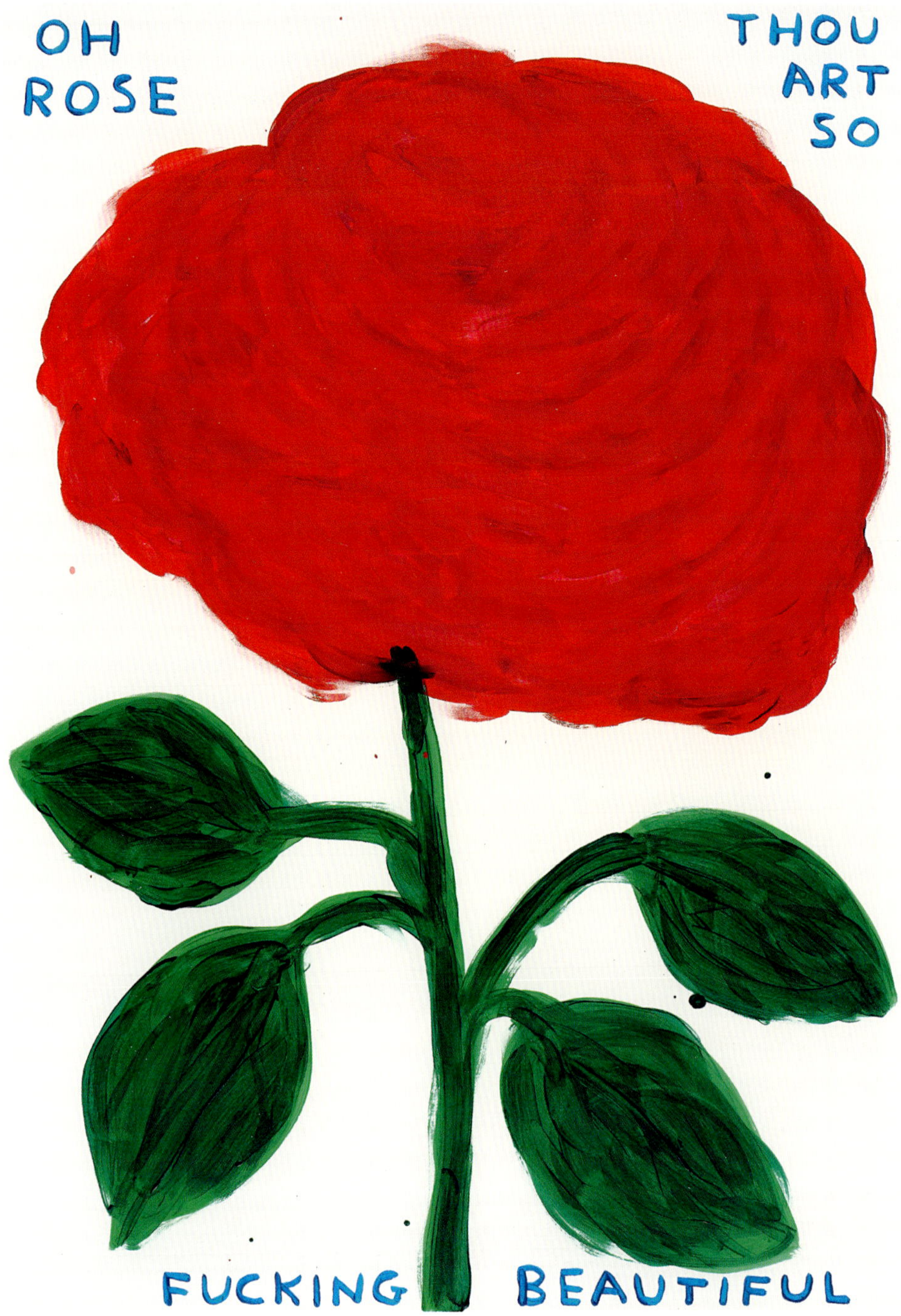

DAVID SHRIGLEY

Untitled, 2022
Acrylic on paper

A luscious bloom in lipstick red, with glossy green leaves, David Shrigley's (b. 1968) rose fills the page, squeezing the words of his text into the corners and pressing them down to the lower margin. Painted with thick pigment in a naive, almost childlike style, it is typical of Shrigley's works on paper, which are bold caricatures of their subjects, ranging from cats to cakes or sex to death, deploying mundane and familiar objects to approach these themes obliquely. Each drawing carries a handwritten caption or comment, introducing a note of surrealism or an element of mordant humor, although Shrigley has emphasized that the words are always those of a character he invents for the purpose of the work. His rose is captioned "Oh rose thou art so fucking beautiful," a phrase that suggests an exaggerated awe, but the audience might also recognize that Shrigley is quoting, in part, William Blake's famous poem "The Sick Rose" (see p.79). The red rose is a symbol of love and passion, often featuring in love poetry, and Shrigley's introduction of the emphatic adjective "fucking" can be seen as a reference to the flower's erotic symbolism. It may even be a subtle equivalent for the central metaphor of Blake's poem, which describes the deflowering of the pure, unsullied rose by the priapic "invisible worm which flies in the night."

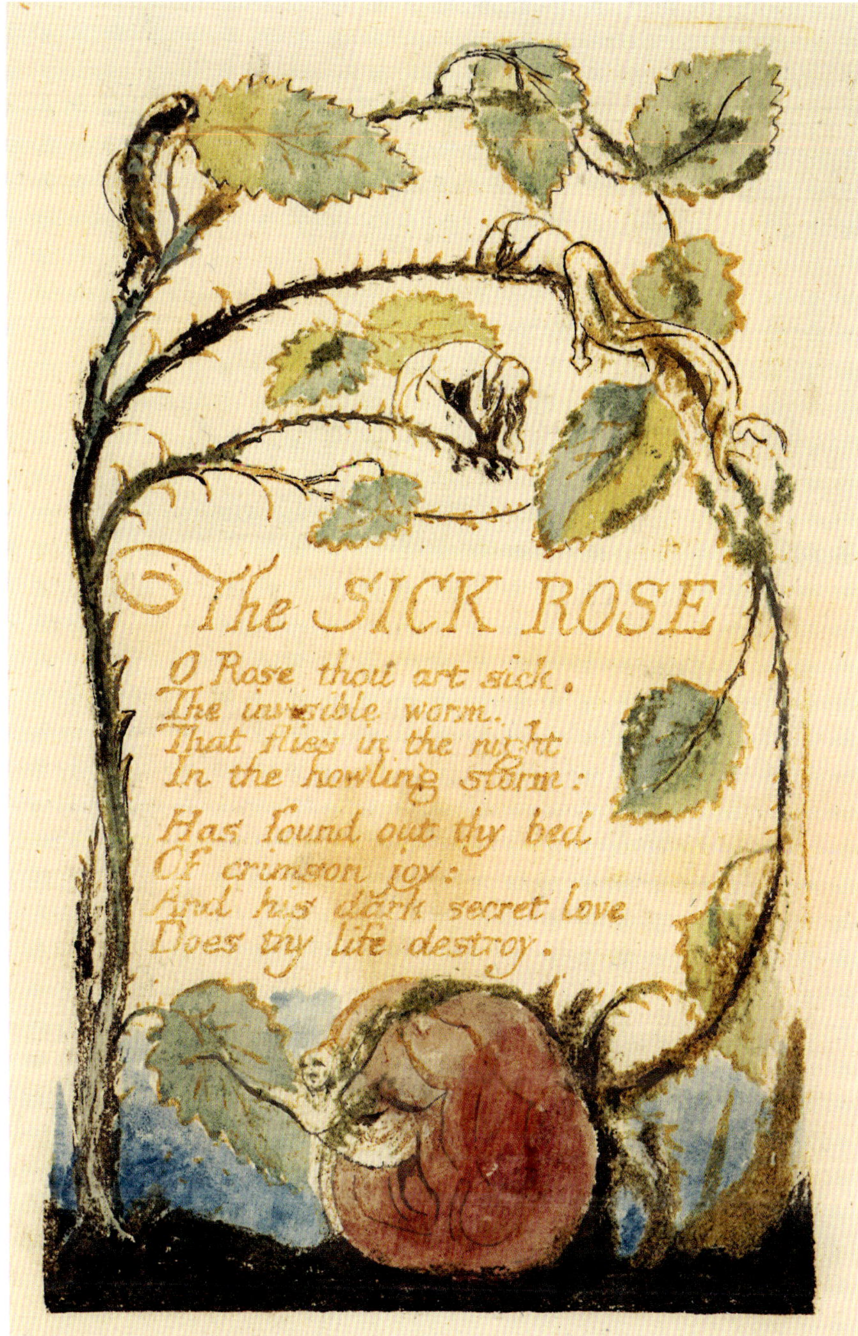

WILLIAM BLAKE

The Sick Rose, from *The Songs of Innocence and of Experience*, 1794
Hand-colored relief etching on paper, 4⅜ × 2⅝ in. /
11 × 6.7 cm
British Museum, London

In 1794, English poet, painter, and printmaker William Blake (1757–1827) self-published his illuminated book *The Songs of Innocence and of Experience*, in which each of his poems is vividly illustrated. One of the best known is "The Sick Rose," shown here. Heavily inspired by children's literature, Blake also challenged and subverted the genre, which was then new. Although appearing to be for children, the book often concerns themes of childhood in a way that is unsettling and, at times, even startling. The image of "The Sick Rose" is as enigmatic as the verse, and both are laden with metaphors. An elongated rose stem lined with sharp thorns frames the page, holding the poem in a protective enclosure. The stem winds around, ending in a large open flower that is weighed down to rest on a patch of dark soil at the bottom of the page. A spectral figure emerges from the flower with outstretched arms, an expression of elation on her face. Two gowned figures lie atop the thorns, in deeply sorrowful poses, while at top left, a worm or caterpillar chews on a leaf. The other leaves are severely yellowing, adding to the rose's sickly appearance, as the title suggests. The rose may have armature, but it is still vulnerable to threat from a potential pest. In eight short lines of verse, words serve as double entendres, where "bed" might refer to flower bed or a marital bed, and sickness could relate to a plant pathogen or a broken heart.

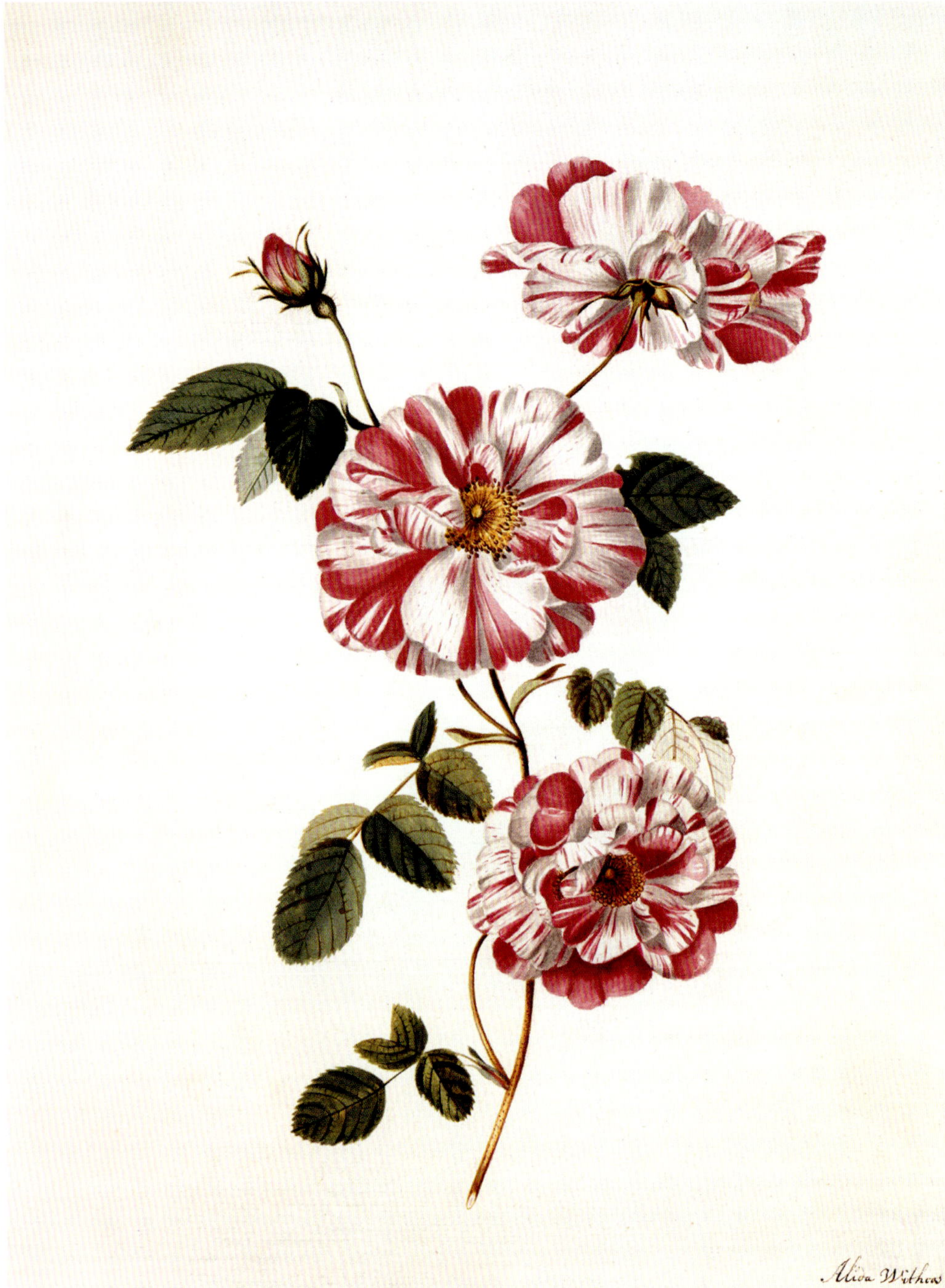

ALIDA WITHOOS

Rosa Mundi, from *Konstbоeck van Simon Schynvoet*, c. 1690–1750
Watercolor, 15⅜ × 10⅜ in. / 39 × 26.5 cm
Wageningen University and Research Library, the Netherlands

Thanks to its semidouble flower and white petals splashed with stripes of pink, *Rosa gallica* 'Versicolor', more commonly known as Rosa Mundi, is perhaps one of the most easily recognizable of the historic roses. Legend says that it was named after Rosamund Clifford, the mistress of English king Henry II, and reportedly first found near her grave. However, it probably does not date back to her lifetime in the twelfth century because the first known description of it was not recorded until 1581. A sport (a mutation in the character of an established variety, usually a variation in color) of an even more ancient rose, the completely pink *R. g.* 'Officinalis', this rose was painted by Dutch botanical artist Alida Withoos (c. 1661–1730) for the *Konstboeck*, or "art book," commissioned by garden designer and art collector Simon Schynvoet during the Golden Age of Dutch art. A collection of 142 botanical studies, the *Konstboeck* features six striking watercolors by Withoos alongside works by other artists, including her brother Pieter. Taught by their father, artist Matthias Withoos, Alida and five of her eight siblings specialized in still lifes, and were part of a network of highly sought-after painters for commissions such as this. Today, Rosa Mundi is one of the most popular of the Old Garden roses, partly thanks to its hardiness. It grows and flowers well, reaching a height of about 4 feet (1.2 m) with little or no maintenance; suckers freely, covering large areas; and mixes particularly well with other plants, such as salvias and geraniums.

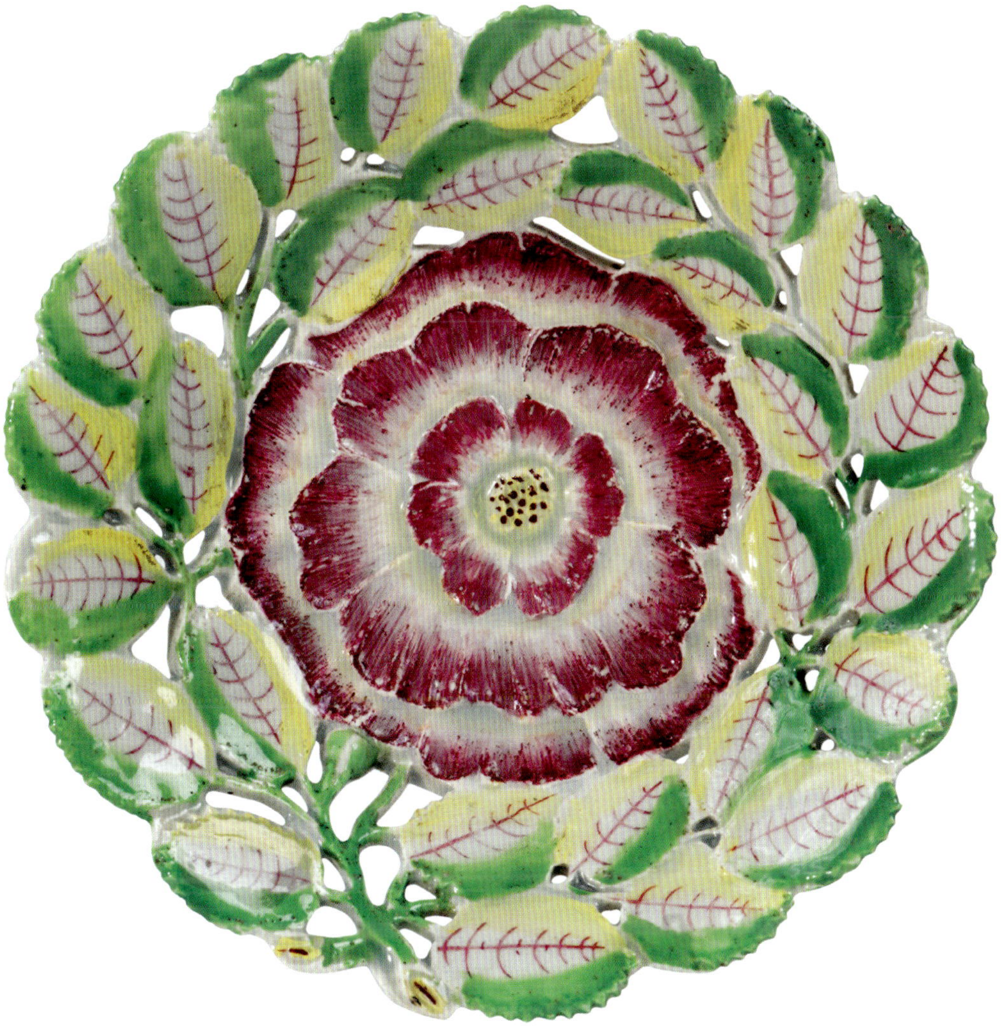

DERBY PORCELAIN FACTORY

Rose basket, c. 1760
Porcelain, Diam. 7⅞ in. / 20 cm
Private collection

A purple-edged rose with its petals completely opened is surrounded by a wreath of white and green leaves in this reticulated rose basket, which was produced in the middle of the eighteenth century for the British Derby Porcelain Factory. Anchored by the colorful central rose, this piece was made by an anonymous potter whose skill is clear in the deft execution of the basket's network of leaves, which appear to grow from the curled stems, creating gaps in the ceramic bowl that give it its height. Reticulated porcelain—the name given to ceramic objects that incorporate intricate latticework, such as this—became popular at the beginning of the eighteenth century as a craze for decorative china objects swept Europe. Each gap between the leaves had to be cut out separately, so not only was the production of this piece time-consuming and laborious, the finished basket was extremely fragile. The Derby Porcelain Factory was founded in about 1747 by Andrew Planché, a French Huguenot who had fled religious persecution to the city of Derby, the center of the new British china industry. After Europeans broke the Chinese monopoly on the production of porcelain when German alchemist Johann Friedrich Böttger discovered how to make hard porcelain in 1707, beautiful objects such as this one became a more affordable luxury for the increasingly wealthy British upper classes. This rose basket was probably intended to be displayed in the formal sitting room of a well-to-do lady in Georgian England.

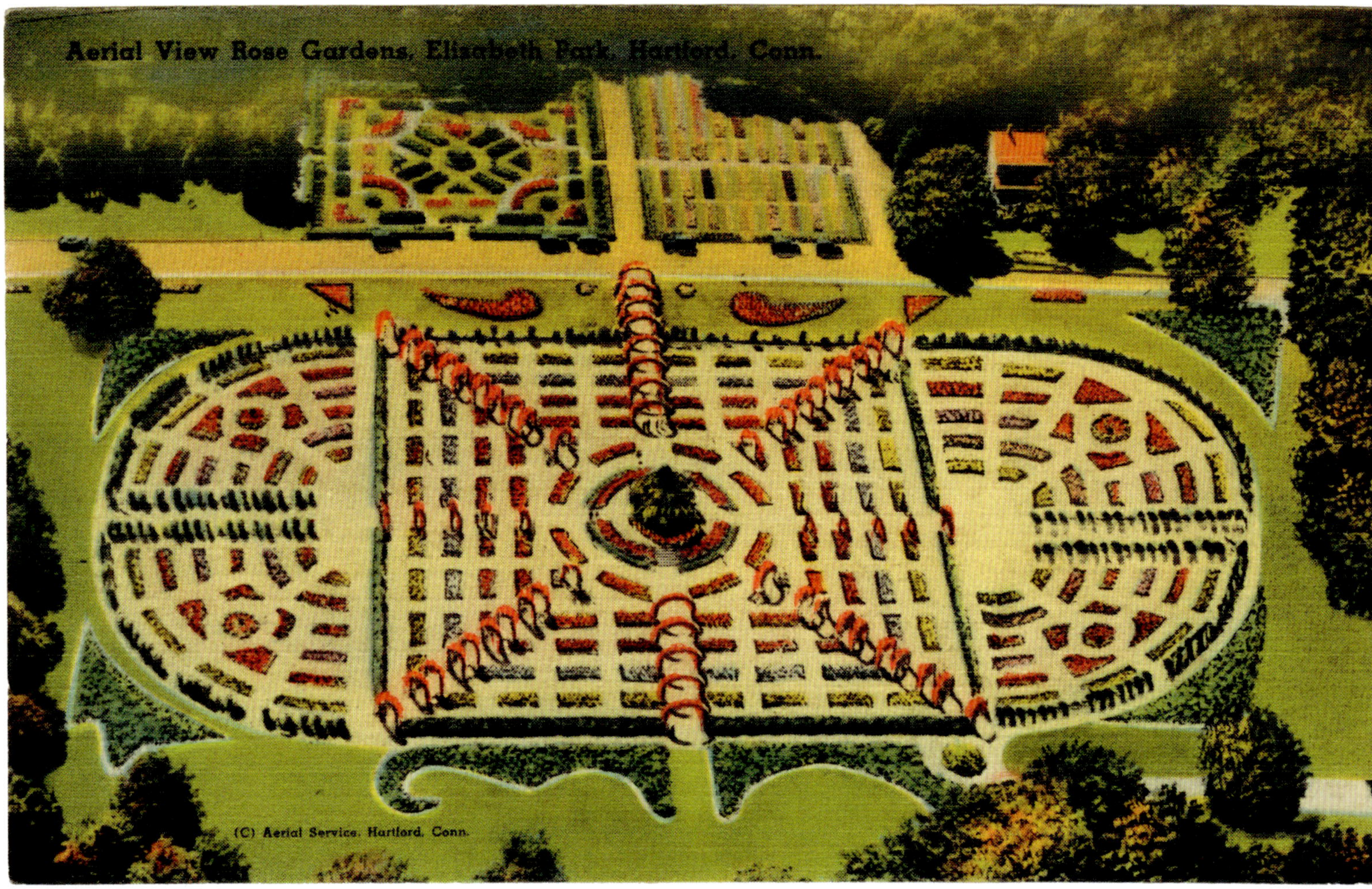

ANONYMOUS

Aerial View of Elizabeth Park Rose Gardens, Hartford, Conn., c. 1930s
Postcard, 3½ × 5½ in. / 9 × 14 cm

This colorized bird's-eye view of a large, formal rose garden was created in the years before World War II, a time when sending postcards of local attractions to loved ones was an essential and delightful part of any trip. This card commemorated a visit to Hartford, Connecticut, where in 1903 horticulturist Theodore Wirth designed Elizabeth Park, a 100-acre (247-ha) public park that is the home of the first municipal rose garden in the United States. It opened in June 1904 with 190 rose varieties arranged in a series of symmetrical beds and pergolas, the rosy arches of which are just visible in this postcard. Wirth's design of the main square, center circle, and eight grass pathways initially covered 1 acre (2.5 ha), but the garden later more than doubled in size with the addition of the north and south gardens, adding another 475 beds. By the 1950s, nearly a thousand rose varieties were growing in the garden, but by the 1970s, it had fallen into disrepair and was threatened with demolition. Connecticut-born artist and aeronautical engineer Helen Kaman spearheaded the campaign to save the park, and it is now known as the Helen S. Kaman Rose Garden in her honor. With more than 15,000 rose bushes and featuring many rare varieties, including Hybrid Tea, Hybrid Perpetual, Floribunda, shrub, and pillar roses, the garden is arranged in beds, each containing a single variety of rose, some of which bloom only in the early summer. The park is also home to the Heritage Rose Park, which contains varieties that date back to before 1867.

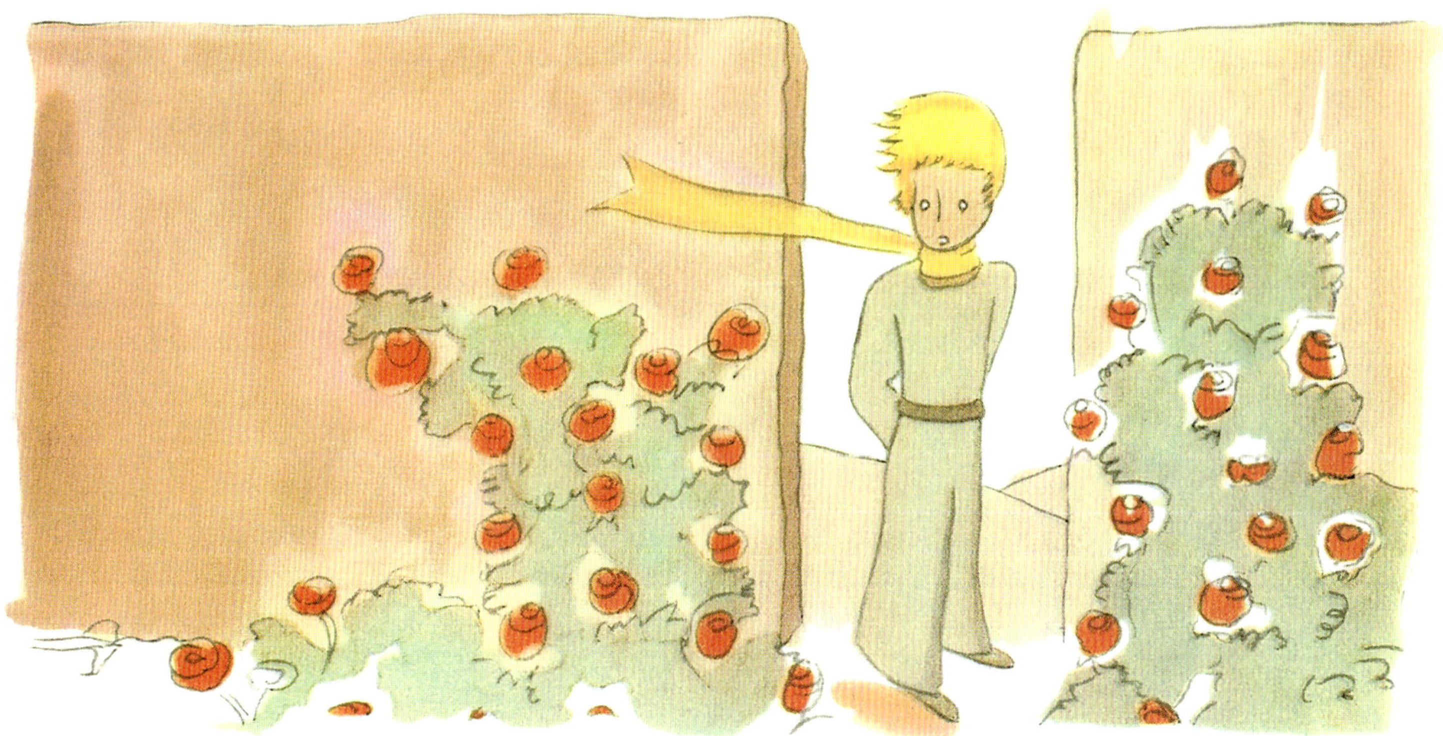

ANTOINE DE SAINT-EXUPÉRY

The Little Prince in the Rose Garden (detail), from *Le Petit Prince*, 1943
Color lithograph, page 9 × 7 in. / 23 × 17.5 cm
Saint Exupéry–d'Agay Estate

A golden-haired boy stands at the entrance of a sunny rose garden in this simple drawing by French writer and illustrator Antoine de Saint-Exupéry (1900–1944). Saint-Exupéry, a lifelong aviator, was a temporary refugee in New York City when he wrote and illustrated *Le Petit Prince* (*The Little Prince*). It was first published in an English translation in 1943, about the time he returned to France as a pilot during World War II. He was declared missing the following year, when his plane disappeared on a reconnaissance mission. The original French version came out in 1946, after the liberation of France. This illustration from *The Little Prince* captures anguish. With his mouth an "O," the boy—a traveler from space who has recently landed on Earth—is stunned. He has learned that the rose on his tiny home planet, Asteroid B-612, is not unique in the universe, as he was led to believe. "I thought I was rich because I had just one flower," the prince cries, "and all I own is an ordinary rose." His solitary rose is just like the thousands in the garden. Saint-Exupéry portrays the rose on B-612 as lovely, vain, somewhat devious, and very demanding. The prince tends her with regular watering and protection against the elements. "She perfumed my planet and lit up my life!" he remembers, now on Earth, realizing his beloved, *his* rose, had been an absolute necessity after all. *The Little Prince* has become one of the best-selling books of all time, with more than 200 million copies sold, and has been translated into hundreds of languages.

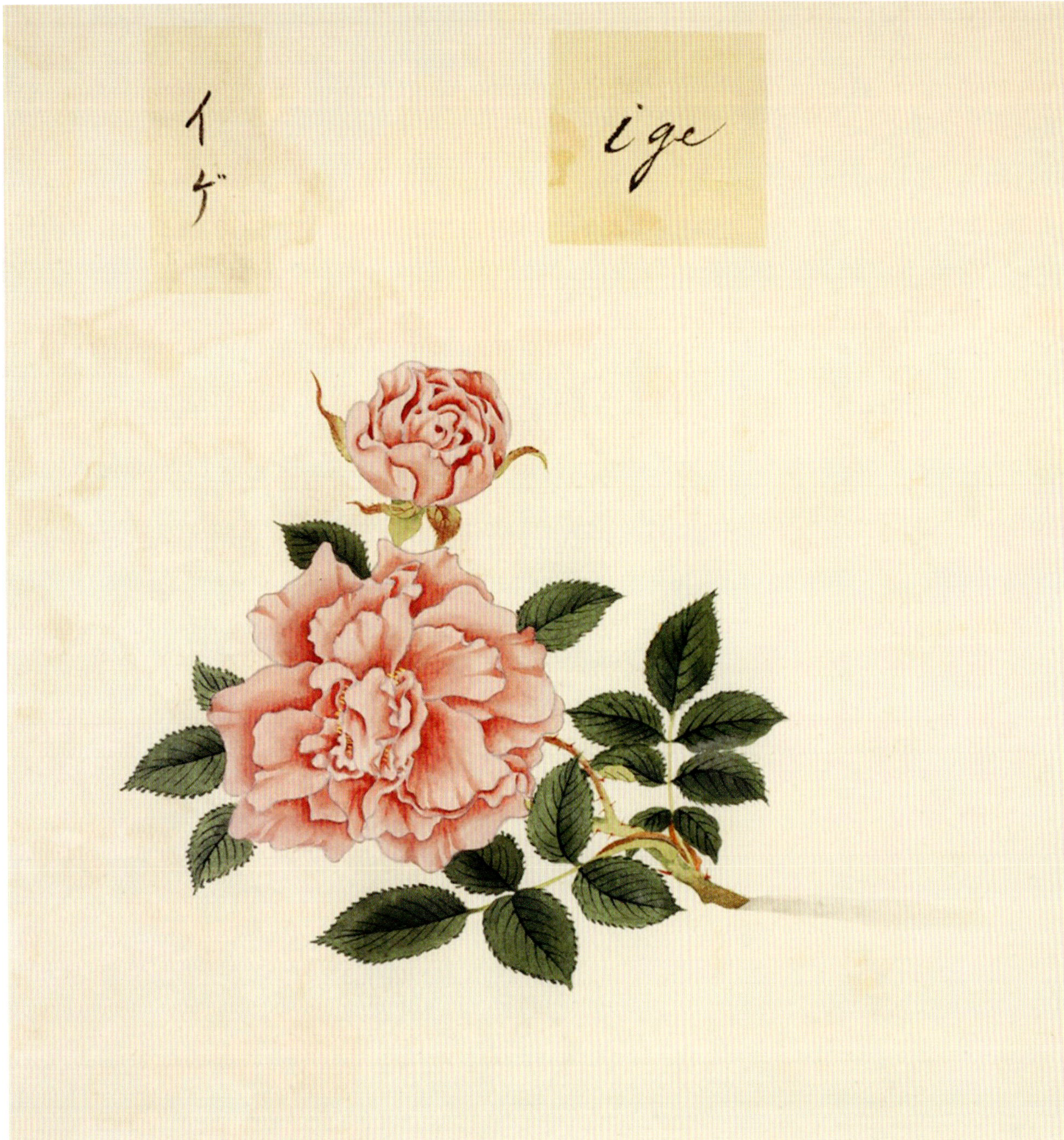

KAWAHARA KEIGA

Rosa, c. 1823–29
Pencil drawing and watercolor on paper
Naturalis Biodiversity Center, Leiden, the Netherlands

The great majority of the roses that gardeners grow today are the result of the hybridization of European varieties with the so-called China roses (see p.35). These two types of rose are different in both appearance and flowering ability; the latter repeatedly flowers constantly if the weather and temperature are favorable, whereas roses native to Europe flower just once each year. At the end of the eighteenth century, the two groups started crossing, initially with the help of pollinating insects. The rose depicted here, with its informal flower and pale pink petals, is probably a type of China rose, and, indeed, it could be a loose representation of the first variety from China that came to Europe in the mid- to late eighteenth century: *Rosa* 'Old Blush', also known as 'Parson's Pink'. This illustration was painted by Japanese artist Kawahara Keiga (1786–1860), and it is probable that the China roses were grown in Japan well before they came to Europe. Kawahara painted this rose in the 1820s, at which time the Japanese government had established an isolationist policy whereby foreigners were allowed to live only on the artificial island of Dejima, where Kawahara was employed as an artist by the Dutch factory there. Only twenty non-Japanese were allowed at any one time. One of those was German physician Philipp Franz von Siebold, who commissioned Kawahara to produce more than a thousand works cataloging plant and animal life, including this China rose.

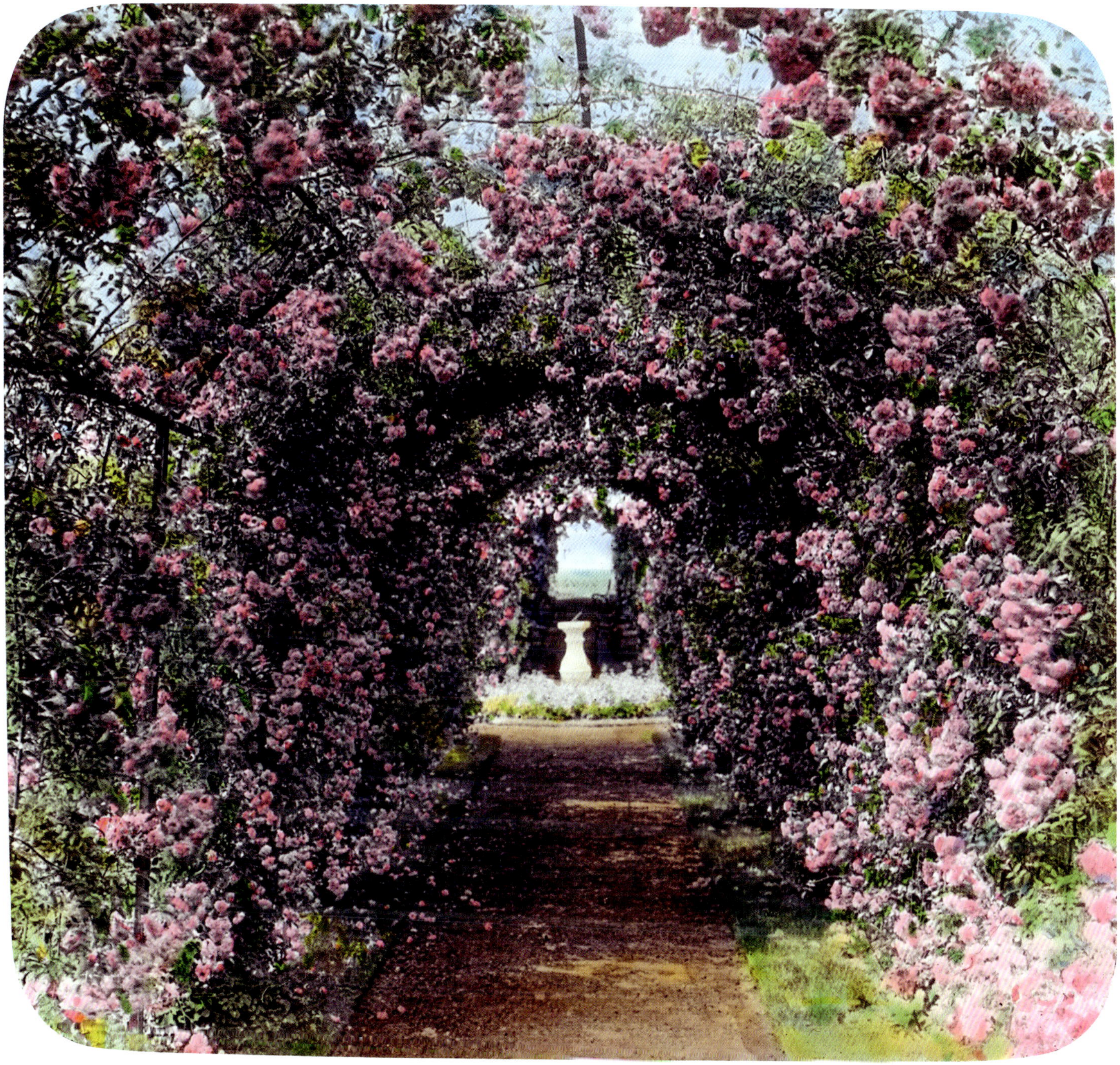

FRANCES BENJAMIN JOHNSTON AND MATTIE EDWARDS HEWITT

Armsea Hall Rose Trellis, 1914
Hand-colored glass lantern slide,
3¼ × 4 in. / 8.3 × 10.2 cm
Library of Congress, Washington, DC

On a summer day in 1914, American photographers Frances Benjamin Johnston (1864–1952) and Mattie Edwards Hewitt (1869–1956) set up camera and tripod at Armsea Hall, the estate of Zelia Hoffman in fashionable Newport, Rhode Island. Hoffman's 'Dorothy Perkins' rambler roses were in full bloom, and Johnston and Hewitt captured the tunnel of flowers, its trellising made by French craftsmen. Johnston's first love had been art, and she studied at the Académie Julian in Paris for two years. In 1888, a family friend, George Eastman of Eastman Kodak fame, gave Johnston her first camera. She took to the medium, becoming a photojournalist and an in-demand portraitist in Washington, DC. Ten years later, she moved to New York City with her business and romantic partner, the recently divorced Mattie Edwards Hewitt, who had honed her skills in her husband's St. Louis photography business. This was the American country place era, a boom time for horticulture. Society women embraced flowers, and rose gardens were a part of the zeitgeist. As a philanthropist, able hostess, and founder of garden clubs in New York and Newport, "Zealous Zelia" Hoffman was an able advocate of growing roses. Of Armsea Hall, she wrote, "When the roses are in bloom from July 27 (as that is almost the date to a day . . .) to the middle of August, they are a wonderful picture which people come from far and near to admire." She wanted professional photographs of her creation, and the Johnston-Hewitt Studio was perfectly positioned to provide them.

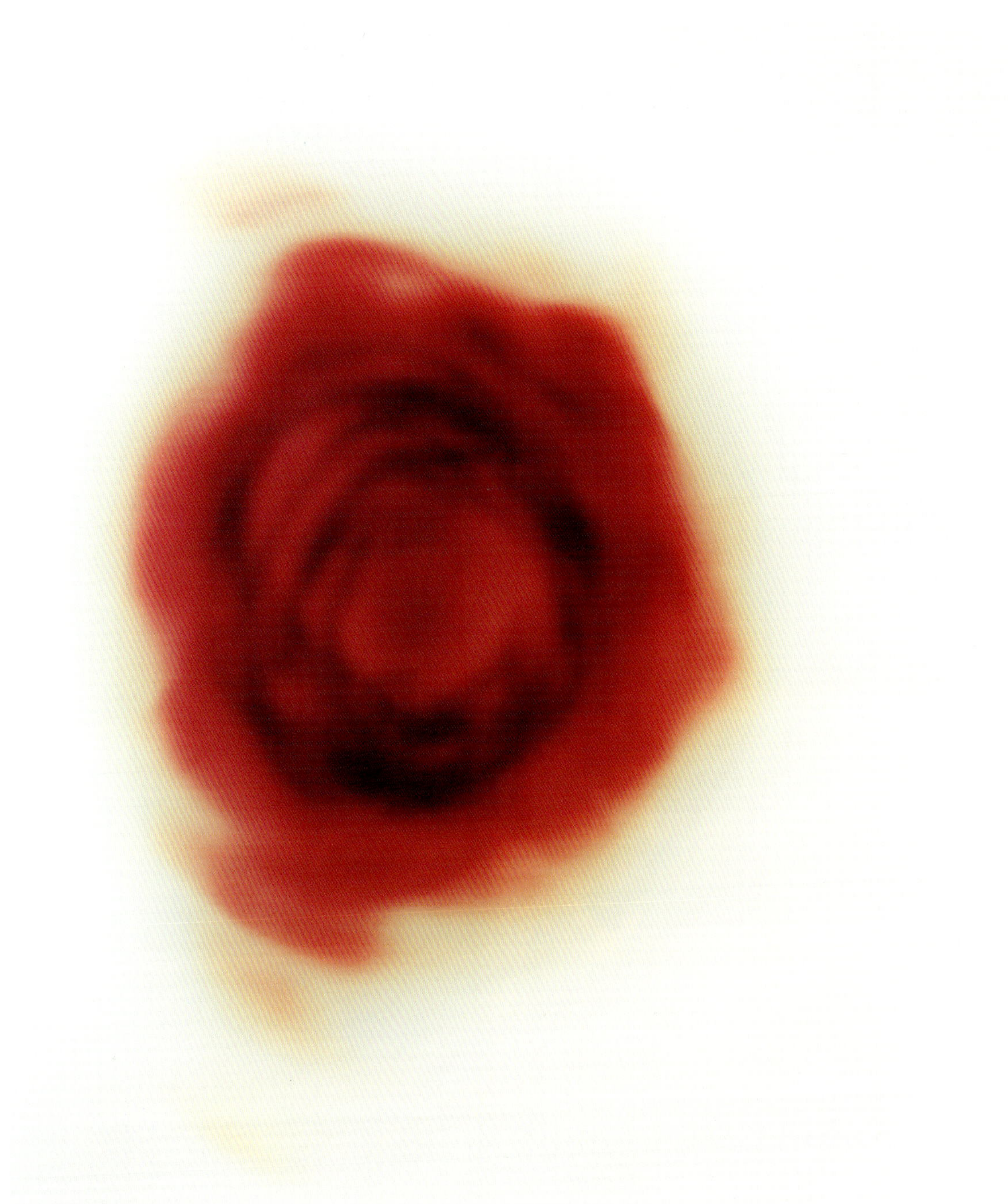

WOLFGANG TILLMANS

Rose, 1999
Chromogenic print, 14 × 12 in. / 35.5 × 30.5 cm
Museum of Modern Art, New York

Created shortly before the turn of the millennium, this image by German photographer Wolfgang Tillmans (b. 1968) was not made with a camera, but by the artist drawing a lit flashlight directly onto photographic paper in the darkroom. Instead of representing a fixed ideal or a sharp picture of a perceived reality, Tillmans suggests what ultimately appears to be the subject of the photograph with its title, *Rose*. As in his other well-known works made without a camera, *Freischwimmer*, *Greifbar*, and *Blushes*, Tillmans encourages viewers to complete the work through their reading of the image as a representation of reality—here, to see the rose. He has said how important it is to his art that he resists the desire to own, control, or interpret the subjects of his photographs—desires that he believes will result in bad art. Instead, he values authentic interest in his work, and he uses the tools of photography as objects positioned between himself and the subjects he photographs, as ways to record what his mind sees. Tillmans, who is inspired by science and news photography, kept a scrapbook of images as a young boy. Although magazines were his first major platform, he has been exhibiting his photographs since the late 1980s. Over the years, he has evolved a signature mode of exhibition for his work, in which photographs printed in a range of sizes and techniques, including using a photocopier, are presented in a nonhierarchical way, pinned and taped to the wall and laid out on tables.

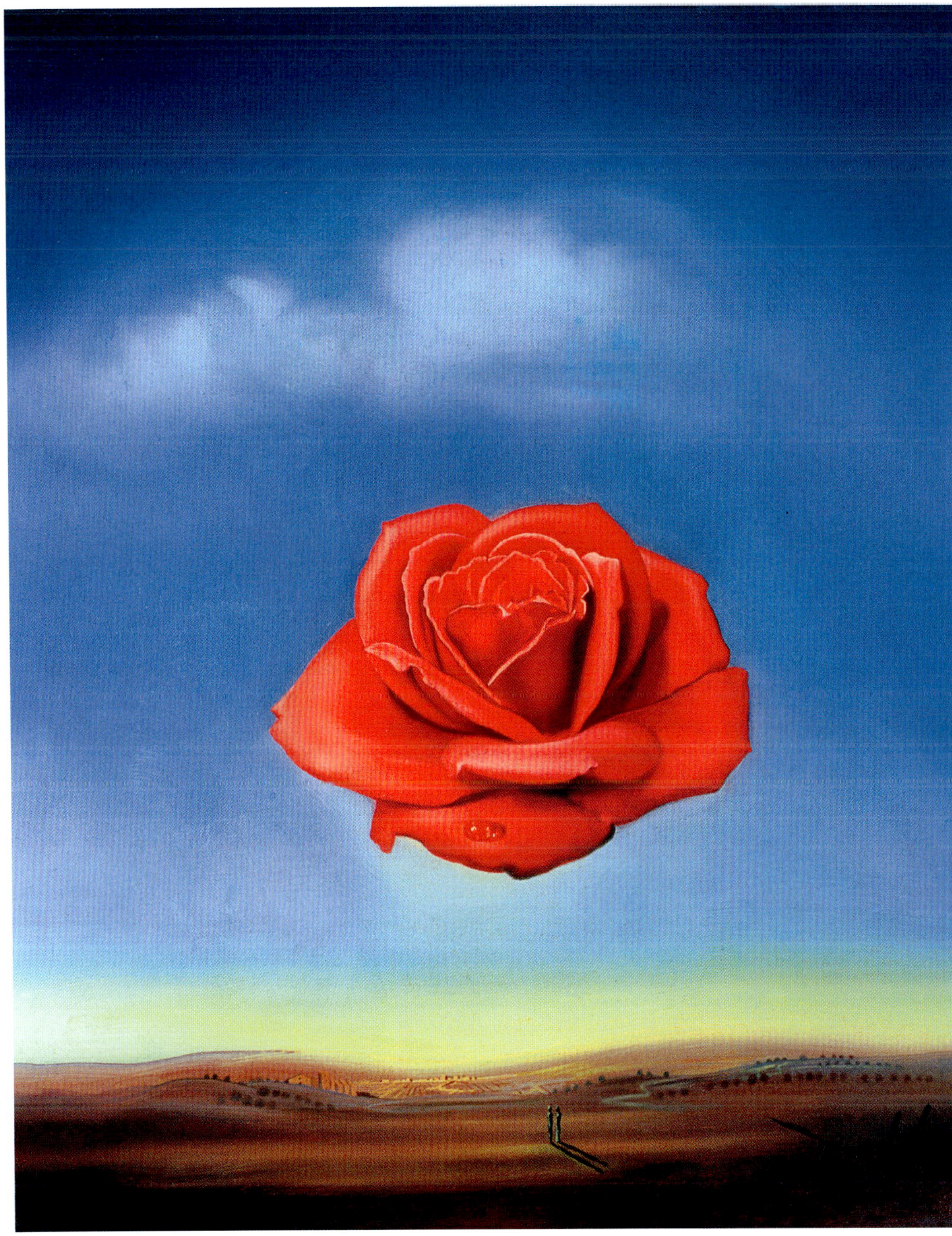

SALVADOR DALÍ

The Rose, 1958
Oil on canvas, 14⅛ × 11¼ in. / 36 × 28 cm
Private collection

Red roses had occasionally haunted Spanish Surrealist artist Salvador Dalí's (1904–1989) dreamscapes before he floated this gigantic version toward heaven. In his earlier paintings, roses had replaced women's heads or wombs. Perhaps even the dewdrop on the petal seen here reflected Dalí's Surrealist methods of self-analysis, recalling childhood tears and traumas. Alternatively, if he were channeling Catholic symbolism, the red rose represented Christ's sacrificial blood, and the dewdrop the tears of his followers. In the 1950s, after the atomic violence of World War II, Dalí turned increasingly to traditional Christian imagery and mysticism, most magnificently when he painted *Christ of Saint John of the Cross* (1951)—in which he slanted the Crucifixion over a watery landscape with fishermen.

This striking rose is similarly transcendent. Beneath it the tiny, praying people are taken from Jean-François Millet's iconic painting *The Angélus* (1857–59), in which a peasant couple working in a field pause to pray at sunset. Around his couple in miniature, beneath the mystic rose and heavenly blue sky, Dalí has spread a Spanish landscape lit by the setting sun. In 1958 he also published his "Nuclear Mysticism" manifesto, which combined elements from quantum metaphysics and religion with his art practice. Both menacing and serene, Dalí's *The Rose* is an apt symbol for the artist's intellectual interests during this period, in which he pondered the tendency of nature and humanity toward unpredictable outcomes.

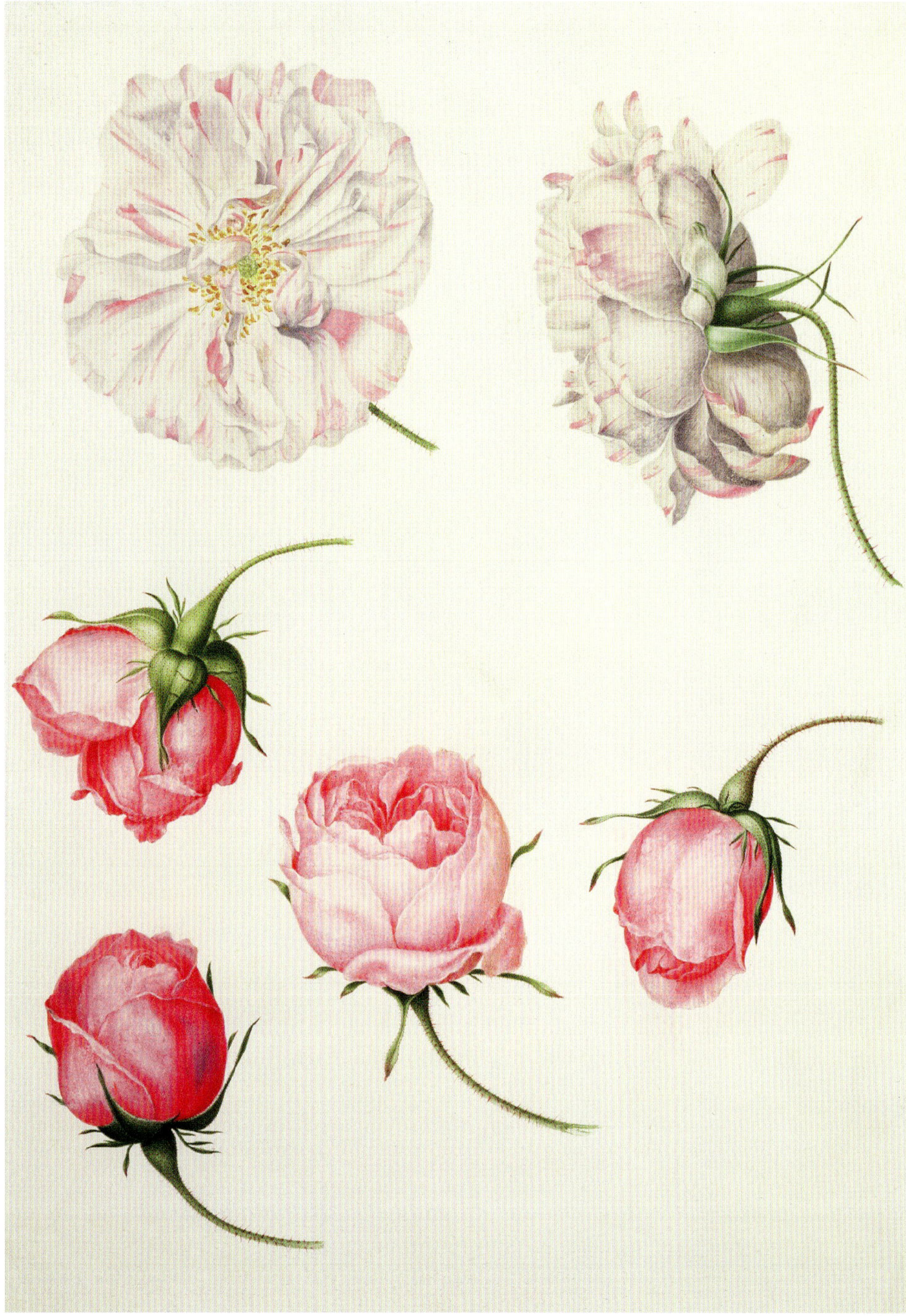

NICOLAS ROBERT

Rosa × Centifolia Variegata and Rosa × Centifolia, c. mid-17th century
Graphite, watercolor, and bodycolor on vellum, 12¼ × 8⅞ in. / 31 × 22.7 cm
Fitzwilliam Museum, Cambridge, UK

The two roses named in this beautiful study attributed to French artist Nicolas Robert (1614–1685) are *Rosa × centifolia variegata* and *R. × centifolia*. Robert, who enjoyed royal patronage from Louis XIV and the French court, had a great love of plants and included scientific details of roots, leaves, and seeds in his intricate botanical watercolors. At the bottom of this study, he delicately paints the completely pink bloom of the original Centifolia rose, now more commonly—and perhaps unkindly—known as the cabbage rose. Its origin is uncertain, and it may have been the result of a chance hybridization in the Netherlands between a Gallica (see p.122) and a Damask (see p.76) in the late sixteenth century. Once it appeared in nurseries, it immediately became popular both in gardens, thanks to its showy, highly scented blooms, and with the Dutch masters, who often used it in still-life arrangements (see pp.10 and 140). Rose breeders continued to cross it with other varieties, and it became the parent of new cultivars collectively known as the Centifolias, distinguishable from other Old Garden roses by their stiff growth, large leaves, and thorny stems. Other varieties have also resulted, not by hybridization but by sporting—when a mutation occurs that results in a different morphological character, often color. This is how *R. × centifolia variegata*, now commonly known as village (or cottage) maid, originated, its petals depicted here with a wonderful combination of creamy white blooms delicately striped with pale lilac pink.

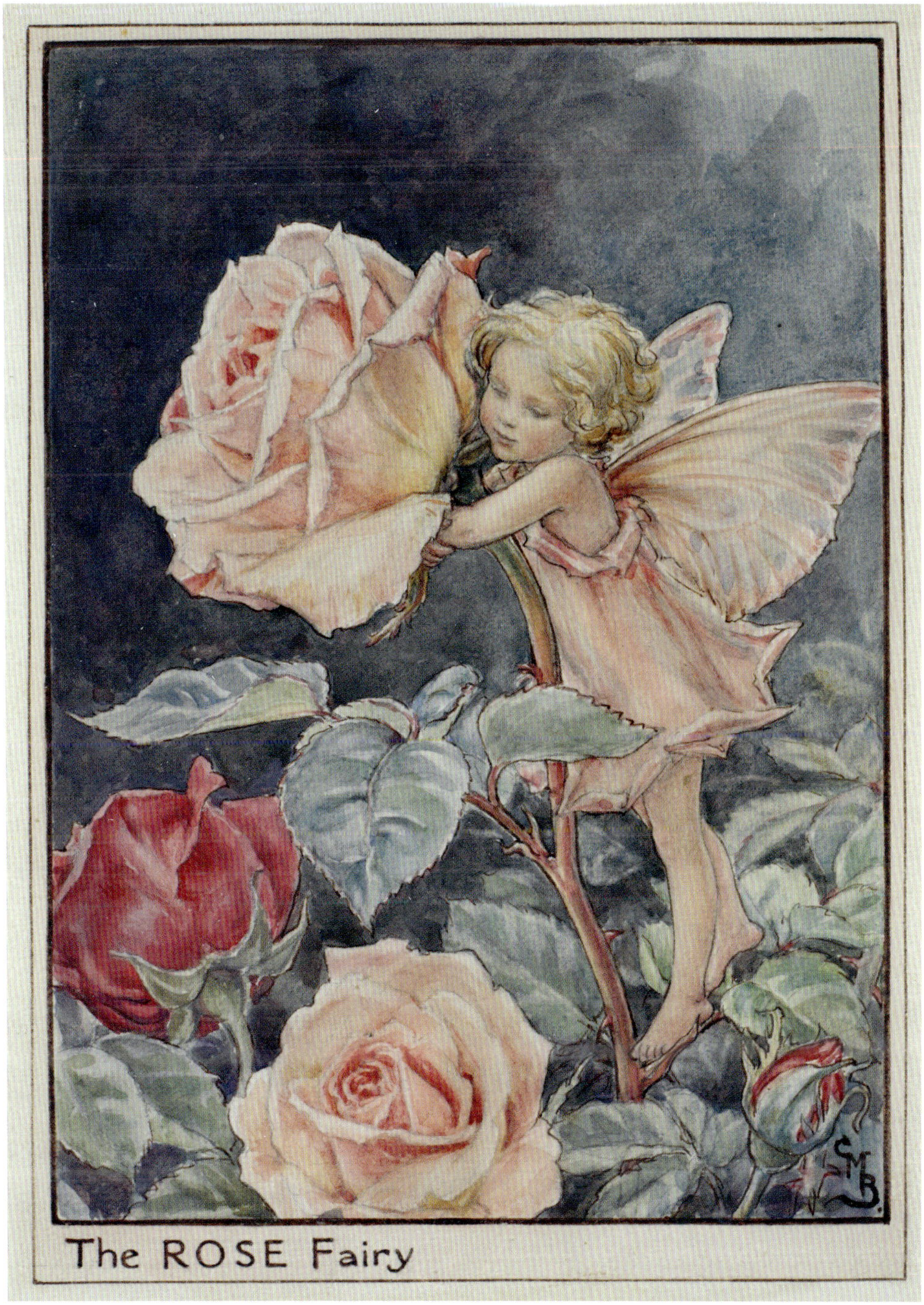

CICELY MARY BARKER

The Rose Fairy, from *Flower Fairies of the Summer*, 1925
Ink and watercolor, 4⅜ × 3 in. / 11 × 7.5 cm
Penguin Random House

With a dress and wings that match the flower, the Rose Fairy balances on a large pink rose, carefully avoiding the thorns as she embraces the bloom. This drawing by English illustrator Cicely Mary Barker (1895–1973) is one of a collection of 170 ethereal illustrations that Barker made for her series of *Flower Fairy* books, which she published from the 1920s to 1950s. Her fairies are tiny—the tallest is just 8 inches (20 cm) in height—and each takes care of their chosen flower. Here, the Rose Fairy tends the rose with an embrace, extending as the flower grows. A poem accompanying the image begins, "Best and dearest flower that grows/ perfect both to see and smell/ Words can never, never tell/ Half the beauty of a rose." Barker went to great lengths to ensure the botanical accuracy of her illustrations; here, the pink and red roses are in full bloom, with a red rose still in bud and thorns clearly visible on the stems. Each fairy's face was based on that of a real child from the nursery school run by the artist's sister, and each illustration began as a loose sketch that Barker then worked into a full illustration. Heavily influenced by the Pre-Raphaelite painters with their "truth in art" mantra, Barker created her first fairies in 1923, collected in *Flower Fairies of Spring*. After the end of World War I, and with the popularity of such works as J. M. Barrie's *Peter Pan* (1904), fairies were an increasingly popular topic for art and literature as people sought to escape the memories of the horrors of war.

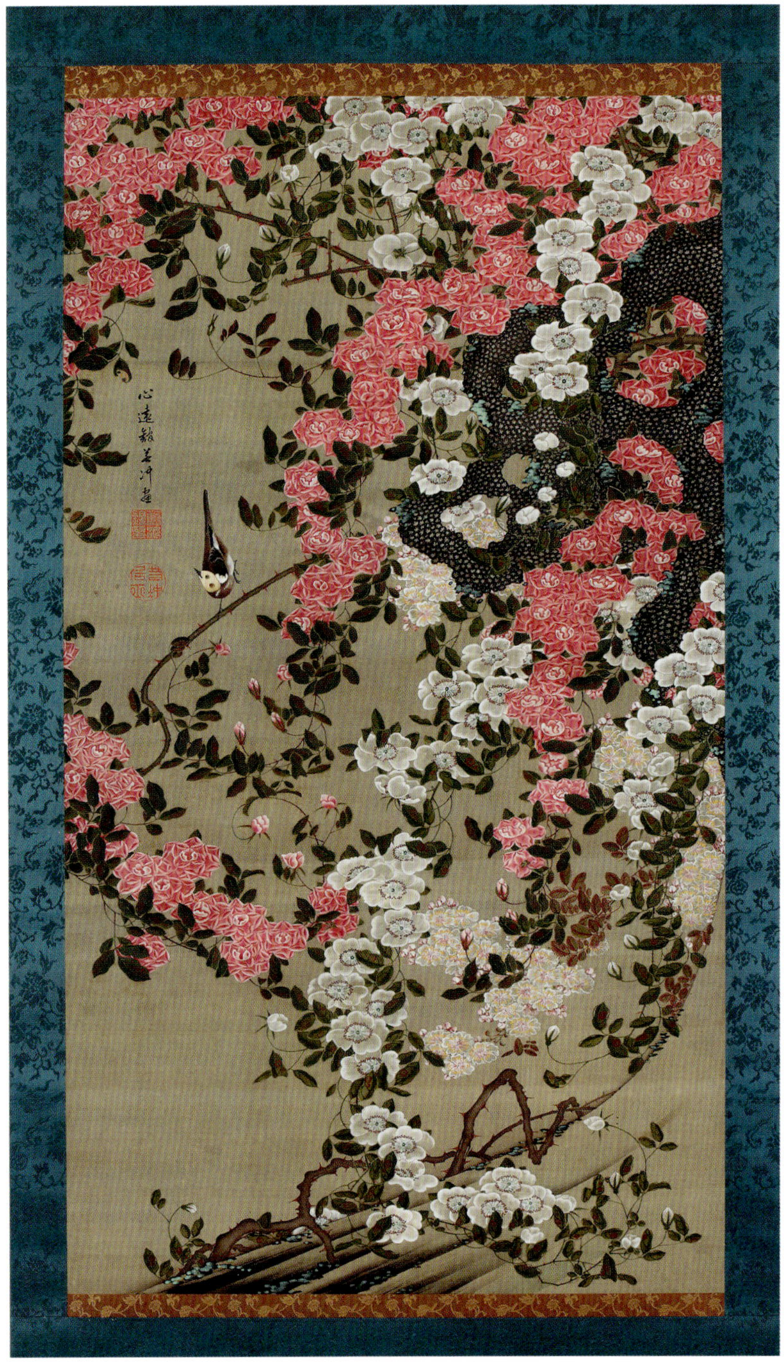

ITŌ JAKUCHŪ

Roses and Small Bird, from *Colorful Realm of Living Beings*, c. 1757–66
Color on silk, 56⅛ × 31⅜ in. / 142.6 × 79.7 cm
Museum of the Imperial Collections, Tokyo

Itō Jakuchū (1716–1800), a Japanese artist of the mid-Edo period, was best known for his superb paintings of traditional subjects from nature, especially chickens and other birds. In about 1757, he began painting his masterpiece: thirty large silk hanging scrolls collectively entitled *Doshoku Sai-e* (*Colorful Realm of Living Beings*) as a depiction of the sublime found in nature and an offering to the Shōkoku-ji temple in Kyoto. Each scroll illustrates flowers, fish, or birds and is a superb example of Japanese art in the eighteenth century. The unidentified pink and white roses here take center stage, cascading down the entire piece of fabric, with the bird—a wagtail—taking a minor role perched on a thorny branch. The dark pink rose is probably the rambler variety commonly known as Seven Sisters, after the seven shades of pink that the blooms can take at any one time, although it has many other names, including 'Bosatsu', 'Grevillei', *R. multiflora* 'Platyphylla', and 'Qi Zi Mei'. The white one is probably *R. laevigata*, based on the large single flowers and trifoliate leaves that can be distinguished in the painting. A native of southeastern China and Vietnam, it has since naturalized widely in the American South, where it is known as the Cherokee rose (see pp.120, 127, and 217). The small multiheaded pale pink rose at bottom right is probably a form of *R. multiflora*, a native of Korea and Japan.

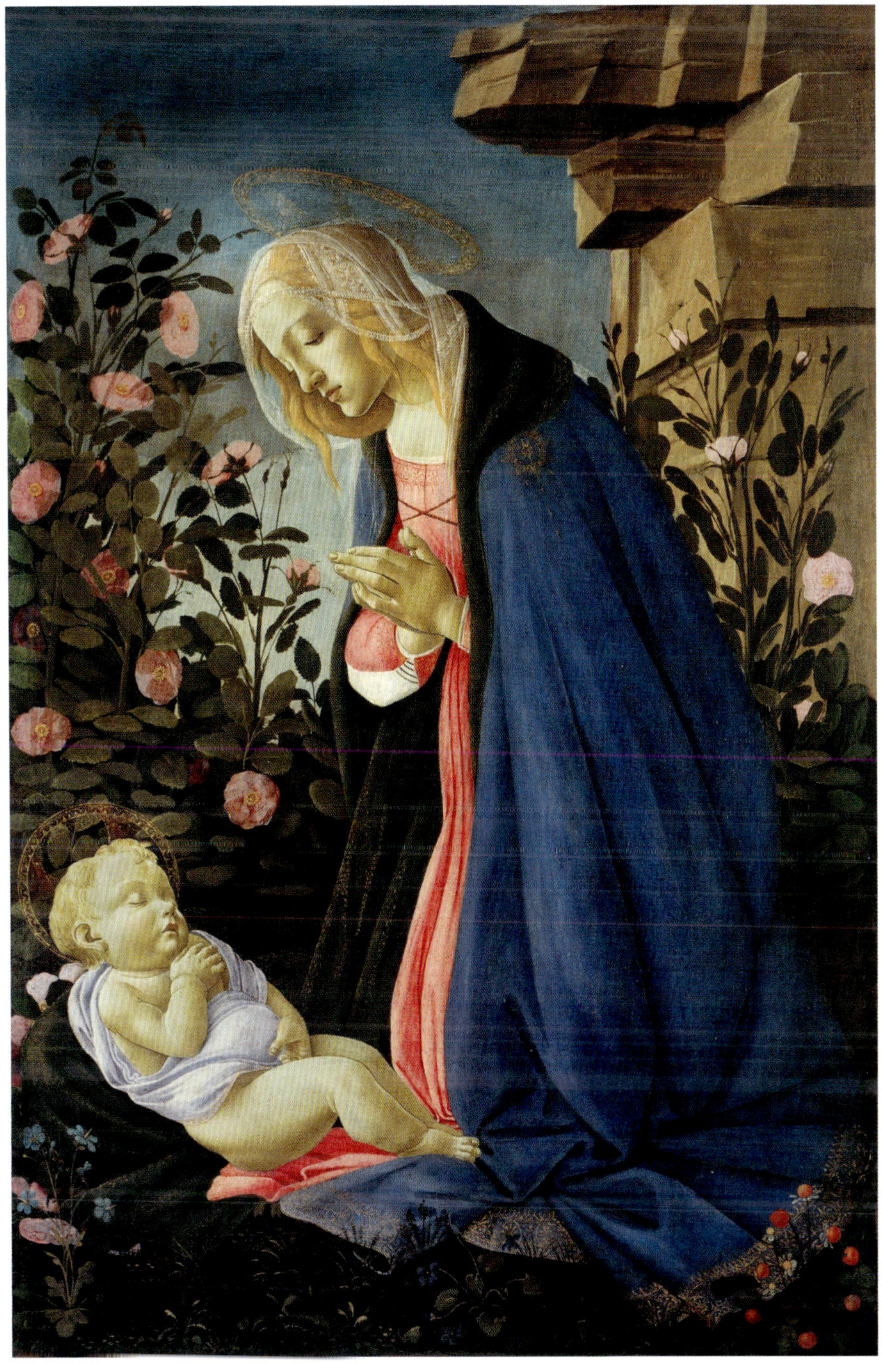

SANDRO BOTTICELLI

The Virgin Adoring the Sleeping Christ Child, c. 1485
Tempera and gold on canvas, 48 × 31⅝ in. /
122 × 80.3 cm
National Gallery of Scotland, Edinburgh

Pink is a common color for roses, but it appears rarely in Renaissance art. Florentine painter Sandro Botticelli (c. 1445–1510), however, associated pink roses both with the Virgin Mary and with Venus. His famous Venus, painted as she was born naked from the waves, was at that moment virginal, and her attendant rushed to clothe her in a pink mantle while the breezes blew blushing roses toward her. The Renaissance humanists who were Botticelli's patrons had created a synthesis between Christianity and the Classical writings they so admired, arguing that Venus was a prototype of the Virgin, since both manifested a kind of divine love. This also made it acceptable for Botticelli to use the same type of feminine beauty—and roses—for them both. Since the Middle Ages, devotional writings about the Virgin had linked her with the flower. She was described as *rosa sine spina*, "a rose without a thorn," and the enclosed space of the rose garden, the *hortus conclusus* (see p.118), became an additional symbol of her virginity. An Old Testament prophecy, declaring that the desert shall rejoice and blossom as the rose, was taken to foretell the birth of Christ to a virgin mother. In Botticelli's painting of her maternal love, Mary prays over her sleeping child. With the rose petals reflecting the color of her dress and their stamens the gold of her halo, the sense of fulfillment is complete.

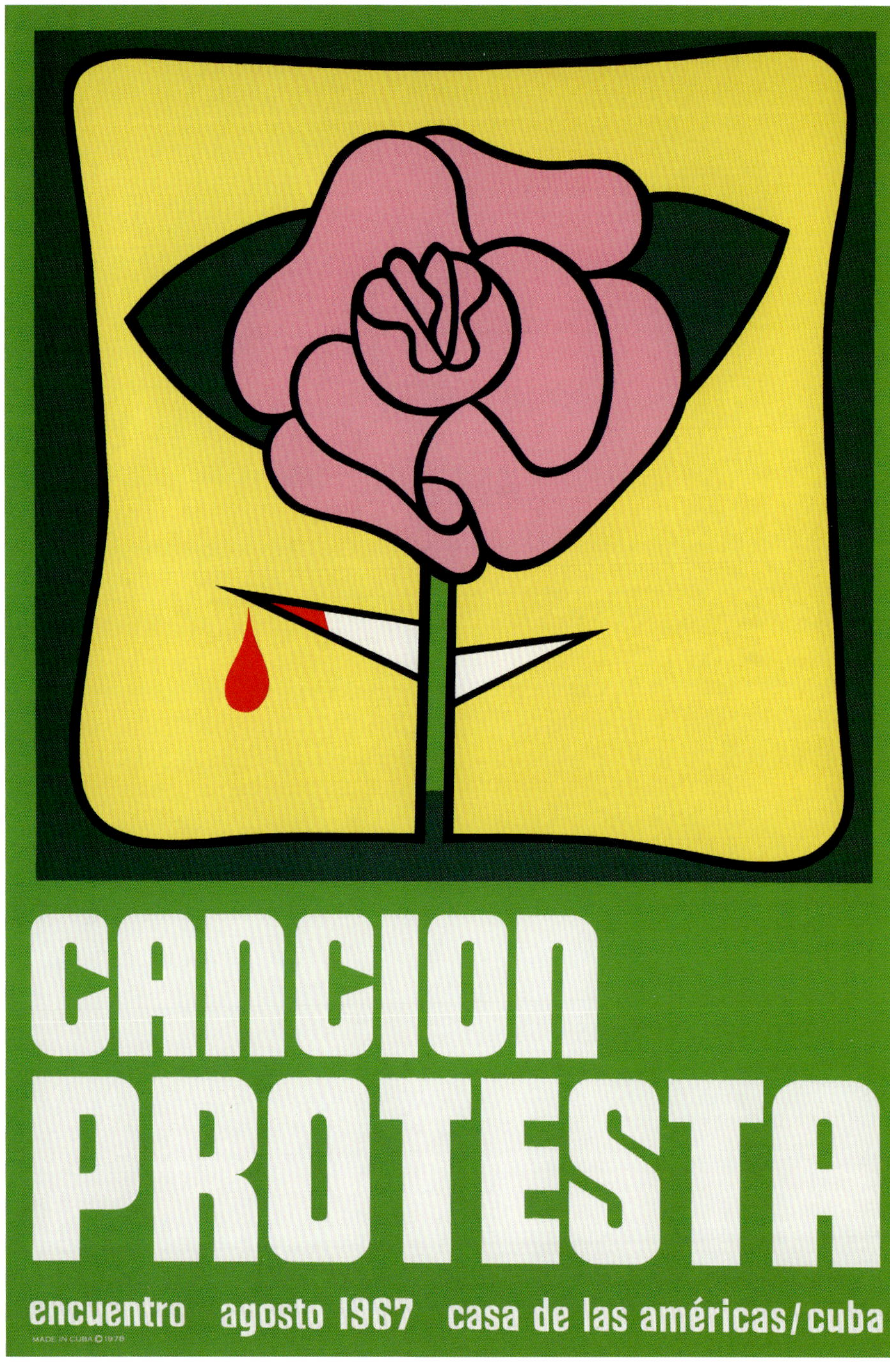

ALFREDO ROSTGAARD

Cancion Protesta, 1967
Silkscreen on paper, 44¼ × 29⅞ in. / 112.4 × 75.9 cm
Museum of Modern Art, New York

The red rose has long been an emblem of the left wing in politics (see p.159), but in this poster, it has been reimagined in pink, perhaps to provide a contrast with the scarlet blood that drips from a thorn. In this context, the rose's thorns represent adversity, sacrifice, and the blood spilled in revolution. The poster was designed by Alfredo Rostgaard (1943–2004), a leading graphic designer in Cuba who was closely involved with promoting the ideas and values of the Cuban Revolution. The Spanish words *canción protesta* translate as "protest song," and this poster was designed to advertise the First International Meeting of the Protest Song, which was held at the Casa de las Américas in Havana in August 1967. In Latin America, the protest-song genre was also known as *nueva trova* (*trova* being derived from the word "troubadour"). *Nueva trova* musicians employed traditional poetics, local instruments, and distinctively Latin American musical idioms as symbols of resistance. In the Cuban context, however, the resistance was not against the government but against the cultural and economic influence of North America and Europe. Ironically, Rostgaard's design has the bold outlines and bright colors we associate with Pop art and the psychedelic posters of the period, but here deployed for a revolutionary rallying cry. This rose, with its bloody thorn, became a widely recognized emblem of Cuban *nueva trova*. The original poster was later reissued and also reproduced on an album cover.

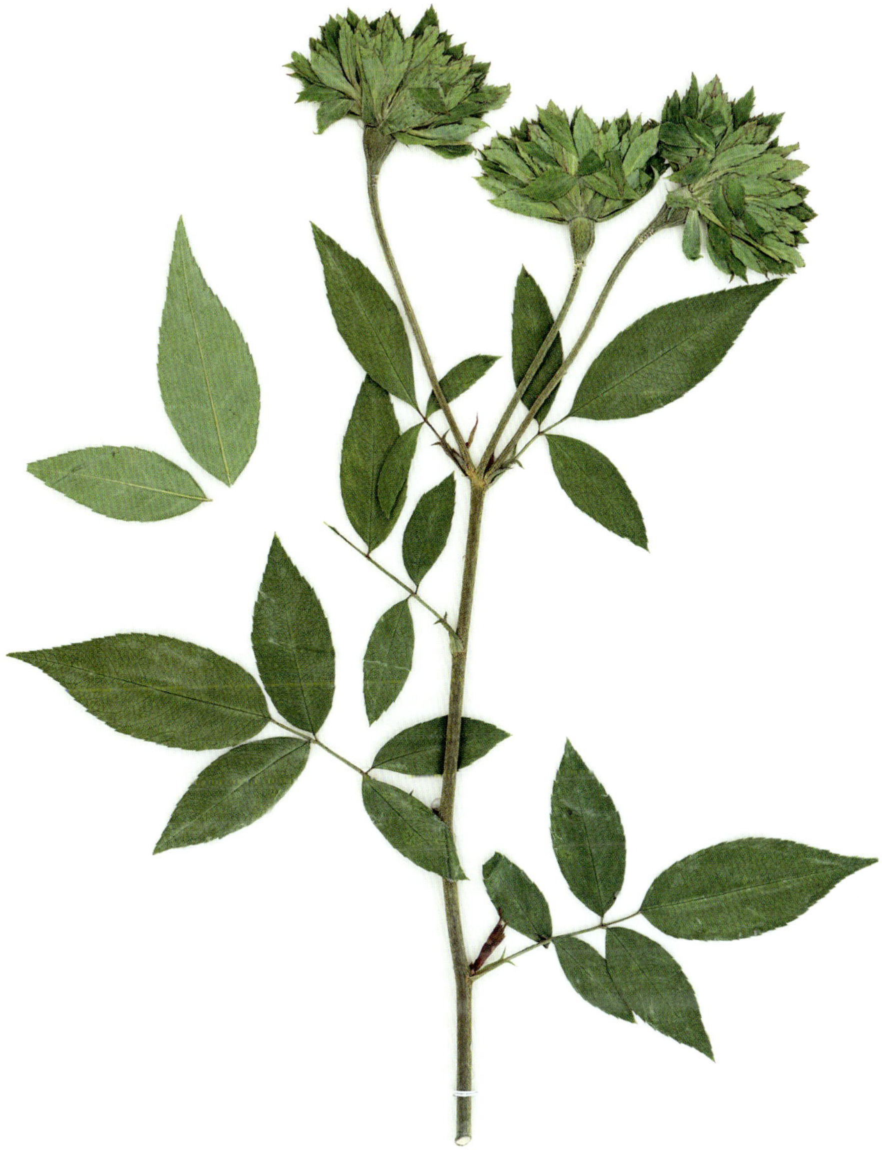

RALPH PORTILLANO

Rosa × odorata 'Viridiflora' (Andrews) Sweet, 2016
Voucher specimen, dimensions variable
William and Lynda Steere Herbarium, New York Botanical Garden

Considered by some to be an unappealing rose because of its drab green flowers, *Rosa × odorata* 'Viridiflora' might not appeal to the typical gardener, but this unusual flower is a survivor. A peppery-scented deciduous shrub—pictured here as an herbarium voucher, collected for the New York Botanical Garden (see p.152) by Ralph Portillano the rose blooms in large clusters in poor soil or shade, its drab green petals often tinged with brown. Its resilience might account for its place in American folklore. According to historians, it has strong links to the nineteenth-century abolitionist movement, when members of the Quaker religion, who were fervently against slavery, planted the roses in their front yards. The presence of the rose signified that the inhabitants were conductors of the Underground Railroad, a series of secret safe houses and routes used by enslaved African Americans to escape northward to freedom. Conductors aided in their escape by providing shelter and food. The precise role of this rose—a mutation from a pink rose named 'Old Blush' (see p.84) that originated in China and arrived in Europe and the United States in the 1790s—in the Underground Railroad has never been verified. The story originated with a Quaker family in Baltimore, Maryland, who claimed the rose and its symbolic meaning were part of their family history. However, the significance of the Underground Railroad itself has never been disputed; current estimates suggest that some 100,000 enslaved people used it to escape before 1850.

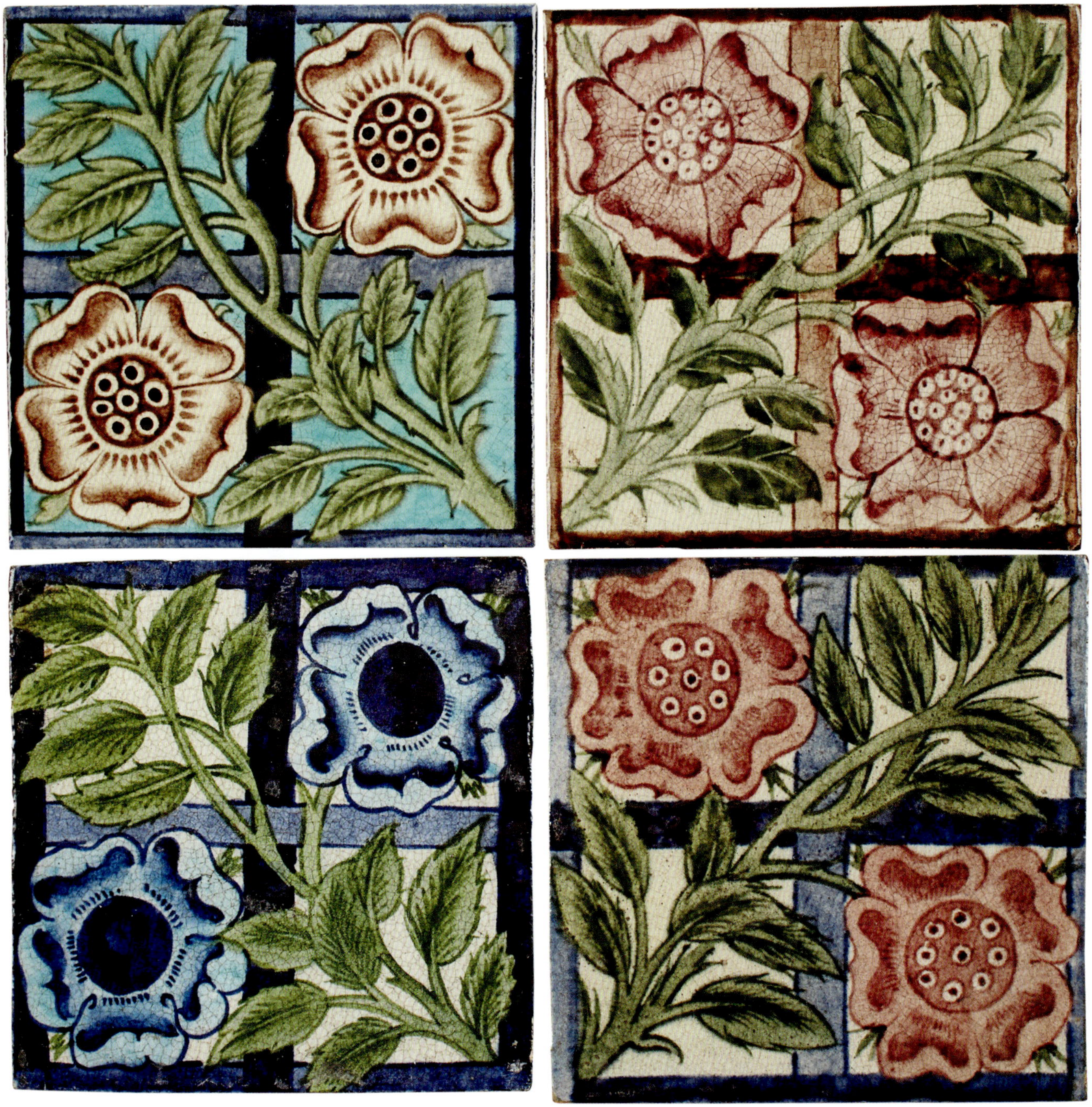

WILLIAM DE MORGAN

Rose and Trellis, c. 1880s–90s
Tin-glazed earthenware tiles,
each 6¼ × 6¼ in. / 15.3 × 15.3 cm
Private collection

These four hand-painted earthenware tiles, in slightly different colorways, are typical of the early products of William De Morgan (1839–1917), a leading English designer and manufacturer of ceramics in the late Victorian period. The stylized floral motif is derived from the Tudor rose, an emblem of royalty adopted by Henry VII, the first monarch of the Tudor dynasty (see pp.33 and 164). This royal symbol was itself based on the simple five-petaled flowers of England's native wild rose. Throughout his career, De Morgan was closely associated with the artistic ideals of the late nineteenth-century Arts and Crafts movement, which rejected industrial production in favor of the handmade and aimed to supply home furnishings that were both beautiful and functional. Much of De Morgan's work reflects the medieval aesthetic that was so characteristic of the style. He was a close friend of William Morris (see p.32), and worked for Morris & Co. before setting up his own business in 1872. Many of his hundreds of tile designs were strongly influenced by Morris's own patterns for textiles and wallpapers, and it seems probable that the *Rose and Trellis* tiles were inspired by Morris's wallpaper *Trellis*, first issued in 1864. The diagonal thrust of the tile's repeat pattern echoes the wallpaper's motif of a rose twining up through a wooden trellis, and the soft reds, blues, and greens used for the tiles are typical of the Arts and Crafts palette of muted tones derived from natural pigments.

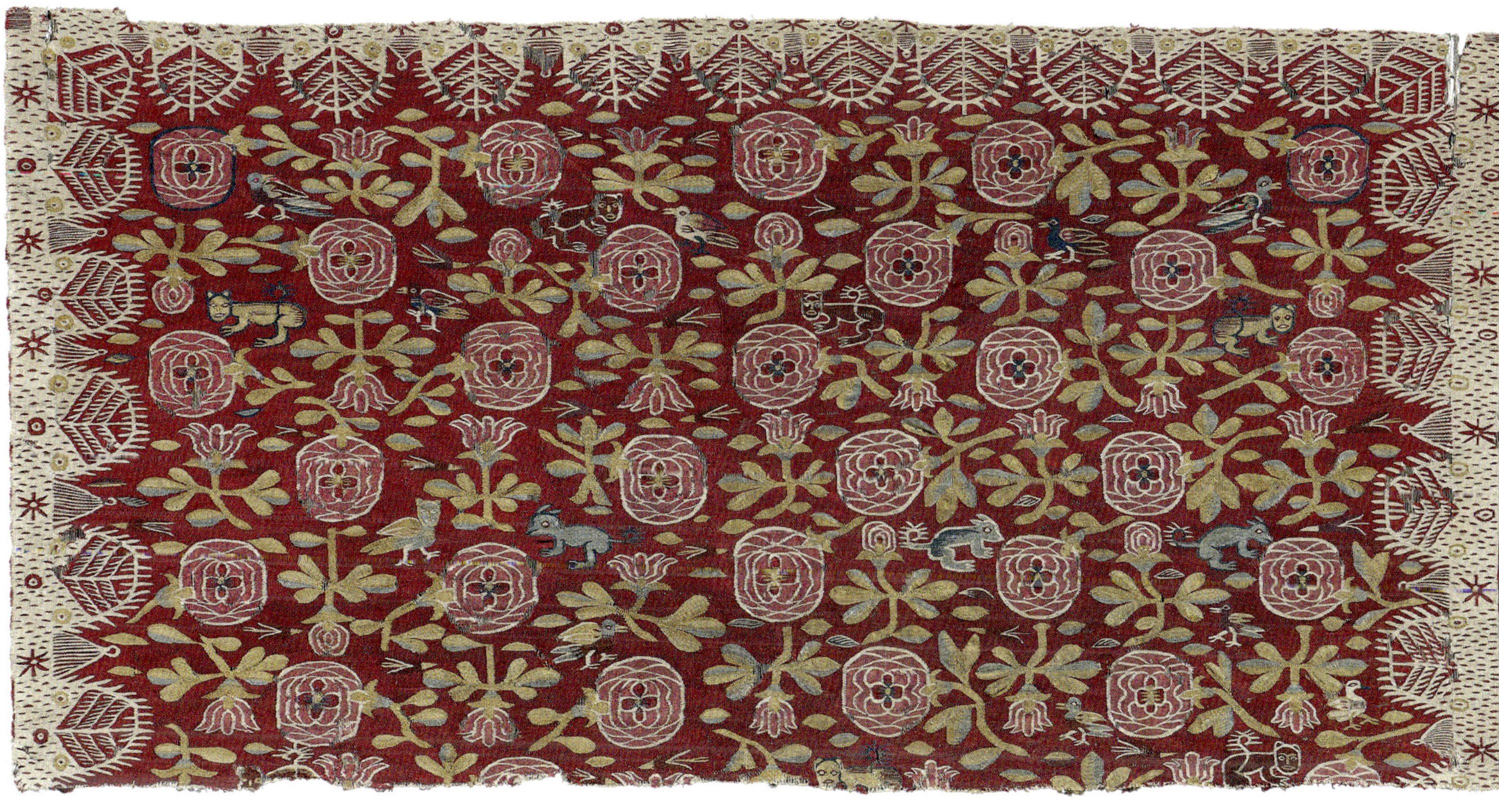

ANONYMOUS

Fragment of a tapestry with roses, c. 17th–18th century
Wool and cotton, 22 × 43½ in. / 56 × 110.5 cm
Museum of Fine Arts, Boston

A dynamic pattern of roses, in bloom and bud, fills this crimson textile fragment, dotted with mythical-looking birds and blue creatures. Woven sometime during the seventeenth or eighteenth century in colonial Peru, the rose-laden design of this textile points to its use as a mantle for a statue of the local saint, Rosa de Lima (see p.50), who had close associations with the flower. The mantle is similar to a number of other Andean tapestries produced during the colonial period that used modern weaving techniques from Spain, but it also shares another, perhaps more exciting, commonality. Other contemporaneous textiles feature the same lively visual motif of a tiny animal hidden amid flowers or foliage, as seen here scattered among the roses, and always depicted as blue. This is not, in fact, a creature from legend, but a small rodent native to the highlands of the Andes, the viscacha. Symbolically and materially important to the Indigenous cultures of the region, the viscacha is related to the chinchilla, but its tall ears make it seem more like a rabbit in appearance. It is especially notable for its soft fur, an exotic hair often used in luxury or ritual objects, such as this mantle. Although this fragment is spun from wool and cotton, its pattern still references the important symbolism of the viscacha. Such textiles, with designs and techniques that hybridized Andean and Spanish cultures, offer striking material examples of the complex exchange between Indigenous societies and the Spanish colonial world.

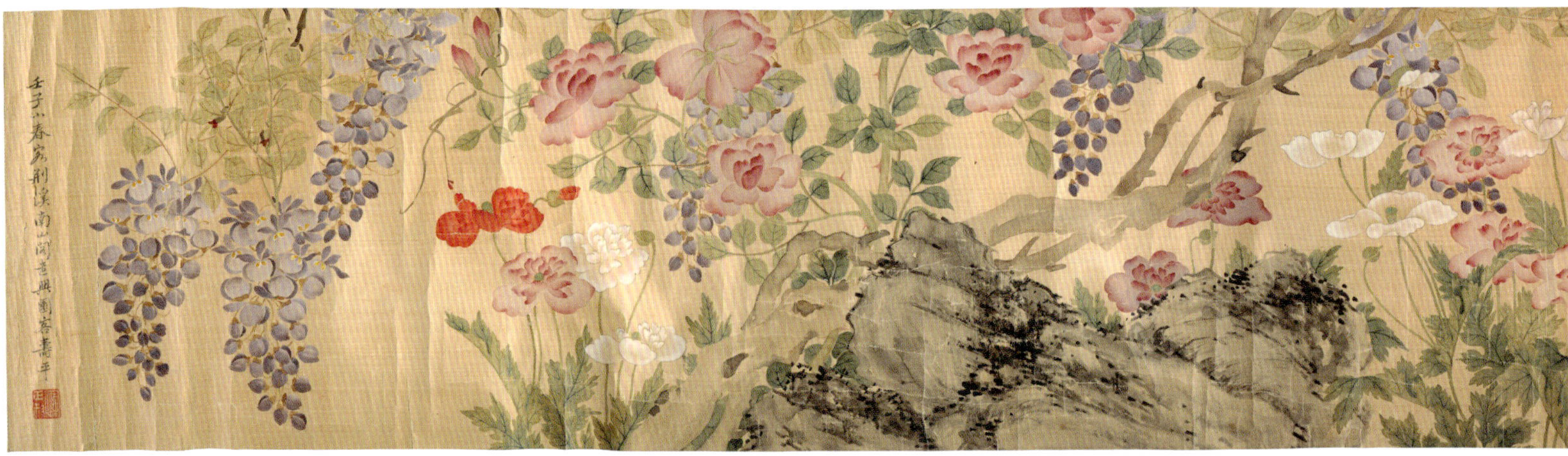
壬子小春寫荊溪南山閣並與園客壽平

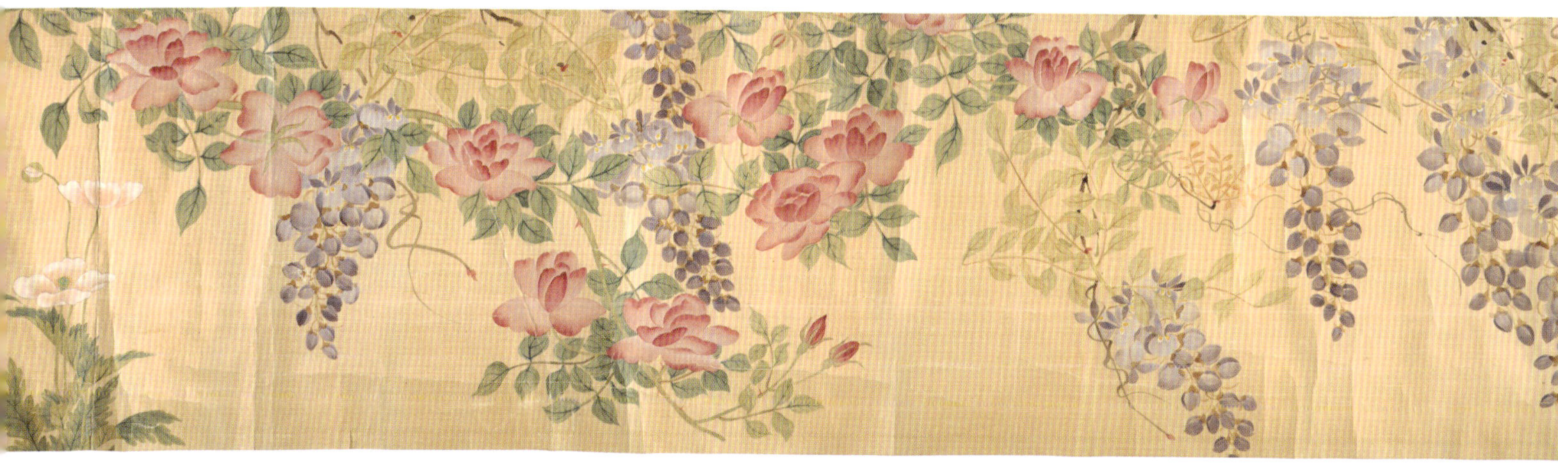

ANONYMOUS

Roses and Wisteria, c. 18th–19th century
Ink and color on silk, 12 × 102½ in. / 30.3 × 260 cm
Metropolitan Museum of Art, New York

Romantic roses alongside hanging blooms of wisteria feature in this exquisite handscroll in the style of Yun Shouping, one of the Six Masters of the early Qing period in China (1644–1911). The inscription states that it was painted in the tenth lunar month of the *renzi* year (1672), and it is signed by an otherwise unknown artist called Yuanke, although both date and signature are probably spurious. Emulated here by the artist, Yun began as a landscape painter but soon turned to flower, animal, and insect painting after admiring the landscapes of the celebrated artist Wang Hui. He popularized the eleventh-century painter Xu Chongsi's *mogu* (boneless) style, which built up forms with colorful washes instead of using outlines, creating a more graceful and unaffected depiction of flowery scenes. After creating the shapes, he applied color on wet paper to produce a sophisticated yet natural image. The Six Masters of the early Qing espoused the orthodox tradition of painting following the classifications of the earlier Ming dynasty (1368–1644) painter and art theoretician Dong Qichang, and Yun earned the respect both of his peers and of later generations as a representative of "literati painting" (*wenrenhua*). The works of the Six Masters—and those inspired by them, such as this rose scene—are generally careful and subtle, in contrast to the dynamic, intense paintings of their more individualist contemporaries.

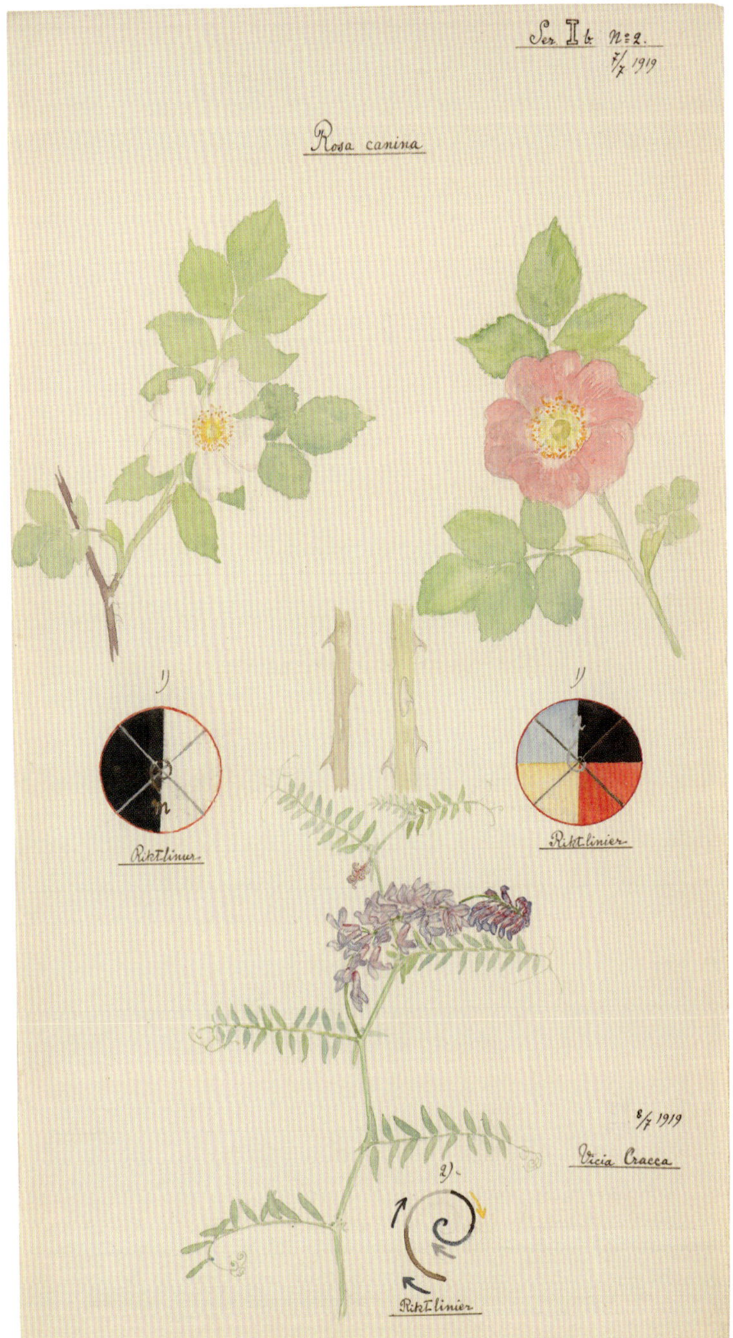

HILMA AF KLINT

Dog Rose Bush, Cow Vetch (Nyponbusken, Kråkvickern), from *Dornach Nature Studies*, 1919
Watercolor, pencil, ink, and metallic paint on paper, 19⅝ × 10⅝ in. / 49.9 × 27 cm
Museum of Modern Art, New York

Pink and white dog roses (*Rosa canina*; see pp.103 and 185) are the focus of this unusual drawing by Swedish artist Hilma af Klint (1862–1944), which forms part of a portfolio of forty-six nature studies. Botanically accurate and delicately brought to life, this work includes dog-rose cuttings, nearly micrographic depictions of their thorns, and a cow vetch in purple flower, with mysterious circles and swirls floating. The study synthesizes many aspects of af Klint's interests, combining scientific illustration, geometry, and color theory. Originally inspired by the nineteenth-century occult philosophies of spiritism and theosophy, she later adopted the ideas of Austrian polymath Rudolf Steiner, whose anthroposophy movement taught that one could comprehend the spiritual world by training the intellect through sensory discovery. In 1913 Steiner opened the Goetheanum, a school of "spiritual science," in the Swiss town of Dornach, which af Klint often visited. She developed a unique diagrammatic language as an attempt to grasp the true nature of the plant in a way that was radically different from botanical observation. Her circles and swirl diagrams were often divided into quadrants and included dynamic colors, which she tagged here in pencil as *riktlinier* (guidelines) to aid in transcendent perception of the roses. An attuned mystic and pioneer of artistic abstraction, af Klint strove through her esoteric botany to understand the spiritual nature of life through the detailed study of the plants—including the roses—around her.

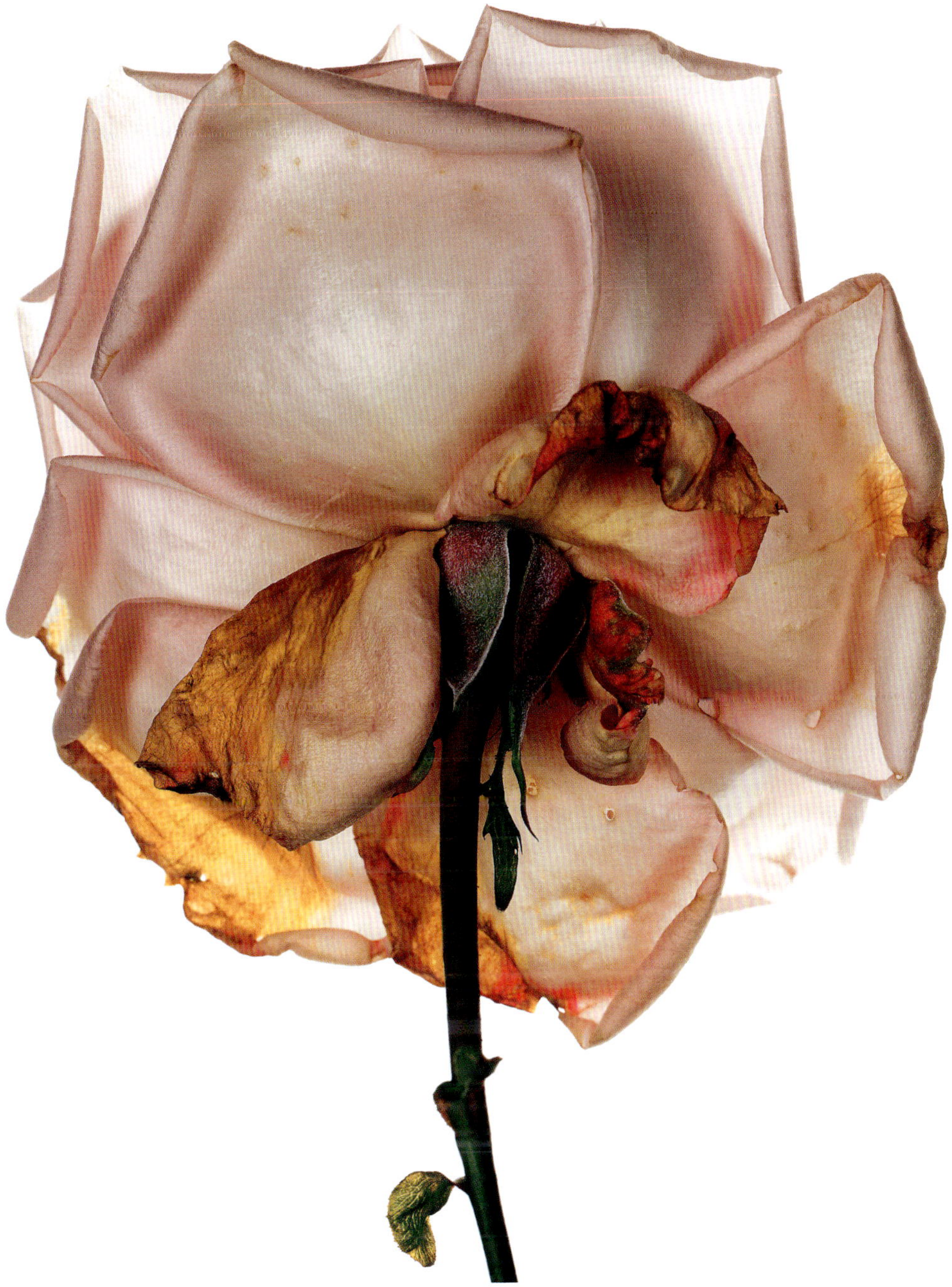

IRVING PENN

Rose: Blue Moon, London, 1970
Dye transfer print, dimensions variable
Private collection

This image, in which an individual bloom is presented in a modern, unfussy way, absent of a bouquet or flower arrangement, is typical of the minimal style of American photographer Irving Penn's (1917–2009) flower—and fashion—photography. The focus is a single stem, shot from behind. Combining glamour, poise, and vulnerability, it is emblematic of Penn's close examination of form, texture, and detail instead of being an attempt to create a perfect representation of an ideal. As discoloration marks the petals, and pink veers to orange, yellow, and brown, Penn finds beauty in imperfection in order to engage completely with his subject. The dignity of the flower recalls that of a slender fashion model, with two leaves evoking a pair of arms and the stem a spine, atop which a voluminous coiffure seems to blossom. The artist began photographing flowers in 1967, when American *Vogue* commissioned him to illustrate the Christmas issue with images of tulips. He went on to shoot flowers for seven consecutive December issues of the magazine, each dedicated to a different species, which were collected in the book *Flowers*, published in 1980. In the 1990s and 2000s, Penn returned to flowers, exploring the beauty of fading blooms to symbolize his own aging. He was not particularly interested in horticulture but rather in the simple enjoyment of form and color, and his flower portraits, like his photographs of people, embrace the uniqueness of each model.

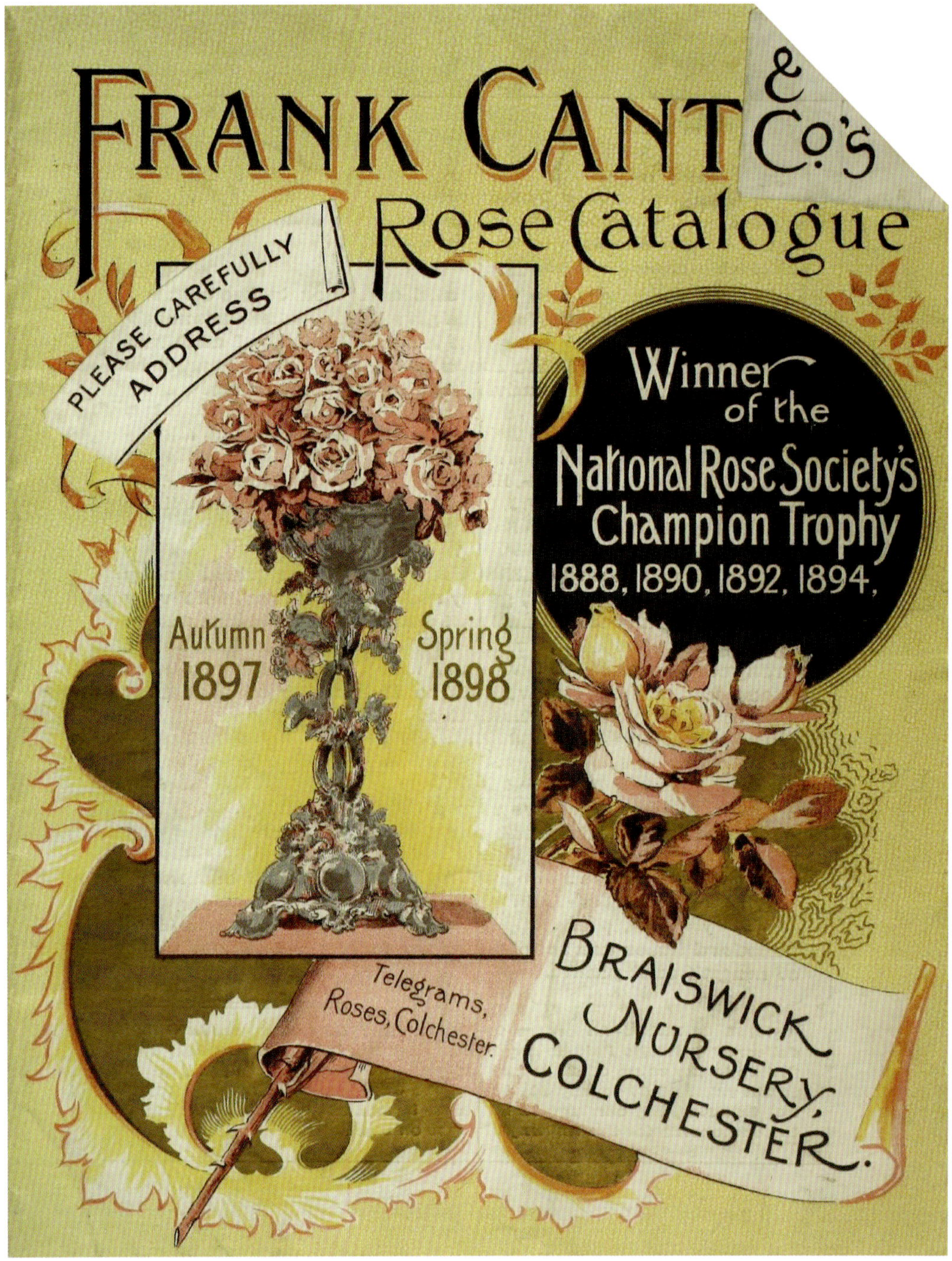

FRANK CANT & CO.

Rose Catalogue, 1897–98
Trade catalog, 12⅝ × 9½ in. / 32 × 24 cm
Cornell University Library, Ithaca, New York

The history of growing roses in Britain was enlivened in the late nineteenth century by two rivalries between uncles and nephews. The first was a rivalry between the members of the Paul family; William Paul, whose nursery was at Waltham Cross in Hertfordshire—and whose classic book *The Rose Garden* went through ten editions between 1848 and 1903—had an eager rival in his nephew George Paul, whose nursery was in Cheshunt. The two spat gentle venom at each other in the gardening magazines for decades. Over in Colchester, the second rivalry concerns the trade publication illustrated here. Benjamin R. Cant, whose catalogs proudly bore the heading "Cant's of Colchester," secured his reputation by introducing into England the celebrated French roses 'Général Jacqueminot' and 'Gloire de Dijon'. However, in about 1880 his nephew Frank Cant left the firm and started up his own nursery on Braiswick Road. He was also successful, both socially and financially; he served a term as mayor of Colchester, and on his death left a small fortune, equivalent to about a million pounds (nearly $1.3 million) in today's money. The back covers of his catalogs, as shown here from the Fall 1897/Spring 1898 issue, displayed some of the many awards he won at exhibitions. But even if Frank Cant were a four-time winner of the National Rose Society's Champion Trophy, his collection of accolades was ultimately dwarfed by the 2,000-odd cups and medals awarded to the original Cant, his uncle Benjamin.

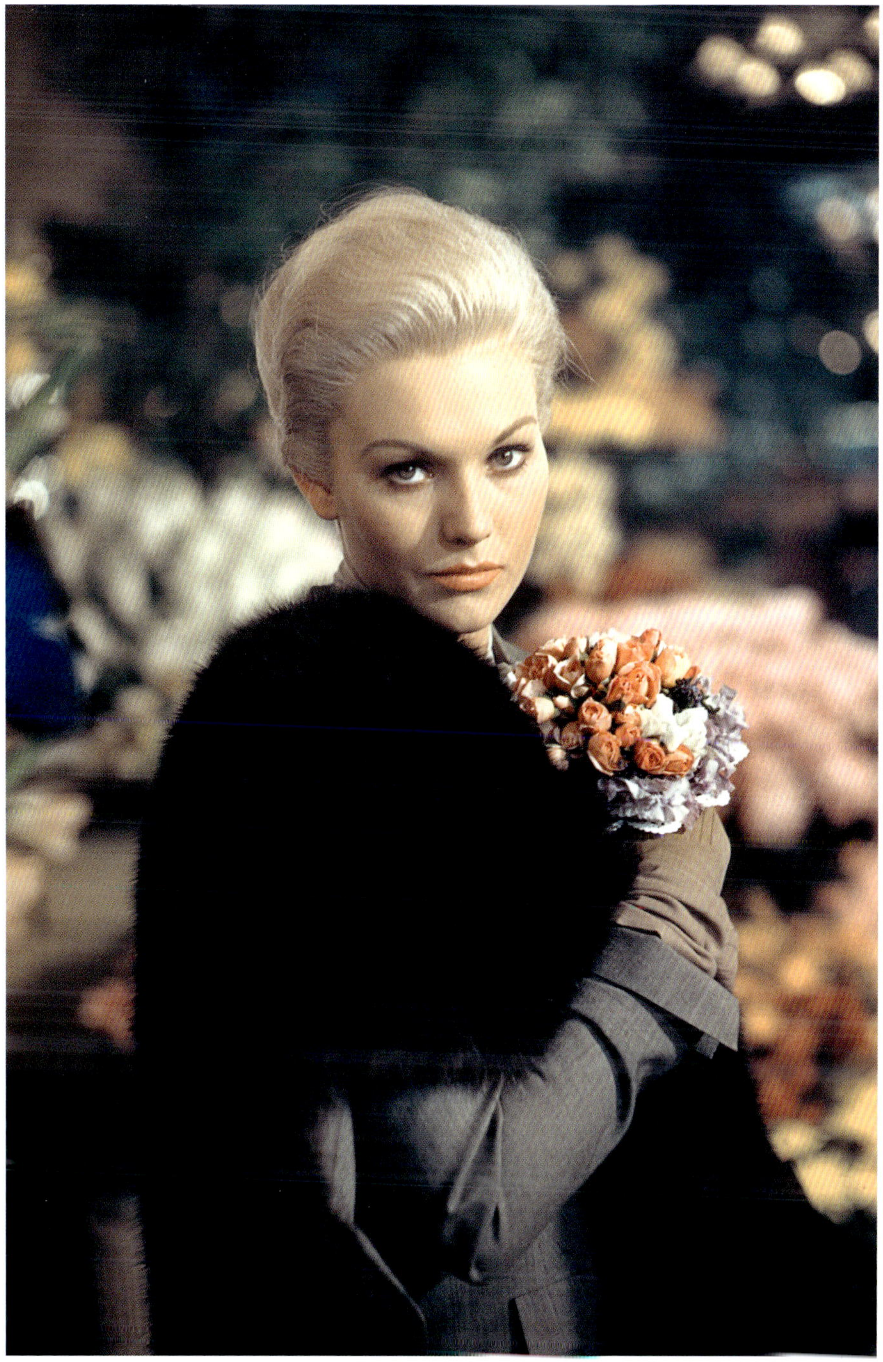

ALFRED HITCHCOCK

Vertigo, 1958
Film still, dimensions variable

Flowers are a central motif in *Vertigo*, an undisputed masterpiece of twentieth-century cinema directed by Alfred Hitchcock (1899–1980). Early in the psychological thriller, Madeleine Elster (played by Kim Novak) enters a flower store overflowing with colorful blooms, where she buys a delicate nosegay containing miniature pink roses. Unbeknownst to her, she is being watched by former police detective John "Scottie" Ferguson (James Stewart), who has been hired by old schoolfriend Gavin Elster (Tom Helmore) to trail his enigmatic wife, whose strange behavior is worrying him. The scene was filmed on location at the former Podesta Baldocchi store, which was founded in 1871 and at the time was one of the oldest operating florists in San Francisco. The fragile-looking roses are intended to represent Madeleine's disposition, and the bouquet appears several times in the movie, most prominently when she is standing under the Golden Gate Bridge, plucking it to pieces and tossing the flowers into the bay before seemingly attempting to commit suicide by jumping in. Later, after Scottie has rescued Madeleine from the water, he witnesses her apparently plunging to her death from a church tower, an event that triggers a nightmare in which the roses swirl around the screen before disintegrating violently in a colorful animated sequence. With the image of the real flowers transforming into a flurry of paper-thin fragments, there is the suggestion that Madeleine was not all she appeared to be.

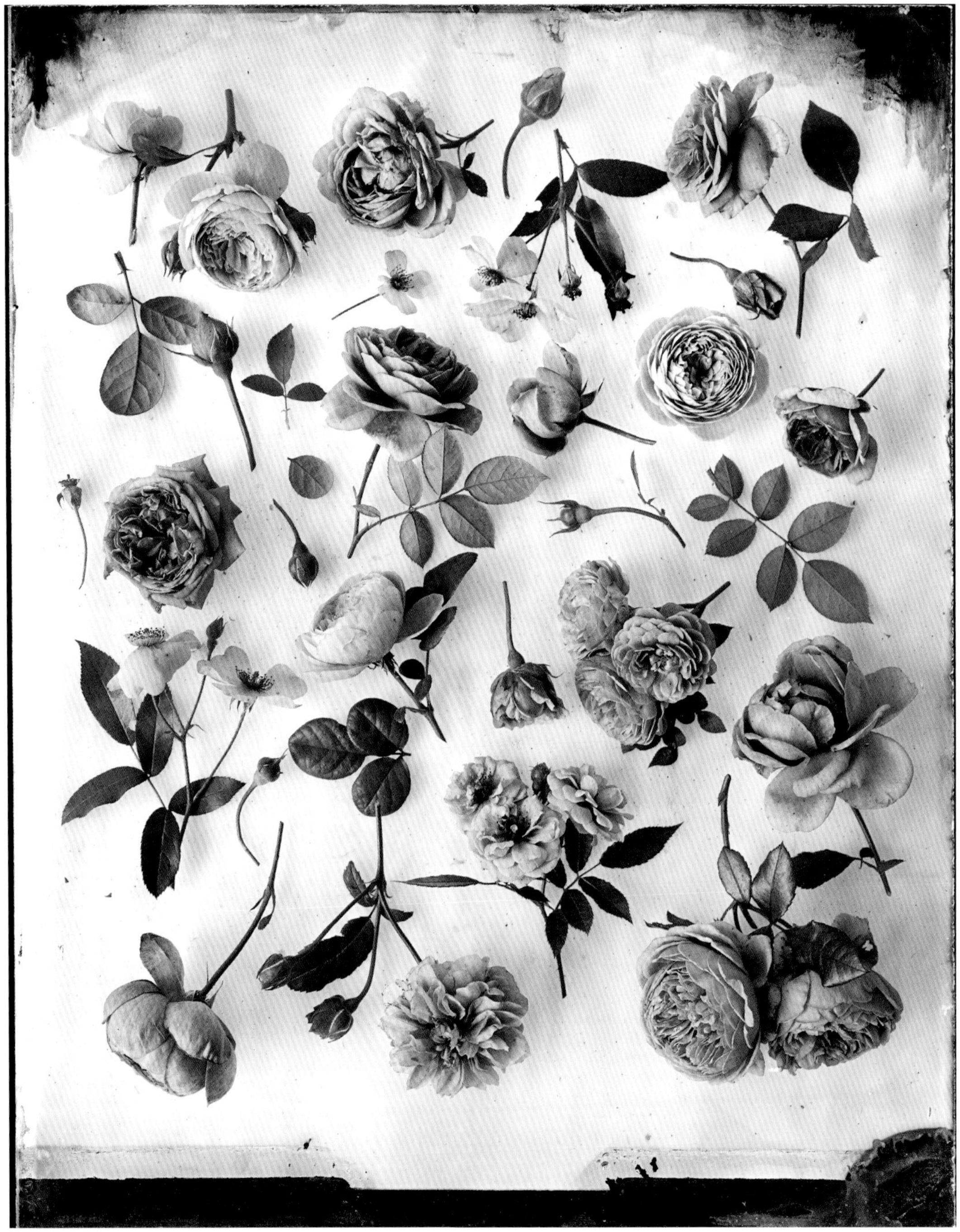

EMILIJA PETRAUSKIENE

Roses Herbarium, from *Silver Herbarium*, 2018
Gelatin silver print on mat paper,
11⅝ × 8¼ in. / 29.5 × 21 cm

Blossoms, leaves, bud, and hips from several varieties of rose, including delicate dog roses and plush cabbage roses, such as 'Queen of Sweden', 'Lyda', and 'William Shakespeare', are laid out for the camera, displaying their morphologies, sizes, and textures from different angles. Lithuanian photographer Emilija Petrauskiene's (b. 1981) *Rose Herbarium* is one of several herbaria she has compiled from flowers, plants, and crops grown in her garden. They are part of her *Silver Herbarium* series, referencing to the light-sensitive silver-nitrate coating on the glass plate where the image is formed. Each image features a carefully planned composition of elements in which the absence of color allows the structure of the plant to become the main focus of interest. A full-time family doctor, Petrauskiene is a self-taught photographer and began using a digital camera as a hobby before switching to large-format analog photography and working with wet-plate collodion negatives. Drawing inspiration from early Flemish painters, as well as classical botanical illustrations, scientific herbaria, and fine-art photographers, including Sally Mann, Petrauskiene recognizes the beauty in all classifications of plants, from flowers to weeds. Her preference for working with negatives over computers is one facet of her embracing of photography as a slow process, akin to gardening, and she views her creativity as a mindfulness practice, which provides a space of relaxation and an antidote to the challenges of working as a professional physician.

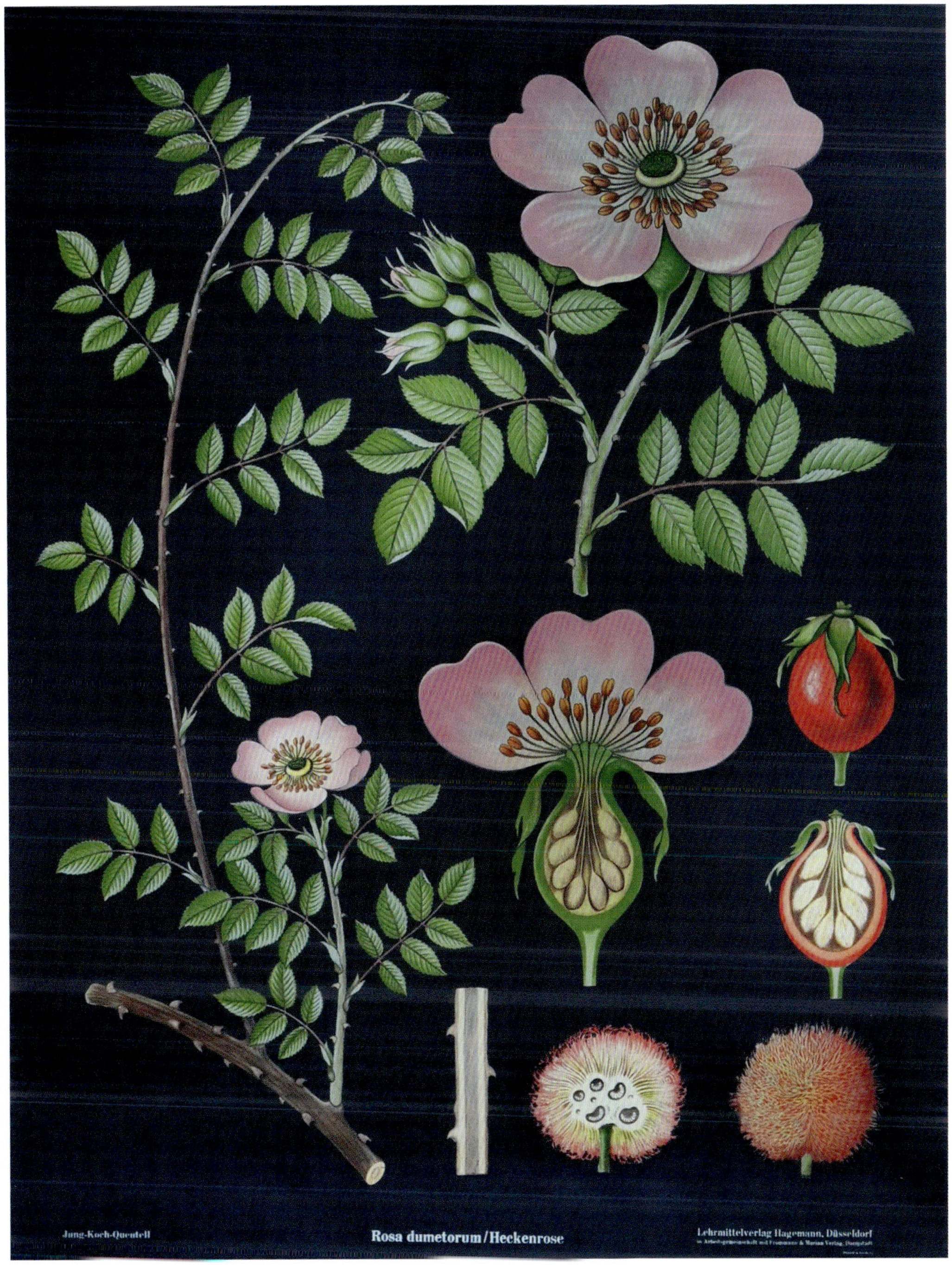

JUNG KOCH QUENTELL

Rosa dumetorum / Heckenrose, c. 1960s
Lithograph print on paper-backed linen, 45¼ × 32⅝ in. / 115 × 83 cm

Studying botany was once a vital part of the school curriculum and early learning. This striking illustrated chart—printed large enough for the whole class to view and attached to a roller so that it could be pulled down and up—was designed by Jung Koch Quentell. The name was derived from the surnames of three men, all scientists, who from the 1950s created educational charts for the German publishing company Hagemann. This chart illustrates *Rosa* × *dumetorum*, a hybrid dog rose (*Heckenrose* in German), so named for the ancient belief—attributed to Roman scholar and naturalist Pliny—that the roots of the plant could help to cure a bite from a mad dog. The formula for this particular hybrid is *R. balsamica* (a shrub rose) × *R. canina* (a climbing wild rose; see pp.98 and 185).

Some students may have bisected a wild rose flower or its brilliant red rose hip, or fruit, depending on the time of year, and compared their findings to the chart. Their teacher may have explained to them that, while its scent is barely discernible to the human nose, the flower is highly successful in attracting pollinators. The shape of the stamens may have been likened to a coronet and the children asked if they knew why this plant is commonly known as a dog rose. Although originally created for scientific and educative purposes, the beauty and strong sense of design of Jung Koch Quentell's charts have led them to become highly desirable for decorating stylish domestic and commercial interiors.

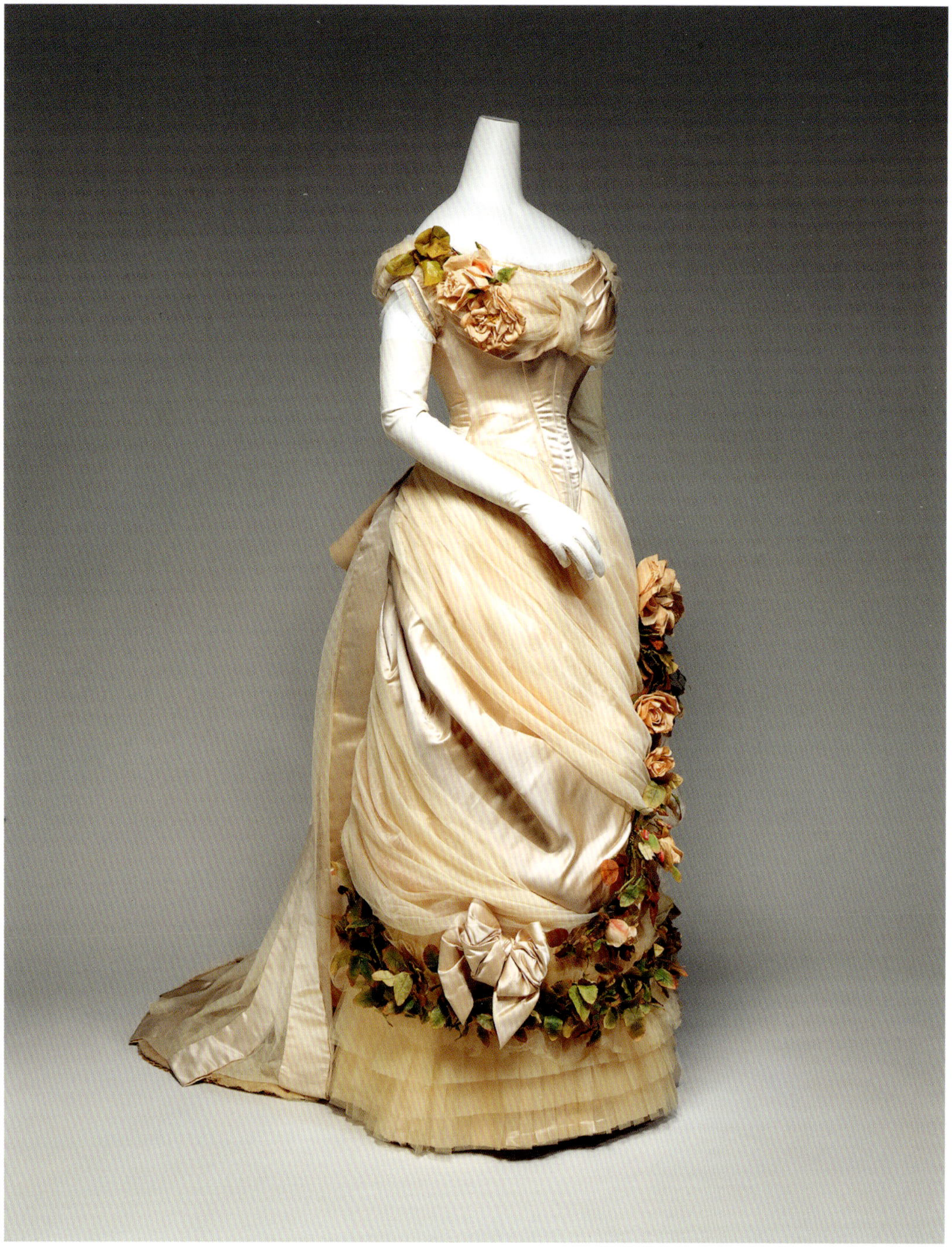

CHARLES FREDERICK WORTH

Evening dress, c. 1882
Silk
Metropolitan Museum of Art, New York

By the mid-nineteenth century, artificial flower making (see p.176) had become a vital ancillary trade to the haute couture houses of Paris. Of all flower makers, those who made roses were the most revered and highly paid. To create both the exquisite corsage of golden roses, set at the décolletage of this evening gown, and the leafy vine of blooms that meanders down and around the hemline of its skirt, required an apprenticeship of at least three to five years. In 1880s Paris, a gown designed by British-born couturier Charles Frederick Worth (1825–1895) was an international symbol of fashionable taste, wealth, and social prestige. His signature adoration and abundant use of silk flowers were evident in his evening-gown designs, which were the most highly priced in Paris and in insatiable demand. Hailed as the "father of haute couture," Worth skyrocketed from his humble origins as a haberdashery assistant to the head of the world's most famous fashion house. By the 1860s, the House of Worth employed 1,000 workers and boasted an enviable client list that included the French empress Eugénie herself. Worth was the first designer to create a label under his own name and to present seasonal collections from which clients could place an order, instead of taking individual commissions. He also embraced new technologies, including the sewing machine and photography to document his collections, which were already distinguished by his lavish use of luxury fabrics and sumptuous floral ornamentation.

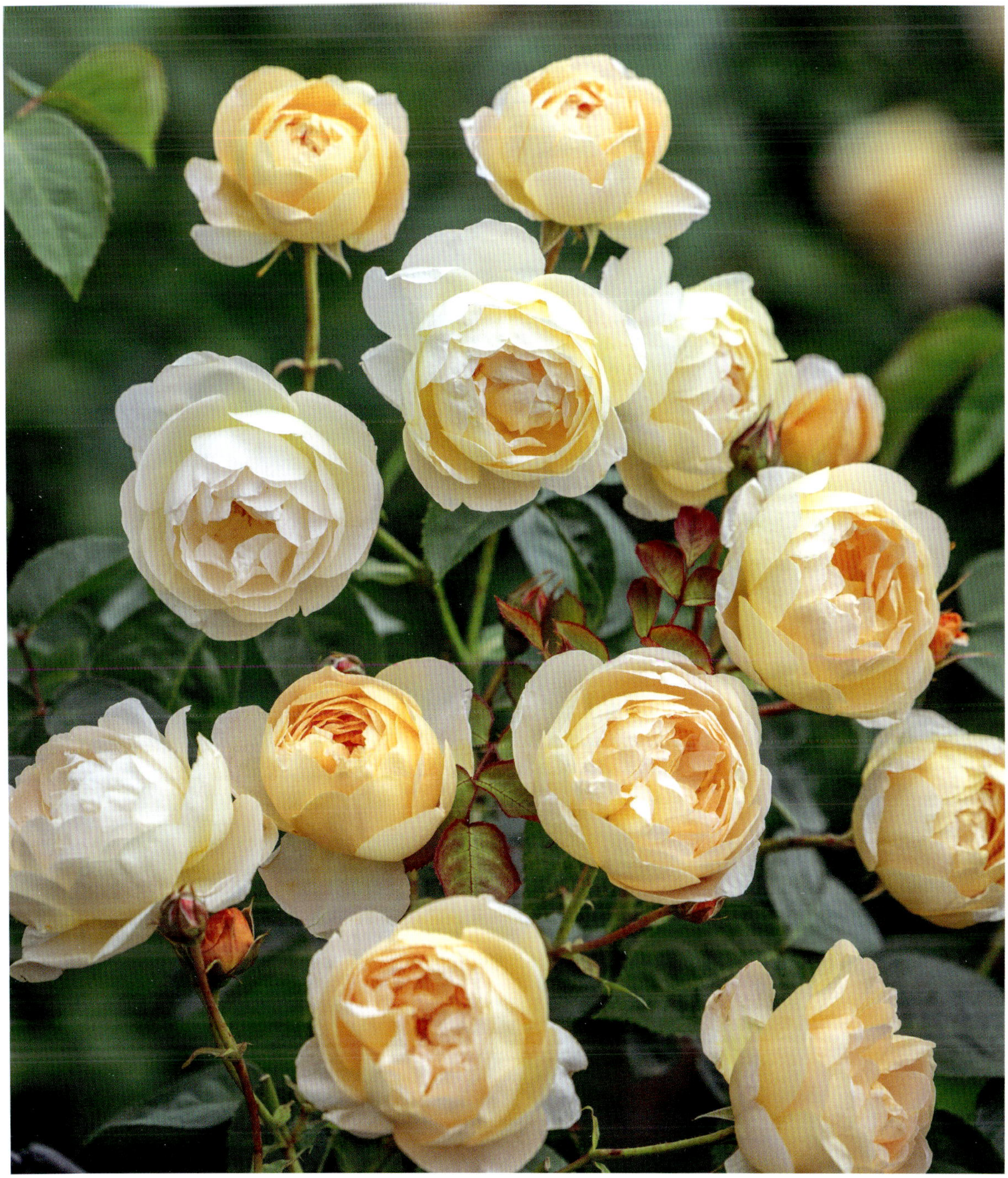

DAVID AUSTIN ROSES

Rosa 'Dannahue', 2023
Photograph by Jonathan Buckley, 20 × 13½ in. / 50.8 × 34.4 cm

Naming new roses is always a challenge. It is such an important part of the success of a rose that some rosarians are known to say, "the easiest way to kill off a new rose is to give it a bad name!" Prestigious English rose breeder David Austin (1926–2018; see pp.14 and 77) always ensured that the names of his roses should roll off the tongue and in some way be associated with Great Britain. Some of his varieties are named after literary characters, such as 'Jude the Obscure', or prominent people in the gardening world, as in 'Gertrude Jekyll' (see p.73), and even in association with charities, such as 'The Generous Gardener' for Britain's National Garden Scheme. Introduced in 2023, *Rosa* 'Dannahue' was the first David Austin rose to be named after a person of color. Captured here in all its apricot splendor by renowned English gardens photographer Jonathan Buckley, the compact, bushy shrub is named in honor of Danny Clarke, a television presenter and garden designer known to viewers in Britain as "The Black Gardener." Clarke is a leader in making British green spaces more diverse and accessible, and £10,000 (about $7,000) of the proceeds made from purchases of 'Dannahue' are donated to Grow to Know, a nonprofit that encourages a more inclusive environment within horticulture to reconnect people with nature and one another. A versatile rose with a zesty fragrance that can grow in sun or shade, in containers or smaller borders, 'Dannahue' mirrors Clarke's message and will thrive in any type of garden, grown by any kind of gardener.

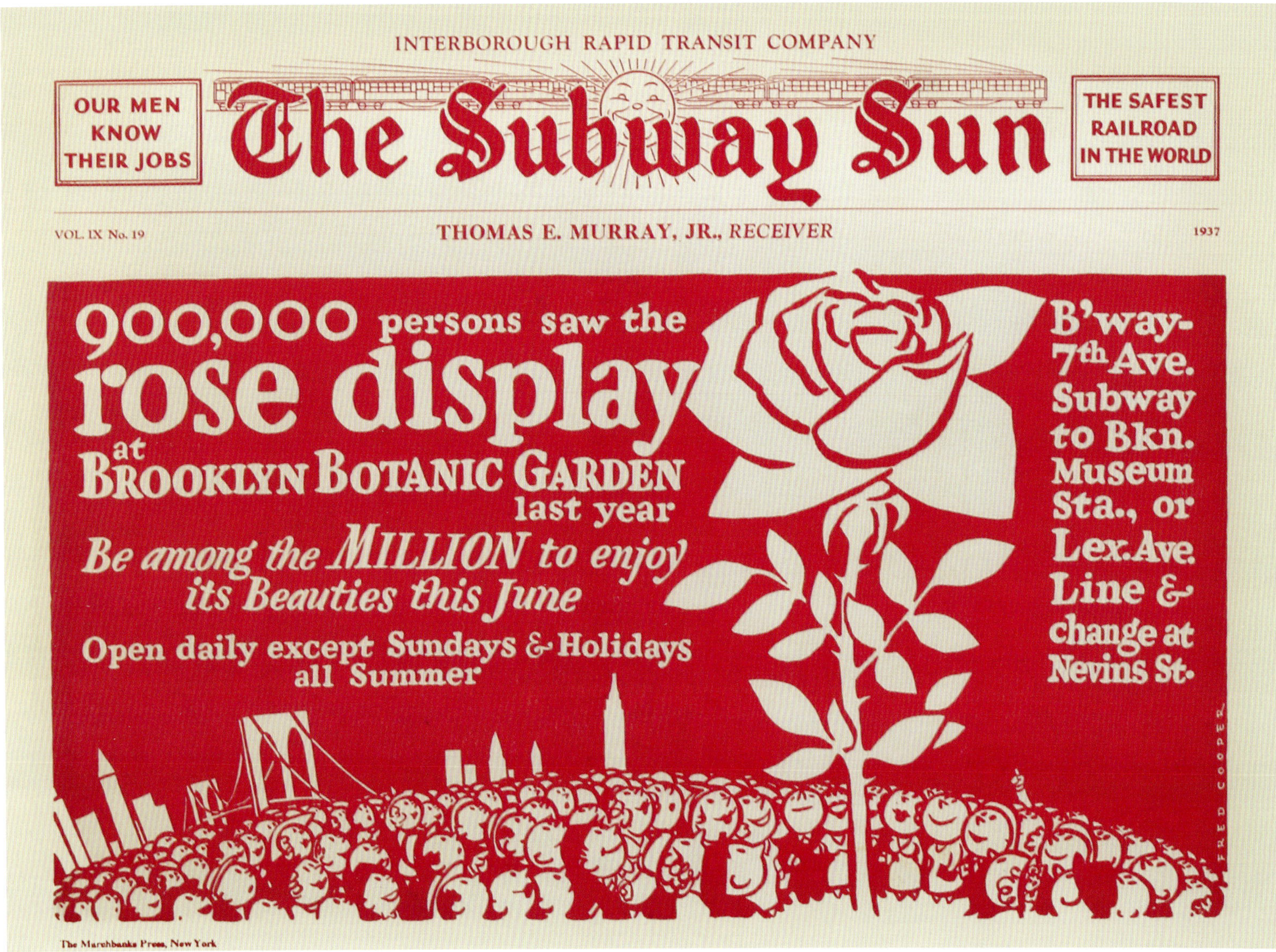

FRED G. COOPER

The Subway Sun, Brooklyn Botanic Garden Rose Display, 1937
Poster, 15¾ × 21⅝ in. / 40 × 55 cm
Private collection

Vying for the attention of weary commuters, *The Subway Sun* is a series of humorous advertisements that first appeared in New York subway cars in 1918. Designed to look like newspapers, the original ads were created by American artist and illustrator Fred G. Cooper (1883–1962). Some of the graphic posters promote the city's landmarks and cultural institutions, such as this cheery pink edition inviting passengers to visit the newly opened rose display at the Brooklyn Botanic Garden. Cooper, who is known for his popular magazine covers, uses his recognizable style to feature the spectacle of roses in peak bloom. With the Manhattan skyline and Brooklyn Bridge in the background, the poster shows a dense crowd of people looking up at a towering, larger-than-life rose stem.

Opened in 1928, the Cranford Rose Garden was designed by landscape architect Harold Caparn, and remains one of the garden's most popular attractions. Caparn's plan carefully considered the combination of modern and historic roses, as well as their sequence of blooms, fragrance, and foliage, ensuring its year-round beauty. Many of the original roses planted in 1927 remain in the garden today. Although *The Subway Sun* does not convey the kaleidoscope of roses cultivated and their environments—from Hybrid Teas to Grandifloras, Floribundas, and ramblers in formal beds framed by lattices, an overlook, and a pergola—it does colorfully insist that everyone should follow Cooper's gigantic rose to the garden.

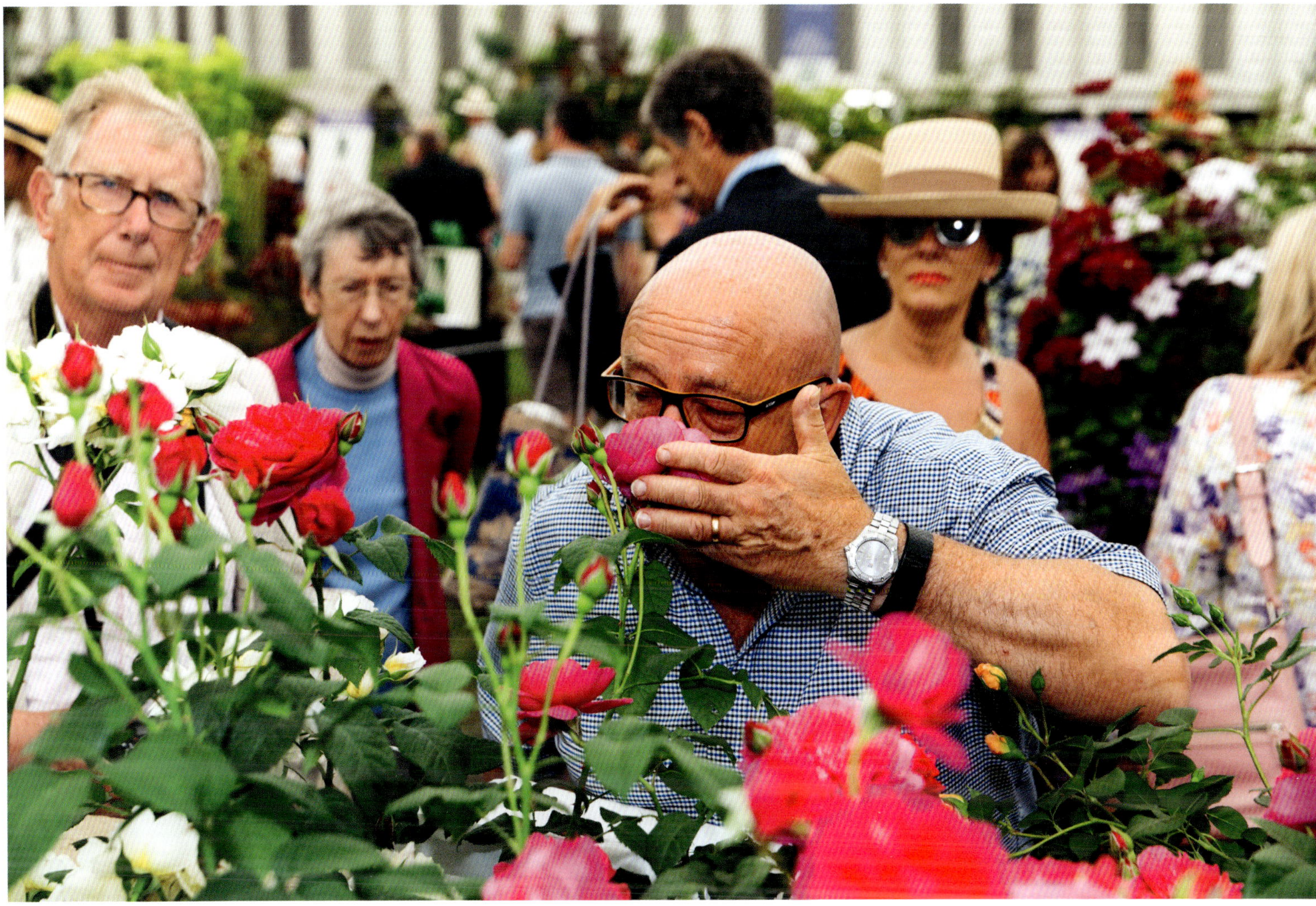

MARTIN PARR

RHS Chelsea Flower Show, London, England, 2018
Photograph, dimensions variable

In attempting to smell an exciting new rose, the man in this photograph has pulled the bloom so close to his face that it resembles a clown's nose. British photographer Martin Parr (b. 1952) shot this humorous picture at the Chelsea Flower Show, a prestigious annual garden show held over five days in May by the Royal Horticultural Society in the grounds of the Royal Hospital Chelsea in London. Parr is one of the best-known documentary photographers of his generation, recognized for his candid and often witty photographs documenting aspects of modern British life, and, in particular, the idiosyncrasies of the nation's social classes. For a photographer interested in capturing the people of Middle England, the Chelsea Flower Show presents the perfect opportunity—up to 168,000 visitors attend the event each year, and the demographic is predominantly white and middle class. Positioning himself near a vibrant pink rose in 2018, he shot a sequence of photographs—none of them posed—of people smelling the disappointingly weak-scented flower. The Chelsea Flower Show is an important barometer of new trends in growing flowers, featuring avant-garde show gardens by leading designers as well as smaller, less ostentatious displays. New roses are launched there by some of the world's leading breeders, and prizes are given to the best. The grower David Austin Roses (see pp.14 and 105) is a regular exhibitor and has accumulated a remarkable forty-four medals over the years, including twenty-eight Gold awards.

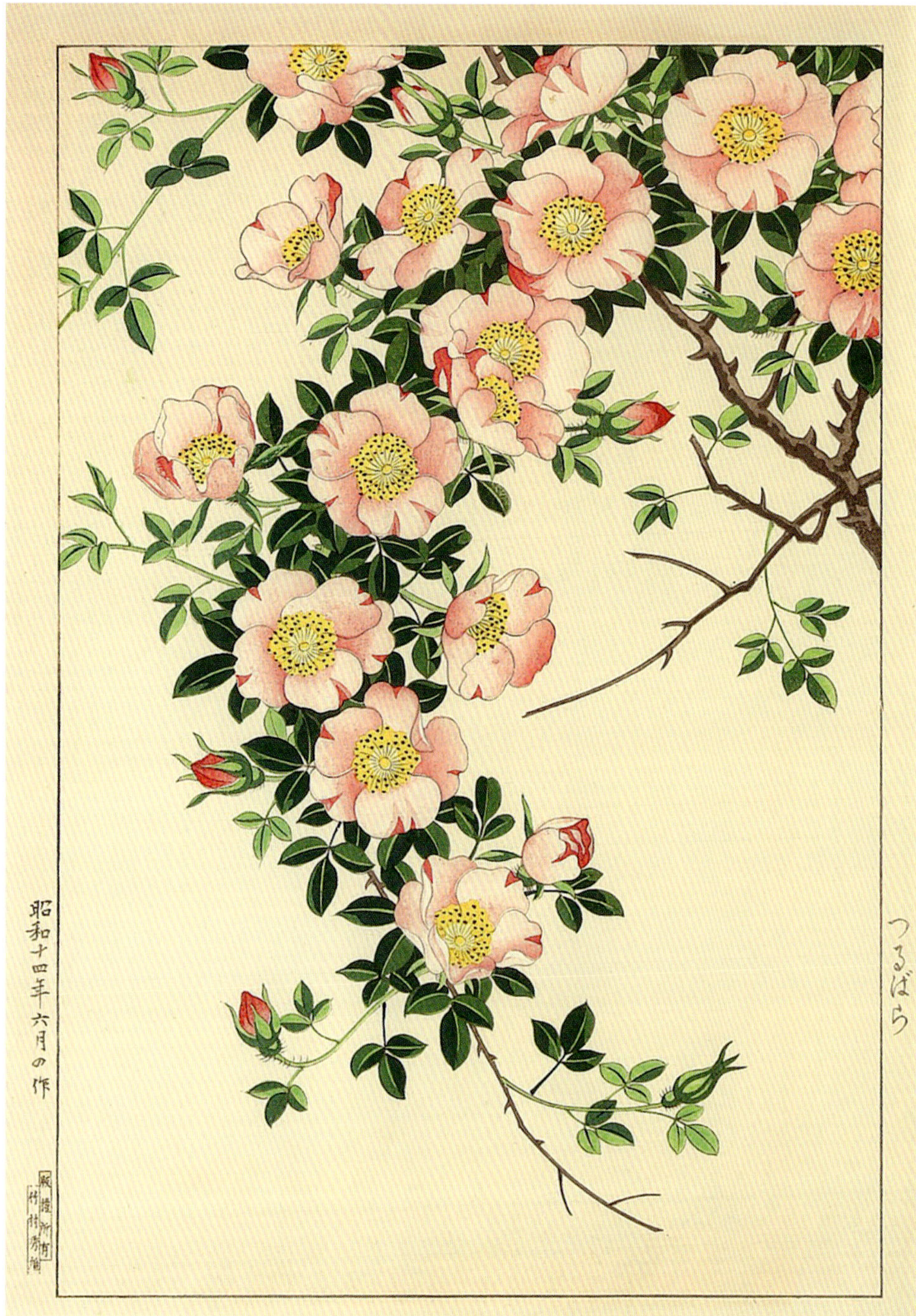

NISHIMURA HODO

Wild Vine Rose, 1939
Woodcut print, 14⅜ × 9½ in. / 36.5 × 24.1 cm
Private collection

Wild roses, such as the one cascading down the center of this woodblock print by Japanese artist Nishimura Hodo (fl. 1930s), are thought to have originated in Asia. These flowers were first recorded in Japan in the eighth century in the *Hitachi Fudoki* and *Man'yōshū* books of poetry, but they are also notably mentioned in the famous epic *The Tale of Genji*, published before 1021. Among the wild roses native to Japan are the multiflora rose (*Rosa multiflora*), the memorial rose (*R. lucieae*), and the beach rose (*R. rugosa*). However, unlike cherry or plum blossoms, or iris, chrysanthemum, and peony flowers, the rose was not a popular traditional floral theme in Japanese art. Japanese depictions of cherry blossom often feature the blossom seemingly cascading onto the page in a graceful sweep, and Nishimura's woodcut print owes much to this aesthetic. The rambling pink rose trailing gracefully down the print is probably a multiflora rose, identifiable by its copious sprays of pink-tinged flowers and hardy arching branches with recurved prickles. The process of Japanese printmaking is a collaboration between the artist, the block cutter, the printer, and the publisher. It is perhaps for this reason that little is known about Nishimura beyond the legacy of his prints. Known as a designer of *kacho-e* (bird and flower pictures) and landscape prints in the *shin-hanga* style of early twentieth-century Japan, he is believed to have been an in-house artist for Yokohama publisher Takemura Hideo from the 1930s until the business closed in 1941.

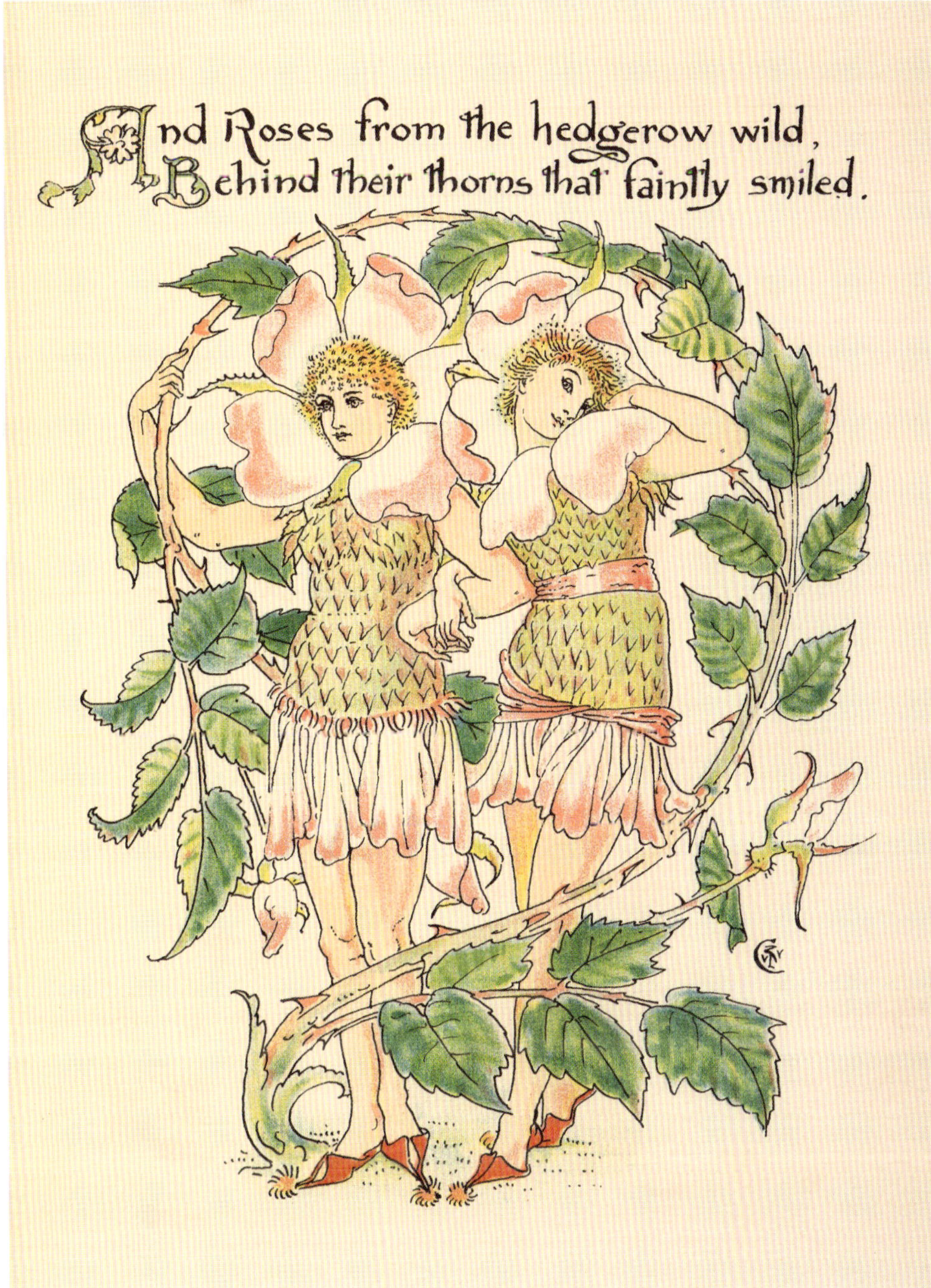

WALTER CRANE

Wild Roses, from *Flora's Feast: A Masque of Flowers*, 1902
Lithograph on paper, H. 10¼ in. / 26 cm
National Library of Poland, Warsaw

Throughout the nineteenth century, an anthropomorphic craze swept through visual culture, and the conflation of people, particularly women, with flowers—especially roses—flourished throughout much of the West (see p.151). Walter Crane (1845–1915) was a British socialist, artist, and illustrator who incorporated his political ideology into floral fables. A student of John Ruskin (see p.168) and follower of William Morris (see p.32), he also romanticized the medieval era as a pastoral idyll. In *Flora's Feast: A Masque of Flowers*, one of his most famous books for children, Crane follows the goddess Flora as she wakes the flowers in her garden for a spring festival. Among the nearly forty flowers depicted, two are roses: a cultivated blushing rose and the wild ones seen here.

Crane's portrayal of the wild rose as two masculine figures with queer undertones was radical. Surrounded by a climbing stem and with faces framed by petals, the entangled couple don short dresses comprising drop-waisted prickly bodices and petal-clustered skirts, accessorized with thorny red mules. Crane championed a similar style in his capacity as vice president of the radical Healthy and Artistic Dress Union, established in the early 1890s, which encouraged both men and women to dress in loose, unstructured clothes instead of the restrictive fashions common to the Victorian era. A departure from the traditional feminization of the rose, Crane's wild roses offer a unique nineteenth-century example of the androgynous floral dress so common in contemporary fashion.

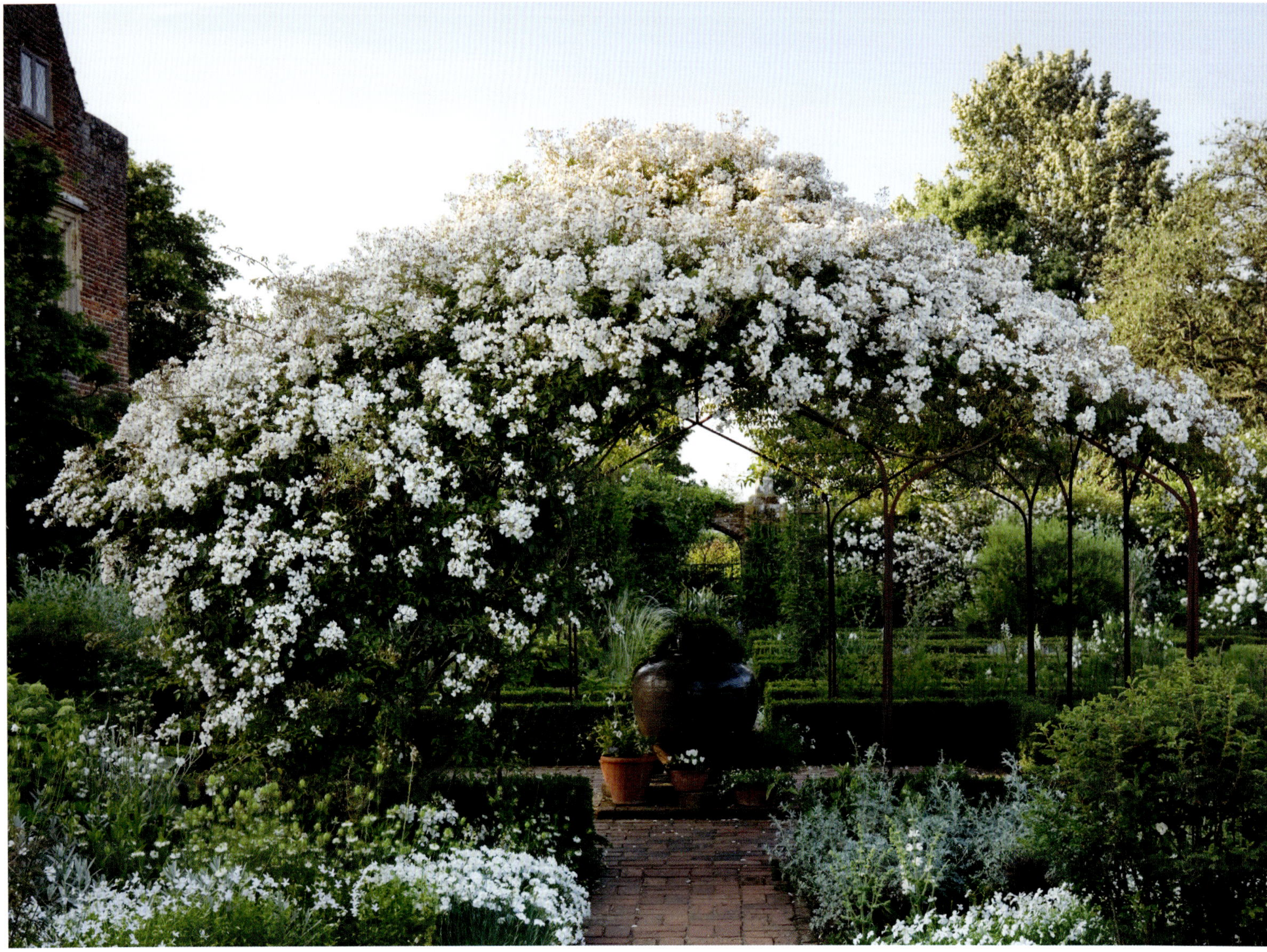

VITA SACKVILLE-WEST AND HAROLD NICOLSON

Rosa mulliganii bower in the White Garden at Sissinghurst Castle Garden, 2013
Photograph by Jonathan Buckley, dimensions variable

Vita Sackville-West's (1892–1962) garden at Sissinghurst Castle in Kent is the epitome of the classic naturalistic English garden. It is full of roses, which form the centerpiece of all her plantings and clamber over the Tudor brick walls that give the castle its architectural framework. In 1930—before Sackville-West, an English writer and garden designer, and her husband, diplomat Harold Nicolson (1886–1968), had even signed the deeds—she planted the creamy white Noisette rose 'Madame Alfred Carrière' by the South Cottage. Over the next thirty years, Sackville-West designed intricate garden rooms and assembled more than three hundred rose varieties in a collection of Old Garden roses that is now recognized as one of the world's finest. Her planting style made use of exuberant profusion and a wondrous sense of color; her catchphrase was "cram, cram, cram." Now managed by Britain's National Trust and overseen by head gardener Troy Scott Smith, Sissinghurst Castle Garden remains world renowned. Sackville-West's most influential experiment was the White Garden, which she designed in 1950 entirely around white flowers of all kinds, complemented by plants with gray and silver foliage, all contained within a grid of mellow brick paths. The centerpiece of the White Garden today is the stupendous rambler, *Rosa mulliganii,* trained over an iron arbor. *R. mulliganii* has sprays of single white flowers, a banana-like scent, hooked thorns, and a spread of nearly 20 feet (6 m). The vigorous, weaving growth of this rose across the arbor has led some to call it "living lace."

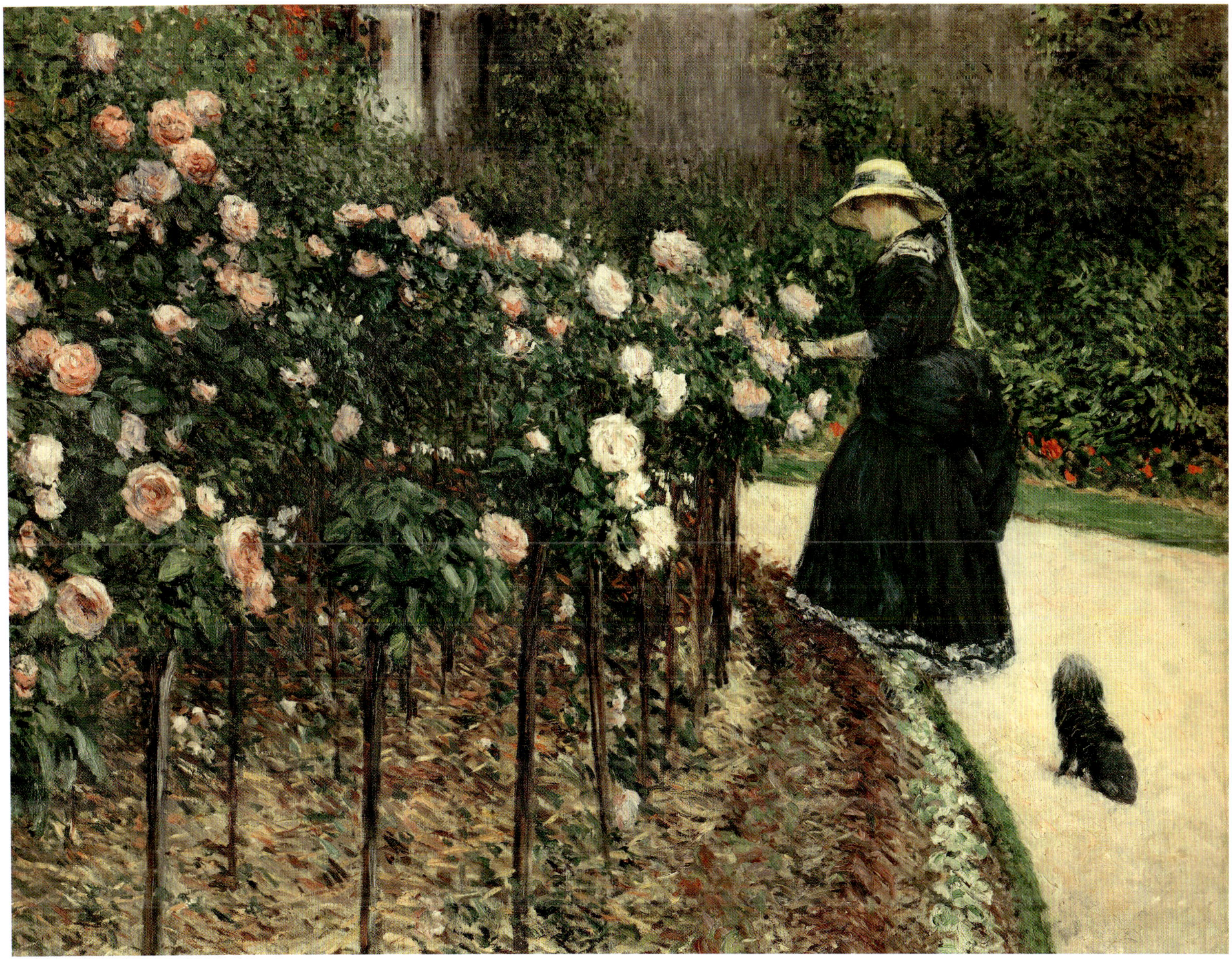

GUSTAVE CAILLEBOTTE

Roses in the Garden at Petit Gennevilliers, c. 1886
Oil on canvas, 35 × 45⅝ in. / 89 × 116 cm
Private collection

In this tranquil summer scene, a woman steps from a gravel path to tend a row of pink standard roses. This is a view of the garden belonging to French artist Gustave Caillebotte (1848–1894), and the unnamed woman is probably his companion Charlotte Berthier. A leading figure in the Impressionist movement, Caillebotte was known best for urban scenes and interiors, until an inheritance from his parents in 1874 left him wealthy, allowing him to buy land in 1881 in Petit Gennevilliers, a suburb of Paris on the banks of the Seine River. Like his close friend and fellow artist Claude Monet (see p.169), with whom he shared planting advice in the letters they exchanged, Caillebotte soon became an avid gardener. Once he was firmly established in Petit Gennevilliers, he largely gave up his Parisian life in favor of rural pastimes. Over the next decade, he expanded the estate and created an extensive garden that was lushly planted with such flowers as irises, chrysanthemums, and roses, as well as vegetables. His garden became a work of art in its own right, and his enthusiasm for horticulture led him to build a large greenhouse for exotics and even to install automatic sprinklers. This passion for gardening was reflected in his paintings, and from the 1880s until his early death he focused almost exclusively on compositions in which the garden's flowers are the chief subject. As seen here, these floral canvases were typically painted with soft, loose brushstrokes, much different from the flatter, smoother style that characterized his earlier work.

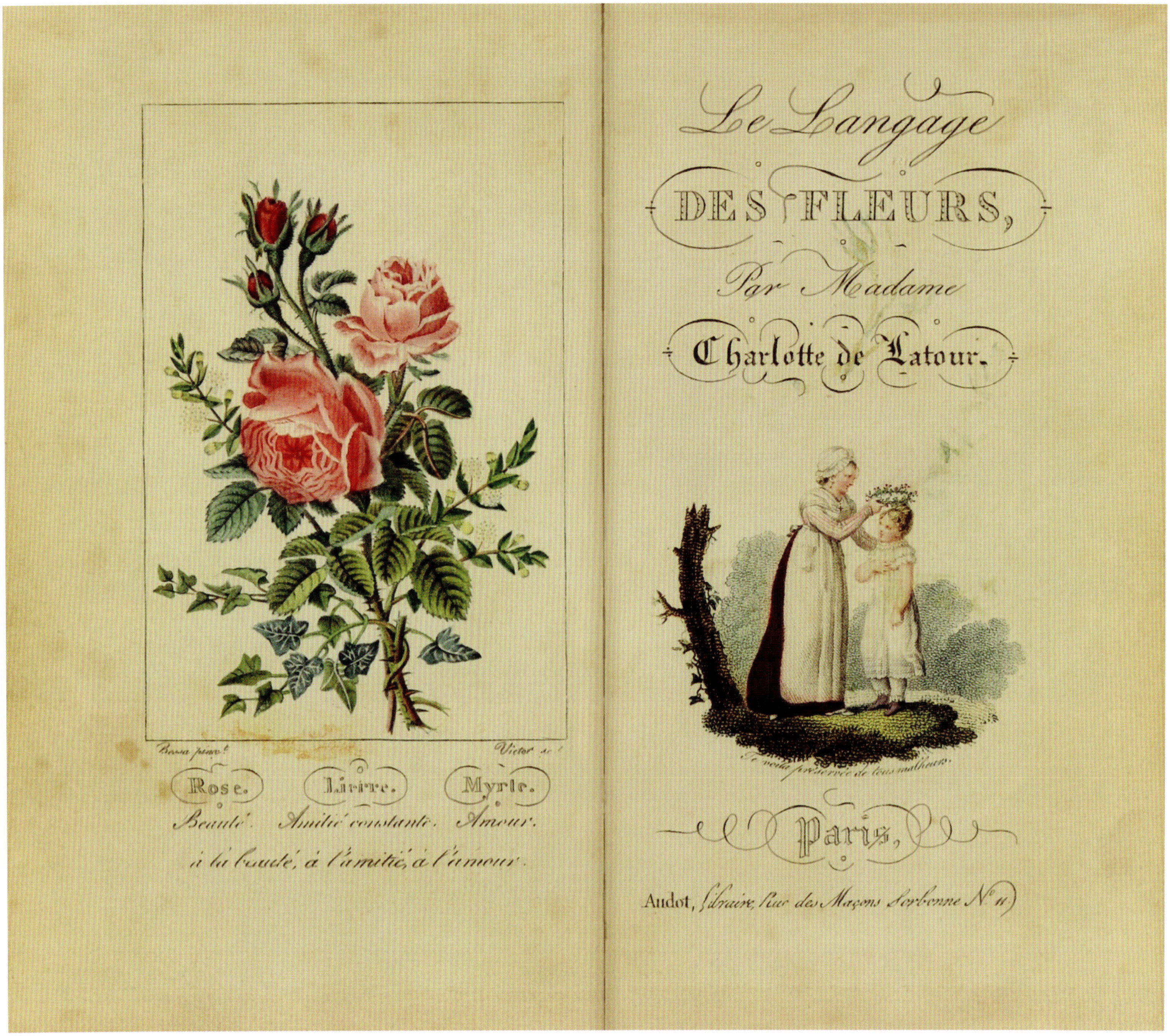

CHARLOTTE DE LA TOUR

Le Langage des Fleurs, 1819
Printed book, 5½ × 3½ in. / 13.5 x 9 cm
Cornell University Library, Ithaca, New York

When words fail, use flowers instead. This is the concept of floriography, which became popular in Victorian England, although the practice of assigning meanings to flowers goes back centuries. Today, flowers are used to convey sentiments and symbolize meaning, with different varieties carefully selected for the occasion, from professing romantic love to sending condolences. As a means of communication, flowers can signify peace, declare war, and send secret coded messages. First published in 1819, *Le Langage des Fleurs* (*The Language of Flowers*) was one of the first popular publications to explore the poetical floral vocabulary, inspiring numerous subsequent flower dictionaries. Coinciding with the popularity of French literature in England, this edition was the work of French author Charlotte de La Tour, the nom de plume of Louise Cortambert. The book is organized by season and includes hundreds of emotions subtly expressed through flowers, describing the plants' history and variations across cultures. Roses are well represented in their own chapter, with different types and colors symbolizing an array of feelings. The plates are based on drawings by French botanical artist Pancrace Bessa (see p.70), and the frontispiece shows a blush-pink double-flowered rose, alongside myrtle and ivy, representing beauty, love, and friendship in the colorful language of flowers. For readers not yet fluent with this lexicon, *Le Langage des Fleurs* includes a long alphabetical list of flower species and their corresponding meanings, and vice versa.

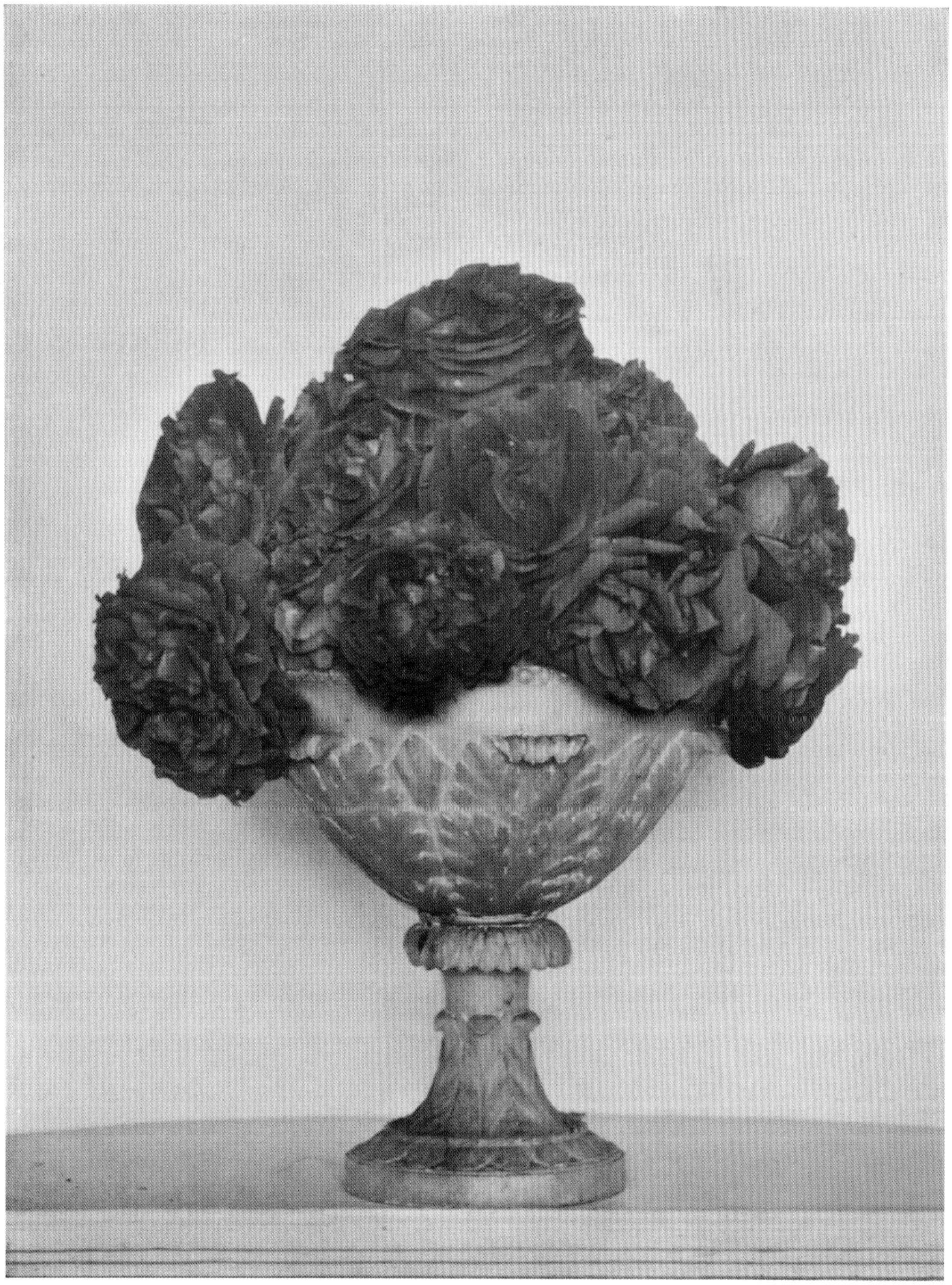

CONSTANCE SPRY

Alabaster Chalice with Red Roses, from *Flower Decoration*, 1934
Floral arrangement in printed book, page 9⅝ × 7¼ in. / 24.5 × 18.5 cm
Photograph by Paul Laib
Private collection

A generous abundance of velvety red roses has been piled into an exquisite chalice in this floral arrangement from the 1930s. The domed design, using a single unnamed rose variety, was created by pioneering English floral decorator Constance Spry (1886–1960) and shot by fine-art photographer Paul Laib for her first groundbreaking book, *Floral Decoration*. Spry was a gardener with a botanist's eye, totally unswayed by fashions, and she grew and chose the materials for her arrangements with the innate instinct of an artist selecting paints. She was especially passionate about roses and devoted to collecting rare and antique varieties, even during the gaudy decades of Hybrid Tea rose breeding in the early twentieth century, when Old Garden roses went out of style. She was said to have one of the best collections of heritage roses in Europe and saved several varieties from extinction. In *Floral Decoration*, Spry references this rose arrangement several times and suggests that red flowers are often best arranged by themselves, with no leaf showing that could dilute their vibrant color. She brings the black-and-white image of this design to life through her description of the roses being cut in several stages of development, some even showing the purple edging of age, to which red roses are prone but which adds immeasurably to their appeal in Spry's eyes. The choice of container was as important to her as the flowers; this alabaster urn is an iconic Spry find, which, at that time, was both unfashionable and affordable.

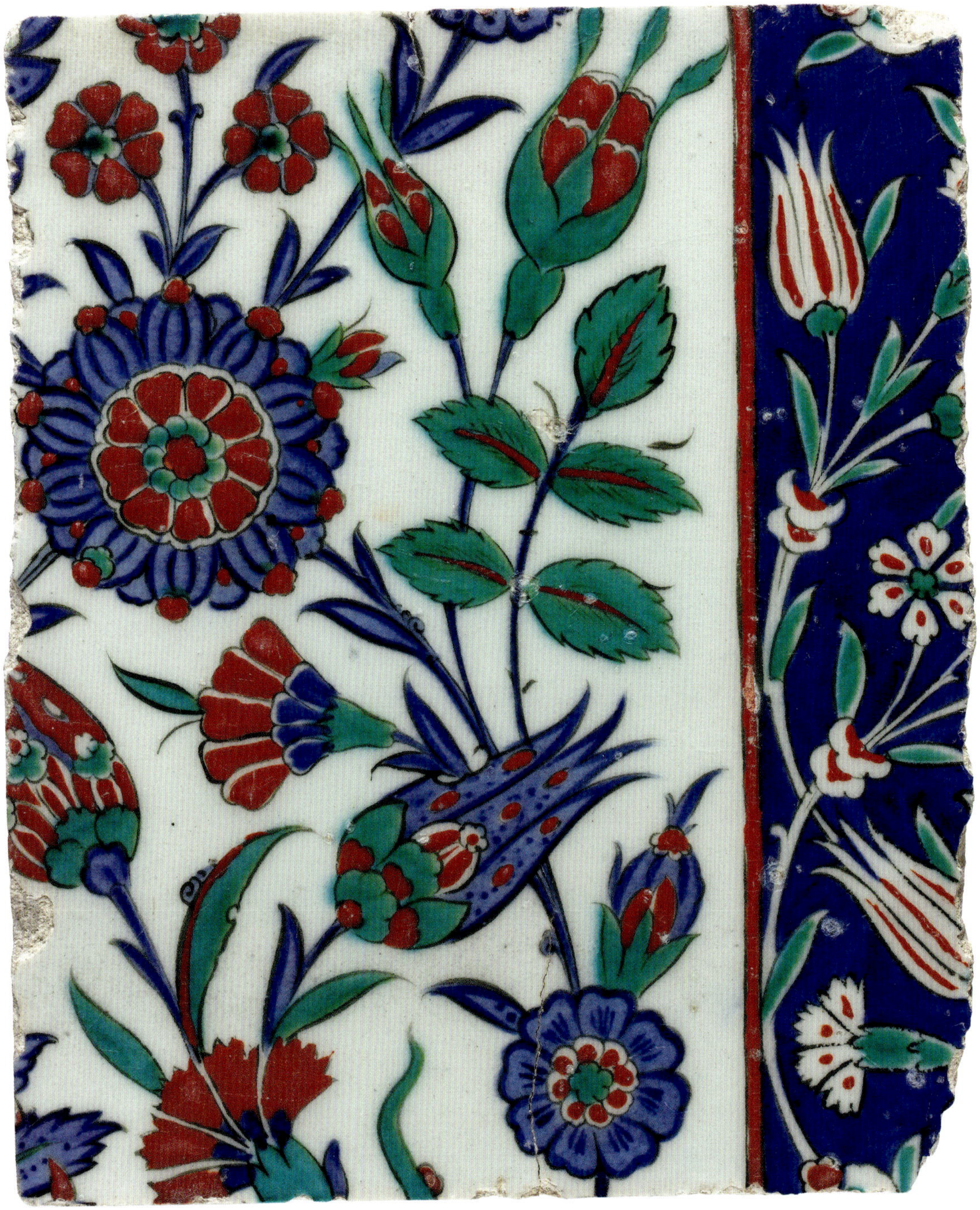

ANONYMOUS

Tile with a garden of roses, tulips, rosettes, and carnations, c. 1560–80
Ceramic, 11 × 9 in. / 27.8 × 23 cm
Musée du Louvre, Paris

The finest Turkish tiles were produced at Iznik, south of Istanbul, and the first reference to blue-and-white ware, made in emulation of Ming porcelain from China, survives in the kitchen accounts of Sultan Mehmet II from about 1470. First, the use of cobalt blue was mastered, followed by a turquoise glaze about 1535, during the reign of Süleyman the Magnificent. Soon afterward came green, and in about 1560, the rich and highly prized red was created from a bright scarlet clay called Armenian bole. This was the complete repertoire of tile colors, but the red glaze was difficult to use, causing such wastage of defective pieces that the Iznik potteries went into decline. However, during the heyday of production, these glorious tiles decorated the most stunning buildings of sixteenth-century Istanbul, including the great mosques Süleymaniye, the Blue Mosque, and Rüstem Pasha, and also Topkapı Palace, where corridors and courtyards offer a feast of tiles in an atmosphere charged with mystery. The flowers and leaves that decorate them have a stylized charm, recognizable without being realistic. The flattened roses resemble the *gol* (Persian for "rose") patterns on Oriental carpets; the tulips have pointed petals and contrasting streaks of color; and the carnations present a dainty serrated profile. Individually, the tiles are exquisite, but when they cover walls, facades, *mihrabs* (niches in the walls of mosques that point toward Mecca), columns, and fountains in lavish profusion, even when the flowery details are lost, the effect is amazing.

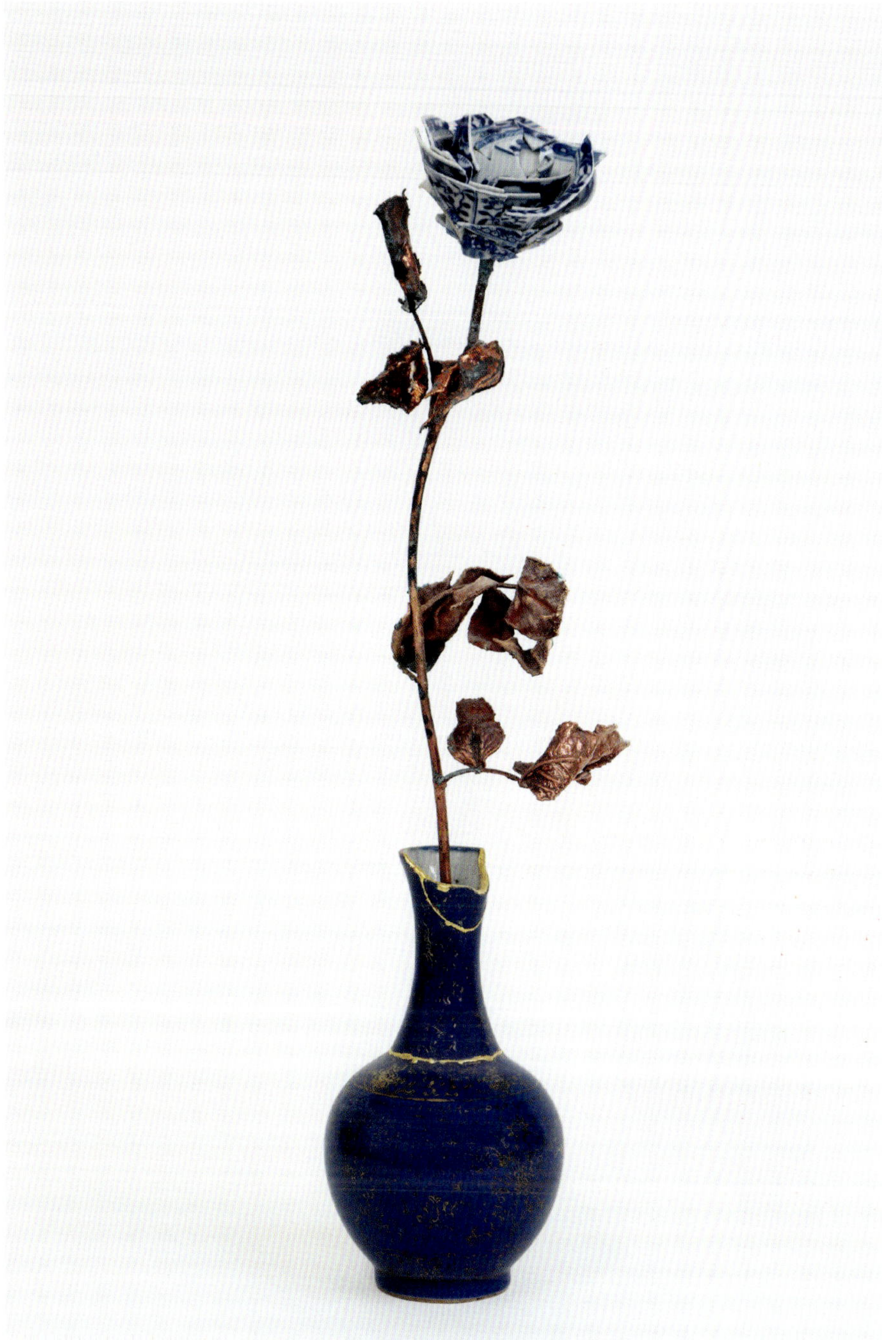

BOUKE DE VRIES

Rose in Kang Xi Vase, 2019
18th-century Chinese porcelain fragments and electro-formed rose stem, 22½ × 6¾ × 5½ in. / 57 × 17 × 14 cm

A tall, unnaturally blue rose stands proudly in a vase, but a closer look adds new meaning to both the cut flower and its decorative container. The head of this rose is actually constructed from pieces of Kangxi porcelain, a famed blue-and-white porcelain pattern dating back to China's Qing dynasty and the Emperor Kangxi, who ruled from 1661 to 1722 and was noted for his promotion of porcelain making. These shards have been glued together to produce the recognizable shape of a rose bloom, but the stalk and leaves are made from copper. The vase is also clearly damaged. This sculpture is the work of Dutch-born, London-based ceramic restorer and artist Bouke de Vries (b. 1960), whose mission is to reclaim works of art that have been condemned as valueless, usually as a consequence of being damaged. De Vries's practice revels in what he calls "the beauty of destruction." Heavily influenced by his forebears, the Dutch flower painters of the seventeenth century, whose paintings both celebrated beauty and implied the inevitability of decay (see pp.10 and 140), De Vries found that his work as an antiques restorer led him to question what gives art value and meaning. Although broken, these pieces of Kangxi porcelain have value, dating as they do back to a golden age of Chinese ceramics. De Vries thus challenges the accepted belief that once a ceramic sustains a crack it becomes worthless, and the sculpture he has assembled from the discarded pieces becomes an object of floral beauty that will take on a new value of its own, while never decaying.

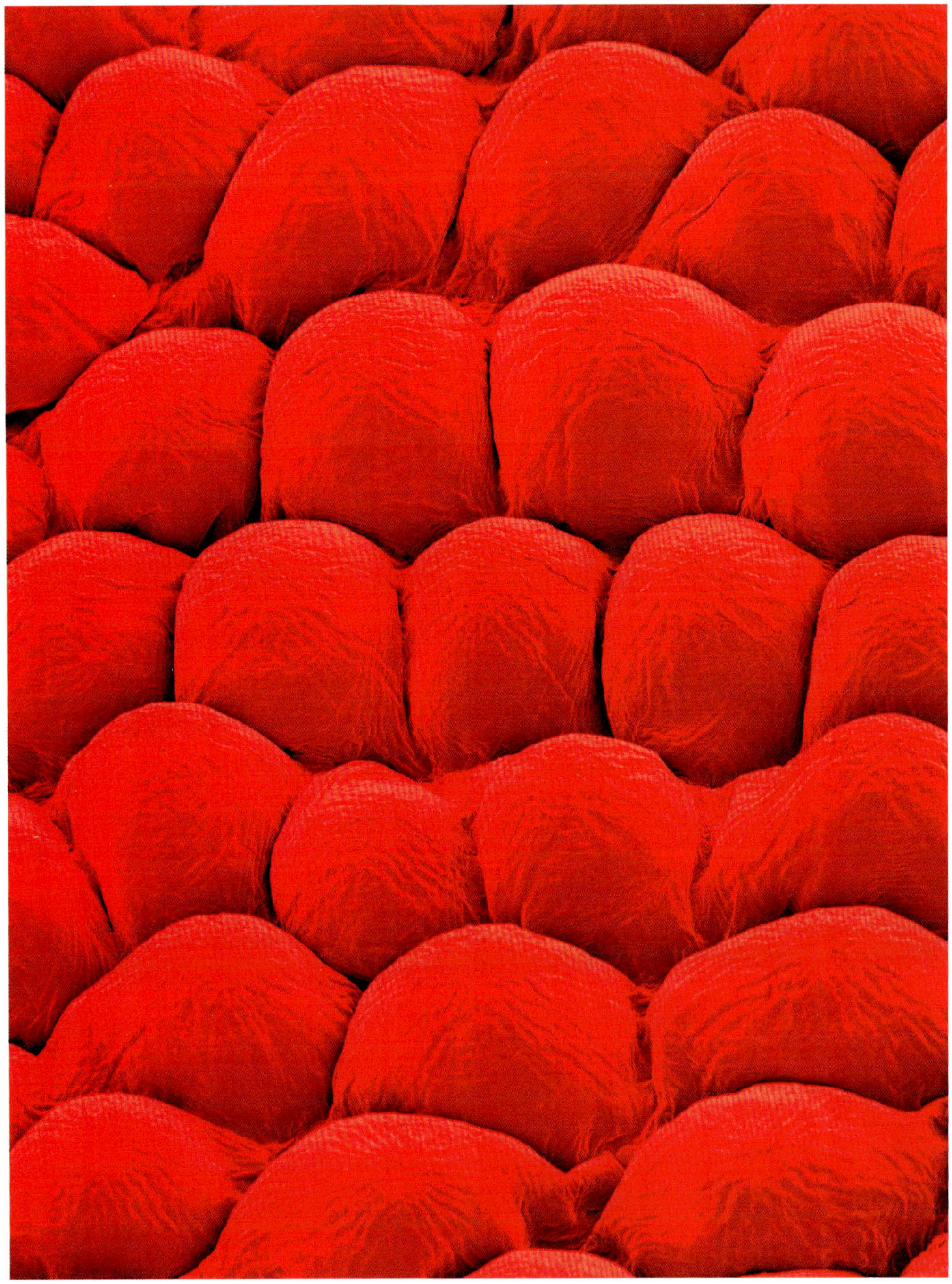

DENNIS KUNKEL

Rose Petal Upper Surface, c. 2010s
Micrograph, dimensions variable

This mosaic of tessellated crimson lumps could be anything, but these forms are actually papillae from the surface of a rose petal. Invisible to the naked eye, the papillae can be seen thanks to scanning electron microscope (SEM) technology, which magnifies the image up to 2,500 times. American scientist Dr. Dennis Kunkel has pioneered SEM technology since the 1970s, applying the skills he originally learned as a microscopist to using SEM to reveal the hidden details of the world. The technology scans the surface of the rose petal with a concentrated beam of electrons that are reflected back to the microscope to create a black-and-white image, which is then carefully colored to achieve the final effect. The papillae are conical projections from the rose's epidermal cells; their surface is often covered with a thin layer of wax to stop the rose from losing too much water through evaporation. It is the tightly packed papillae that make a rose's petal so soft to the touch. SEM technology has helped to revolutionize the microscopic world, allowing for a better understanding of botany and a deeper appreciation of nature. However, such insights take time: to produce this image, the rose petal had to be carefully dried—to avoid it being damaged while it was scanned in a vacuum—then covered with a layer of gold to conduct the electrons. It is difficult, meticulous work—but the results are often stunning.

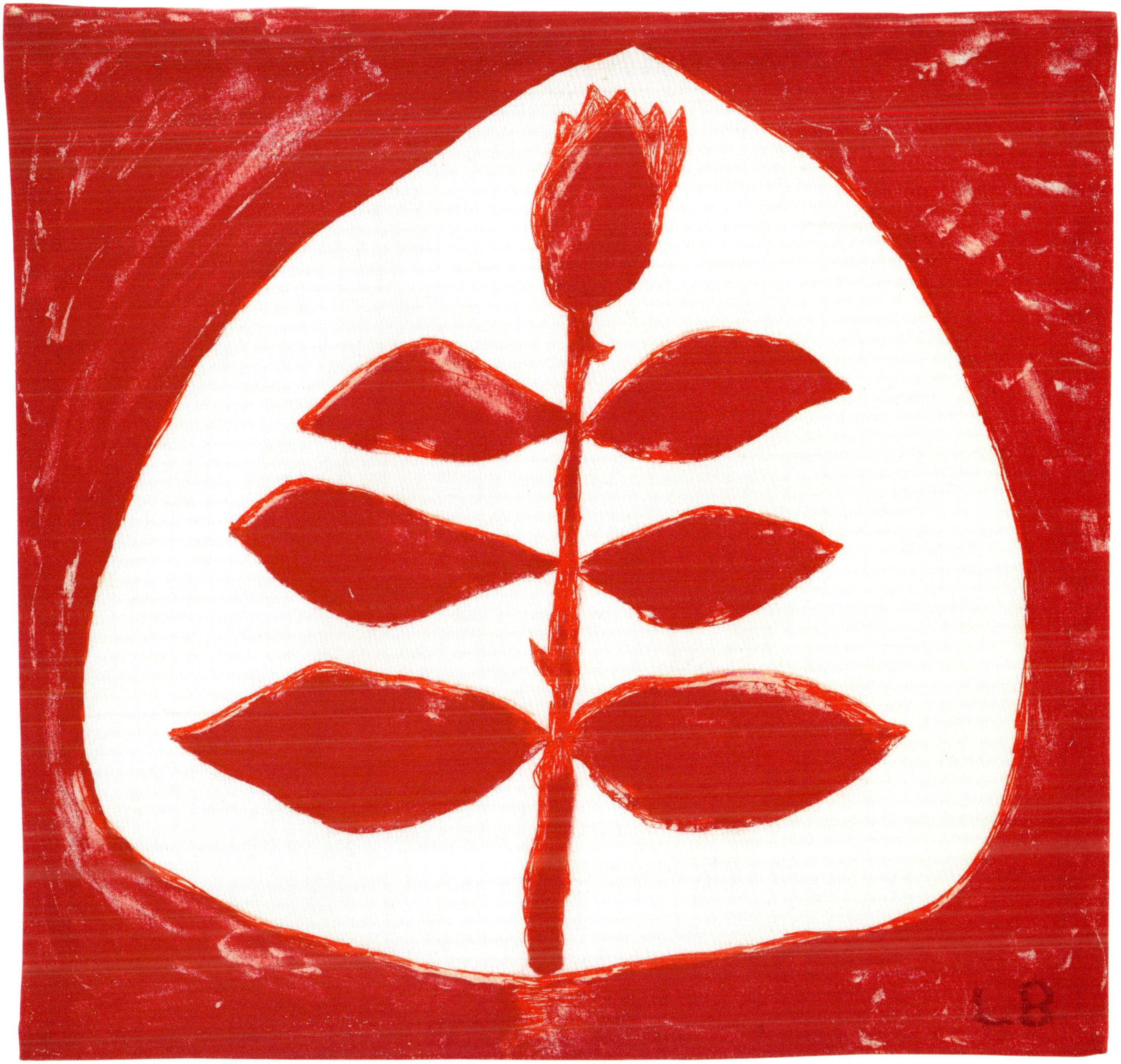

LOUISE BOURGEOIS

Rose, 2002
Lithograph, 12¼ × 12¾ in. / 31 × 32.3 cm
Museum of Modern Art, New York

One of the most innovative American artists of the contemporary period, Louise Bourgeois (1911–2010) is perhaps best known for her monumental abstract and often biomorphic sculptures. Among her most recognizable pieces is a series of large bronze and steel spiders, although she worked in a variety of media and subjects. Born in Paris, Bourgeois studied at the Sorbonne and École des Beaux-Arts before moving to New York. Although she is less well known for her works on paper, her early work often employed printmaking techniques, which continued to be an important part of her creative process and to which she returned later in her career. Her work explored themes of trauma, death, childhood, motherhood, memory, nature, and the human body, some of which can be found in this deceptively simple yet visually powerful lithograph of a single rose stem. The impression, from an edition of twenty, is starkly graphic and intensely red, an important color to the artist that she used again and again. The rosebud is just beginning to open; it is enclosed in a white halo, and the three pairs of opposing leaves give the whole the appearance of a leaf, with the rose itself serving as veins. (Here, the artist defies botanical accuracy; in nature, rose leaves are alternate.) Bourgeois used three shades of red to make this striking image of a rose in its classic color. In the lower right, nearly obscured by ink, her initials are embroidered in red thread, alluding to her childhood experiences of weaving for her family's tapestry business.

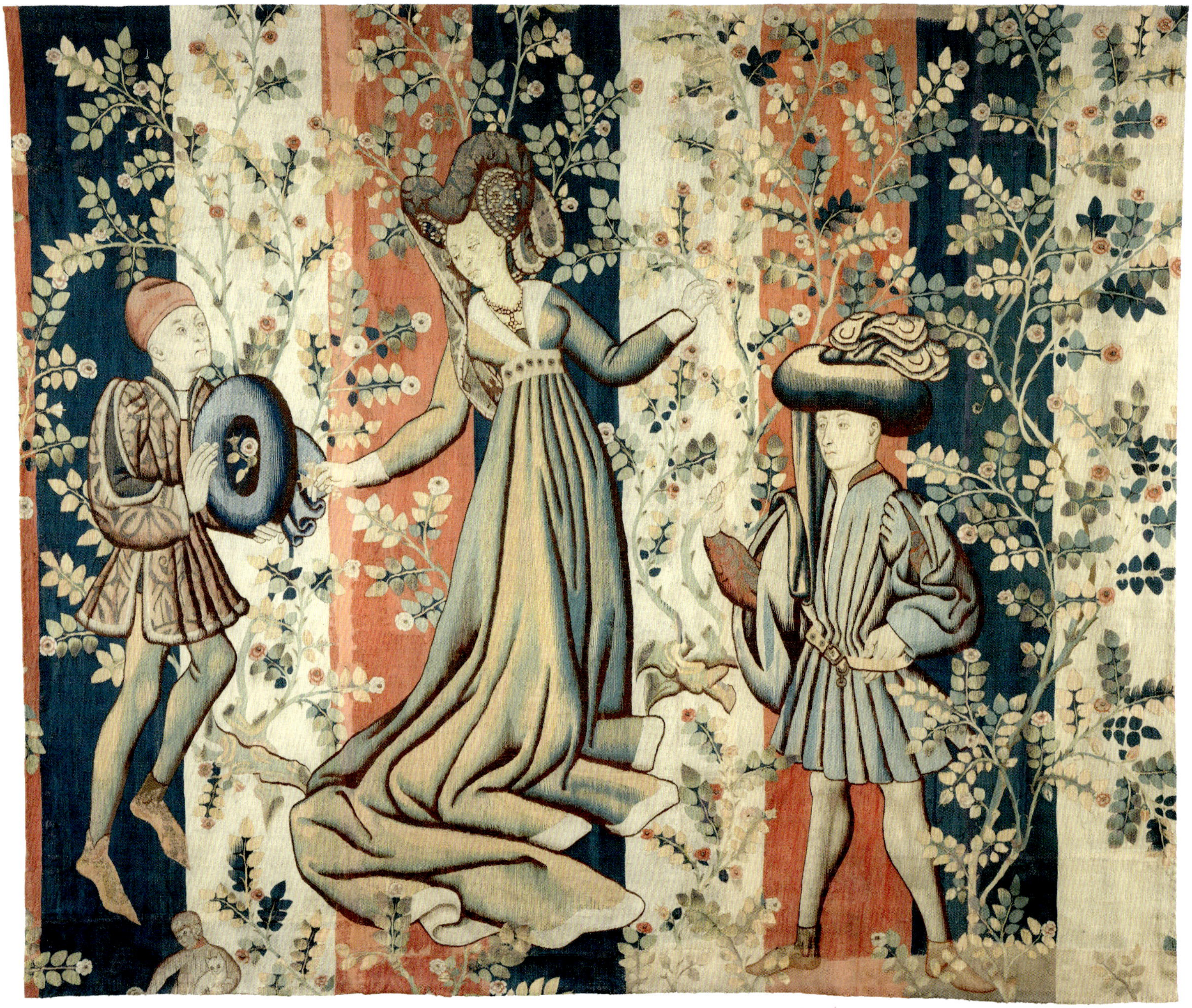

ANONYMOUS

Courtiers in a Rose Garden: A Lady and Two Gentlemen, c. 1440–50
Wool warp; wool, silk, and metallic weft yarns, 9 ft. 5¾ in. × 10 ft. 8 in. / 2.9 × 3.3 m
Metropolitan Museum of Art, New York

Against a flat background of white, red, and blue stripes, two male courtiers wearing richly embroidered pleated tunics surround a female figure in an elaborately jeweled and veiled headdress, her skirts billowing with the heft of a mountain range, filling the lower half of this tapestry by an unknown artist. She turns her body in pleasing *contrapposto,* theatrically posing the dilemma of her choice between the suitors. On the right, she grips with her left hand one of the many thorny rose branches that surround the trio, perhaps hinting symbolically at the risks of sexual attraction in the medieval imagination. But it is the figure on the left, revealing a rose concealed within his hat, who seems closest to capturing the lady's affection. Flemish and Netherlandish tapestries from this period were prized throughout Europe, and this expensively made example, rich with metallic threads that glisten to accentuate the jewels and winding roses as equivalent charms, was possibly made for the king of France, Charles VII. The tapestry's subject matter may also have been inspired by the Old French poem *The Romance of the Rose* (*Le Roman de la Rose*; see p.206), one of the most widely reproduced secular poems of the Middle Ages. The chivalric romance of a pilgrim courting a lady within a *hortus conclusus,* or walled garden, and the plucking of a rose—a traditional symbol of love and the Virgin Mary—is echoed closely here.

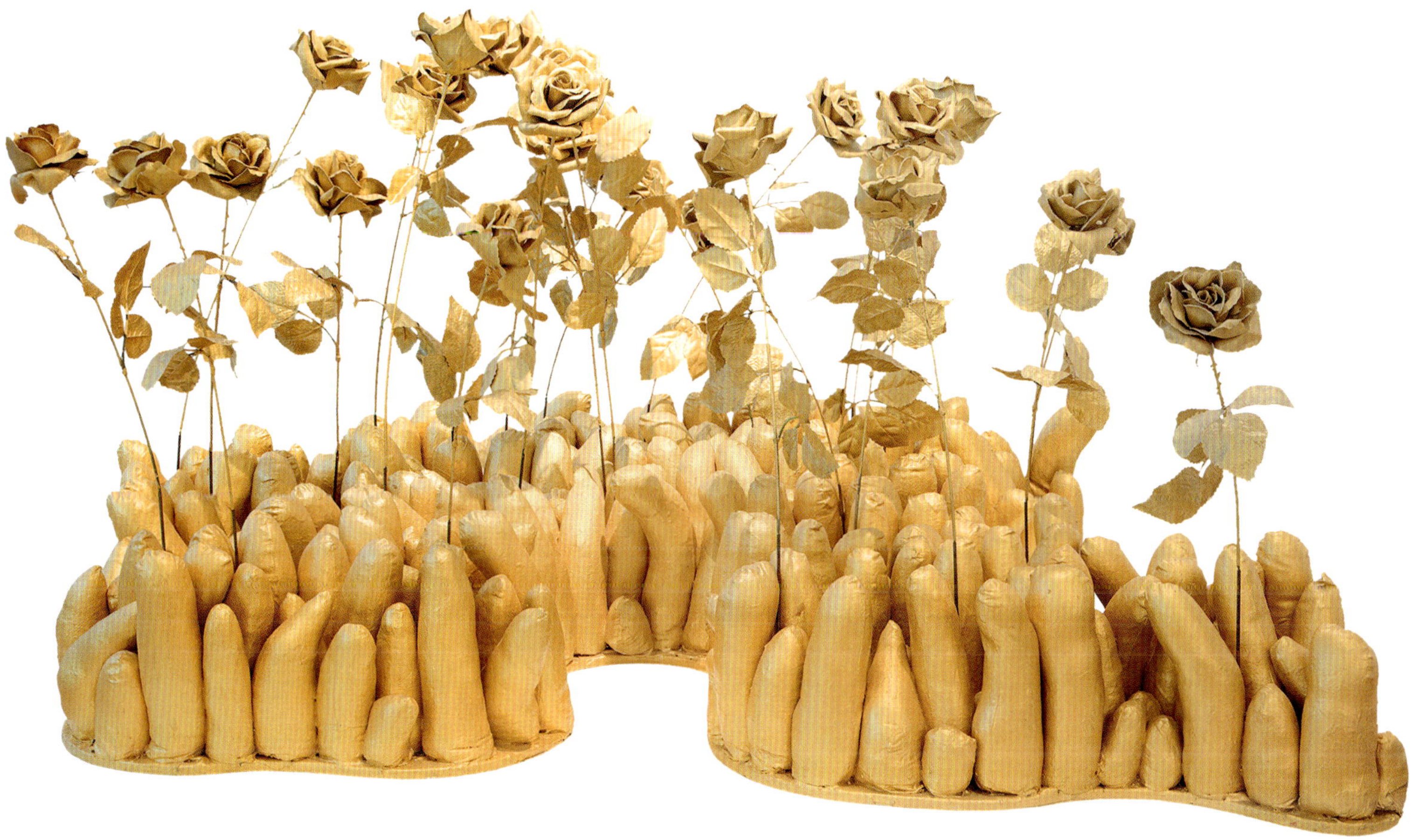

YAYOI KUSAMA

Rose Garden, 1998
Sewn and stuffed fabric, metallic paint, and artificial roses, 27 × 49 × 51 in. / 68.5 × 124.4 × 129.5 cm
Private collection

It is no surprise that floral imagery recurs throughout the work of artist Yayoi Kusama (b. 1929), given that she grew up immersed in her family's plant nursery in Matsumoto, Japan. Kusama's early work often featured more realistic depictions of botanical forms, while her later work became more conceptual and somewhat abstract, although flowers, plants, and gardens remain a central theme. *Rose Garden* is among numerous mixed-media soft sculptures that Kusama started to create in the 1960s. Composed of stuffed, sewn fabric, it shows a compact garden, as if floating on a small, self-contained island. Two dozen single stems of artificial roses arise from a cloud of pillowy protrusions. Amid a juxtaposition of materials and forms, these biomorphic shapes are both confounding and intriguing. Kusama called such repetitions of organic forms "obliterations," and they were common in her work. However, while these fingerlike protuberances nearly swallow up the roses, the flowers, far from obliterated, bloom atop towering stems, defeating the soft, unarmed forms with their armor of thorns. Gilded with gold paint, *Rose Garden* recalls an *ikebana* arrangement, with the protrusions fixing the flowers in place like a *kenzan*, the device used in this traditional Japanese art. For Kusama, flowers represent transformation, and these fragile but strong roses seem to suggest the possibility of beauty arising from unpleasant situations. The sculpture embodies both the avant-garde spirit of her early work and its often cosmic nature.

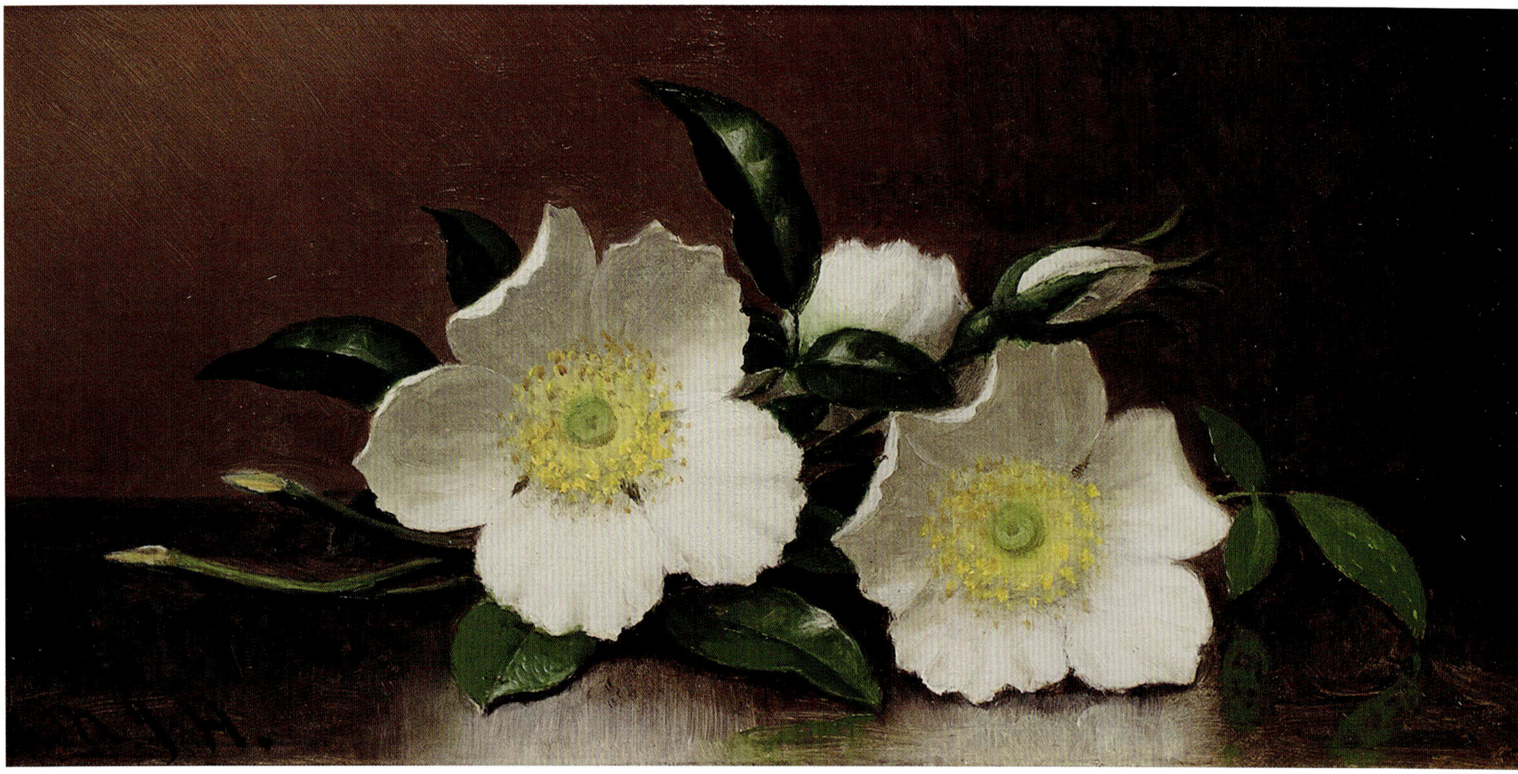

MARTIN JOHNSON HEADE

Cherokee Roses, c. 1880s
Oil on canvas, 10⅛ × 18⅛ in. / 25.7 × 46 cm
Private collection

The original wild roses are native only to regions north of the equator, and while some can be found in North America, those usually have small pink flowers and an upright habit, different from the white ones seen here. The Cherokee rose (*Rosa laevigata*) is, in fact, indigenous to the subtropical regions of southeastern China, where it will scramble up into trees (see pp.127 and 217). It is not certain when it was introduced into the southern United States, but it probably came with early settlers in the seventeenth century. The plant integrated into its new environment so well that early botanists assumed it was a native species—erroneously attributing it as a symbol of the Cherokee people, whose ancestral lands comprised the Southeastern Woodlands of the United States—and it was adopted as the floral emblem of the state of Georgia in 1916. Painted here by American artist Martin Johnson Heade (1819–1904), the Cherokee rose is easily recognizable with its large white flowers and glossy, often trifoliate leaves. Both it and the red rose were favored subjects of Heade, who was known for his landscapes and his portraits of tropical and native flora. This quiet still life shows only a few blossoms, but the Cherokee rose is very vigorous. With the aid of its substantial backward-pointing thorns, it can reach a height of 30 feet (10 m) into trees or, if unsupported, make a completely impenetrable "heap," a far cry from Heade's elegant floral grouping.

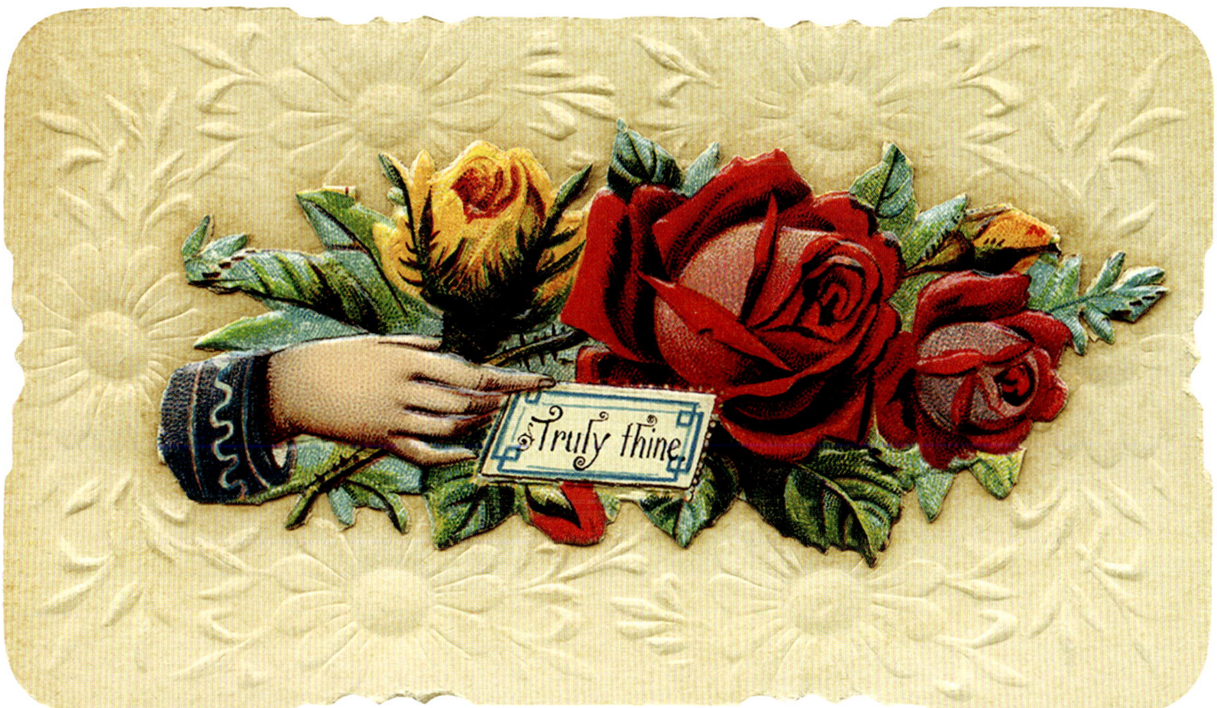

ANONYMOUS

Calling card, late 19th century
Chromolithography and embossing on die-cut card

This charming card shows a woman's hand proffering a bunch of roses and a card with the message "truly thine," set against an embossed pattern of daisies. The card was probably designed with the popular Victorian language of flowers in mind (see p.112). This floriography assigned meanings to flowers and plants and was often used as an elaborate means of communicating feelings or responding to messages. The red rose is traditionally a symbol of love, and the yellow rose sometimes represents friendship, but according to some nineteenth-century sources, red and yellow roses together imply joy, happiness, and excitement. The daisy, meanwhile, was an emblem of innocence, loyalty, and the ability to keep a secret; combined on this card, the implicit message of these flowers is one of affection and trust. The calling cards of the eighteenth and early nineteenth centuries were relatively plain and simple, bearing only a name and perhaps an engraved border. But as printing technology advanced, it became possible to produce elaborate, highly colored designs, such as this one, inexpensively. Just as floral motifs could be used to convey complex messages, the ritual of leaving a calling card was itself part of a strictly regulated system of social etiquette for the bourgeoisie and upper classes in the West. The fundamental rule was that a visitor would not expect to see someone in their home unless that person had first called, left a card with a servant, and received a response to indicate that their visit would be welcome.

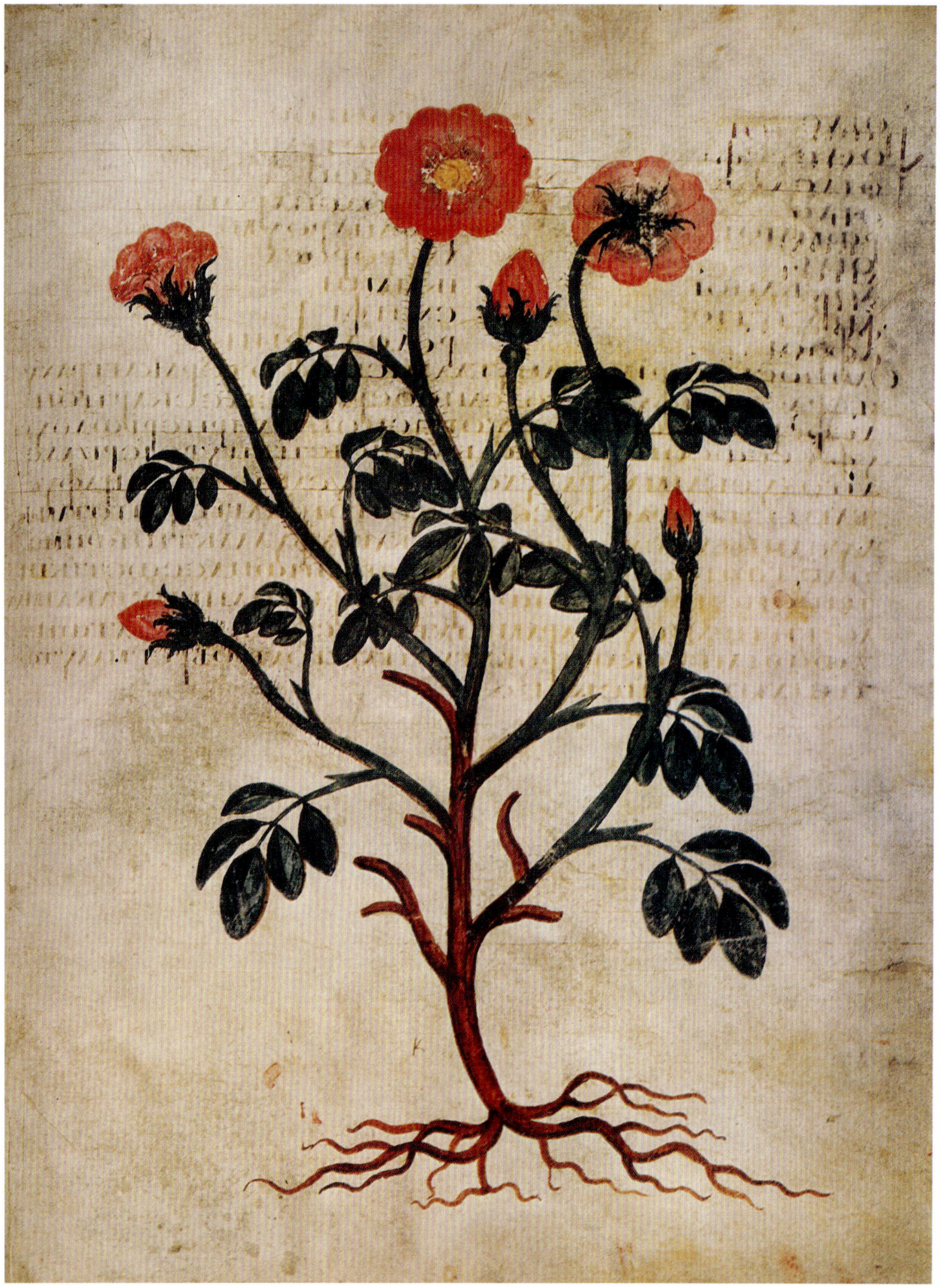

PEDANIUS DIOSCORIDES

Rose, 6th century CE
Parchment, 15 × 12 in. / 37 × 30 cm

Greek physician Pedanius Dioscorides (c. 40–90 CE) completed the Classical herbal *De Materia Medica* (*The Materials of Medicine*) in about 65 CE. Considered one of the most significant treatises on medical botany and pharmacology, it had a long-lasting influence, and while the original unillustrated papyrus scroll no longer survives, numerous manuscript copies were made. The oldest and most famous of these is the Juliana Anicia Codex, made about 512 CE in Constantinople for the imperial princess Anicia Juliana. The elaborately illuminated manuscript contains nearly four hundred paintings of plants and details of thousands of medical uses. Intricate, naturalistic depictions of flora, such as this rose, were atypical of the Byzantine era, suggesting that it may have been drawn from life instead of from a copy. Still, it is difficult to positively identify the species shown, given pre-Linnaean plant-naming conventions and the fact that the original papyrus was not illustrated. Dioscorides's original text calls the red-flowered plant *rhódon*, and later scholars identified it variously as *Rosa* × *centifolia*, *R. foetida*, and *R.* × *damascena*—but it is more likely *R. gallica*, the rose of antiquity (see p.188). Painted on parchment, the double-flowered rose is shown with two open blooms, one partly open, and three in bud, with a hint of thorns on the stem and calyx, the structure that encloses the unopened flower. Commonly referred to as the apothecary rose, it was regarded by the ancient Greeks and Romans as one of the most important medicinal botanicals.

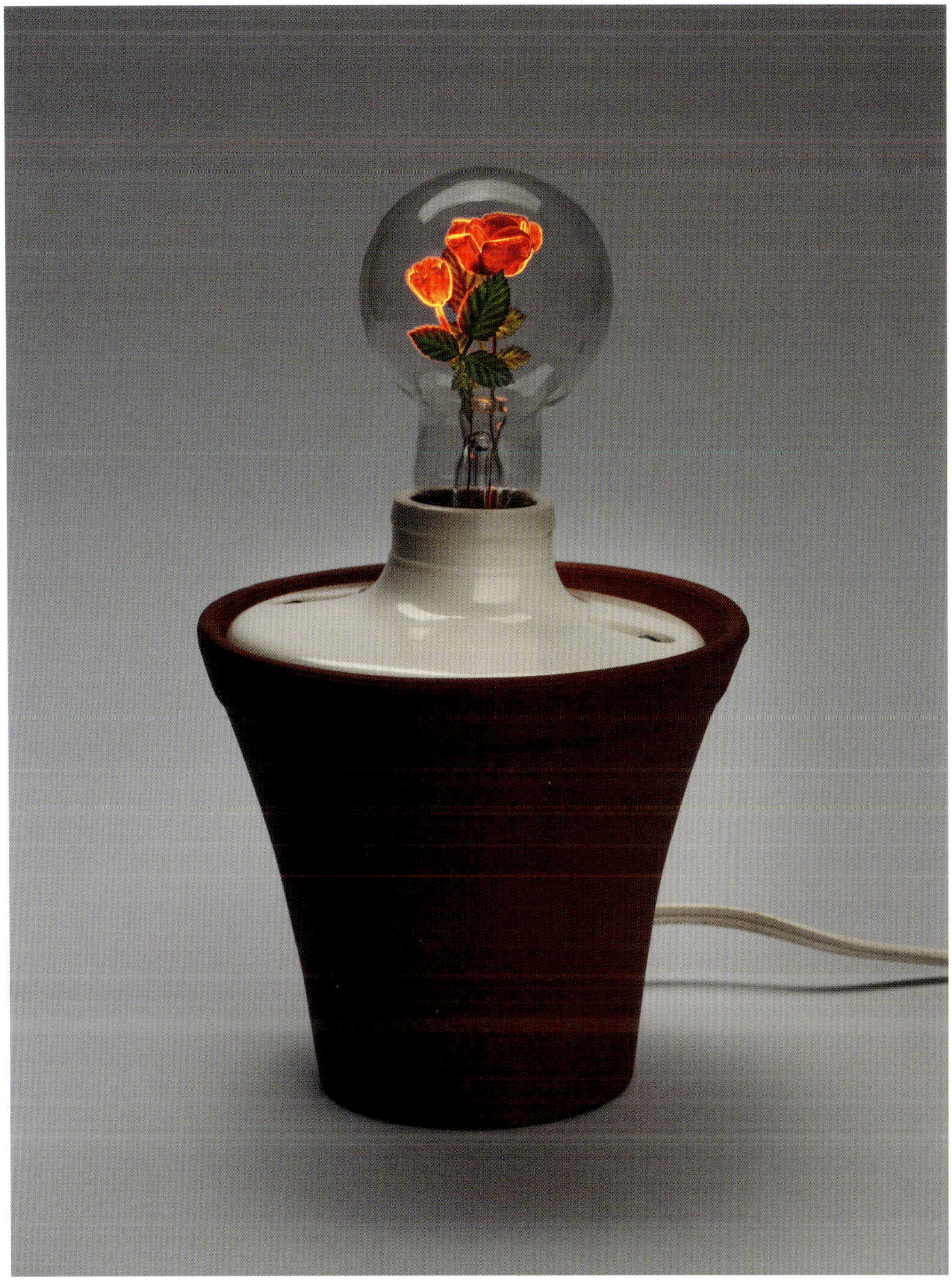

DAN FLAVIN

Roses, 1962–66
Terracotta flowerpot, white plastic light sockets, and Aerolux Flowerlite, 8½ × 5¼ in. / 21.6 × 13.3 cm
Museum of Modern Art, New York

A beautifully crafted rose-shaped light bulb filament emits a gentle red glow from within its glass encasement, which is set in a terracotta flowerpot. Punning on the notion of planting a bulb, this humorous assemblage of commercially available materials was created by American artist Dan Flavin (1933–1996) in an edition of ten. It is one from a series of three related flowerpot works, with titles that are variations on Barbara Rose, the art critic for whom the series is named. The "Flowerlite" light bulb that Flavin used for the edition was produced by the Aerolux Light Corporation, known for its novelty bulbs containing filaments in the shape of birds, animals, religious symbols, and—most popularly—flowers. Manufactured from the 1930s to the 1970s and selling for about twenty cents, these bulbs contained low-pressure gas (neon or argon, or a mixture of the two), which reacted with the phosphor coated filaments, causing them to illuminate in different colors when electrical current was passed through them. Flavin started his career making drawings and paintings, but in the late 1950s he began to create assemblages using detritus collected from the streets near his studio. This rose-themed sculpture, from one of Flavin's early series to incorporate light, connects to his most notable installations made solely from fluorescent lamps and fixtures arranged in different configurations, bathing exhibition spaces in radiant light and color.

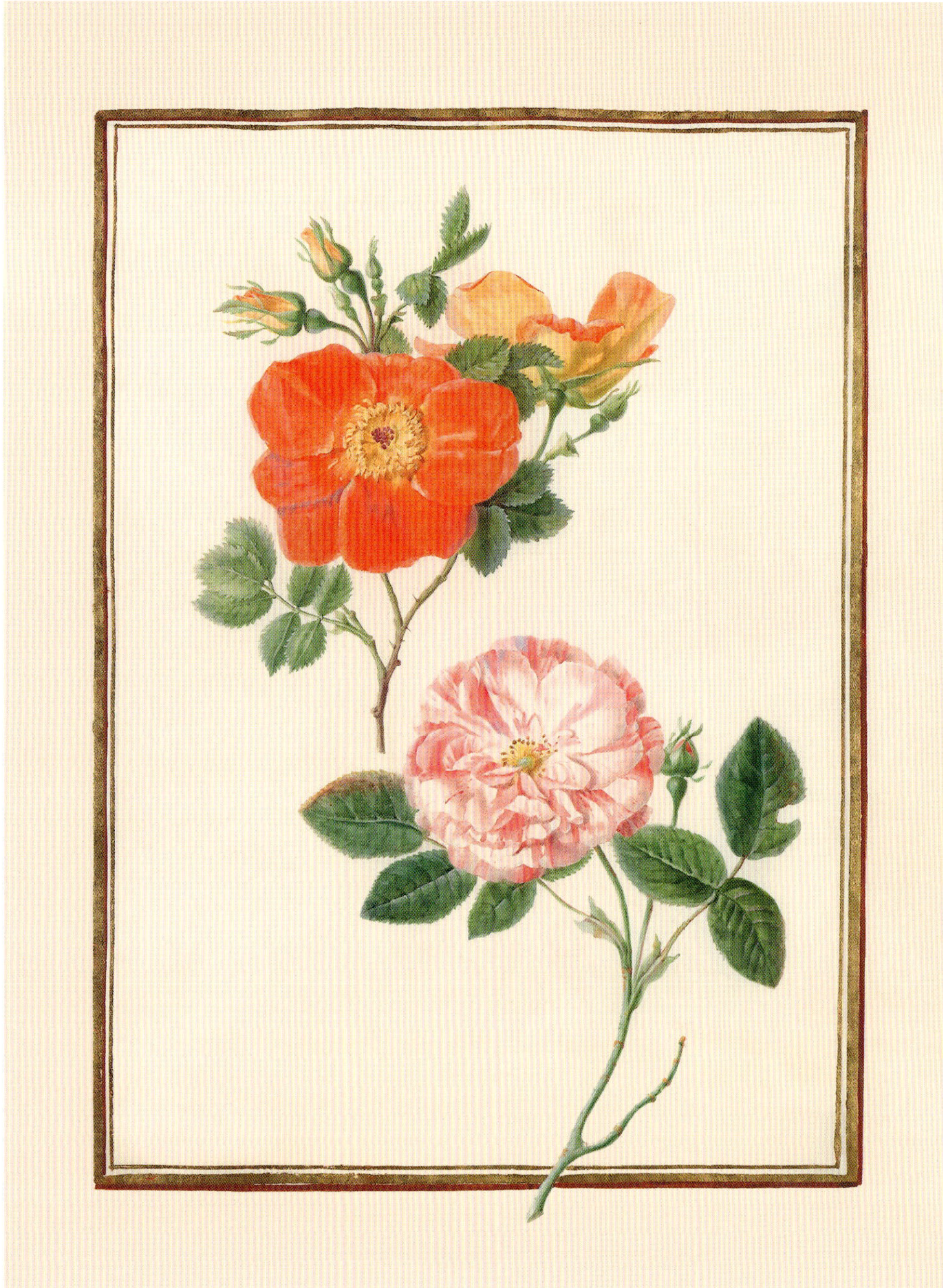

MADELEINE BASSEPORTE AND HER CIRCLE

Double and Single Roses, c. 1750
Watercolor over pencil on vellum, 15¾ × 12¼ in. / 40 × 31.1 cm
Morgan Library & Museum, New York

Two energetic sprays of roses open on a single page, extending ever so slightly over the decorative border into the viewer's space. The regal flowers dominate, with the foliage knowing its place and thorns a mere suggestion. Buds at different stages of maturity assure a succession of blooms. In gardening, art, and life, timing is everything, and French botanical artist Madeleine Françoise Basseporte (1701–1780) had impeccable timing. Selected by Louis XV as the official painter of the King's Garden in 1741, she was the first woman to be so honored by a king of France. She suited the requirements perfectly, having had long training and an apprenticeship with Claude Aubriet, her predecessor. Basseporte also won the favor of the king's official mistress, Madame de Pompadour (see p.36), who may be responsible for commissioning these roses as part of a series of botanical drawings of her gardens at Bellevue. The king further developed the grounds at Versailles, amassed a large botanical collection, and commissioned extensive gardens to surround his new getaway palace, the Petit Trianon. Not only was natural history a passion of the Enlightenment, but it also opened up more artistic spaces for women, who were often barred from academic training. Louis XV kept Basseporte busy. She taught botanical painting, and her pupils included the king's daughters and other up-and-coming women of art and science. Basseporte painted for the king until the end of her long life, when she was one of the highest-ranking botanical artists in all of France.

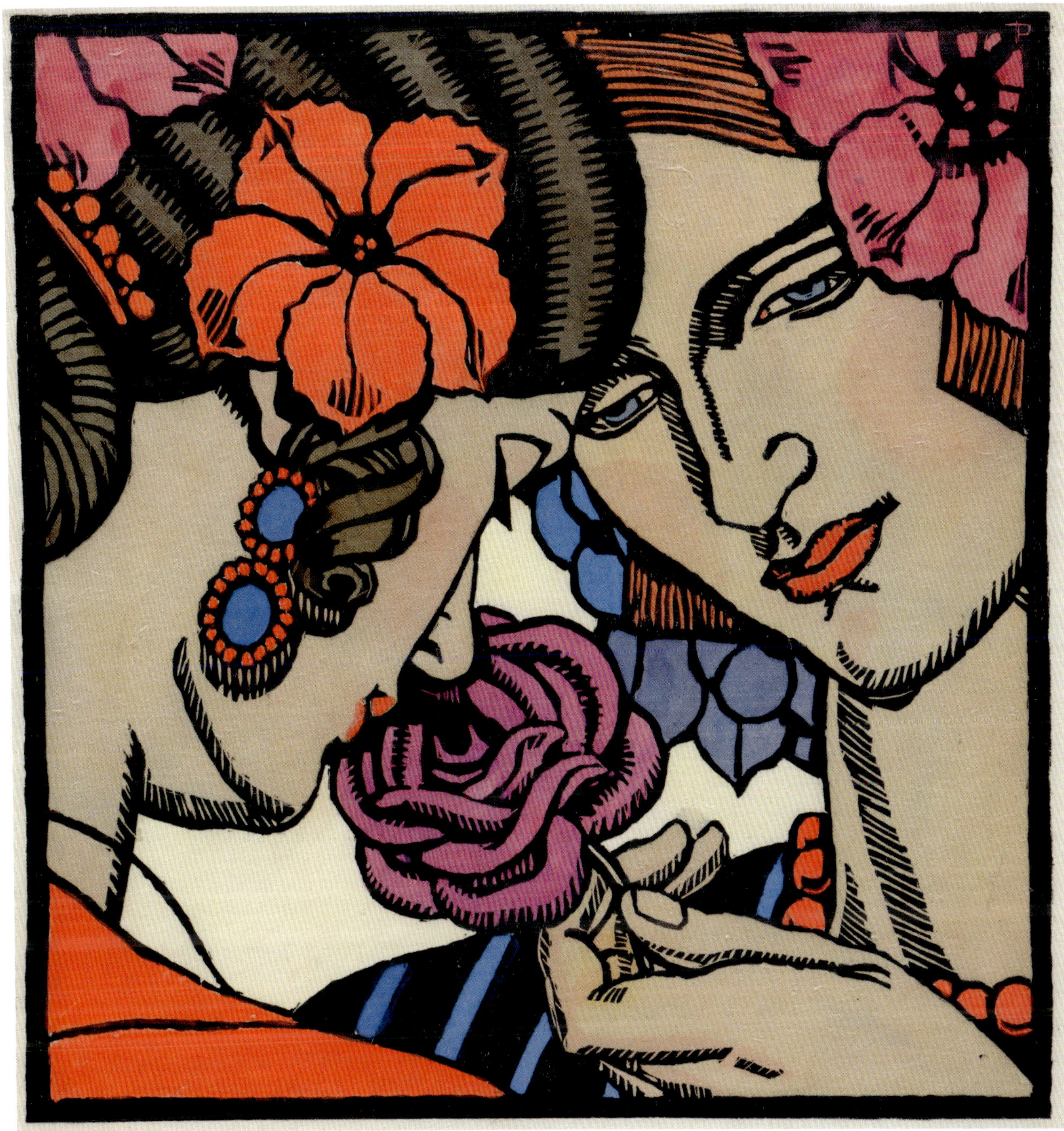

THEA PROCTOR

The Rose, 1927
Woodcut, printed in black ink, hand-colored on thin ivory laid tissue, 8¾ × 8¼ in. / 22.1 × 20.8 cm
Art Gallery of New South Wales, Sydney

A deep pink rose at the height of its fragrant powers is the subject of this woodcut, which celebrates an intimacy between two women focused on the bloom's scent and beauty. Australian artist Thea Proctor (1879–1966) based the work on her original cover art for the July 1927 issue of *The Home* magazine, and this is one of the most famous pieces of cover art in Australian history. Proctor often used as her model her cousin Hera Roberts, who was also an artist and who holds the bloom at right. Both Roberts and Proctor were forward-thinking women who brought a cosmopolitan outlook to Australian magazines that broke down the cycle of male-dominated imagery. They promoted the fashionable and independent femininity of the 1920s through their illustrations and by contributing articles on interior decoration and fashion. As seen in the vibrant color, close-up composition, and graphic approach of this work, Proctor was influenced by traditional Japanese ukiyo-e woodcut prints. She was an expert draftsperson, which, when combined with her color sense and acute eye for composition, gave her works a contemporary yet timeless look. A cofounder of the Contemporary Group—a modernist artist collective established in Sydney in 1926—Proctor promoted printmaking as an exciting and acceptable form of artistic expression. Intimate yet strong, floral but not flouncy, *The Rose* uses the timeless symbol of beauty to make a statement about a new kind of modern woman.

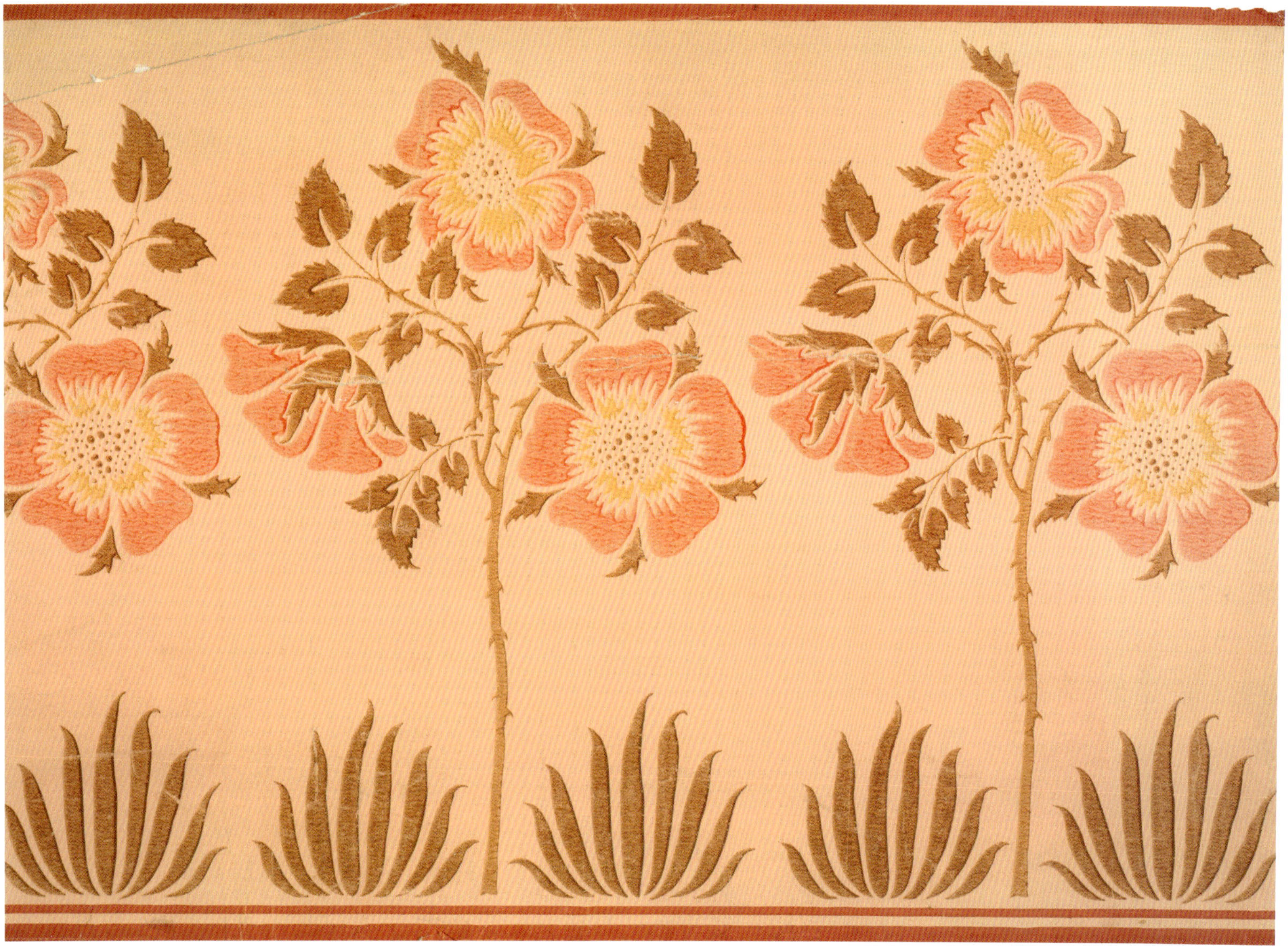

C. F. A. VOYSEY

The Standard Rose, c. 1901
Color woodcut on paper, overall 24¾ × 32⅛ in. / 63 × 81.6 cm
Victoria and Albert Museum, London

This row of roses is a study in pink. The flattened blooms stand open to reveal their yellow centers, their thorny stems emerging between clumps of low-growing abstracted plants against a pink background, with a stronger pink border on the top and bottom. Designed by prolific English architect and designer Charles Francis Annesley Voysey (1857–1941), this woodcut print was intended to edge wallpaper in the home. The flattened flowers are a signature of Voysey's style. Often, he would repeat different patterns, but for this border he simplifies the image by reiterating a single standard rose over and over again. Standard roses are grafted to produce a single stem and resemble, as here, a miniature tree more than a flower. In nineteenth-century Victorian England, wallpaper borders were popular as a means of breaking up a single pattern, and they were often used in hallways and on staircases. Voysey repeatedly featured birds and flowers in his designs, including roses, which were a widespread symbol of love in Victorian Britain. A key member of the influential Arts and Crafts movement, he later became better known for his house and furniture designs. His early career, however, was spent designing wallpaper and textiles, first for Jeffrey & Co., which produced William Morris's designs (see p.32), and—after considerable success—for the company he started himself. Voysey became so successful that, by 1896, *The Studio* magazine hailed him as being to wallpaper "what Wellington was to the boot."

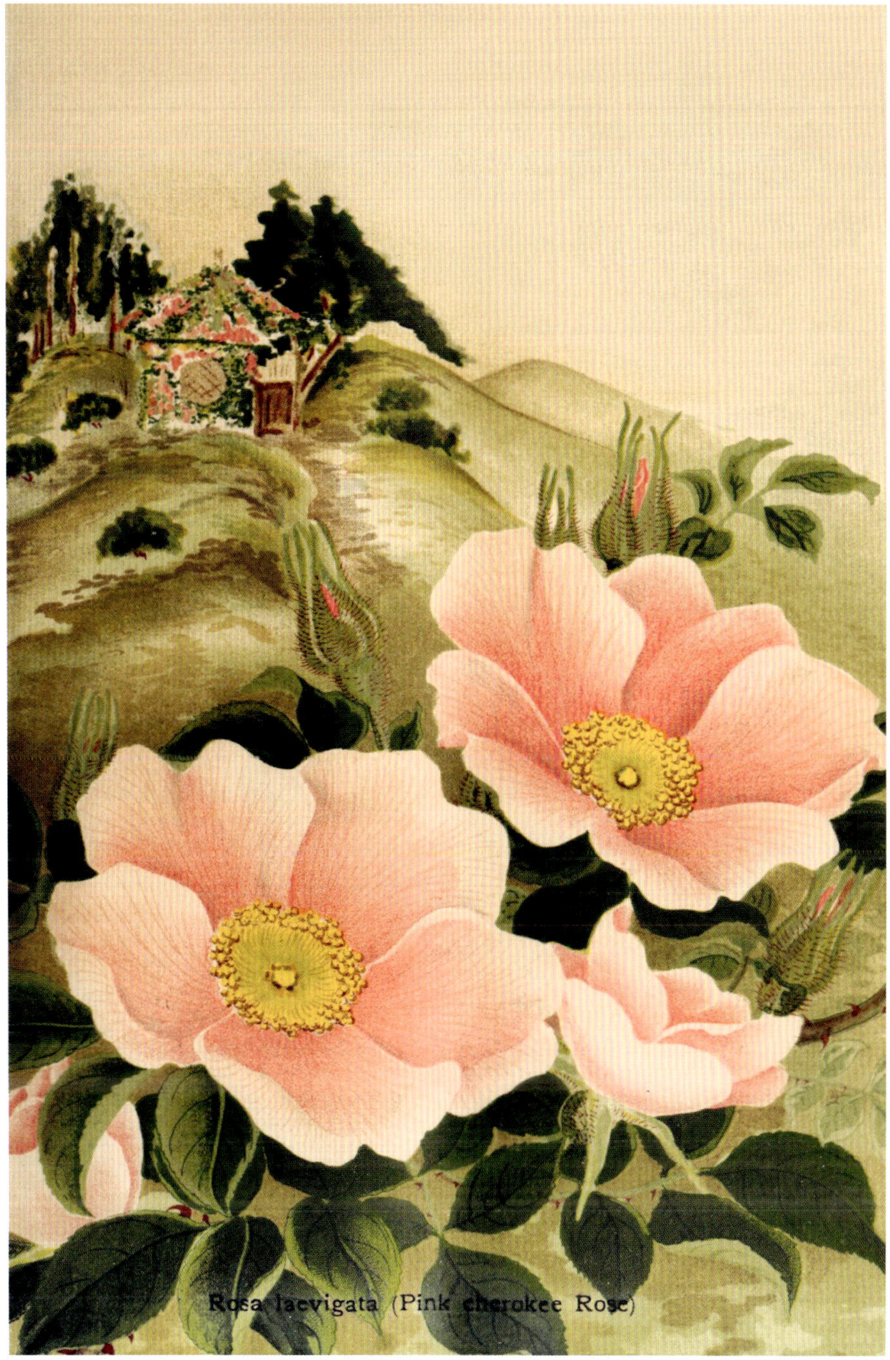

YOKOHAMA NURSERY CO.

Rosa laevigata (Pink Cherokee rose), from Descriptive catalogue of flowering, ornamental trees, shrubs, bulbs, herbs, climbers, fruit trees, 1909
Lithograph, 10½ × 7¾ in. / 27 × 19.6 cm
LuEsther T. Mertz Library, New York Botanical Garden

To appeal to a high-end market, a nursery can present its offerings as works of art. Made from a woodcut, this was one of seven color prints in the Yokohama Nursery Company's hundred-page descriptive catalog from 1909. Pink Cherokee roses bloom in the foreground of a hilly Japanese landscape with a walled flower garden at the summit, an image that would have caught the eye of any sophisticated grower. Instead of the white petals typical of this species (see pp.120 and 217), Yokohama Nursery's selection was bright pink in bud, softening to pale blush as it matured. For this novelty, the buyer would pay a premium. Founded in 1890 by four nursery professionals in the eponymous port city south of Tokyo, the Yokohama Nursery established itself as a leading purveyor of Asian plants to gardeners and botanical institutions across North America, Europe, and the British Empire. By 1909, it had branch offices in San Francisco, New York, and London. The catalog was vast, with trees, shrubs, perennials, bulbs, seeds, and, of course, roses. Where does a Cherokee rose fit among these Asian plants? *Rosa laevigata* is actually native to China and southeastern Asia. By the mid-eighteenth century, it had made its way to England and then to the American South, where it naturalized and acquired its folkloric associations with the Cherokee people. The plant-savvy entrepreneurs of the Yokohama Nursery Company correctly placed it with other roses of Asian origin.

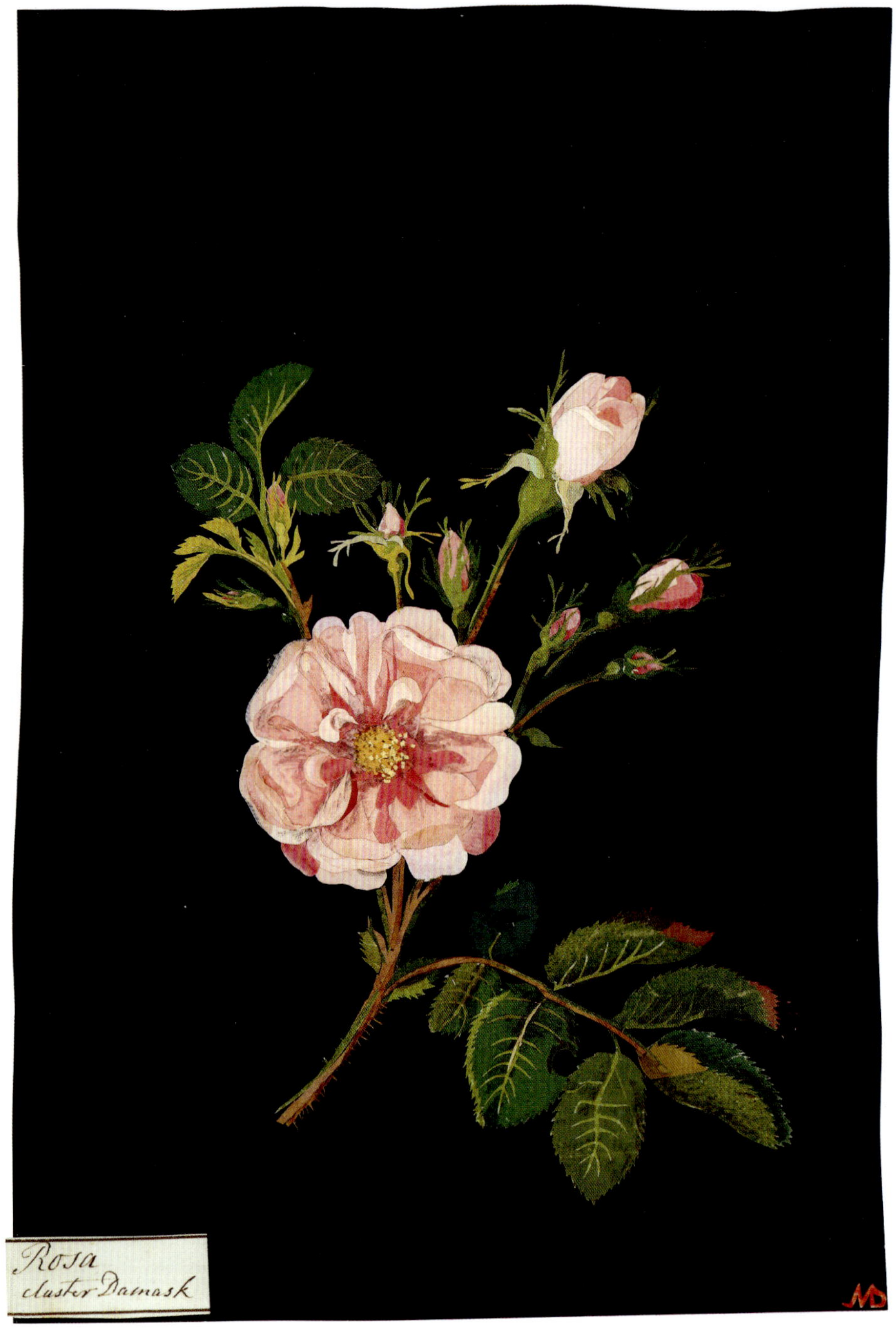

MARY DELANY

Rosa cluster Damask, 1780
Collage of colored papers, with bodycolor and watercolor, on black ink background, 13¼ × 8⅞ in. / 33.5 × 22.6 cm
British Museum, London

This sprig of sweet-scented Damask roses, its many pinks and greens set dramatically against a jet-black background, represents a radical innovation in the representation of flowers. In her early seventies, British artist Mary Delany (1700–1788) innovated a new form of the then fashionable art of decoupage, or paper cutting, to produce uncannily realistic versions of common and exotic blooms. Delany was a "bluestocking": a woman involved in the world of intellectual pursuits, a skilled musician, a writer of a six-volume autobiography, and a correspondent with eminent figures, including Irish satirist Jonathan Swift. In 1771, upon noticing the resemblance between a fragment of colored paper and a fallen geranium petal, Delany set about creating a paper florilegium of 985 collages, which she kept in the ten volumes of her *Flora Delanica*. A sort of reverse dissection, Delany's technique for "paper mosaicks" combined fragments of colored tissue and paper, sometimes supplemented with watercolor, layered to create impressions of shading; these were glued to inky-black supports, a tonal choice that broke with the botanical tradition of depicting plants against a pale background. Using a scalpel, she made her cuts directly into paper, based on close observation and without tracing or sketching, to produce botanically accurate representations labeled with the plant's Linnaean and common names. She rendered countless sharp thorns, veined leaves, and stamens laden with pollen before her failing eyesight led her to abandon the project in 1784.

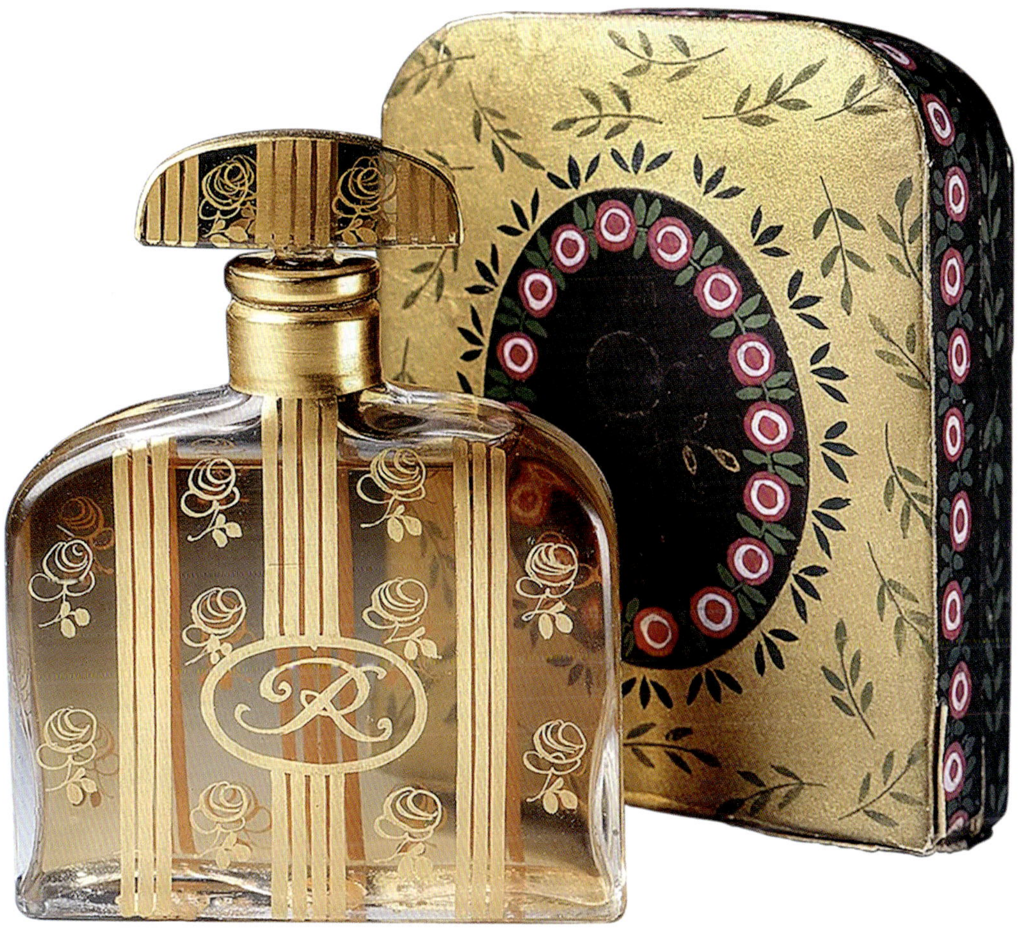

PAUL POIRET AND PAUL IRIBE

La Rose de Rosine, 1912
Perfume bottle, dimensions variable
Private collection

A swirled letter R for "La Rose de Rosine" is surrounded by stylized roses on a perfume bottle with a distinctive stopper. On the box in which the perfume comes, the rose motif is even more abstract, with colored circles and flat silhouettes of leaves. Launched in 1912, La Rose de Rosine was the signature rose-scented fragrance of the new French perfume company Parfums de Rosine, named after the elder daughter of its founder, Paul Poiret (1879–1944). The same motif appeared on the labels for clothes by Poiret, the French fashion designer who liberated women from the corset and set the blueprint for the modern fashion industry. Hailed in Paris as "*Le Magnifique*" and in the United States as the "King of Fashion," Poiret was the first designer in fashion history to expand into the perfume industry, in 1911. Parfums de Rosine was an immediate success, thanks to the marketing skills of Poiret, who employed perfumers (including Henri Alméras, who later created the iconic Joy for Jean Patou) to create the fragrances and used distinctive bottles and packaging. This example was designed by famed French Art Deco artist Paul Iribe (1883–1935), who took inspiration for its Baroque shape from the gown worn by the Spanish infanta in Diego Velázquez's celebrated painting *Las Meninas* (1656). If the story sounds familiar, it is: Coco Chanel would use many of the techniques Poiret pioneered to greater effect when she launched her first fragrance, Chanel No. 5 (see p.47), in 1921.

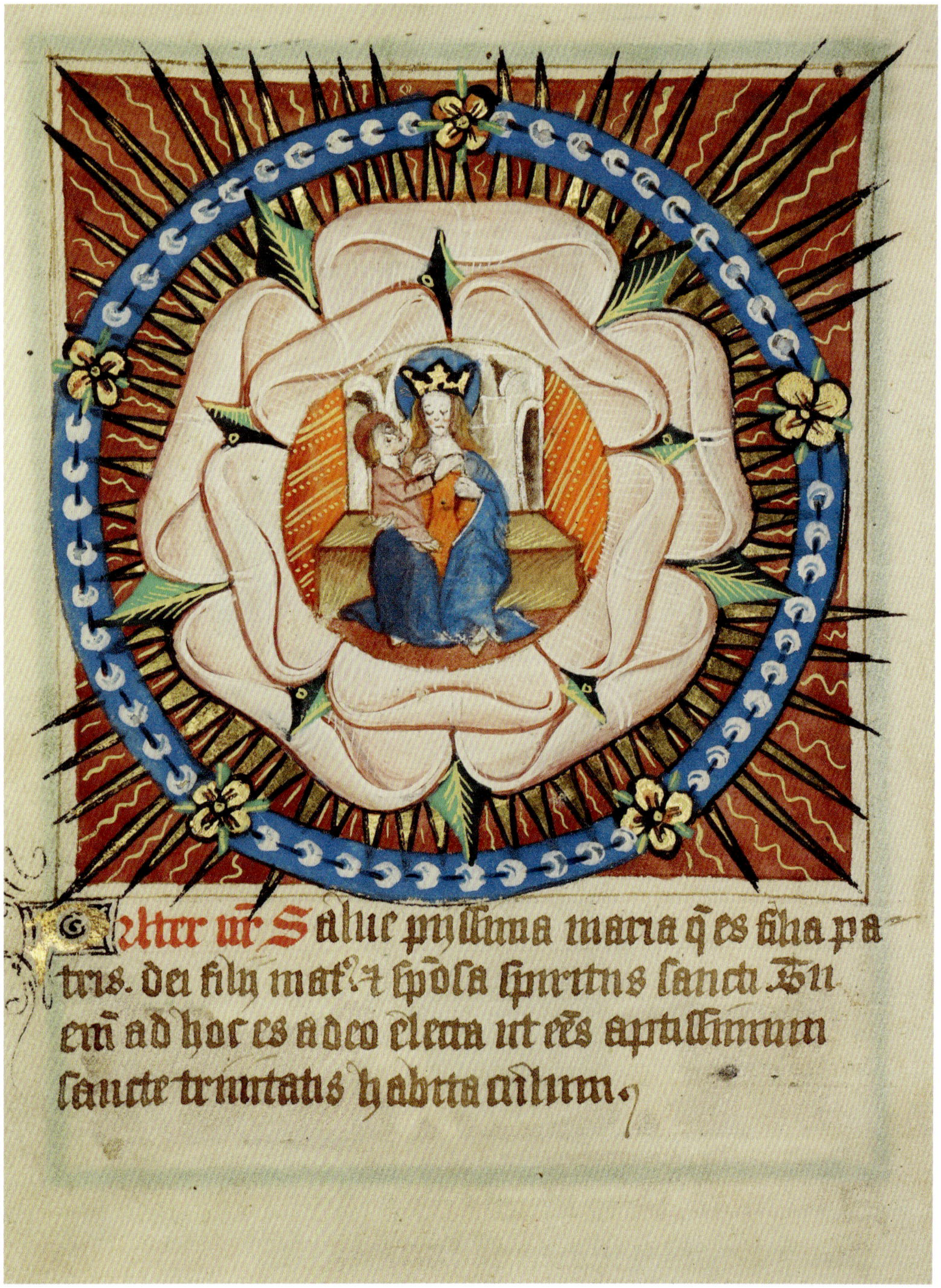

ANONYMOUS

The Virgin and Child in a White Rose, from *Life of Christ*, c. 1480–90
Tempera colors and gold leaf, 6¾ × 4⅝ in. / 17 × 11.9 cm
J. Paul Getty Museum, Los Angeles

White roses have long signified purity, even in antiquity, and therefore became linked with the virginity of Jesus Christ's mother. Mary was venerated by early Christians, and her cult achieved an upsurge in popularity from the twelfth century onward, mainly thanks to the teachings of Saint Bernard of Clairvaux, a Cistercian abbot who was an advisor to five popes and motivated the Second Crusade (1147–49). Bernard's teachings dispelled all doubt of Mary's virginity, making her into the personification of the Church and the main intercessor for human salvation, to whom prayers should be addressed. On this page from an illustrated *Vita Christi* (*Life of Christ*), a devotional prayer book of miniatures depicting the Christ cycle, the prayer salutes the Virgin as "daughter of your son," and she is shown enthroned as Queen of Heaven, similar to the description of Dante Alighieri's white rose of Heaven in *The Divine Comedy* (c. 1308–21). Further rose symbolism accrued during this period, and cathedral stained-glass windows in a circular shape were known as rose windows (see p.131), where multicolored lights represented a vision of Heaven. This manuscript illustration, created in East Anglia, in late fifteenth-century England (not long before the English Reformation denied all Catholic imagery), reflects further rose symbolism. The five petals and thornlike leaves represent the five wounds of Christ's Crucifixion and the Crown of Thorns. Then comes redemption, with the rose being encircled by the blue ring of Heaven and the golden rays of the sun.

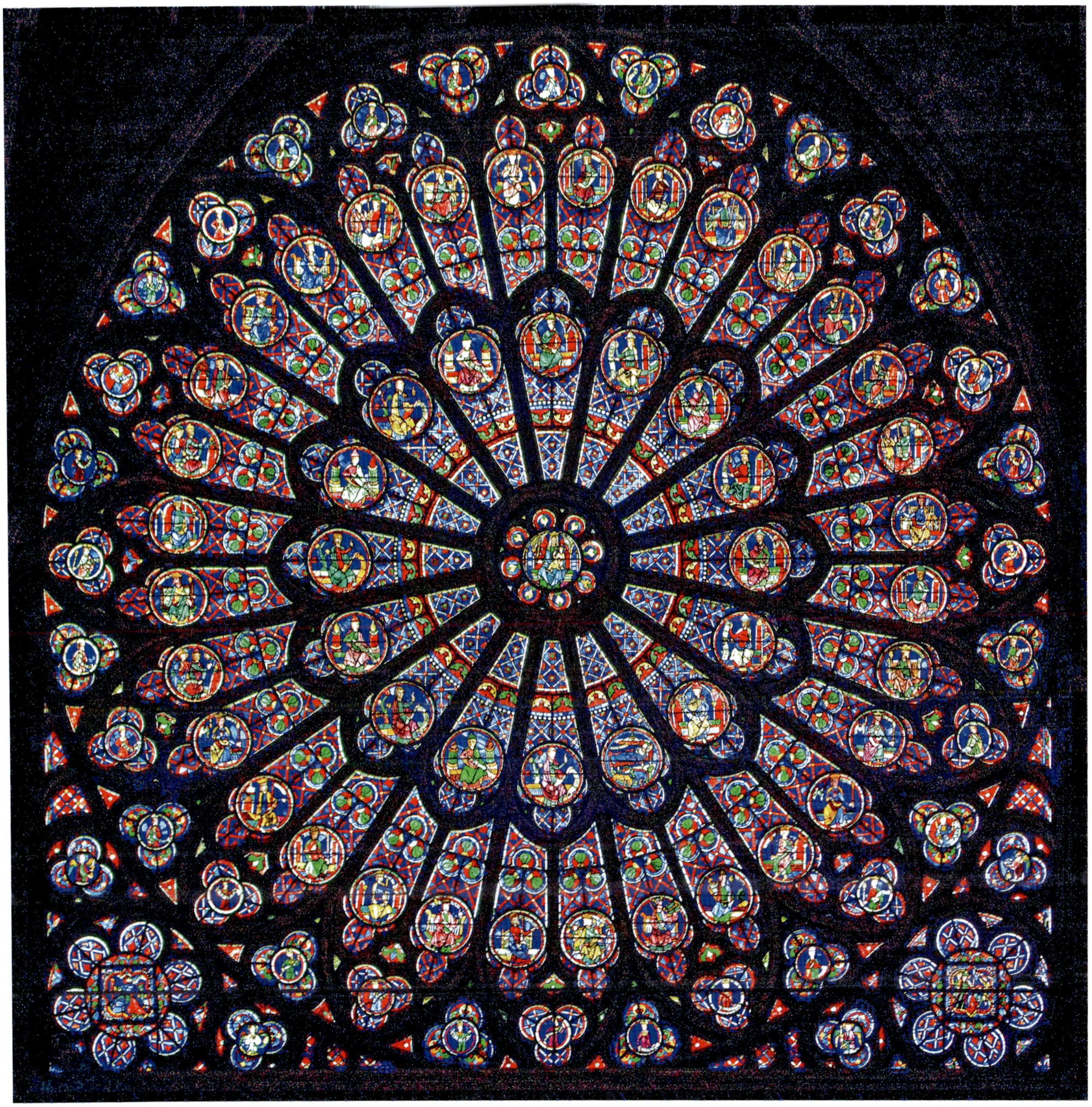

ANONYMOUS

North rose window of Notre-Dame de Paris, c. 1250
Stained glass, D. approx. 42 ft. / 13 m
Notre-Dame Cathedral, Paris

When rose windows are used in the transept ends in Gothic cathedrals and churches, one of those windows is frequently dedicated to the Virgin Mary, who is also known by the title "the Mystical Rose." A marvel of French Gothic architecture, the north rose window of Notre-Dame Cathedral in Paris portrays in its central image Mary enthroned, holding the Christ Child. Surrounding them are images of kings, judges, and prophets of the Old Testament, with sixteen prophets in the inner circle of medallions and thirty-two kings and judges in the outer circle. The window is the only one of the three rose windows at Notre-Dame to retain its original thirteenth-century stained glass, having survived the fire that destroyed much of the structure in 2019. Originally known as a wheel window, in which glass shapes spread out from a center point, these openings became known in the seventeenth century as rose windows for their resemblance to a flower with its petals. The circle with spokelike mullions and delicate tracery of stone looks like an open rose. The rosette window is the shape of the openwork bay in the wall, while the rose itself is the stained glass. To prepare for such a large circle of glass piercing a wall that carries the weight of tons of stone above it, the opening is interwoven with a visually subtle but physically sturdy web of stone tracery. This breaks the glass into smaller shapes in a regular way, distributing the weight of the window evenly across the circle.

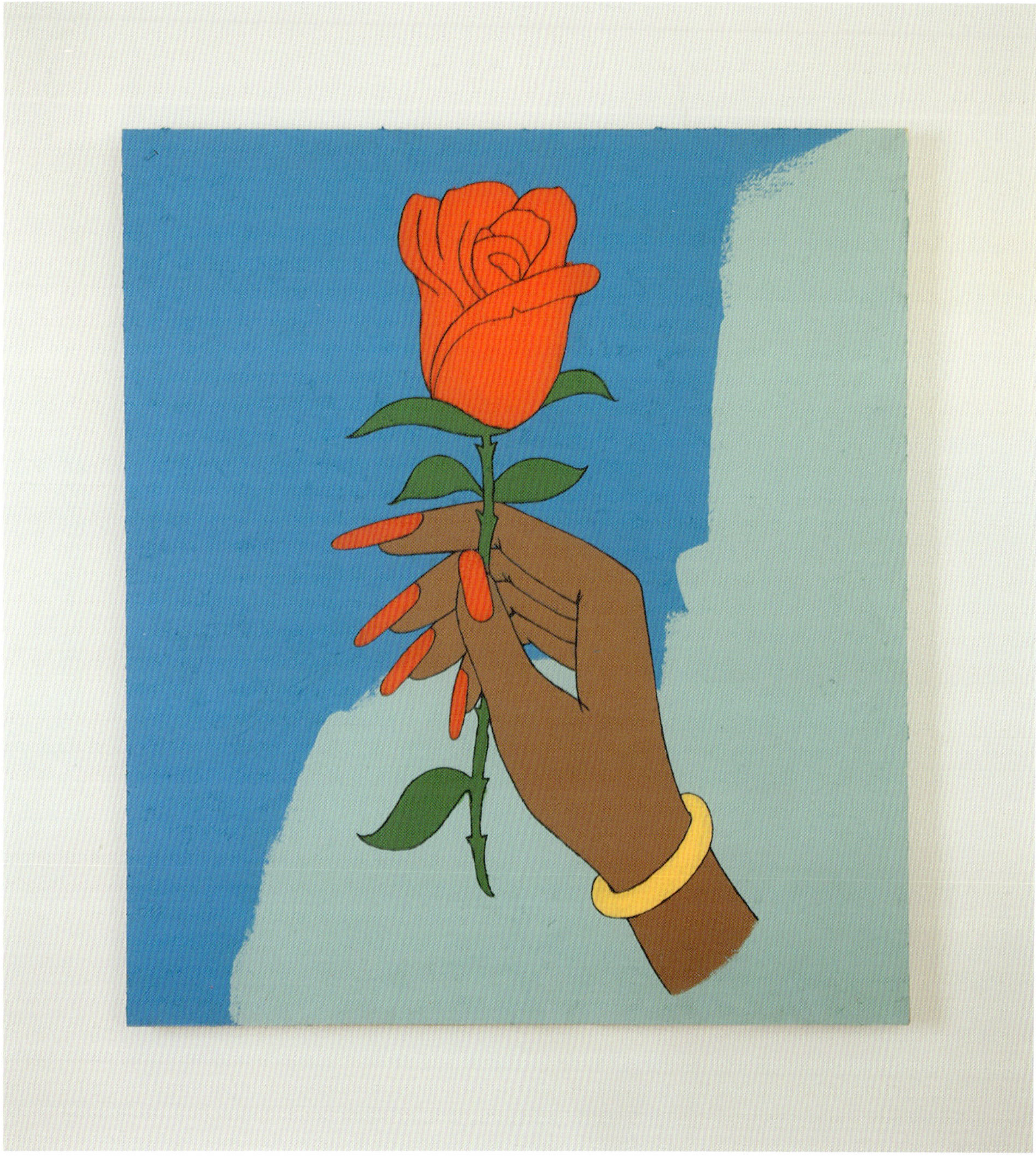

AWOL ERIZKU

Switch – Roy Woods, 2016
Acrylic on wooden panel, 42¼ × 38 in. / 107.3 × 96.6 cm
Private collection

Against a two-toned blue background, an immaculately manicured Black hand holds an upright red rose in an ambivalent gesture. Is the hand accepting the rose, or giving it? Taking inspiration from the neon sign used to advertise nail salons in Los Angeles, this ambiguous painting is the work of Ethiopian-born, Los Angeles–based multidisciplinary artist Awol Erizku (b. 1988). The artist often uses floral motifs to play with symbolism and representation, perhaps most famously in the flower-laden photographs of Beyoncé from 2017, which announced the star's second pregnancy and garnered 10 million likes on Instagram. One of the most famous icons of love and beauty, the bright red rose is central to a nail salon's promise of an attractive manicure, although Erizku reveals, "What drew me to the ubiquitous nail salon at first was the fact that I never had seen a version with a Black hand." *Switch – Roy Woods* comes from a series of conceptual paintings and sculptures, each of which reimagines the instantly recognizable sign with a Black hand, giving it new meaning in the contemporary political climate of the United States. The works use sound (every title is taken from the song that Erizku was listening to when he finished the piece), as well as different colors, nontraditional media, and varying degrees of abstraction, to strip both hand and flower of their original significance. A new kind of floriography, Erizku's work deconstructs the traditional use of the rose in visual culture, finding exciting ways to define this age-old symbol.

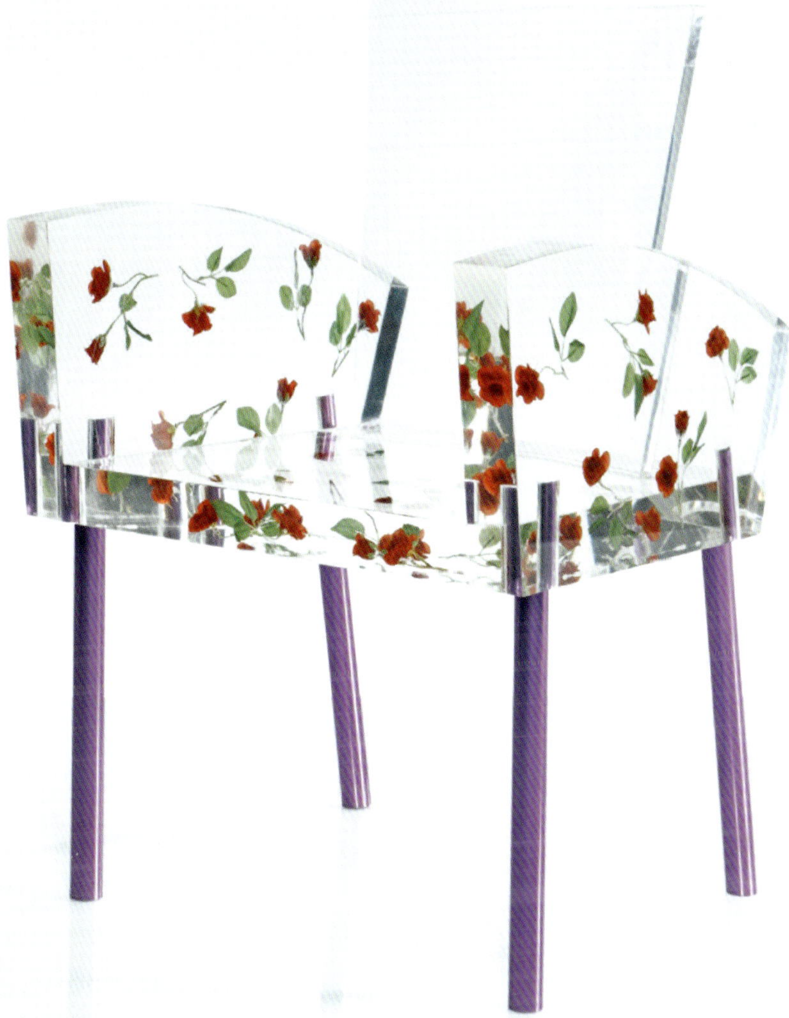

SHIRO KURAMATA

Miss Blanche, 1988
Acrylic, artificial roses, and anodized aluminum,
36¼ × 24¾ × 24 in. / 92 × 63 × 61 cm
Private collection

This remarkable artwork also happens to be a functional chair, with vibrant red artificial roses floating within thick panels of clear resin, their contrasting too-green stems and leaves dancing playfully among one another inside the seat and arms. Japanese designer Shiro Kuramata (1934–1991) fused whimsical ideas with a technical edge, such as his *Glass Chair* (1976), which was constructed solely from glazed panels, and *How High the Moon* (1987), made from a cloudlike mesh. *Miss Blanche* is perhaps his most recognizable piece. The chair encases run-of-the-mill artificial roses in slabs of resin, creating a seat that appears clear as glass, accented by the red blooms and balanced on slim tubular legs of anodized aluminum. The piece is named after Blanche DuBois, the central character in Tennessee Williams's 1947 play *A Streetcar Named Desire*, perhaps as a riff on her illusory personality and veneer of propriety. Like Blanche, the chair is also a cunning sleight of hand; despite being an apparently simple one-liner, the idea was deviously difficult to produce. Abandoning the original concept of using real roses, which the resin burned black, Kuramata made the limited-edition chairs individually in one-off molds. Each required partly pouring a resin slab, inserting the fake roses with tweezers before the liquid hardened, and, finally, submerging them in an upper layer. They are held in perpetual suspension, resulting in a chair graced with seemingly effortless floral animation.

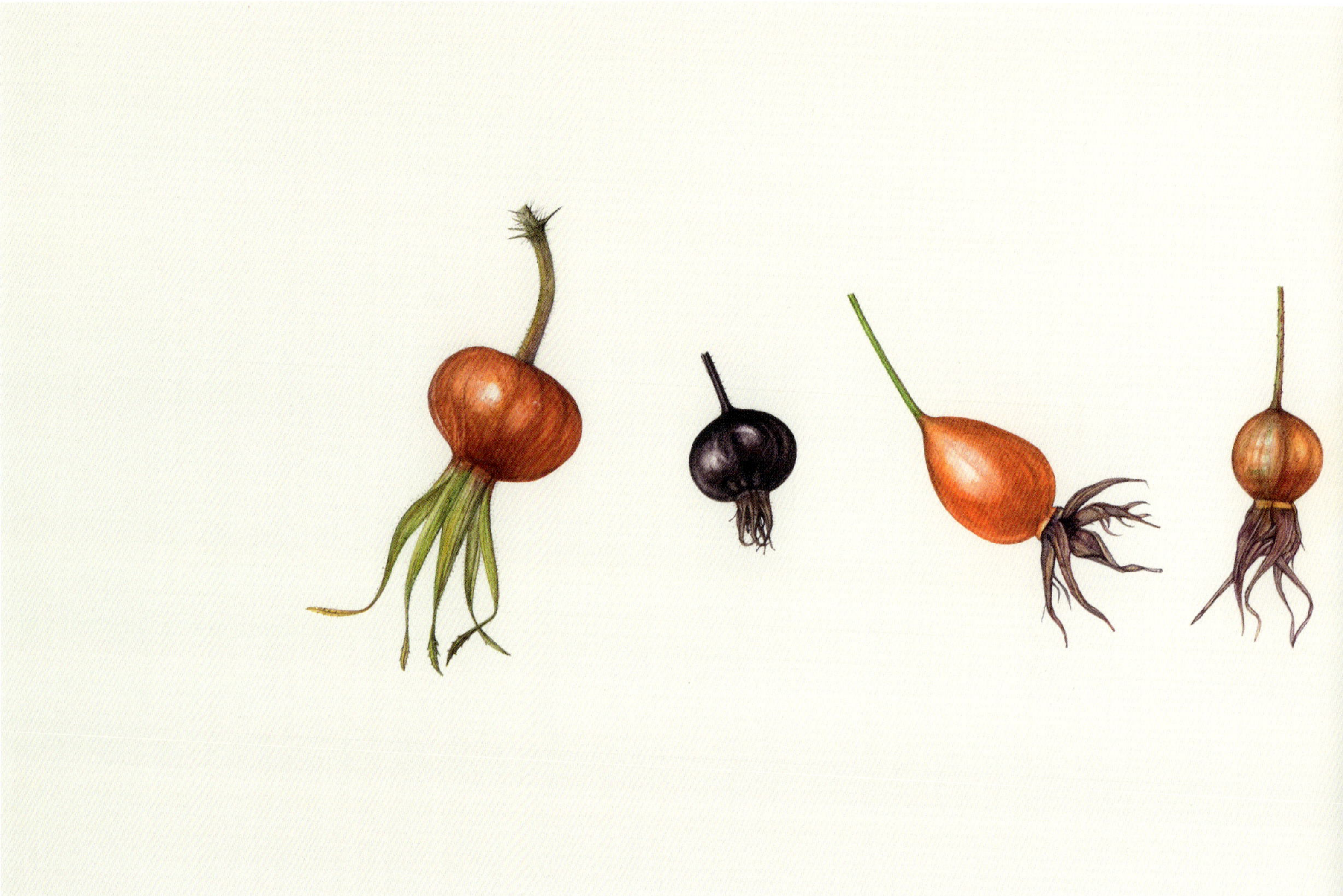

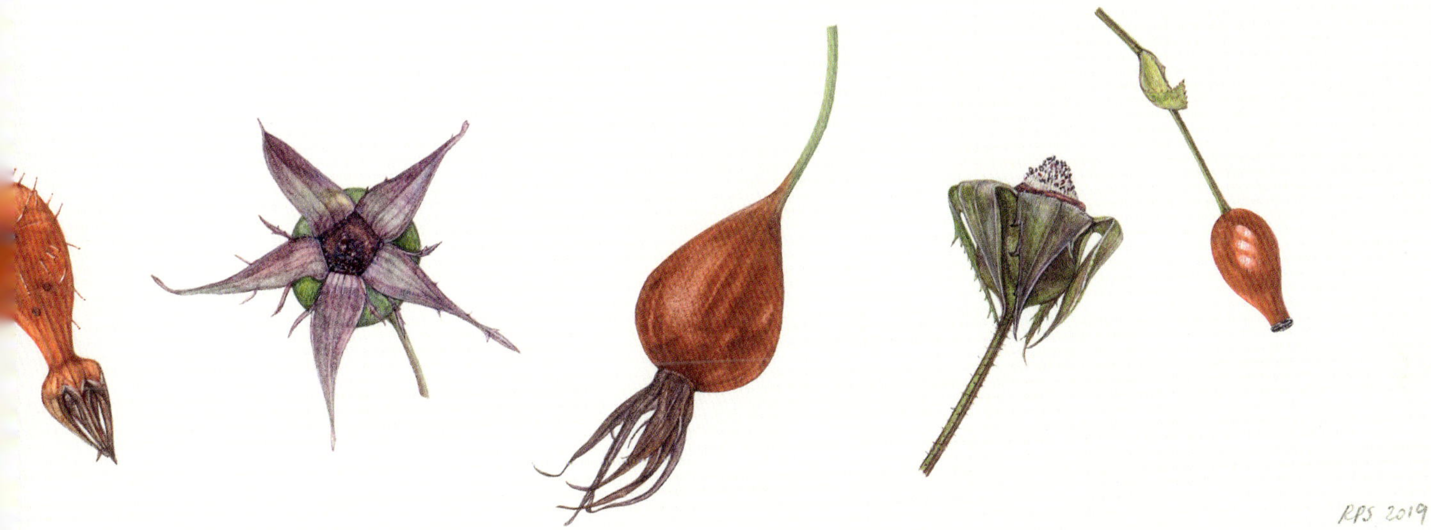

RACHEL PEDDER-SMITH

Rose Hips, 2019
Watercolor on paper, 13 × 22½ in. / 33 × 57 cm
Private collection

Lined up in a row, these nine shapes—each one unique—appear at first glance to be perhaps a series of planktonlike creatures that might be found in the ocean, with swollen bodies and multiple legs. In fact, the shapes are rose hips, each depicting the fruit of different species of *Rosa* at varying stages of development. Rose hips contain the seed of the rose plant, whether a wild hedgerow rose or a Hybrid Tea rose. These oft forgotten but integral parts of the plant start to form in late spring or early summer after pollination, and reach maturity in fall. Their bright colors, which range from reds and oranges to pinks and deep purples, ensure the fruit attracts birds to eat, digest, and spread the seed widely, continuing the cycle of rose reproduction. This illustration is by award-winning English botanical artist Rachel Pedder-Smith (b. 1975), whose work at the Herbarium of the Royal Botanic Gardens, Kew, established her reputation for stunning, botanically accurate drawings. Her images have been widely reproduced, including as fabric designs for iconic London department store Liberty, bringing the garden into the home. The late fashion designer Vivienne Westwood even used Pedder-Smith's "Floral Eve" pattern for a dress in her Summer 2013 collection. Pedder-Smith often sets her illustrations against a white background to emphasize the unique qualities of each specimen. Here, while the bottom of one rose hip appears as a purple flower, others show their hairy stems and rounded forms, filled with the seed that will create the next generation.

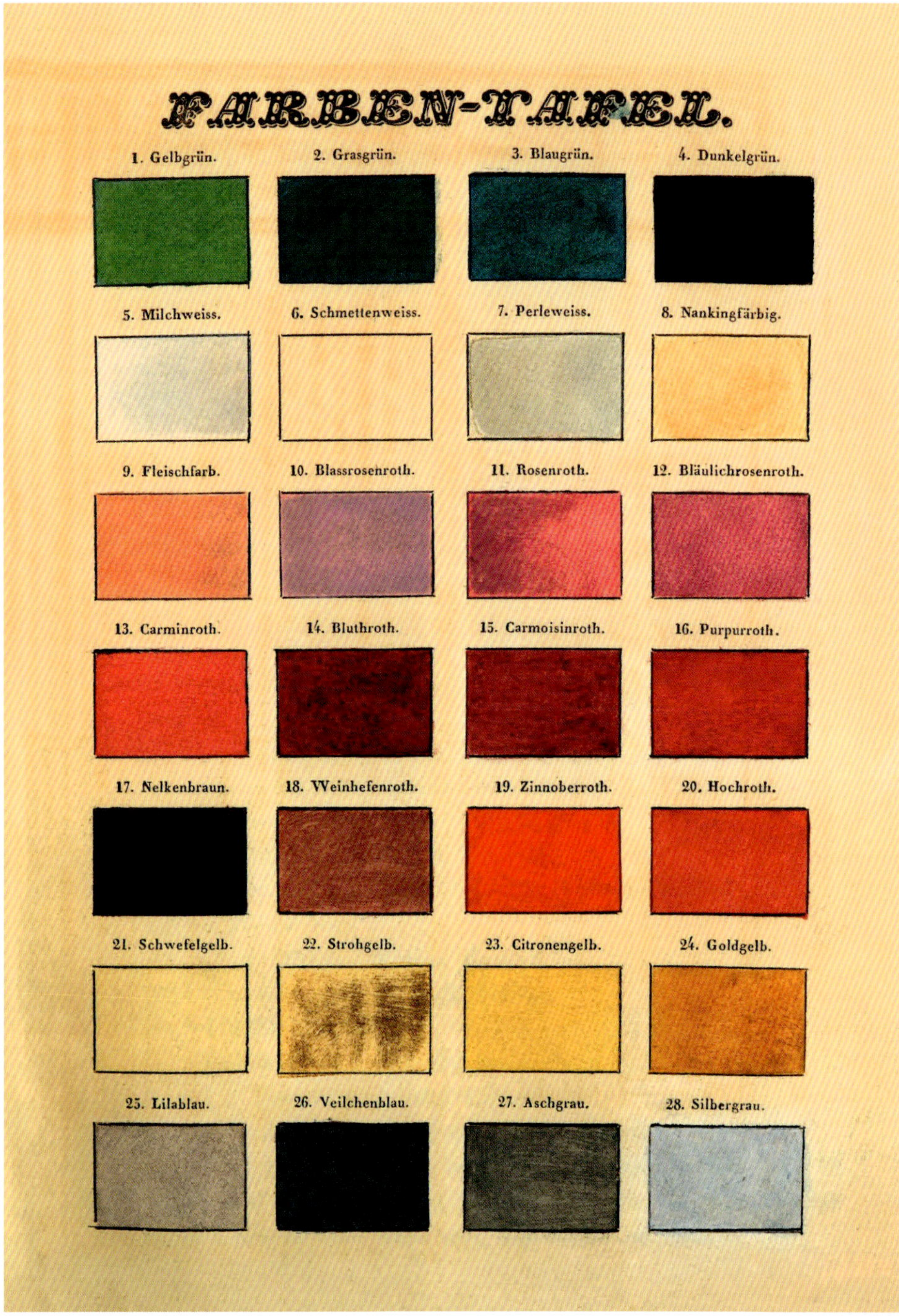

CLARENCE NICKELS

Color chart, from *Cultur, Benennung und Beschreibung der Rosen*, 1845–46
Lithograph, H. 8⅝ in. / 22 cm
Cornell University Library, Ithaca, New York

This gridlike artwork seems more likely to be associated with the twentieth-century Bauhaus movement in Germany than with a mid-nineteenth-century instructional book on horticulture. However, this color chart comes from *Cultur, Benennung und Beschreibung der Rosen* (*The Cultivation, Naming, and Description of Roses*), an eccentric guide written by German rosarian Clarence Nickels in the 1830s and subsequently reprinted. A self-proclaimed rose lover, Nickels had spent twenty years simultaneously growing and learning about roses. He believed that a written description of a rose was a more useful tool for gardeners trying to identify a specific variety than an illustration, so his book included few images other than this chart. The first color chart ever published to help categorize the different hues of roses, the graphic includes a range of shades with specific descriptors, such as *weinhefenrohe* (wine-yeast red) and *fleischfarb* (meat-colored). Once the color of the real-life flower was matched to the chart, a gardener could locate and read the extensive description of each rose to identify the species. Nickels's approach was in stark contrast to that of other European botanical authors in the nineteenth century. For them, a book about roses—of which hundreds were published, particularly in England to satisfy the Victorians' insatiable love of roses—was essentially a pictorial guide, its success determined by the quality of the illustrations of each flower and not the writing that accompanied the pictures.

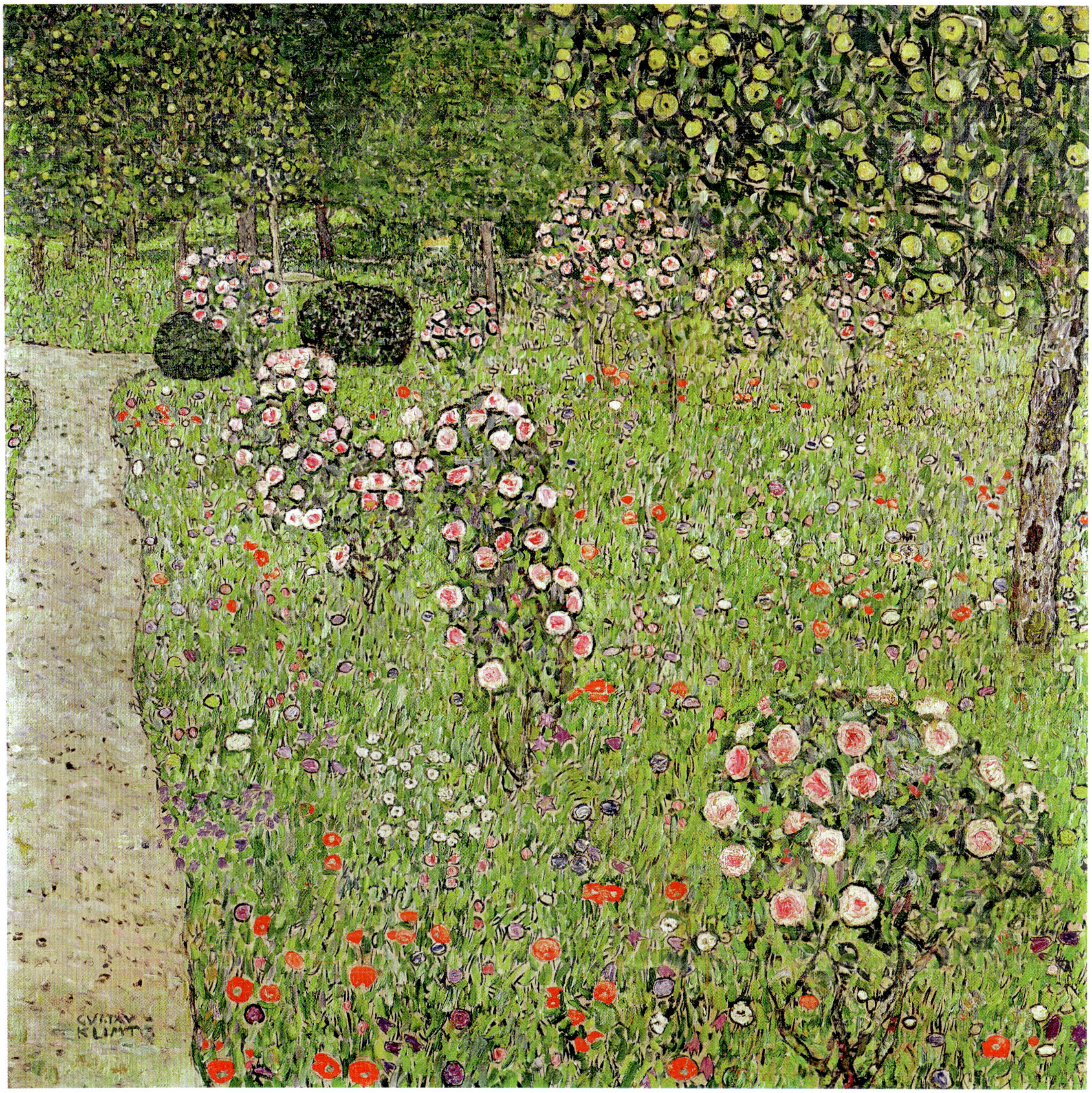

GUSTAV KLIMT

Orchard with Roses, c. 1911–12
Oil on canvas, 43¼ × 43¼ in. / 110 × 110 cm
Private collection

One of Austrian painter Gustav Klimt's (1862–1918) many extraordinary landscapes, this is an image of abundance, the trees heavy with ripe, golden fruit; the grass studded with poppies, daisies, and other wildflowers; and the standard rosebushes bright with warm-hued pink blooms. Painted in a semi-pointillist style, with dashes, dots, and blobs, it forms an all-over jeweled pattern of pulsating color, a vivid tapestry of flowers and foliage that hovers between figuration and abstraction. All the elements exist in a flat plane; there is no perspective, no sense of recession, except for that suggested by the narrow path to the left that leads the eye to the darker trees in the middle distance. Klimt is known best for his dazzling decorative portraits of wealthy women and mythical female figures, but from 1900 onward he also made landscape paintings every summer, when he escaped the bustle of "hot, dreadful" Vienna—where, he said, "everything is parched"—for the verdant countryside. He spent these summer vacations on the Attersee, a tranquil lake in northwestern Austria. These landscape pictures were painted for his own pleasure, not as commissions, but they quickly found an admiring audience and were sought after by avid collectors. *Orchard with Roses,* in common with the other landscapes of his later career, is in an unusual square format, a reflection of the artist's fascination with photography.

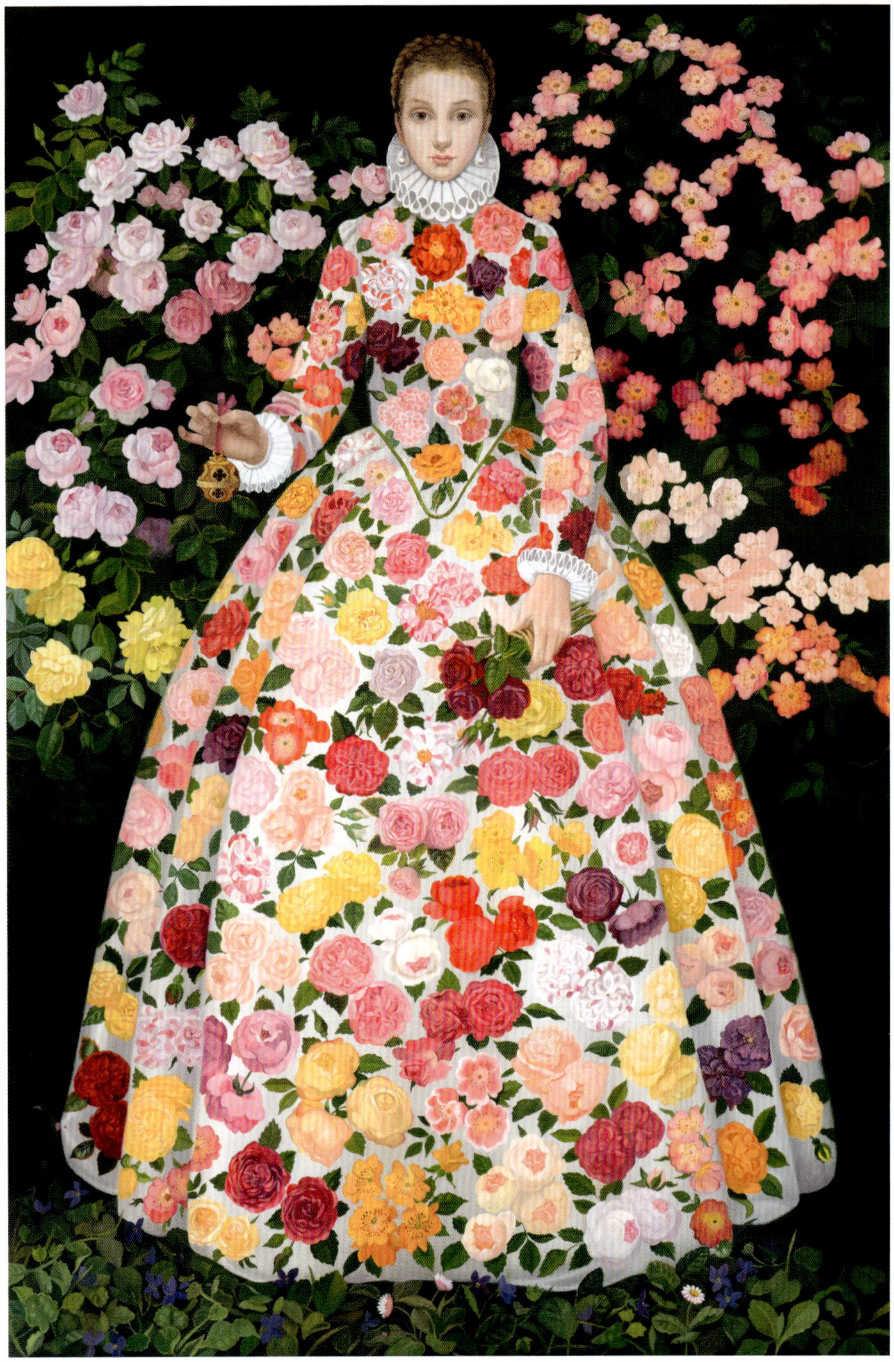

LIZZIE RICHES

The Rose Dress, 2021
Oil on canvas, 54 × 34 in. / 137.2 × 86.4 cm
Private collection

This striking, nearly life-size painting by English artist Lizzie Riches (b. 1950) was specially commissioned by French perfume house Fragonard, which is based in Grasse (see pp.47 and 143). A modern reinterpretation of a traditional Elizabethan portrait, *The Rose Dress* shows a young woman dressed in an ornate sixteenth-century-style gown, complete with collared ruff and large teardrop earrings of shiny pearls. Most significantly, the subject appears surrounded by roses: her dress is completely covered in representations of the flowers, she holds a bunch of purple roses in her left hand, and the background is filled with rosebushes of four different species and colors. Each rose on the dress is unique, from wild red roses to Centifolias in full bloom in an array of hues. The overwhelming impression is one of joy and color. Both these emotions are intentional, as is the reference to the Tudor period. Riches fell in love with the sixteenth century when, as a child, she visited the Jacobean mansion Hatfield House in Hertfordshire and saw the "rainbow portrait" of Queen Elizabeth I (see p.33). Riches's contemporary portraits are almost always rooted in the Tudor and Elizabethan periods, and they have gained her a cult following over the years. *The Rose Dress* debuted in the *parfumerie*'s Christmas 2022 collection, appearing on the box of a diffuser named De Tout Coeur, which promised to fill a home with the scent of a bouquet of flowers.

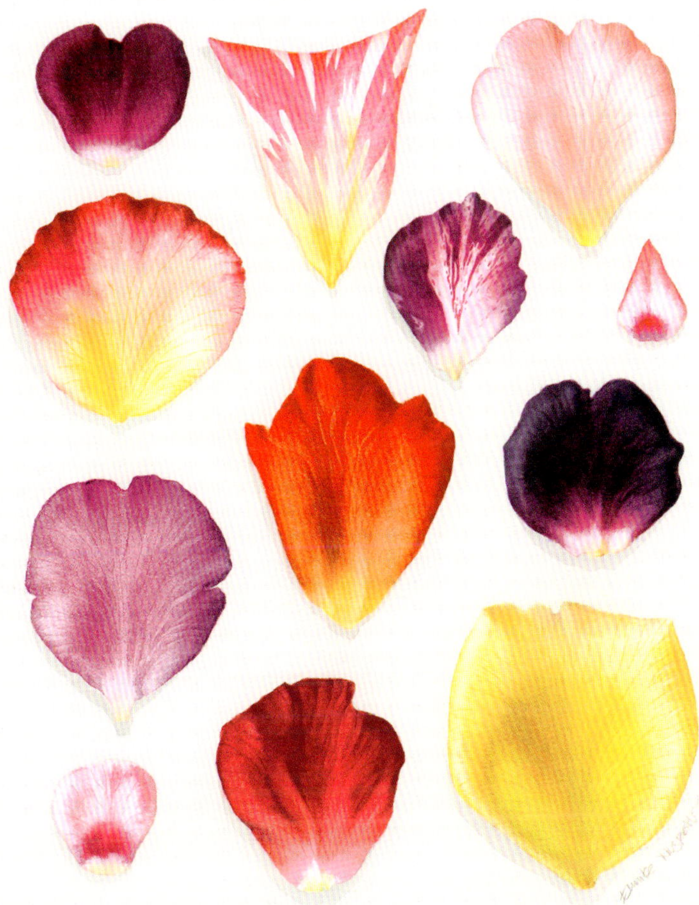

EUNIKE NUGROHO

Rose Petals, 2014
Watercolor on paper, 12⅝ × 10⅜ in. / 32 × 26.5 cm
Private collection

Scarlet and burnt orange, violet and pale pink, lemon-yellow and deep magenta—the vivid colors of these twelve photorealistic rose petals may seem like fiction, but they all occur in nature. Indonesian botanical artist Eunike Nugroho (b. 1981) collected these petals from Sheffield Botanical Gardens in northern England after identifying them as botanical elements that would enable her to play with multiple colors in one composition. Before beginning this work, she compared the colors of each petal with her own radial color charts and wheels, allowing her to find the synthetic paint colors that would best match the petals' natural vibrancy. A skilled illustrator of the natural world, she won a gold medal at the Royal Horticultural Society's Botanical Art & Photography Show in 2023 for her six paintings of plants in the genus *Hoya*, also known as waxplants, which are native to Indonesia. She chose this group of climbing or scrambling plants with their clusters of handsome, waxy flowers because she was inspired by their resilience and diversity. Her presentation was the first time that Indonesia and Southeast Asia had been represented in the RHS's annual exhibition. Nugroho's attention to detail shines through in the preparation and execution of her work. Here, in intricate watercolors, she showcases not only the different tones between the rose petals but also their shape, their form, and the fine detail of their veins.

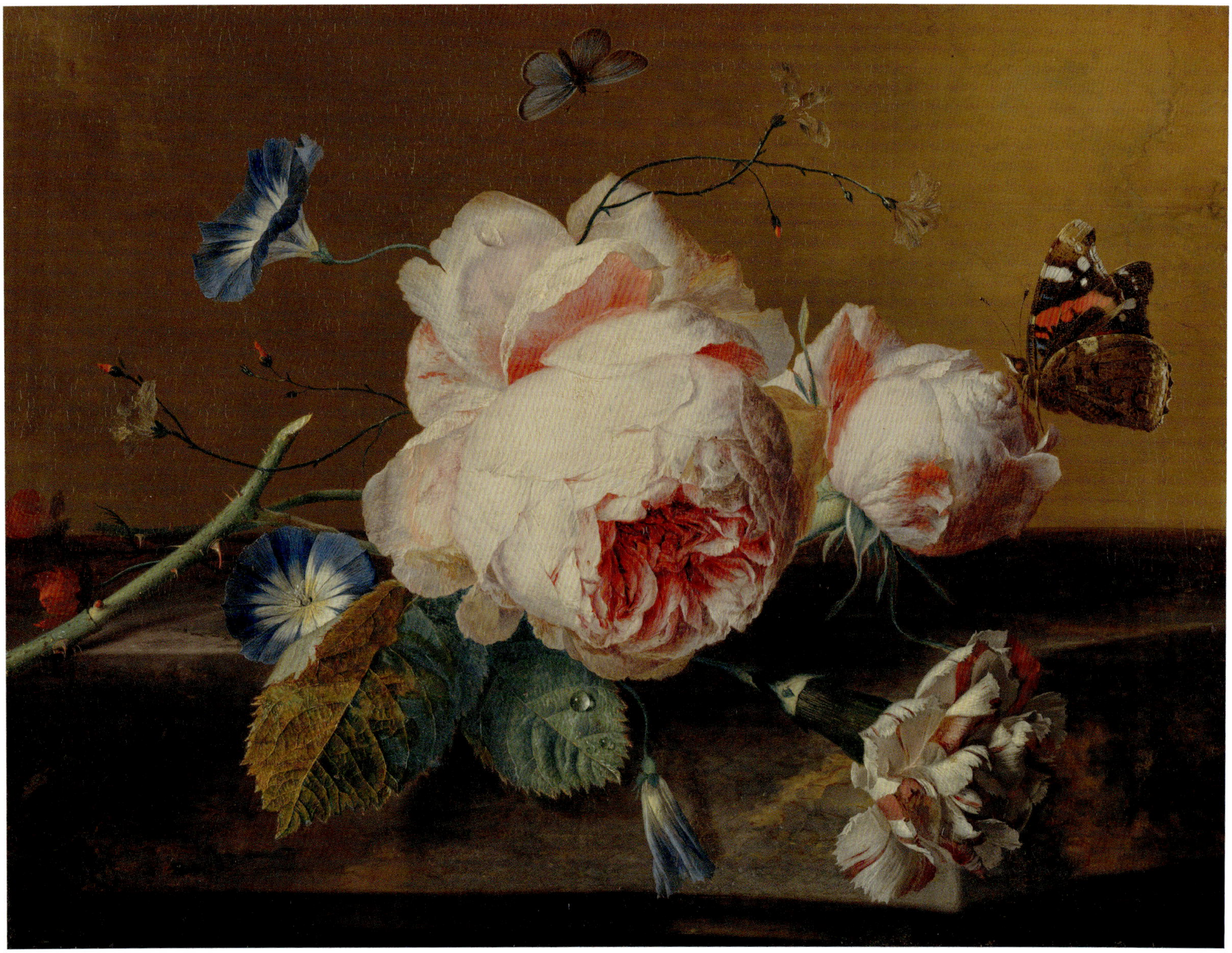

JAN VAN HUYSUM

Flower Still Life, date unknown
Oil on copper, 8¼ × 10⅝ in. / 21 × 27 cm
Mauritshuis, The Hague, the Netherlands

A sumptuous blowing rose commands the viewer in this small work painted by Jan van Huysum (1682–1749), a native of Amsterdam and one of the great Dutch flower painters. In the tradition of the time, art was a family business, and Van Huysum came from a family of prominent artists during the Dutch Golden Age of the seventeenth and eighteenth centuries. It is commonly believed that he was trained by his father, who was also a still-life painter. The majority of Van Huysum's paintings are large, elaborate, elegant compilations of flowers painted from life over time, which makes this smaller work on copper an unusual one. Blue morning glories at left and wispy blooms of *Dianthus* at the top of the composition act as delicate foils for the central pink roses in luscious bloom.

Two butterflies flutter around the flowers, which are draped casually on the edge of a marble table, as if ready to be arranged in a large vase or container—a visual device that Van Huysum also used in his larger compositions. This perhaps signaled an overflowing excess of flowers, an abundance associated with wealth. Although compact, this delightful assemblage acts as a powerful symbol of the owner's comfortable lifestyle; it forms a pair with a fruit still life, and both were acquired by Sir Robert de Neufville, Postmaster General in Leiden. Although a date of about 1736 is commonly associated with the work, it probably comes from an inventory of de Neufville's estate following his death in 1735.

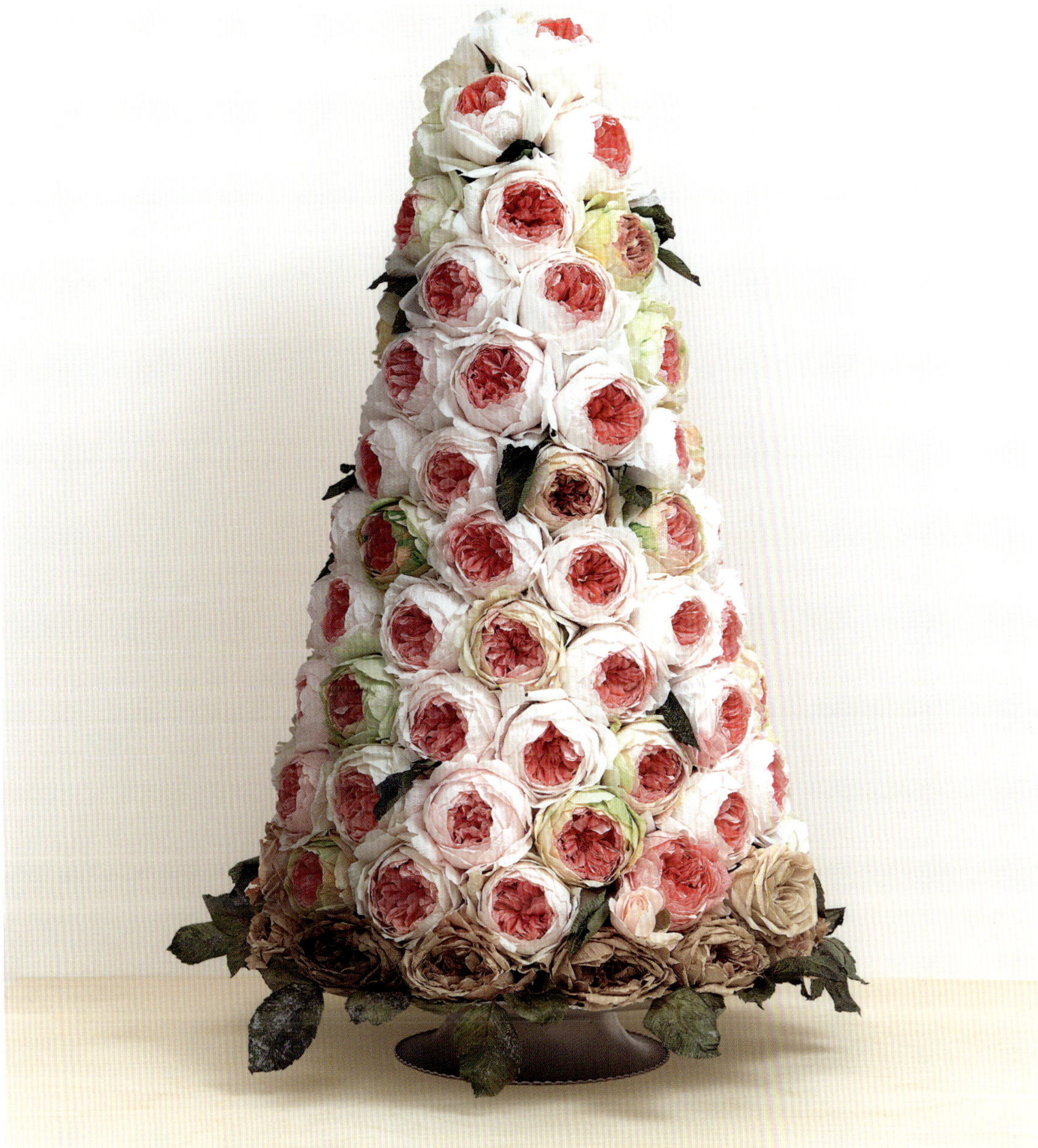

TIFFANIE TURNER

Croquembouche, 2022
Italian crepe paper, stain, glue, floral wire, chalk, chicken eggshells, hat stand, metal platter on pedestal, and epoxy adhesive, 26½ × 15 in. / 67.3 × 38.1 cm
Private collection

This conical tower of delicate pale pink roses looks good enough to eat for a reason. The pile of flowers—most in full bloom, but others clearly dead or decaying and one still in bud—is an homage by American artist Tiffanie Turner (b. 1970) to the French confection known as the *croquembouche*. The sculpture, standing about 2 feet (60 cm) tall, is made from materials that are unexpected for either sculpture or floristry, including Italian crepe paper, a hat stand, and eggshells. Having trained as an architect, and heavily influenced by seventeenth-century Dutch paintings of flowers, Turner uses her structural knowledge to create a sculpture that is simultaneously beautiful and doomed; the dead roses at the bottom hint at a decay that will inevitably spread to the rest of the blooms.

Turner's roses take hundreds of hours to make and each is unique, the petals hand-stained to be as realistic as possible. The artist welcomes flaws and blemishes as part of her attempt to reproduce in permanent form things that usually go unnoticed. This sculpture marks a departure for her because she is better known for the heads of single flowers, often roses but also chrysanthemums and buttercups. The *croquembouche* after which the artwork is named is an edible tower of filled choux buns, typically served as a wedding cake in France and Italy. Turner's version turns the celebratory cake on its head, both capturing the transitory nature of life and providing a warning about what she sees as the political decline of her homeland.

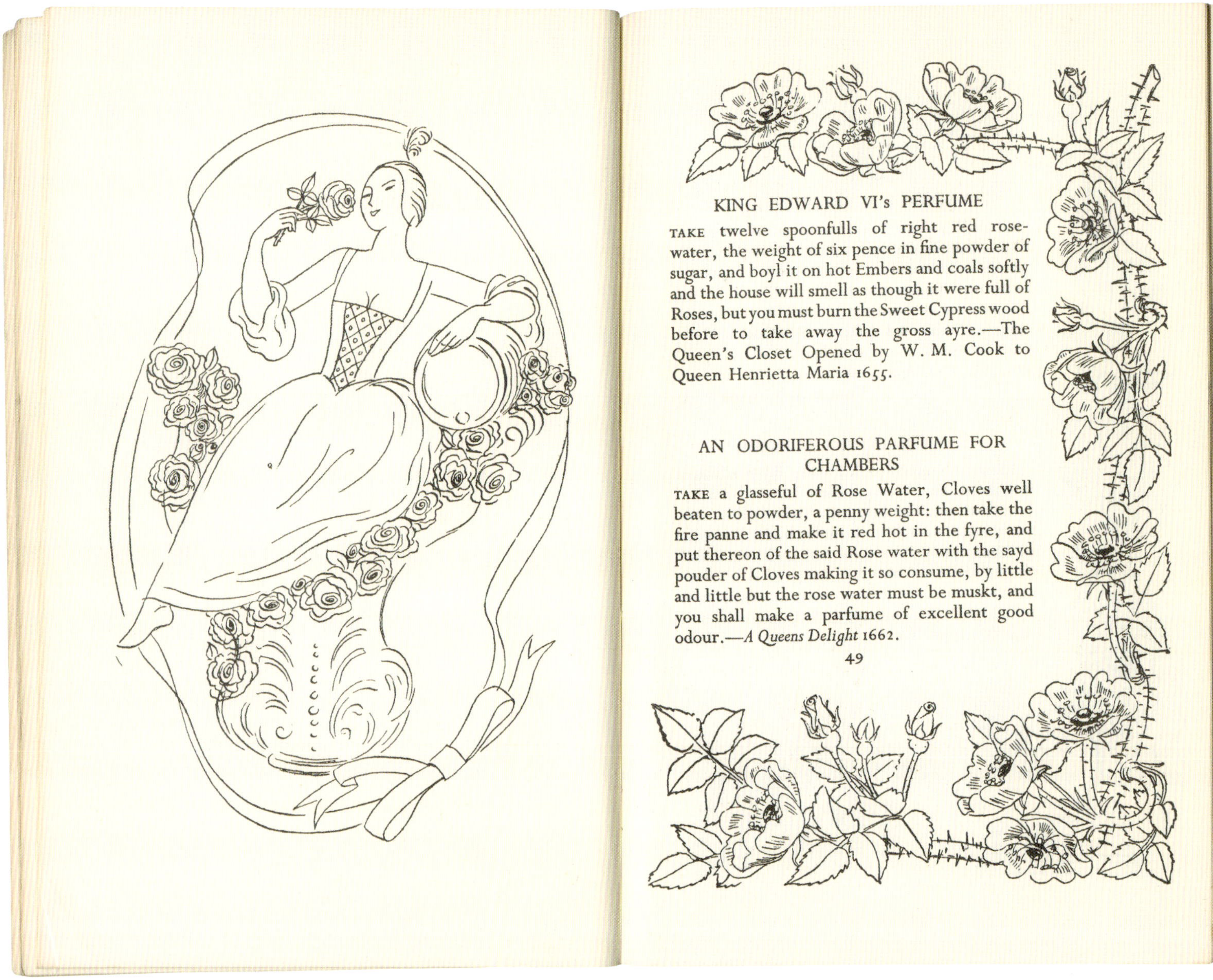

KING EDWARD VI's PERFUME

TAKE twelve spoonfulls of right red rose-water, the weight of six pence in fine powder of sugar, and boyl it on hot Embers and coals softly and the house will smell as though it were full of Roses, but you must burn the Sweet Cypress wood before to take away the gross ayre.—The Queen's Closet Opened by W. M. Cook to Queen Henrietta Maria 1655.

AN ODORIFEROUS PARFUME FOR CHAMBERS

TAKE a glasseful of Rose Water, Cloves well beaten to powder, a penny weight: then take the fire panne and make it red hot in the fyre, and put thereon of the said Rose water with the sayd pouder of Cloves making it so consume, by little and little but the rose water must be muskt, and you shall make a parfume of excellent good odour.—*A Queens Delight* 1662.

49

ELEANOUR SINCLAIR ROHDE AND HELEN KAPP

Rose Recipes, 1939
Printed book, page 8⅜ × 5⅜ in. / 21.2 × 13.7 cm

On opening this book one can almost smell the velvety sweetness of roses. Eleanour Sinclair Rohde (1881–1950), a British medievalist-turned-garden historian, compiled and edited more than eighty recipes primarily from sixteenth- to eighteenth-century sources for her *Rose Recipes*. Born in India and educated at St. Hilda's College, Oxford, Rohde was appointed tutor in history at the college after graduation. She then turned some of her prodigious energies to garden research, marrying her passions for history and horticulture. Instead of targeting academics, she wrote for an audience of amateur and professional gardeners. Her work covered a range of subjects, but she is best known for reigniting interest in the herb garden in Great Britain and beyond. Rohde was no armchair gardener: she grew unusual herbs and vegetables at her Surrey home, Cranham Lodge, where she also ran a commercial nursery. *Rose Recipes* treats the rose as an herb—that is, a useful, not merely ornamental, plant. She organized her finds into recipes for scent and those for taste, and had whimsical illustrations made by British artist Helen Kapp (1901–1978). From its pages, a reader can concoct rose jelly, honey, candy, vinegar, or wine, or learn how to make sweet bags to induce slumber, pomander balls to freshen the air, and perfumes and rose waters to anoint one's clothing, gloves, or skin. Some recipes are practical ("Rose Hip Marmalade"), others curious ("Odiferous Candles against Venome and the Plague"), but all are interesting for their personality and provenance.

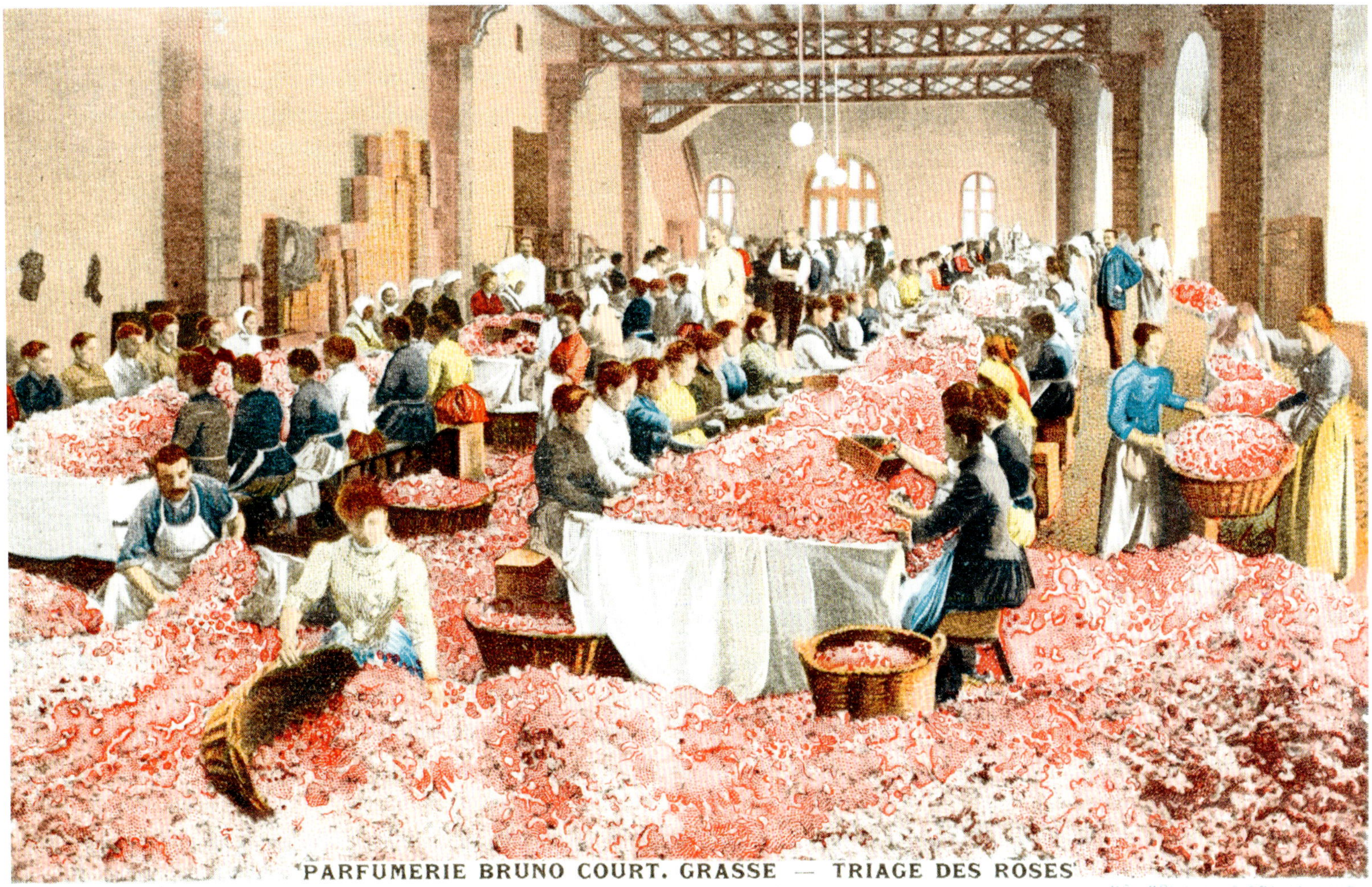

ANONYMOUS

Parfumerie Bruno Court, Grasse, Triage des Roses, early 20th century
Postcard, 3½ × 5½ in. / 9 × 14 cm

A hand-colored photograph from the turn of the twentieth century shows dozens of women working at tables piled high with pink Centifolia roses in the Bruno Court perfume factory in the south of France. Men carry sacks of flowers freshly picked from the surrounding farms to add to their backlog, while foremen supervise. This was the *triages des roses*, the all-hands-on-deck necessary to pick, process, and preserve the flowers before their blossoms begin to decompose and their scent vanish. (It takes approximately 4 tons/3,628 kg of rose petals to make 2¼ pounds/1 kg of essential oil.) The season of bloom for Centifolias is short—about four weeks beginning in May. Bruno Court, founded in 1812, opened its perfumery in Grasse, the global capital of the fragrance industry then and now.

Grasse is an ancient town in southern Provence, near the Mediterranean coast. Its terroir—soil, sun, and situation—make it perfect for cultivating fragrant flowers, and *Rosa × centifolia* (*Rose de Mai* in French) became its hybrid of choice. Although sometimes called the Provence or cabbage rose, the Centifolia is a hybrid that is thought to have been developed by Dutch plantsmen in the sixteenth century (see pp.24 and 180). Its rounded shape, abundant petals, and rich pink color made it a favorite prop for portraits and still-life paintings, but perfumers—"noses" in the parlance of the trade—soon recognized its olfactory significance. The essence of Centifolias became a "note" in many famous scents, Coco Chanel's Chanel No. 5 among them (see p.47).

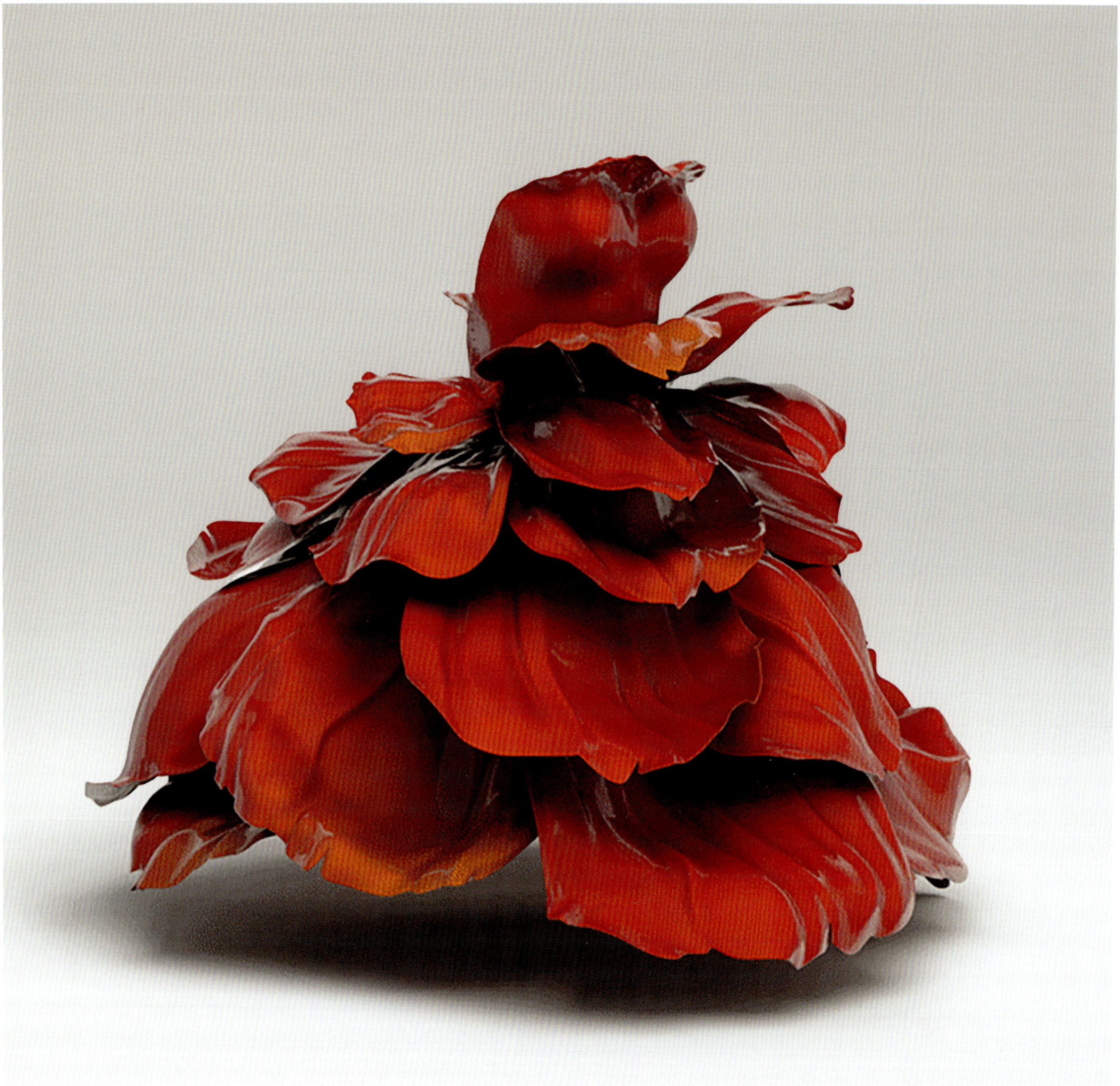

ZAC POSEN

Rose dress, 2019
21 plastic rose petals, 3D-printed and attached to a wearable titanium cage

This glittering evening gown resembling an upturned red rose wowed the crowds in New York when it was worn by British fashion model Jourdan Dunn at the 2019 Met Gala. Born from the imagination of American designer Zac Posen (b. 1980), it was inspired by a 'Mr. Lincoln' rose—a dark red Hybrid Tea cultivar bred in 1964 by Herbert Swim and Weeks Rose Growers and named in honor of President Abraham Lincoln. The organic-looking ensemble is constructed from twenty-one overlapping 3D-printed petals, each of which weighs about a pound (450 g) and took hundreds of hours to print at the specialty manufacturer Protolabs' state-of-the-art facility in North Carolina. Next, they were primed, sanded, and sprayed with a special iridescent automotive paint to give them a lustrous appearance, and were then attached in layers to a titanium frame engineered by General Electric Additive. The frame was designed to fit Dunn's body and weighed about 30 pounds (14 kg). Posen spent five months developing and refining his design, which responded to the Met Gala's theme and the Costume Institute's exhibition that year, *Camp: Notes on Fashion*, exploring the influence of camp culture on clothing and dress across the centuries. Regarding his tribute to the red rose, Posen declared to the *New York Times*, "Mother Nature can be simple, abstract, challenging, and definitely campy, in ways that are surprising . . . You cannot beat Mother Nature. You can just homage."

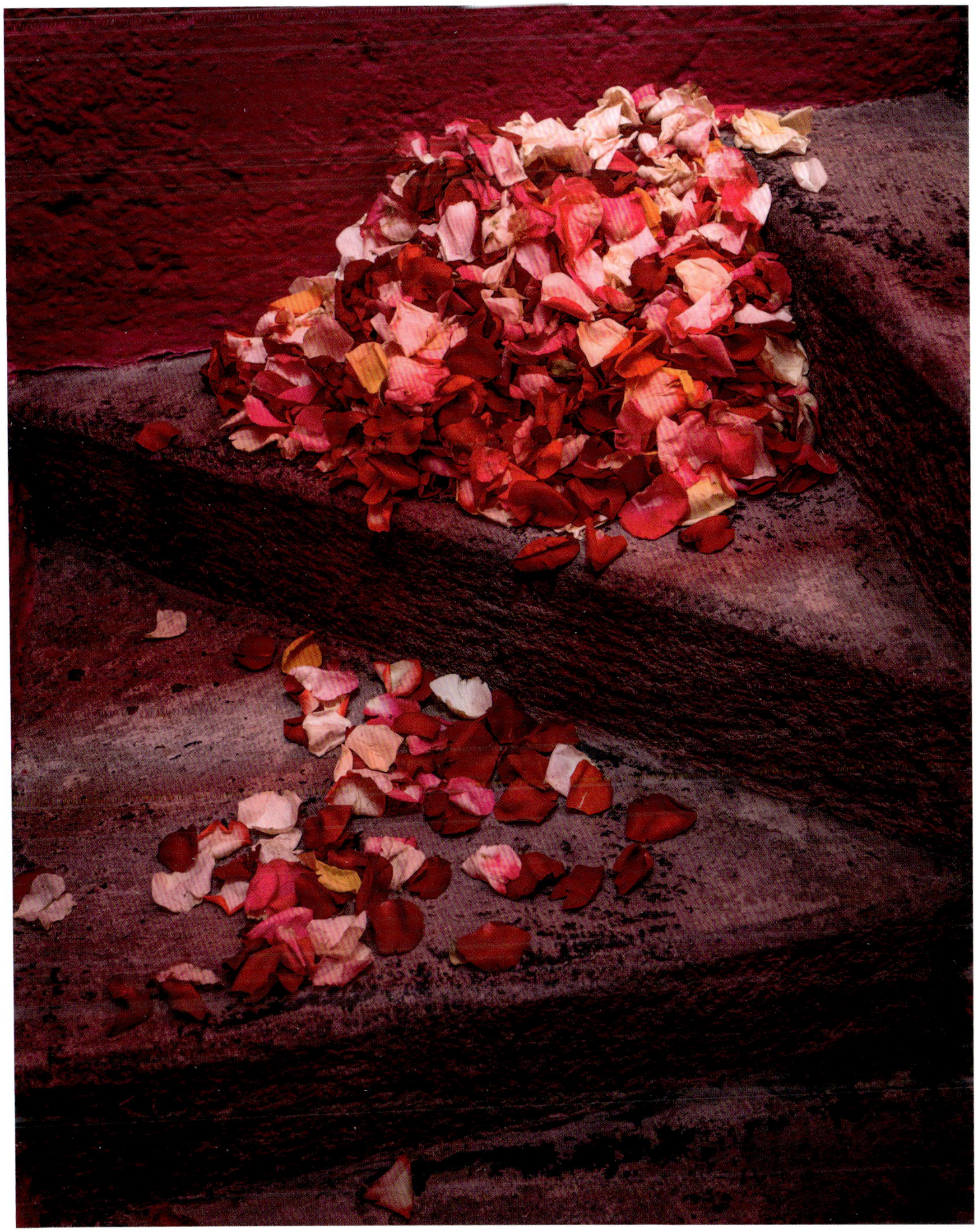

CIG HARVEY

Rose Petals, 2020
Archival pigment print, 40 × 30 in. / 101.6 × 76.2 cm

Some of the many colors incarnated by roses—a species that has between thirty and thirty-five thousand different cultivars—feature in a mound of fallen petals, tumbling down a turned staircase, itself displaying different hues of red, purple, and brown. Although many of British photographer Cig Harvey's (b. 1973) works represent flowers, she has stated that they are not necessarily *about* flowers. Instead, she uses flowers for their metaphorical and symbolic potential, and her depictions of blooms address themes of living and dying, as well as an awareness of the body through the senses. Early in her career, inspired by such photographers as Sally Mann and Andrea Modica, Harvey worked predominantly in black and white to create street photography. In 2003 she began to shift her work toward a deeper analysis of her relationship with color, prompted by an interest in the ways in which different hues can affect the body and mind, and especially examining how they can seduce. Her work explores the connections between trauma and color, following in the footsteps of artists, including Josef Albers and Derek Jarman, who used color and wrote about its relationship with suffering and healing. Initially working intuitively, using the camera as a witness, Harvey then moves into a phase of deep thought about the meanings her images might convey. She embraces the feminine associations of roses and other flowers but also emphasizes their darker side, focusing here on the way they disperse and decay.

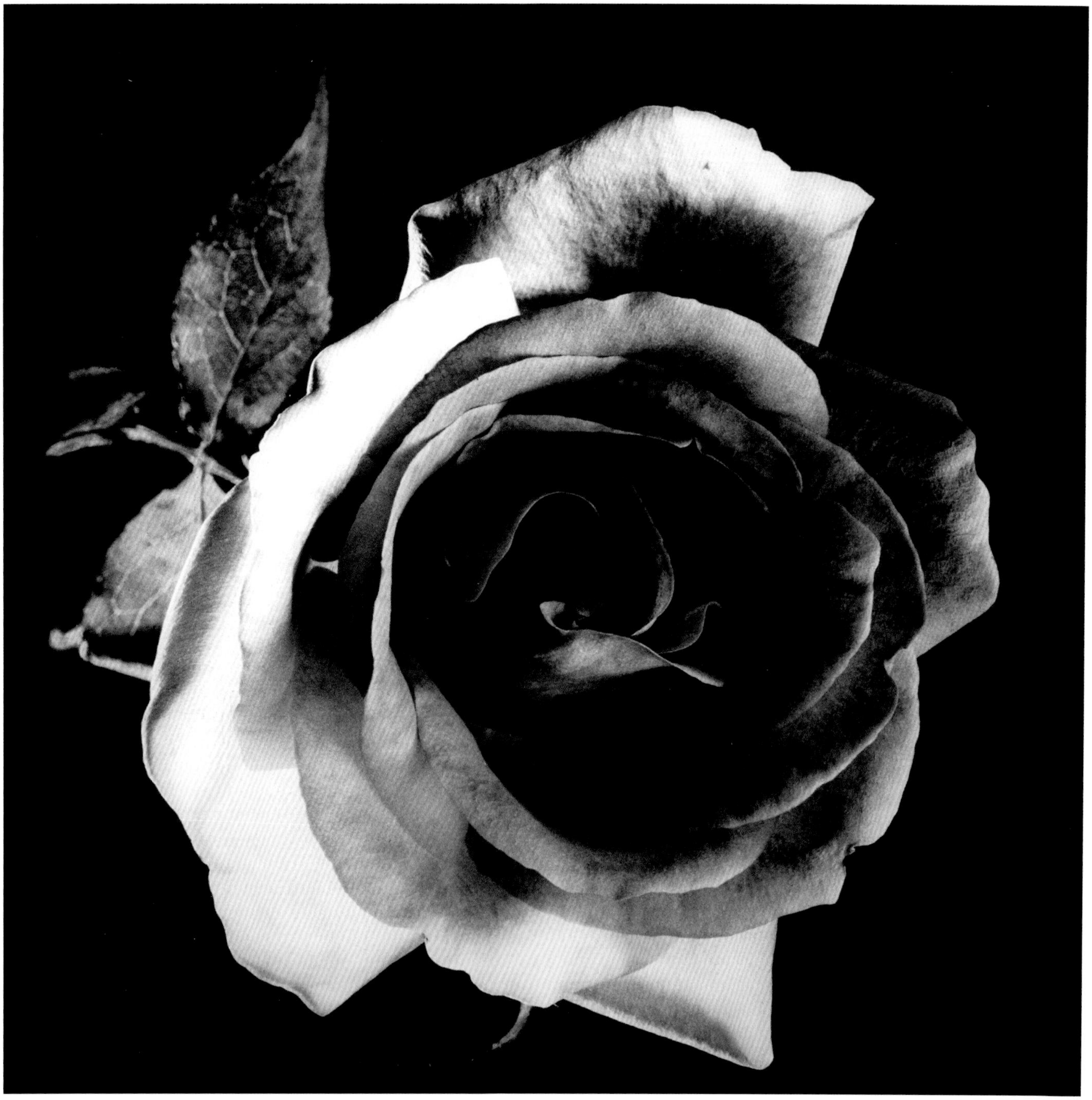

ROBERT MAPPLETHORPE

Flowers (Blue Rose), 1987
Toned gravure, 18¾ × 18½ in. / 47.5 × 47 cm

The blue rose is a fabled flower. Its genetic impossibility in nature has been used to symbolize the unattainable, the longed-for, and the transformative power of love in legends. In Chinese folklore, a princess who is forced to marry sends her suitors on a futile mission to secure this impossible bloom, but, when she is presented with a white rose by the lowly gardener's son—a man she trusts and loves—the rose appears blue to her. This is echoed in Nikolai Rimsky-Korsakov's fairy-tale opera *Sadko* (1898), in which the power of love is so great that a white rose is perceived as blue, a representation of love's divine power. American photographer Robert Mapplethorpe (1946–1989) perhaps draws on these legends in his sensual, stylized photograph *Flowers (Blue Rose)*. Mapplethorpe studied painting, drawing, and sculpture in the early 1960s, but photography would become his principal medium by 1972. He is best known for strikingly realized, intimate nude portraits and still lifes in black and white, but his flower images demonstrate classical aesthetic concerns of form, composition, and beauty. This meditation on a blue rose is both a straightforward depiction of a stunning flower in bloom and a challenge to engage with its fulsome sensuality, laid open for the pleasure and inspection of the viewer. The color, a symbol not only of the transformative power of love but also of all that is unattainable, may speak to the cultural and political contexts in which Mapplethorpe existed and made art—contexts hostile to his expressions of love, sexuality, and the erotic.

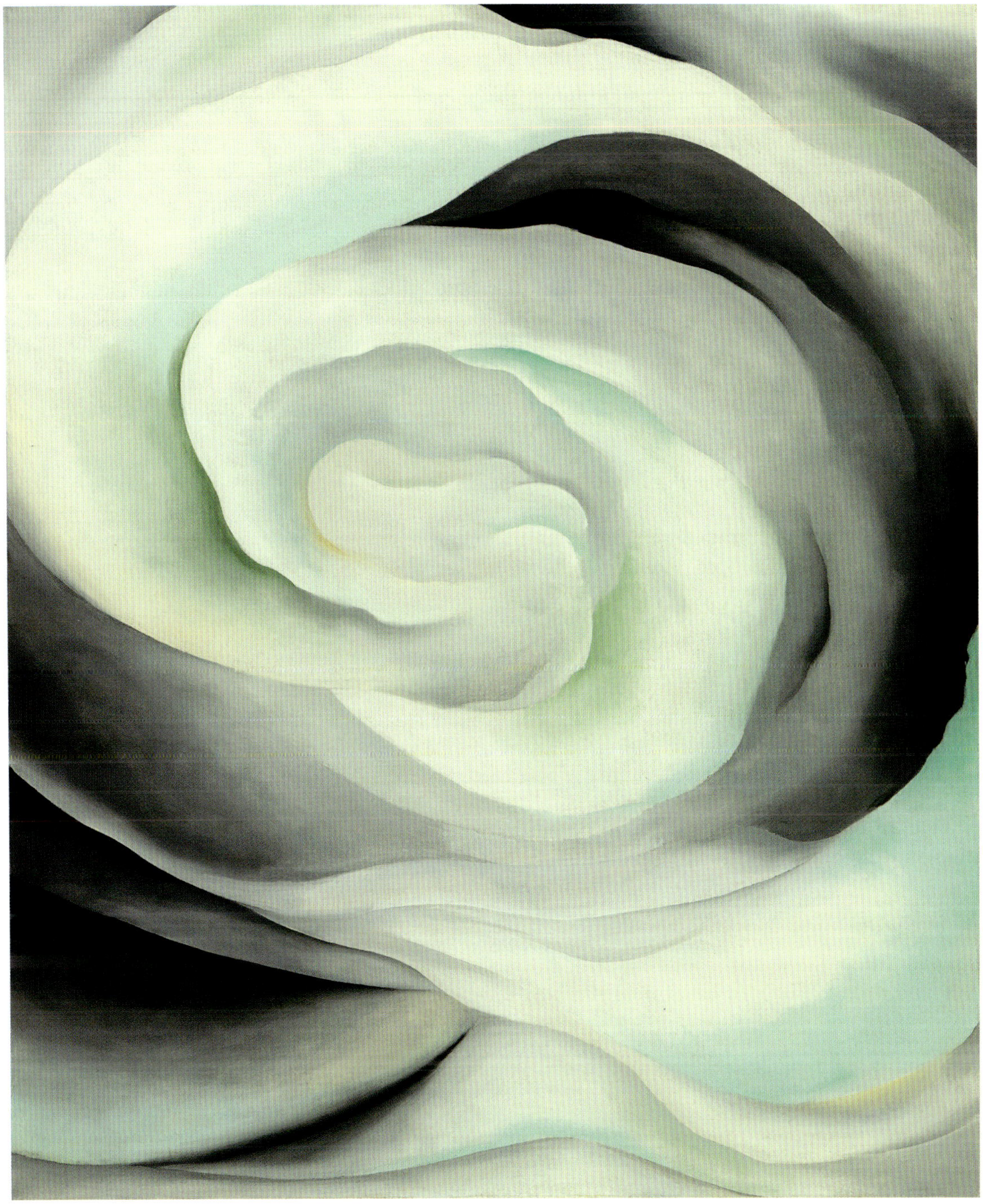

GEORGIA O'KEEFFE

Abstraction White Rose, 1927
Oil on canvas, 36 × 30 in. / 91.4 × 76.2 cm
Georgia O'Keeffe Museum, Santa Fe, New Mexico

In the signature flower paintings of American artist Georgia O'Keeffe (1887–1986), the entire canvas is devoted to a single bloom in extreme close-up. This method of zooming in with great intensity on the part of the flower where its reproductive organs lay evokes natural, but also startling, affinities with human emotions and even anatomy. Photography—with its same power of capturing imagery at dramatically close range—was an important influence on O'Keeffe's particular fusion of realism and illusion. She moved to New York in 1918, meeting and then marrying, in 1924, gallerist and photographer Alfred Stieglitz, with whom she experimented as a model. In the 1920s she started painting flowers that reflected the artistic and emotional inspiration of her relationship with Stieglitz, and which required technical perfection. Her method was to make dense, invisible brushstrokes on extremely fine canvas, creating a surface far richer than the photography they emulated. Here, she captures the shadowy white curves of the rose petals as they swirl toward the center of the abstracted flower. Many of O'Keeffe's subsequent flower paintings have rich and subtle colors, but with this rose the very lack of color seems to add to the power of the image and the mysterious depths between the petals. By naming the painting "abstraction," O'Keeffe suggests reaching for all that a rose might signify: "By emphasis," she said of her art, "we get at the real meaning of things."

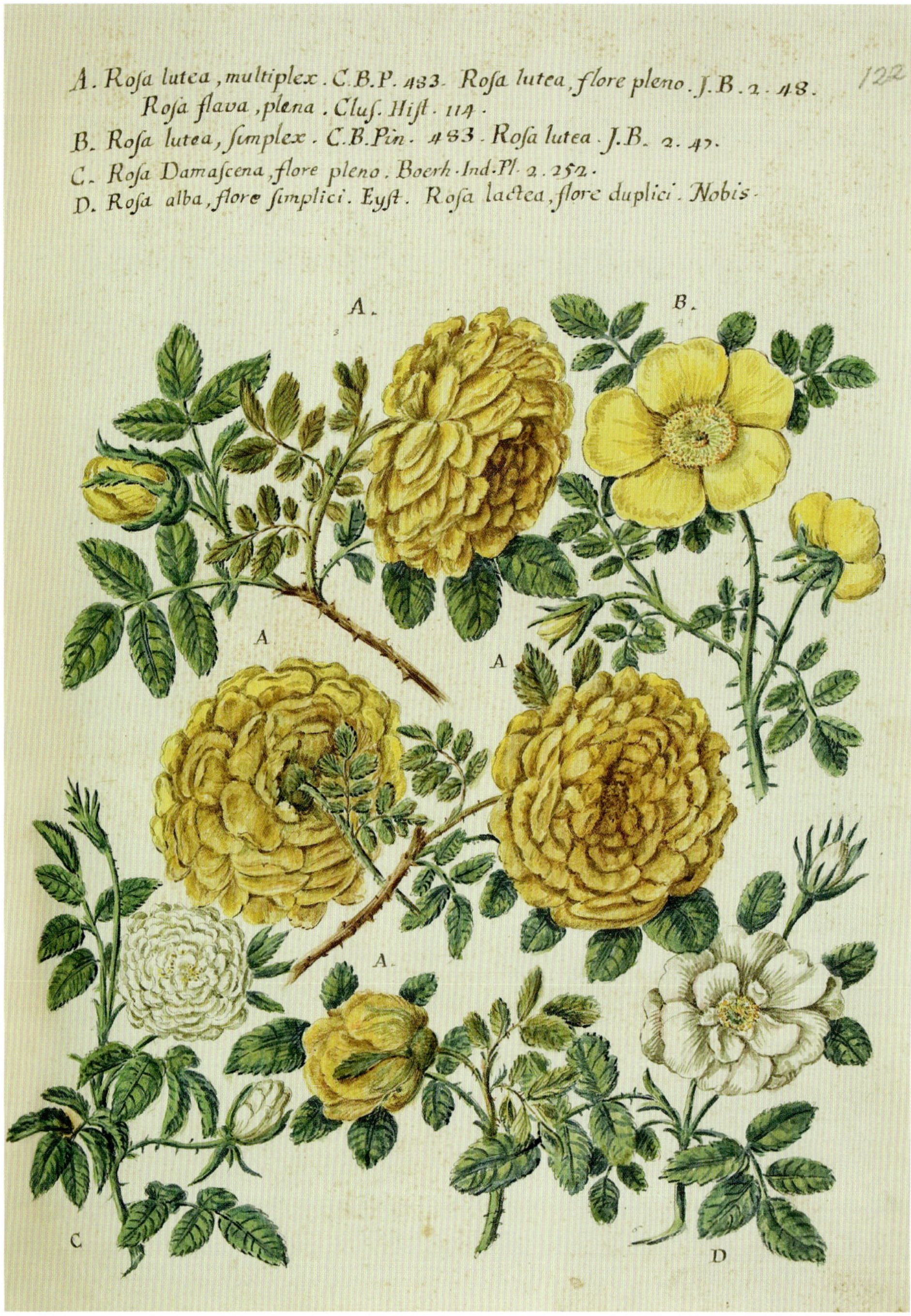

GIOVANNI BATTISTA MORANDI

Rose varieties, from *Hortulus botanicus pictus sive collectio plantarum*, 1748
Illuminated manuscript, 16 × 10⅞ in. / 40.7 × 27.6 cm
Veneranda Biblioteca Ambrosiana, Milan

This engraving of two yellow and two white roses of different varieties belongs to a comprehensive herbal (an identification guide to plants and their medical uses) created in the eighteenth century by Italian botanist and painter Giovanni Battista Morandi (d. 1761). Long before roses became the centerpiece of domestic gardens and floral displays, the flowers were valued mainly for their medicinal and pharmaceutical properties. These four varieties—*Rosa lutea multiplex*, *R. l. simplex*, *R.* × *damascena*, and *R. alba*—are among engravings made by Morandi from his original botanical drawings for *Hortulus botanicus pictus sive collectio plantarum* (*A Painted Botanical Garden or Collection of Plants*). The artist reproduces each rose with great accuracy, showing their thorny stems and carefully highlighting variations in leaf shape and color. This degree of illustrative accuracy became a hallmark of herbals during the Renaissance, when it was made possible by the introduction of metal engraving plates to replace the older woodcuts. During the sixteenth century, a new generation of scientists—Morandi among them—began to study the structure and traits of plants for their own sake instead of as a mere branch of medical science. The medicinal qualities of roses are now overlooked, but for centuries these plants were used to treat such conditions as inflammation, aging, stress, and diabetes. While little is known of the career of the Milan-based Morandi, it is thought that he helped to create the botanic garden in Turin, which opened in 1729.

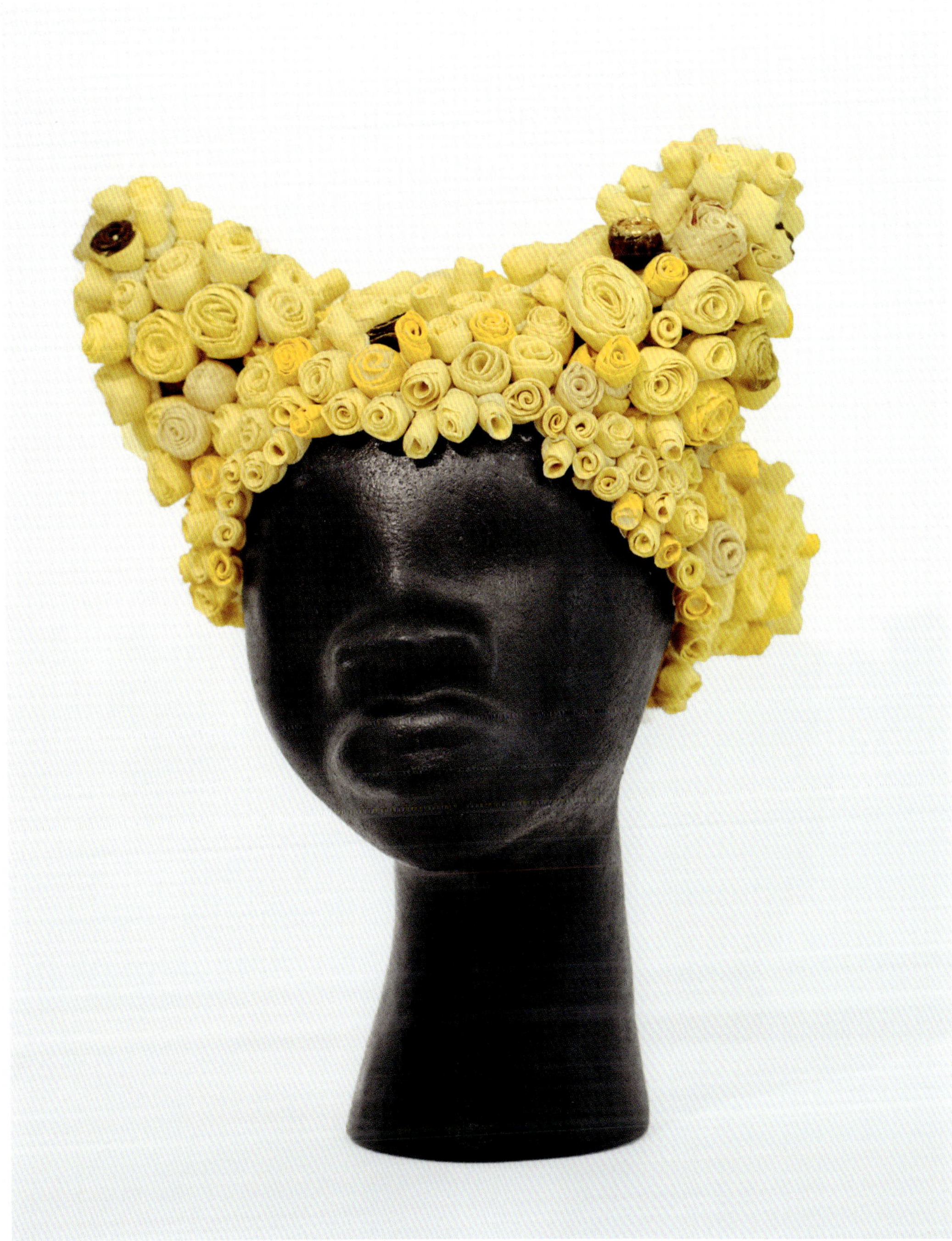

SIMONE LEIGH

Birmingham, 2012
India ink and epoxy on terracotta and porcelain,
15¼ × 11½ × 11½ in. / 38.7 × 29.2 × 29.2 cm
Private collection

This bust of a Black woman, with her mouth closed and her eyes absent, is topped with a magnificent display of budded roses in various shades of yellow. Placed close to its neighbors, each rose is handcrafted in ceramic and, therefore, unique, varying in size and brightness. Tightly curled rosettes in the place of hair have become a signature motif in the work of American artist Simone Leigh (b. 1967), and her figurative sculptures work as an ethnography of the experience of Black women. *Birmingham* has particularly poignant associations; its title connects it to the Birmingham Church Bombing of 1963, a racially motivated terror attack by white supremacists that killed four Black girls in the Alabama city. This serene bust asserts its existence calmly yet powerfully, acting as a monument to and affirmation of those lost girls' lives. Although Leigh's practice includes media beyond sculpture, such as video and installation art, she initially trained as a ceramicist. By using terracotta and porcelain, she connects to the legacies of anonymous African potters who came before her—often women—and also combines references to various folk and visual traditions from Black American and diasporic histories. Leigh's consistent use of the rosette and painstaking creation of each individual flower speak to the diversity in curl shape and type of Black women's hair. The striking yellow rosettes contribute to this regal portrait, their aggregated spiked shapes mimicking the golden form of a crown.

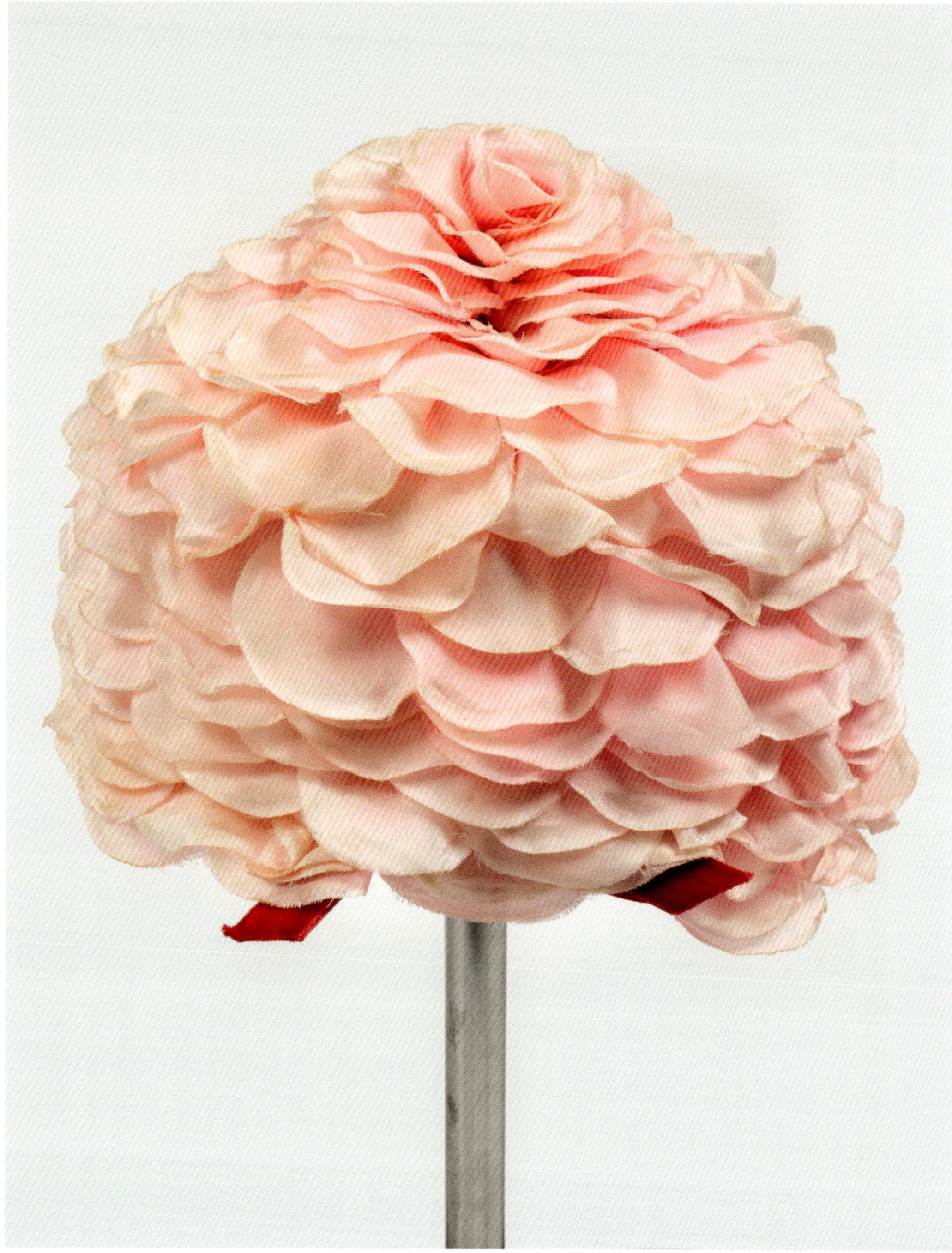

SALLY VICTOR

Hat, c. 1960
Silk and velvet
Museum at the Fashion Institute of Technology, New York

Together with Lilly Daché and Mr. John, American hatmaker Sally Victor (1905–1977) became known as one of the "Big Three" of New York's millinery industry during the mid-twentieth century, when hats were a must-have accessory for smartly dressed women. Victor became one of 1950s America's most successful businesswomen, with clients including Judy Garland, Irene Dunne, Queen Elizabeth II, and Eleanor Roosevelt, but her name is less well known than that of Daché and John. She started her career making hats for Macy's department store in 1924 and became successful enough to open her own studio on 53rd Street in 1934. A member of the Edward C. Blum Design Laboratory at the Brooklyn Museum, Victor drew much of her inspiration from art exhibitions and architecture. She was celebrated for the innovative concentric and collapsible hat constructions she developed throughout her career, merging traditional techniques of hatmaking with synthetic materials and new approaches to shape. Instead of decorating a hat with artificial roses, the milliner created a rose form for this piece, which comprises synthetic fiber rose petals over a pillbox hat with a dark pink velvet bow at center front. A whimsical piece of camp, the hat may have been inspired by Surrealist Salvador Dalí's (see p.87) painting *Woman with a Head of Roses* (1935), which also became the cover of the June 1939 issue of American *Vogue*. Furthermore, Victor may have taken visual cues from the aesthetic convention of likening the female body to a rose in full bloom (see p.61).

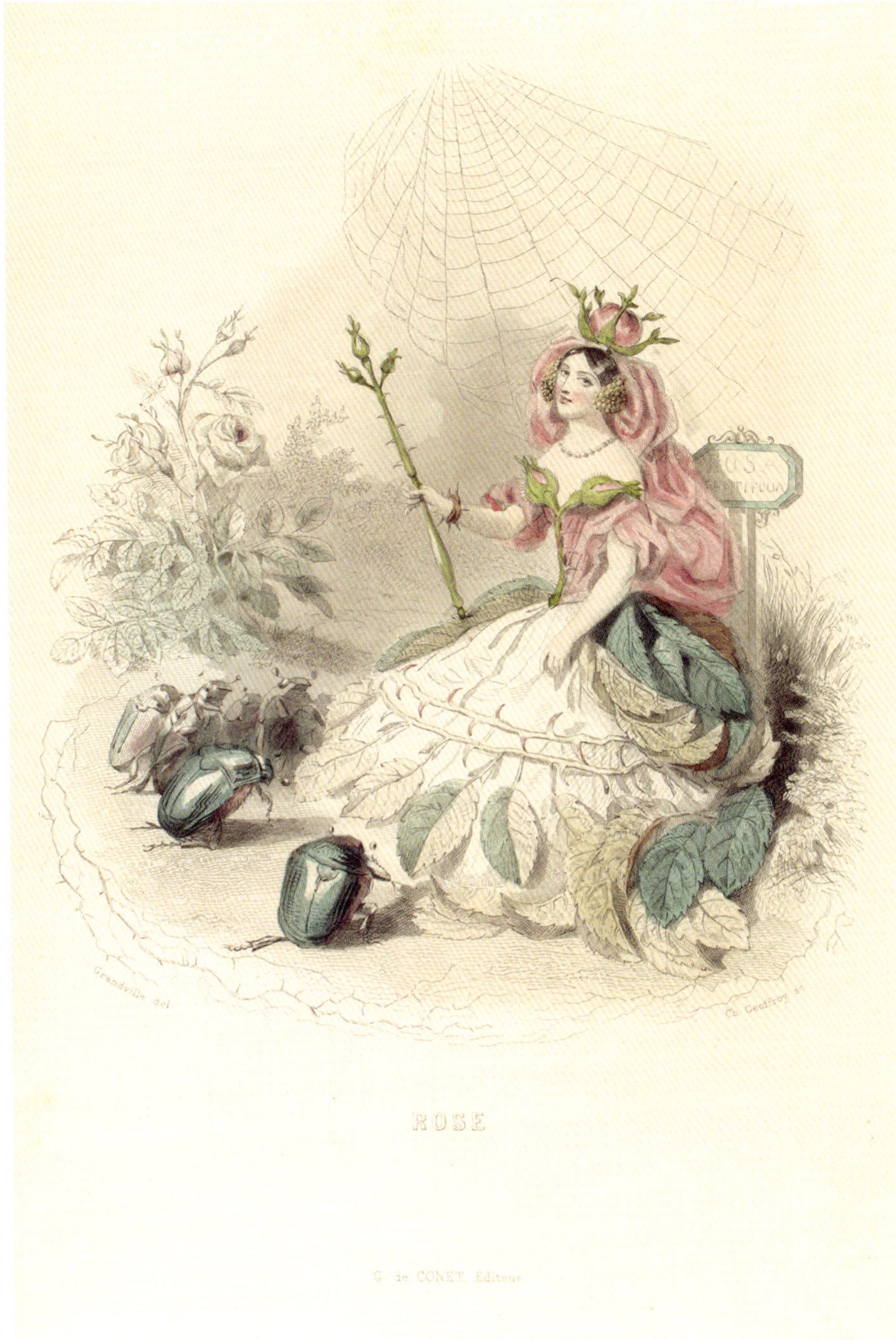

J. J. GRANDVILLE

Rose, from *Les fleurs animées*, 1847
Hand-colored wood engraving, 11 × 6⅞ in. / 28 × 17.6 cm
New York Public Library

In the second-century work *The Adventures of Leucippe and Clitophon*, Greek writer Achilles Tatius personified the rose as female and crowned her queen of the flowers. Visual comparisons between roses and the female body have flourished ever since. The feminization of the flower was particularly popular in the mid- to late nineteenth century, as exemplified in the lively illustrations in *Les fleurs animées* (*The Flowers Personified*) by French artist and caricaturist Jean Ignace Isidore Gérard (1803–1847), better known by the pen name J. J. Grandville. Here, Grandville depicts her rosy royal highness, the cultivated rose, enthroned and costumed in splendor, attended by glittering beetle subjects, and clasping a rosebud scepter of supersharp prickles. Her millinery comprises a rose in full bloom held like a precious jewel by enlarged sepals, while her ceremonial gown, portrayed in the fashionable Victorian full-skirted style of the time, features rose-petal puff sleeves. Female nipples had often been likened to rosebuds, and the queen's bodice is bisected by a prickly stem with two fecund buds, one placed over each breast. A leafy polonaise, or overdress, rambles to the hem and the skirt is encircled by briars. Grandville portrays the flowers in the rosebush beside her from tightly furled to bursting buds and full blooms, a life cycle commonly likened to female sexual awakening. Laden with double meaning, Grandville's queen offers a window into Victorian sensibilities, illustrating the continued use of the rose as an evocative symbol of royalty and femininity.

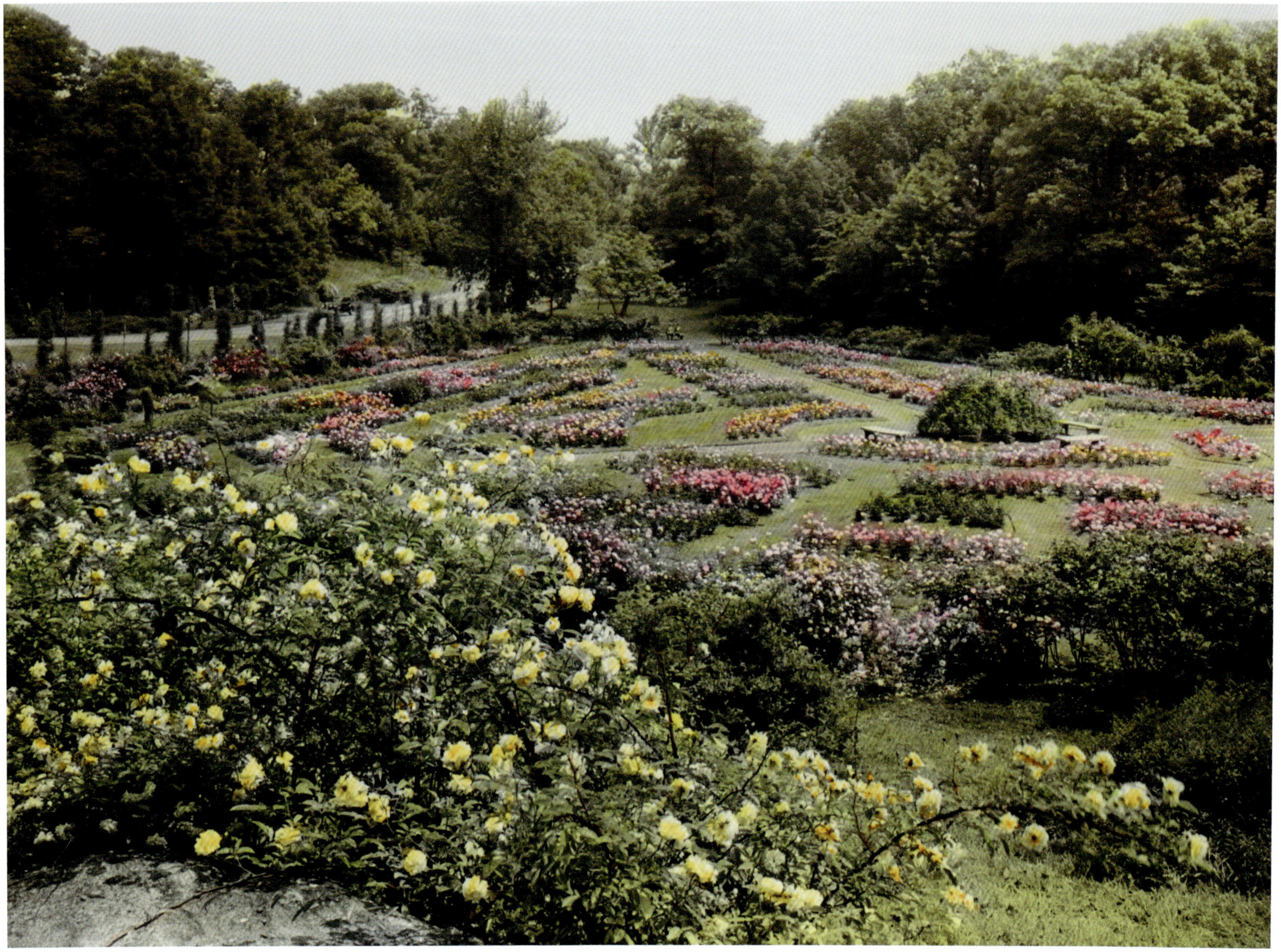

FLEDA GRIFFITH

Rose Garden, New York Botanical Garden, 1937
Hand-colored lantern slide, 4 × 3¼ in. / 10.2 × 8.3 cm
LuEsther T. Mertz Library, New York Botanical Garden

In 1916, celebrated landscape architect Beatrix Farrand designed a rose garden for the New York Botanical Garden. Following the natural lines of the Bronx garden's topography, Farrand's design features an asymmetrical triangular configuration, surrounded by iron lattice fencing and rows of roses radiating from a central gazebo. However, the plan was only partly constructed in 1917, when World War I iron shortages interrupted the project. It was not completed until 1988, when Farrand's original designs were rediscovered and shown to David Rockefeller, whose support made the completion a reality. It was named after his wife, Peggy Rockefeller, who adored roses, and is one of the few remaining intact gardens designed by Farrand, a founding member of the American Society of Landscape Architects and a leading force in her profession. On June 5, 1937, American photographer Fleda Griffith (1878–1962) captured this shot of the garden just as it was entering peak bloom. Griffith was the first woman staff photographer at the New York Botanical Garden and a botanical watercolorist, creating numerous scientific illustrations for the Garden's publications. She regularly delivered public lectures about the Garden, and her talks were accompanied by hand-colored lantern slides that she painted herself, such as this one. Today, the Peggy Rockefeller Rose Garden is one of the most popular horticultural displays at the Garden, featuring a colorful tapestry of more than three thousand roses representing more than seven hundred species, hybrids, and cultivars.

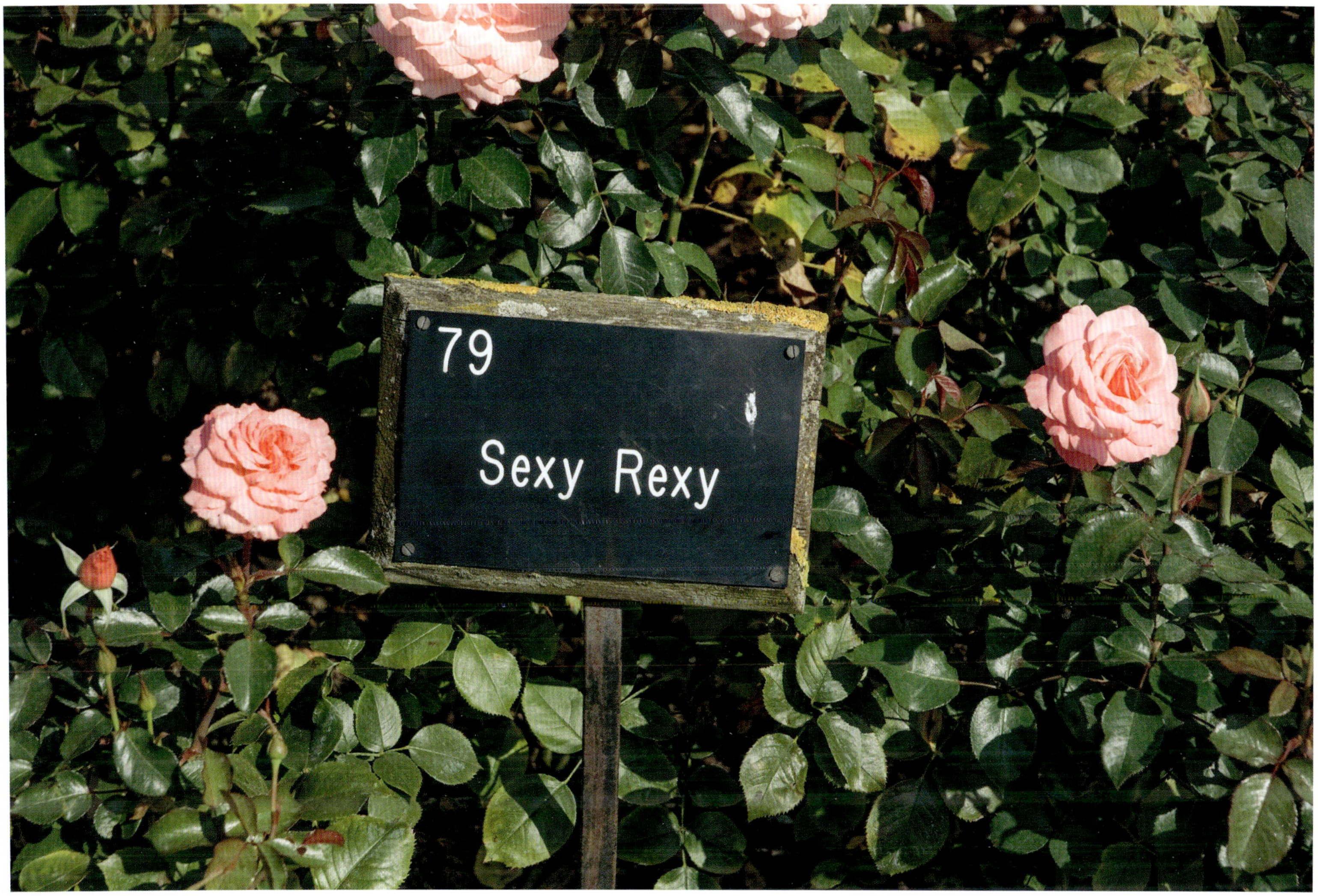

RUSS MCCLINTOCK

Sexy Rexy, from *24 Rose Varieties*, 2012
Photograph, dimensions variable

The names given to roses can be as rich and intriguing as the plants themselves, ranging from celebrities and historical figures to eccentric, evocative, and even cheeky monikers. For his series *24 Rose Varieties*, American photographer Russ McClintock documented some of the more unusual roses found in English gardens, including at Waddesdon Manor, Blenheim Palace, and Queen Mary's Garden in Regent's Park (see p.161). Finding the names more fascinating than the roses themselves, McClintock focused his lens on individual plant labels, including 'Teasing Georgia', 'Pretty Polly', 'Village Maid' (see p.88), and 'Sexy Rexy'. The last, shown here, is an award-winning Floribunda rose bred by Sam McGredy IV and initially introduced into New Zealand in 1984. It produces soft pink, heavily scented, double flowers and boasts glossy green foliage. McGredy was one of the world's greatest rose breeders, renowned for his vigorous and brightly colored plants and awarded many international prizes, including the Golden Rose at the Hague Rose Trials an unprecedented four times. He was bold not only in his breeding goals but also in his choice of names, which include 'Picasso' and 'Elizabeth of Glamis', named after and with permission from the late British Queen Mother. 'Sexy Rexy' was named after a female friend, who was reportedly flattered by the gesture. Despite receiving criticism from the British Rose Growers' Association, which declared the name to be in poor taste, this fun rose remains an international favorite.

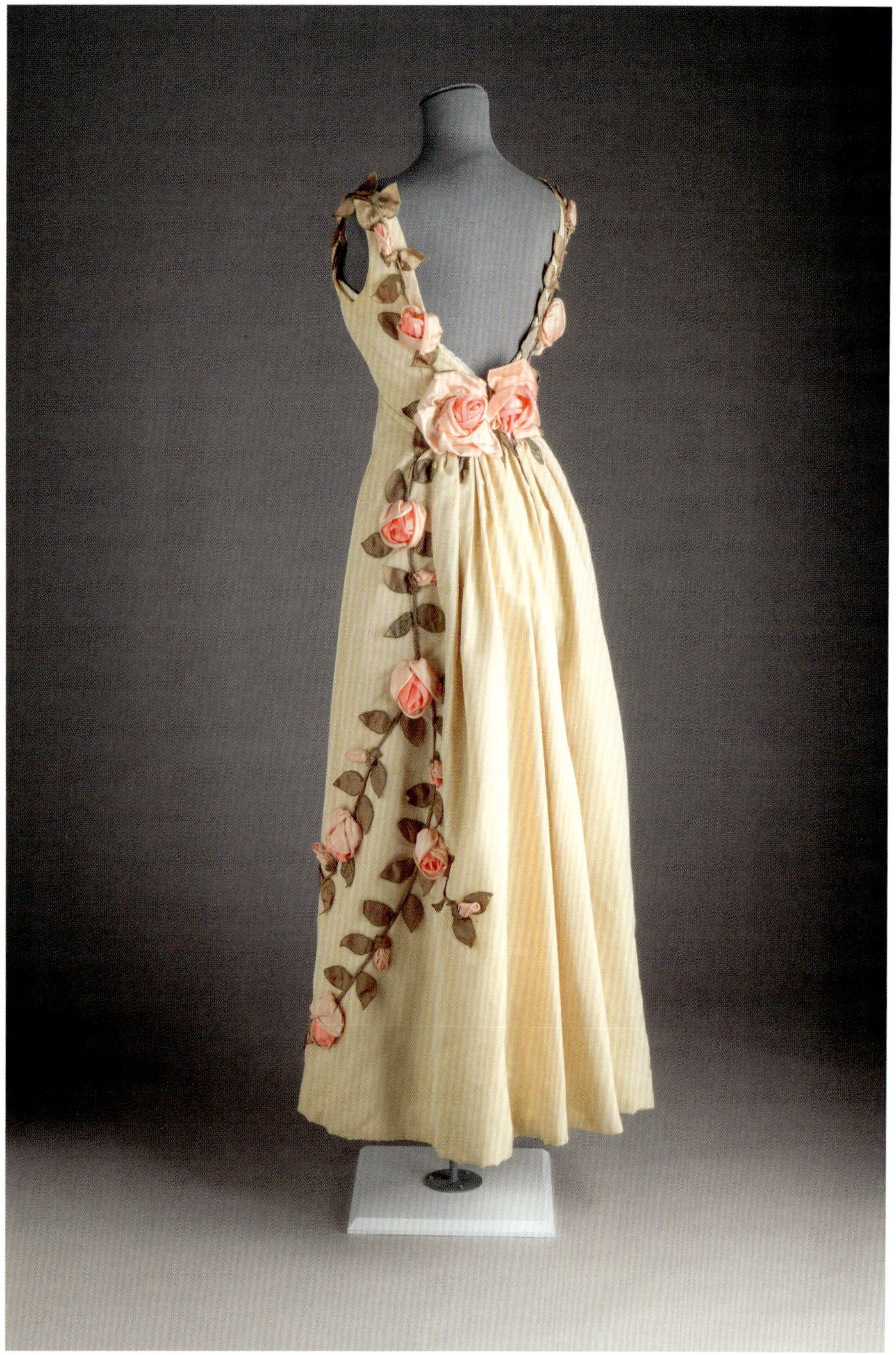

ANN LOWE

Dress, 1966–67
Silk, tulle, linen, metal, and elastic
National Museum of African American History and Culture, Washington, DC

Ann Lowe (1898–1981) is one of American fashion's most important designers and one of the first African Americans to obtain commercial success in the field. Between 1968 and 1972, she ran Ann Lowe Originals on New York's prestigious Madison Avenue, and rose-decorated gowns were among her signature styles. The granddaughter and daughter of enslaved women, whose freedom was purchased by her grandfather, Lowe descended from dressmakers. In the 1910s, her mother and grandmother developed a thriving dressmaking business in Montgomery, Alabama, that catered to affluent white clients. As a child, Lowe re-created flowers from the garden using scraps of leftover fabric and made her own dress patterns by age ten. She moved to New York in 1917 to attend the S. T. Taylor School of Design, although she was made to work in a room alone because of racist segregation. In the 1920s, she worked for such leading designers as Hattie Carnegie. She opened Ann Lowe Gowns in the 1950s, specializing in evening dresses and debutante gowns for American socialites, most famously designing Jacqueline Bouvier Kennedy's wedding dress in 1953. When the French rose 'Madame Ferdinand Jamin' was imported into the United States in the early twentieth century, it was renamed 'American Beauty'. With its brilliant color, blooms of up to fifty petals, and long, straight stem, it was an immediate hit. Here, the designer abstracted the rose to create decorative vines that trail across the gown, leading to its nickname as Lowe's "American Beauty" dress.

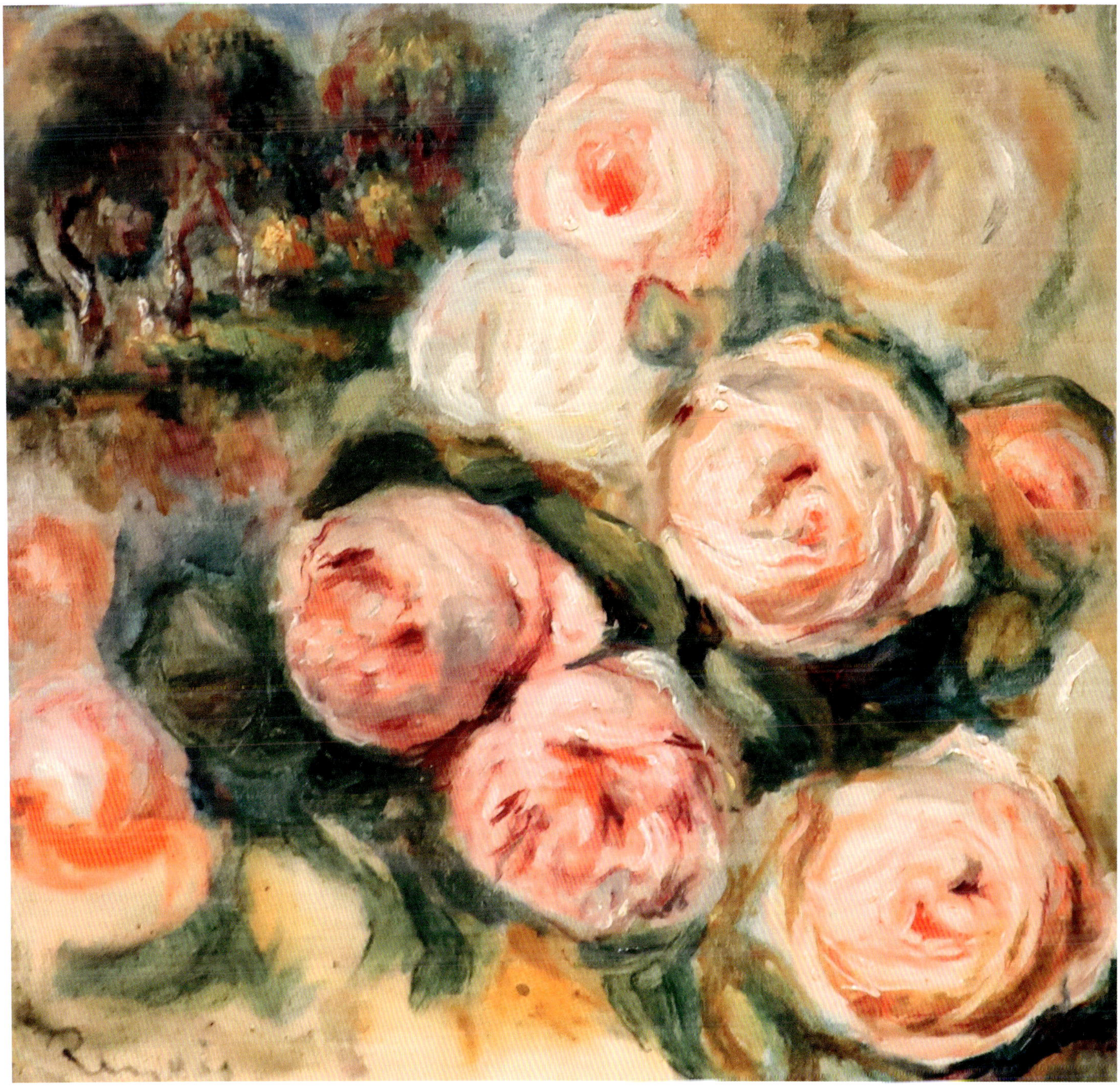

PIERRE-AUGUSTE RENOIR

Roses and Landscape, date unknown
Oil on canvas, 10¾ × 10¾ in. / 27.3 × 27.3 cm
Phoenix Art Museum, Arizona

These romantic roses, full bodied and fleshy pink, loll and sprawl as if about to spill from the frame. Painted by French artist and Impressionist pioneer Pierre-Auguste Renoir (1841–1919), they are the floral equivalent of the artist's portraits of young women with plump bodies and creamy skin, or his scenes of leisure and pleasure in sun-dappled landscapes. The bold, fluent brushstrokes and soft blending of color evoke a mood of calm relaxation. This is one of two paintings with the same title, each dominated by a bouquet of voluptuous roses with a small landscape squeezed into a corner of the canvas. Here, we see trees in full leaf in a sunlit glade, the whole picture a celebration of natural abundance, warmth, and sunshine. In the 1890s, Renoir developed rheumatoid arthritis and, to alleviate his pain, moved in 1907 to the warmer climate of Cagnes-sur-Mer, a town on the Mediterranean coast, where this picture was probably painted. The arthritis made it more difficult for him to hold a paintbrush, and this may explain the shift in his style during this period, when his characteristically smooth, precise marks became much looser and freer. But this stylistic change can also be attributed to his attitude to the subject matter. He felt that his still-life pictures allowed him greater scope to experiment than his figure paintings did: "When I paint flowers, I feel free to try out tones and values and worry less about destroying the canvas."

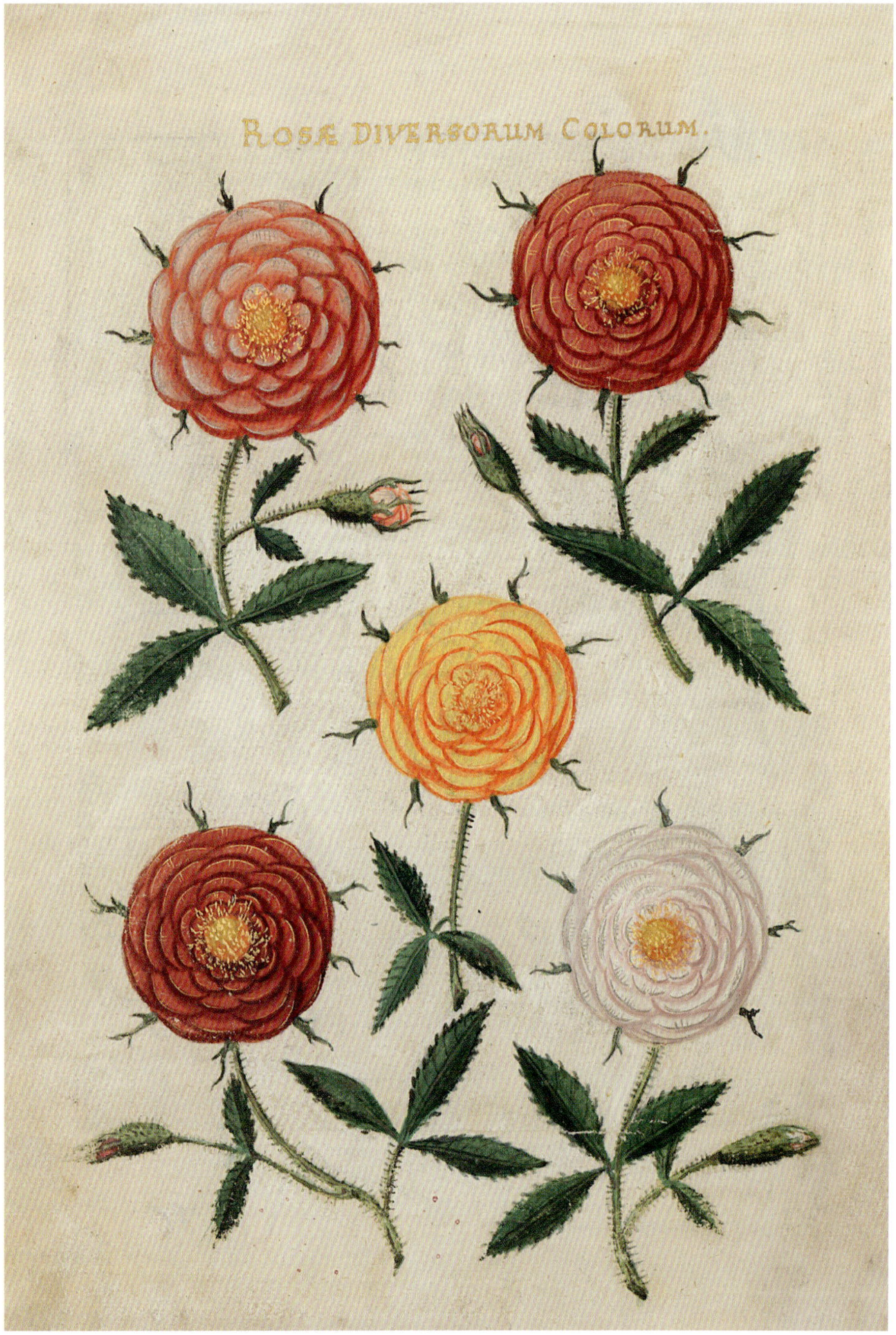

ANONYMOUS

Folio 16v, from *Historische Beschreibung der fürstlichen Kindtaufe Fräulein Elisabethen zu Hessen*, 1598
Illuminated manuscript, 11⅜ × 7⅞ in. / 29 × 20 cm
Munich State Library

Six roses decorate this page from a kind of medieval German scrapbook, filled with allegorical and natural illustrations. In 1596, in the German principality of Hesse-Kassel, the Landgrave Maurice celebrated the christening of his daughter Elisabeth with four days of tournaments and games. An account of the celebrations was published in 1598 by Wilhelm Dilich, a rising artist and publisher in Kassel. Some 150 pages include engravings of marching bands, people costumed as figures from classical mythology, and miscellaneous exercises in horsemanship. The copy that Dilich presented to the Landgrave, and which he himself had hand-colored, is notable not so much for the official content of the work, as for the additional decorations that have been painted throughout by an anonymous hand. These were added wherever there was a blank space on the paper, probably before 1606, the date of the volume's binding. One previously empty page contains a drawing of five *Rosae diversarum colorum* (roses of various colors), each with some three dozen petals. These, while drawn in a stylized manner, may well represent specimens of the new varieties of Gallica rose that were beginning to appear in the Netherlands, and which were named "Centifolia" roses in memory of the hundred-petaled roses described by Theophrastus and Pliny in Classical times (see pp.24 and 180). However, the inclusion of a yellow specimen may suggest that the roses were purely a fantasy because there are no naturally yellow Gallicas.

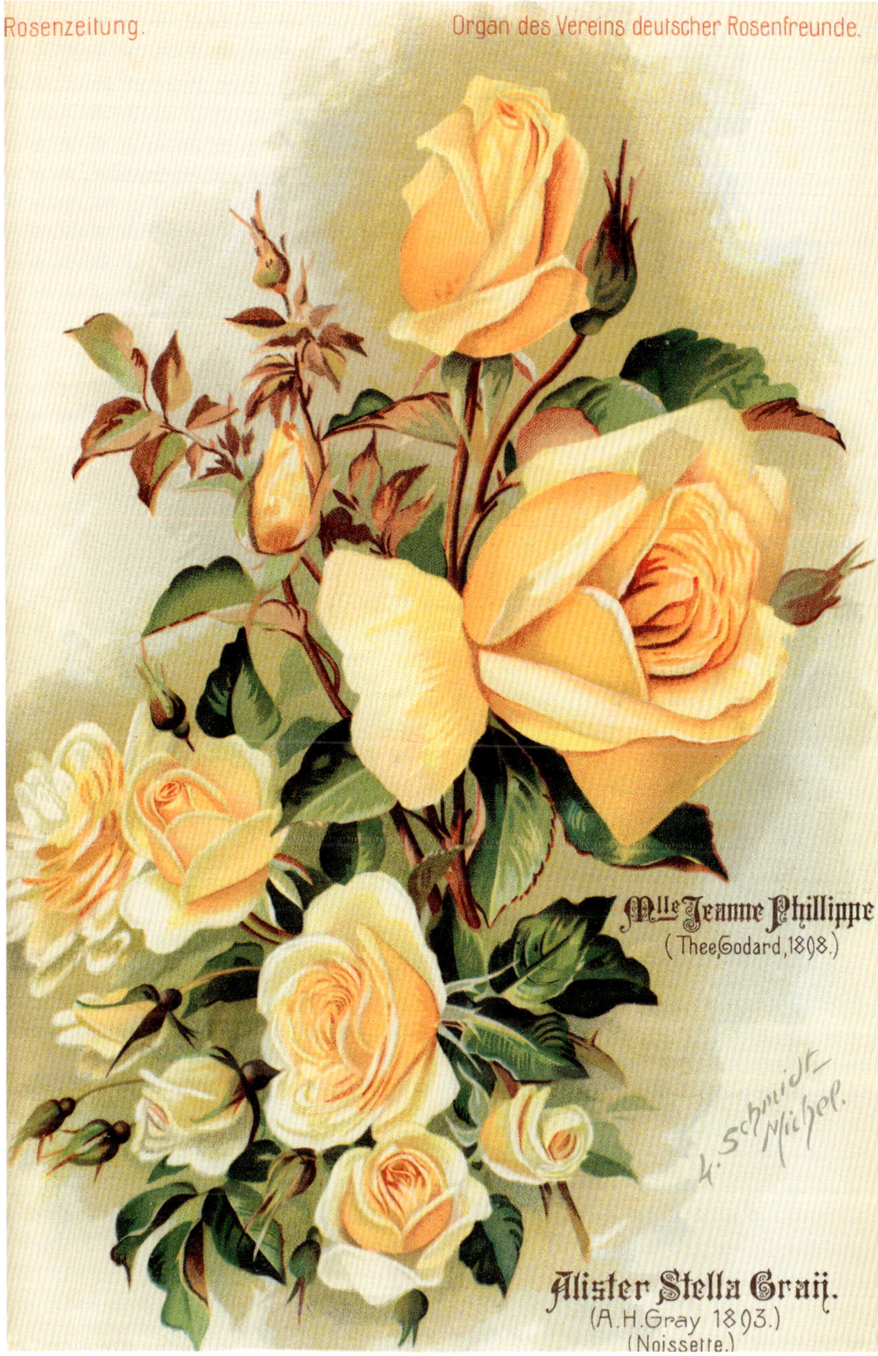

LENA SCHMIDT-MICHEL

Mlle Jeanne Phillipe (Thee, Godard, 1898) and Alister Stella Graii (Noissette, A. H. Gray, 1893), from Rosen-Zeitung, 1903
Chromolithograph, 10⅜ × 7⅛ in. / 26.4 × 18.2 cm
Royal Horticultural Society, Lindley Library, London

Rosen-Zeitung was the magazine of the *Verein Deutscher Rosenfreunde* (German Rose Association), founded in 1886. This popular periodical ran until 1933 and was revived briefly in the 1990s. During its heyday, each issue opened with a specially commissioned chromolithograph depicting a rose cultivar. German artist Lena Schmidt-Michel (d. 1944) was assigned to depict the apricot blooms here, one a Tea rose and the other a Noisette, for the 1903 edition. For each commission, Schmidt-Michel would paint a rose in watercolor, and the work would be printed as a chromolithograph by printer and publisher J. Lintz. The magazine was edited by the Trier rose grower Peter Lambert (see p.57), who helped establish the great rose garden at Sangerhausen in Saxony-Anhalt. At the beginning of the twentieth century, Lambert bred a new range of roses called Lambertianas (the name has been largely abandoned, but these roses were the ancestors of the Hybrid Musks); his most famous rose was 'Schneewittchen' (snow white), which was marketed in the English-speaking world as 'Iceberg'. In this plate, Schmidt-Michel has depicted two roses. The first is the Tea rose 'Mademoiselle Jeanne Philippe', raised by Antoine Godard in 1898. The second is the far more famous 'Alister Stella Gray', bred in 1893 by rosarian Alexander Hill Gray and named after his son. Schmidt-Michel worked on so many commissions for *Rosen-Zeitung* that, in 1906, one of the roses illustrated was named 'Lena Schmidt-Michel' in her honor.

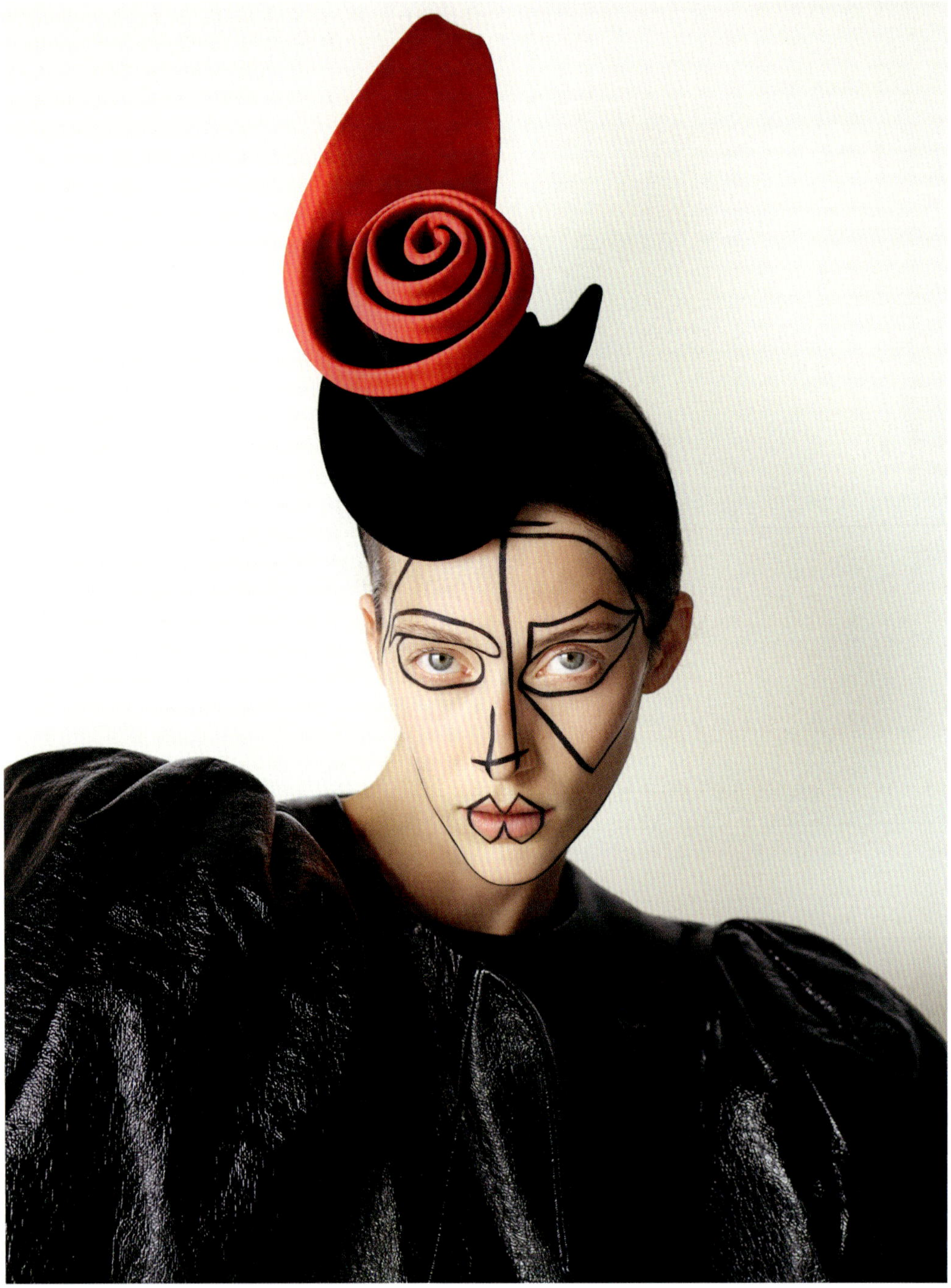

STEPHEN JONES

Rose Royce, from *Contours* collection, 1996
Velvet and satin
Photograph by Ben Toms, modeled by Alana, 2017

This little black velvet top hat—dubbed *Rose Royce*—is edged in what English milliner Stephen Jones (b. 1957) calls "a twirl of lipstick pink." The luxurious satin swirls upward from the crown until it is transformed into a rose. Jones's strategic placing of the deepest pink edge, verging on red, creates the impression of the flower in full bloom. In a career that has spanned more than four decades, Jones has called this hat his favorite of all the thousands he has designed. Collaborating with some of the world's greatest fashion designers, such as John Galliano at Dior, Vivienne Westwood, and Thom Browne (see p.227), Jones creates sensational hats for catwalk shows while also consistently designing pieces for private clients. Jones explains that *Rose Royce*—which he made for his collection *Contours* in Spring/Summer 1996, the first year he worked for Dior—appears effortlessly simple but was extremely difficult to construct. The storied fashion house allowed Jones, whose aesthetic had grown out of the Punk and New Romantic movements in London in the late 1970s and 1980s, to experiment with materials and forms previously unavailable to him in the DIY scene. Using millinery techniques from the 1950s, when women still wore hats daily, he was able to play around with the construction of this shape for his own collection until it became this striking stylized rose.

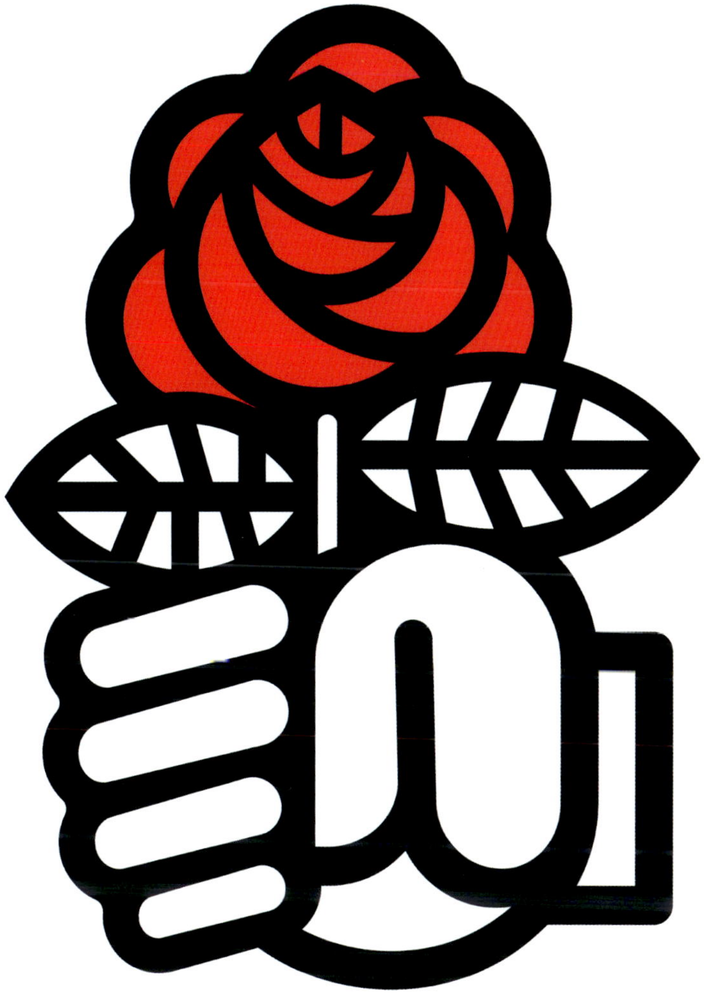

MARC BONNET

Fist and Rose, 1969
Logo, dimensions variable

Created in 1969 by French graphic designer Marc Bonnet (b. 1949) for the French Socialist Party, the rose-and-fist emblem made its street debut in January 1970 with several thousand posters pasted up around Paris. It was adopted as the party's official logo in 1971. Bonnet, who studied in Paris and Zürich, was neither a political activist nor a branding expert, and was more likely to create Art Deco–style flower motifs for textiles than political imagery. At first he turned down the commission to create the logo, but he was eventually convinced to take it on by his friend and former employer Yann Berriet, a communications director with close ties to the party in Paris. Bonnet produced a series of flower-themed options, influenced by 1960s counterculture and the flower power movement initiated during the Summer of Love in San Francisco in 1967. From a fist holding a carnation, a bouquet, and a single rose, the last won out. It is a powerful image that combines the rose as a symbol of the flourishing life promised by socialism with the fist, which represents the struggle, activism, and solidarity needed to achieve such an outcome. In 1979 the image was adopted by the Socialist International group, and the design has since been iterated in numerous versions, adapted by socialist groups across Europe and in Africa, Asia, and the Americas, with changes to skin tone and the number of hands, or with the leaves colored green as an acknowledgment of environmental issues.

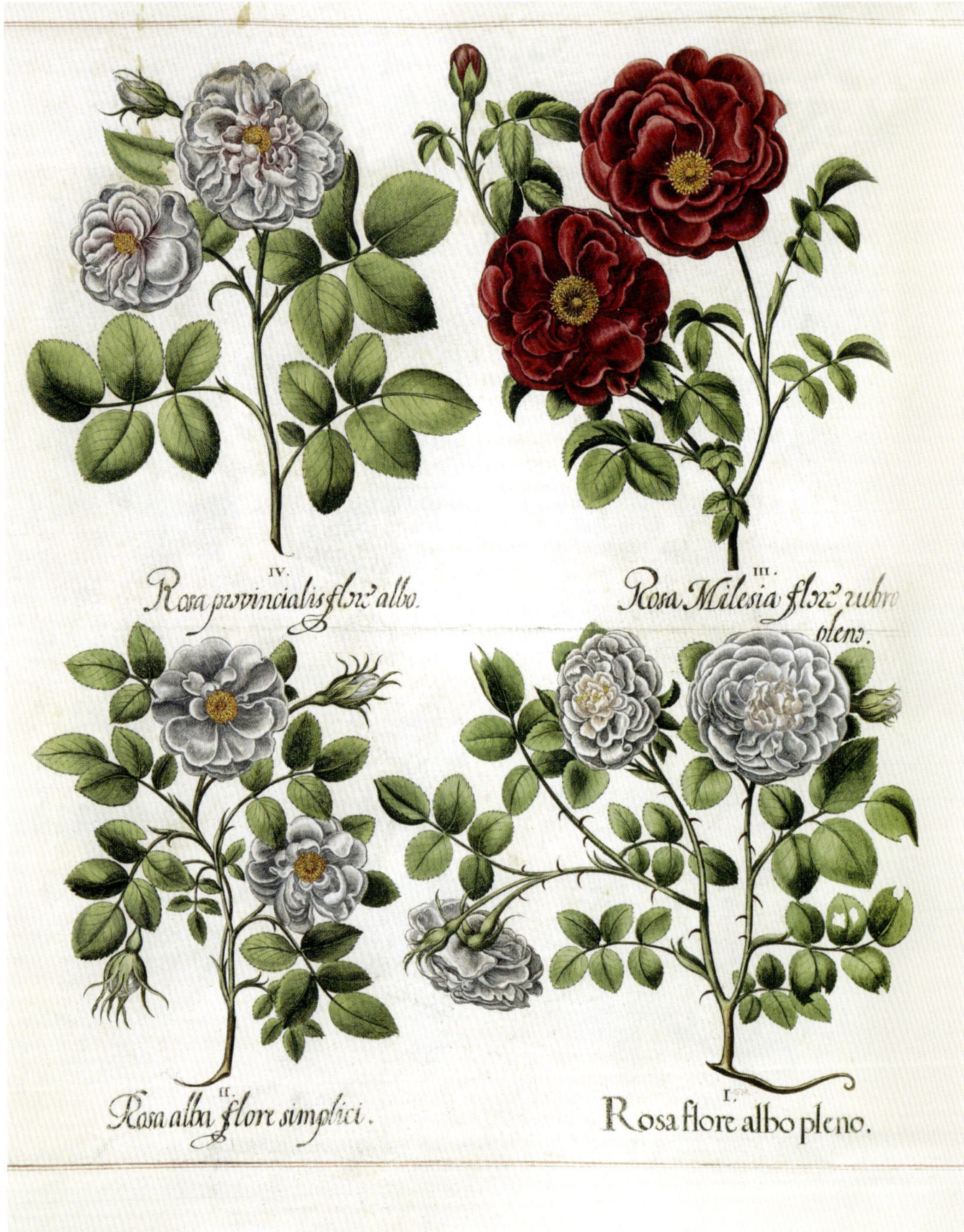

BASILIUS BESLER

I. Rosa flore albo pleno, II. Rosa alba flore simplici, III. Rosa Milesia flore rubro pleno, IV. Rosa provincialis flore albo, from *Hortus Eystettensis*, 1613
Hand-colored copperplate engraving, 20¾ × 17 in. / 52.7 × 43.2 cm
British Library, London

This illustration of four roses, three white and one red, is from arguably the greatest early florilegium, *Hortus Eystettensis*, published in 1613. As opposed to an herbal (a utilitarian compendium aimed at displaying the medicinal and botanical characteristics of a plant), a florilegium showcases aesthetic qualities, so the plant is depicted true to nature and life size. Most of these floral anthologies described plants from a single place or garden. About 1606, Johann Konrad von Gemmingen, Prince-Bishop of Eichstätt in Bavaria, commissioned the Nuremberg apothecary Basilius Besler (1561–1629) to manage a group of illustrators who would create copper engravings of plants from his garden. The prince, an enthusiastic botanist, sent boxes of flowers weekly to Nuremberg for Besler's artists to engrave, and Besler also grew the same species in his own garden to ensure that the freshest flowers were available for his work. The *Hortus Eystettensis* was arranged in three sections according to the seasons: one for spring, one for summer, and a third devoted to late summer, fall, and winter. These roses were featured in the spring section, and it is of historical interest to appreciate which species were grown in a prominent European collector's garden in the early seventeenth century. The *Hortus* was published in a large royal-folio format to display the plants true to size, and the prince's deluxe version featured all the plates in color. Tragically, the prince died before Besler's magnificent tribute to his prized garden was published.

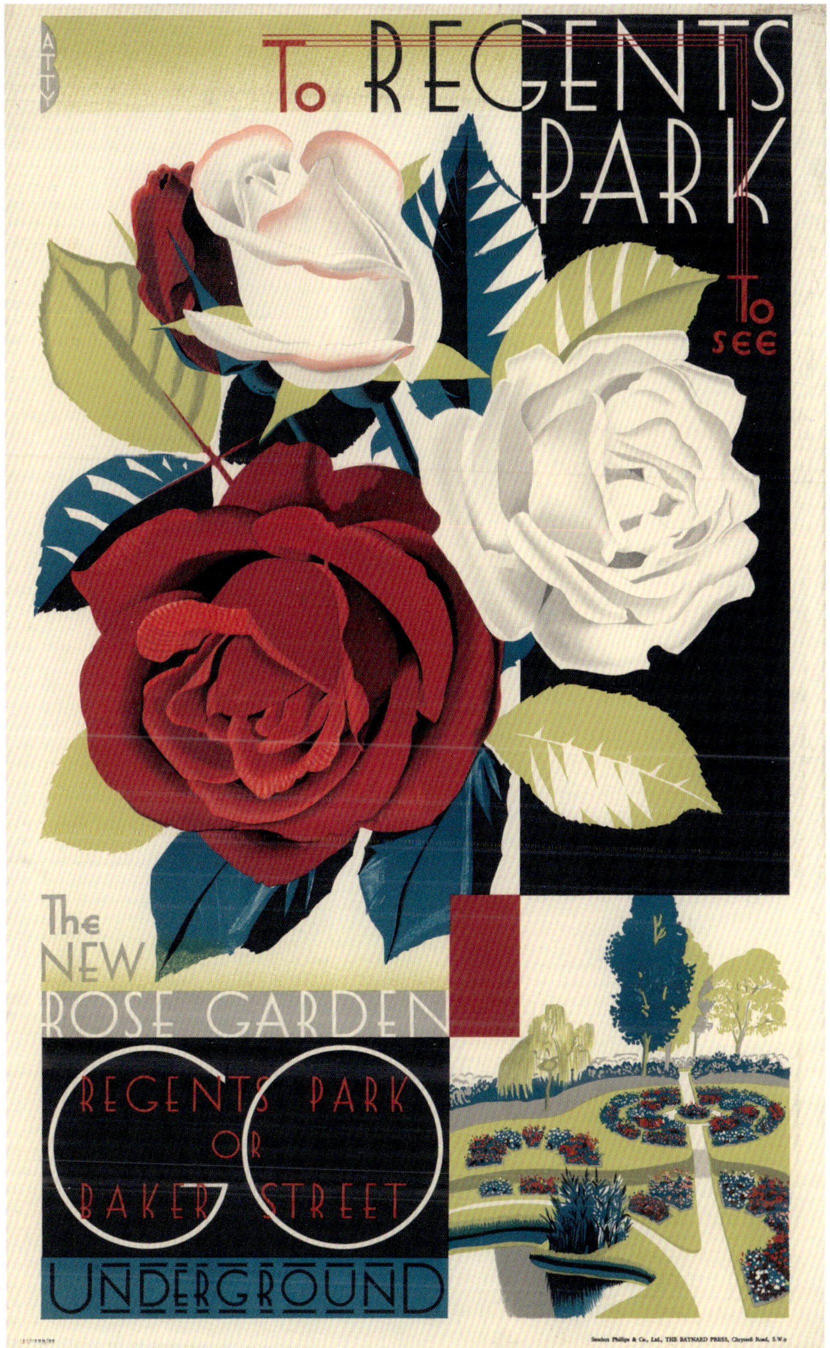

DORA M. BATTY

To Regent's Park to See the New Rose Garden, 1932
Color lithograph, 40 × 25 in. / 101.6 × 63.5 cm
London Transport Museum

Queen Mary's Gardens (named after the wife of King George V) can be found at the heart of Regent's Park in north London. They were opened to the general public in 1932, and in the same year the garden's first superintendent began planting a rose garden, which was completed in 1934. The London Underground's design department was quick to commission a poster alerting people to this new visitor attraction. The chosen artist was Dora M. Batty (1891–1966), an experienced designer and illustrator who had already produced many transportation posters, several of them promoting London's parks and gardens with their floral displays of all kinds, from bluebells and foxgloves at the Royal Botanic Gardens, Kew, to crocuses in the royal parks. Here, velvety red and white roses in bud and full bloom dominate the design, with a vignette illustrating the winding paths and curving beds of the new rose gardens included below. With its narrow, elongated sans-serif lettering and asymmetric blocks of black and red, the poster has a bold Art Deco–style elegance. In a witty touch at bottom left, Batty drew the letters G and O as perfect circles on a black ground to suggest the tubular tunnels of the London Underground railroad. Today, the gardens hold London's largest collection of roses—more than twelve thousand plants in eighty-five beds, a dazzling mix of old and new varieties.

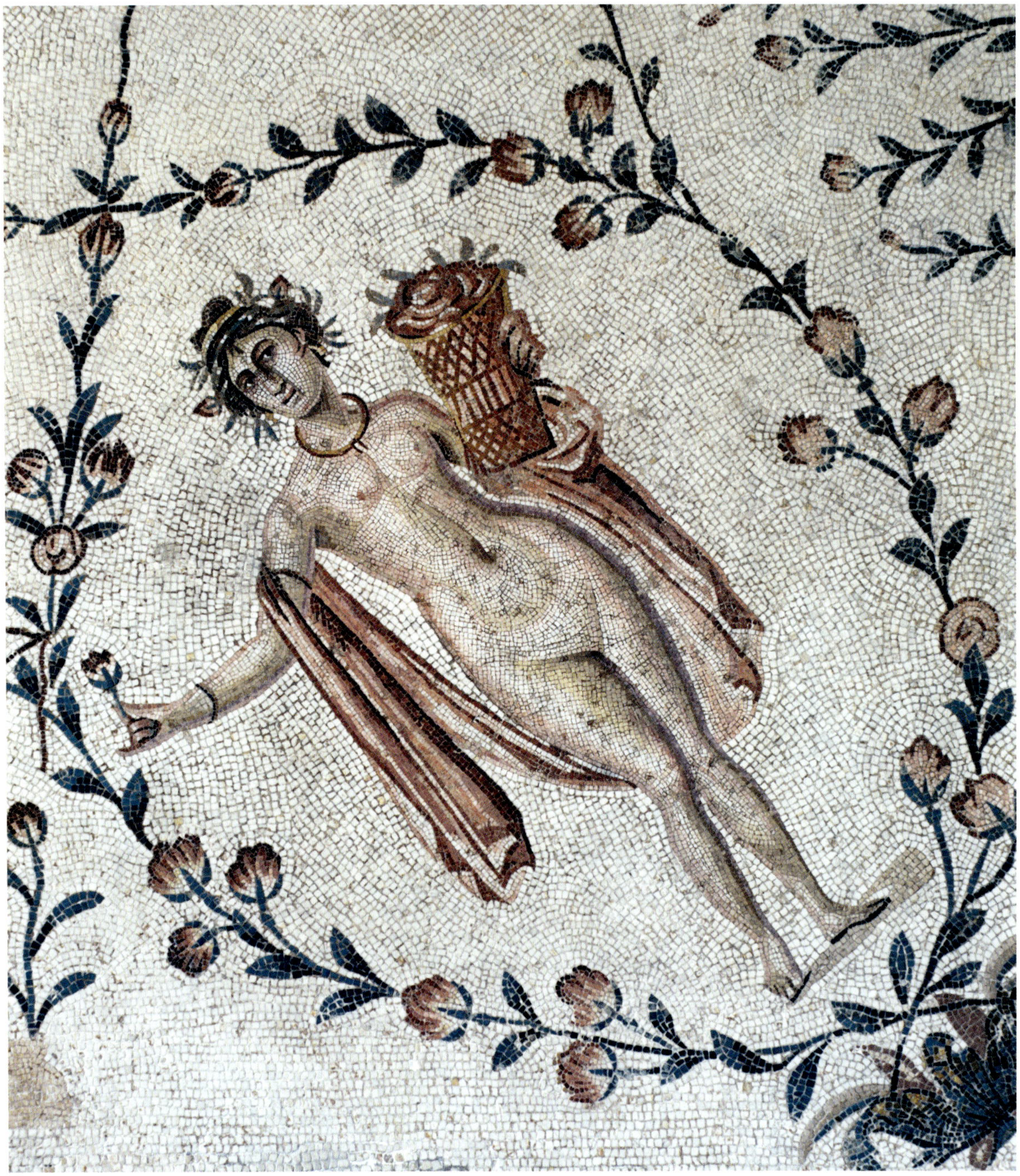

ANONYMOUS

Figure of Spring (detail) from the Triumph of Neptune, late 2nd century CE
Stone mosaic, overall 16 × 16 ft. / 4.9 × 4.9 m
Musée National du Bardo, Tunis, Tunisia

This striking work is just a corner from an enormous mosaic found in the private baths at the Roman settlement Caput Vada (present-day La Chebba) in Tunisia in the 1880s. The central part of this mosaic depicts a story from a myth known as the Triumph of Neptune, which tells of the god of the sea's conquest of the nymph Amphitrite. Neptune first set eyes on Amphitrite while she was dancing at low tide with her sisters on the island of Naxos; he immediately fell in love with her and asked for her hand in marriage. Amphitrite, afraid of his turbulent character, refused his advances and fled. To win her love, Neptune sent gifts of pearls, coral, and other treasures, but still she refused. Finally, Neptune turned to the dolphin Delphinus, who spoke so movingly on behalf of the sea god that Amphitrite gave in and agreed to marry him. Each corner of the mosaic includes an allegorical vignette tied to the four seasons, and this one shows spring as personified by Flora, the Roman goddess of flowers, who is often illustrated surrounded by roses, as here. Flora represents love, beauty, and profusion, and is, perforce, a goddess of youth. Here, Flora is holding a cornucopia, a symbol of fertility, probably tied to the newlyweds' future happiness. She is often surrounded by other figures expressive of the seasons and the natural world, such as birds, animals, and flowers, and this is a striking example of the use in the ancient world of roses as icons of love and abundance.

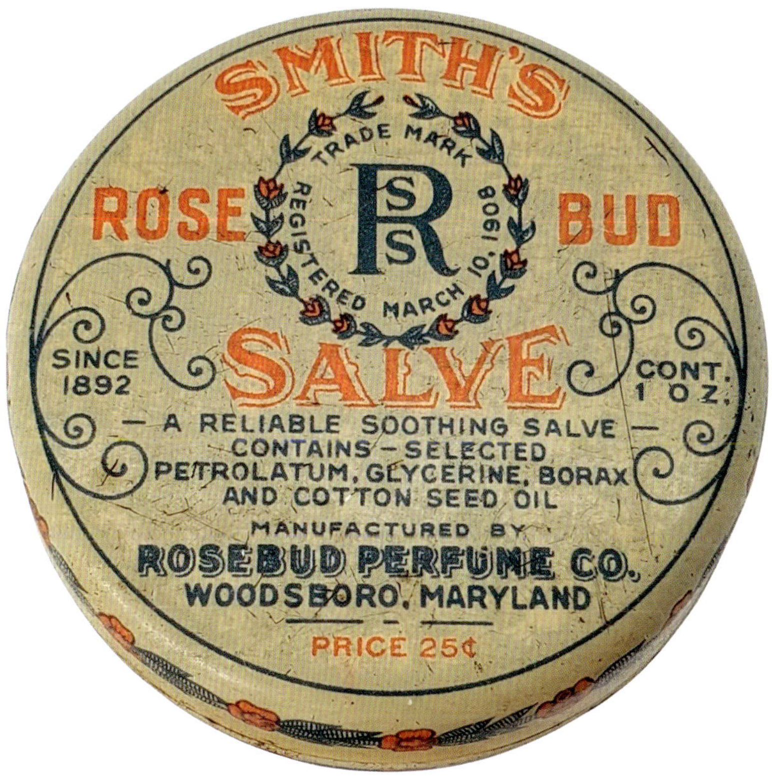

ROSEBUD PERFUME COMPANY

Smith's Rosebud Salve, after 1914
Glycerine, borax, cotton seed oil, and petrolatum in tin, ⅞ × 2⅛ in. / 2.2 × 5.4 cm
National Museum of American History, Smithsonian, Washington, DC

The Rosebud Perfume Company was founded in 1895 by Dr. George F. Smith, a formulating pharmacist from Woodsboro, Maryland, where the business headquarters are still located. In 1908, prompted by customer requests to produce an all-purpose salve for minor skin irritations, Smith drew on the distillation process used in perfume, tweaking it to create a balm that maintained the moisturizing properties of essential oils while preserving their perfume. Originally named Smith's Balsam of Rose Buds, the soothing, emollient concoction found a ready local market, where it could be used to treat chapped lips and skin, minor burns, and insect bites. The business expanded through savvy marketing, placing advertisements in county newspapers, and creating an innovative national mail-order business with tens of thousands of sales agents going from door to door to sell the company's wares, including the renamed Smith's Rosebud Salve. The salve is still made today, and its major ingredients remain cotton seed oil in a base of petrolatum with a sweet rosy aroma. The fragrance and extract of roses have been used in cosmetics for thousands of years, but none has been so successful or long-lasting as this one. It has been produced in several alternative scents over the years, including strawberry, menthol and eucalyptus, brambleberry, and tropical ambrosia, but the original rosebud salve in the blue-and-white decorative tin has endured and is still the best-known, best-selling product made by the Rosebud Perfume Company.

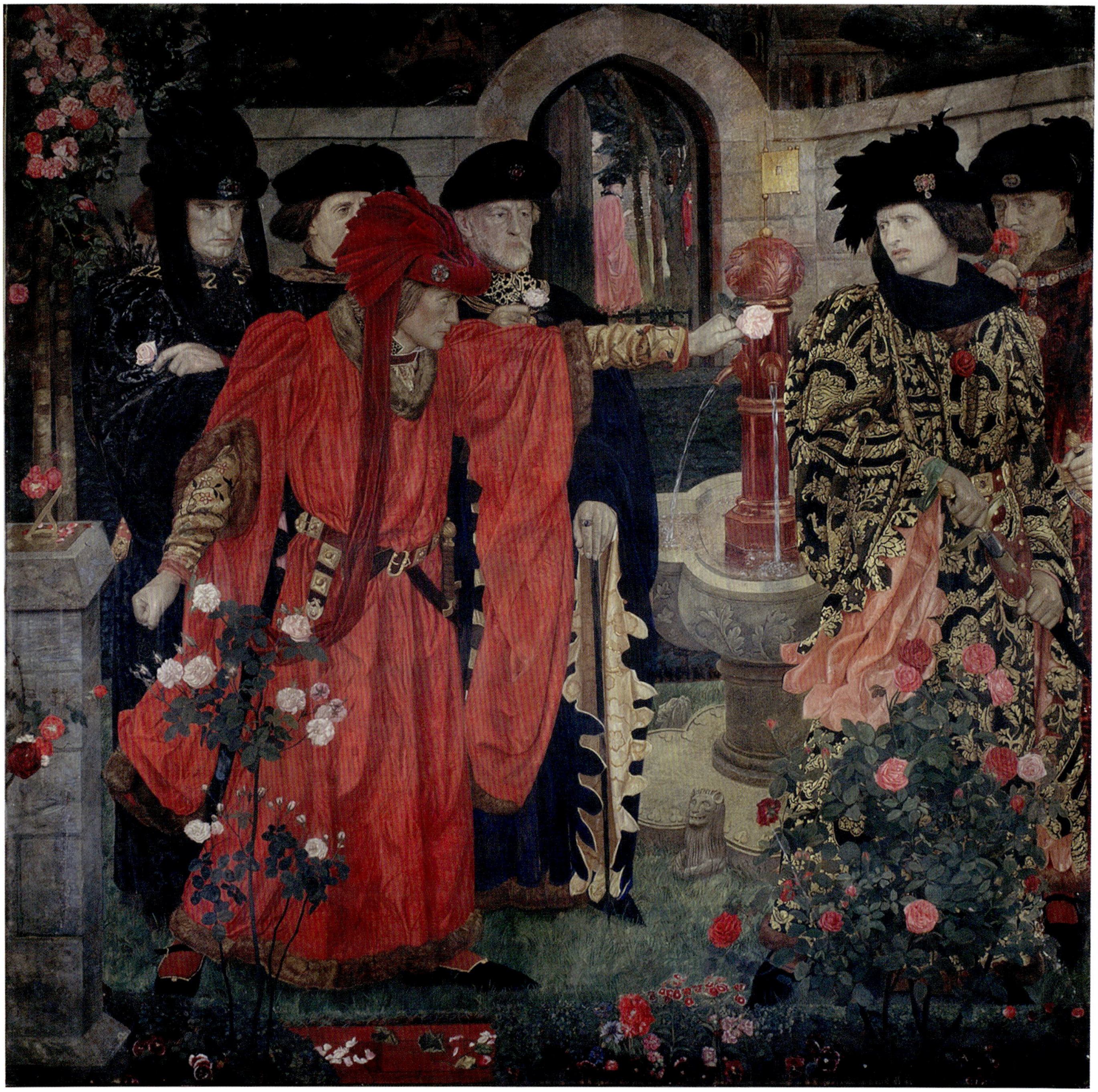

HENRY ALBERT PAYNE

Plucking the Red and White Roses in the Old Temple Gardens, 1910
Oil on canvas, 81 × 83 in. / 205.7 × 210.8 cm
Parliamentary Art Collection, London

The Wars of the Roses, a medieval dynastic struggle for the British monarchy, started because King Edward III had too many sons, and the direct bloodline of the House of Plantagenet died out in 1399. Two rival families emerged in the quarrel for the crown: the House of Lancaster, which initially gained the throne and united under the emblem of the red rose; and the House of York, whose insignia was the white rose, arguing a senior claim. Part of a series of paintings for the Houses of Parliament that show important moments in English history, this work by British artist Henry Albert Payne (1868–1939) depicts a turning point in the first reign of Henry VI (1422–61), as dramatized in William Shakespeare's history plays about the Lancastrian king, written around a century later. At left, Richard of York stands in London's Old Temple Gardens and thrusts his family's white rose aggressively forward, stating his claim that would initiate war. The Lancastrian-allied Duke of Somerset draws his sword to "dye [York's] white rose in a bloody red." Henry VI was eventually deposed by York's son, Edward IV, although the wars did not end until the new King Henry VII won the crown at the Battle of Bosworth Field in 1485, founding the Tudor dynasty. Henry VII was Lancastrian, but he married Elizabeth of York, daughter of Edward IV, and combined their red and white roses into one flower, the Tudor rose, an emblem that came to signify peace and prosperity for a finally unified England (see p.33).

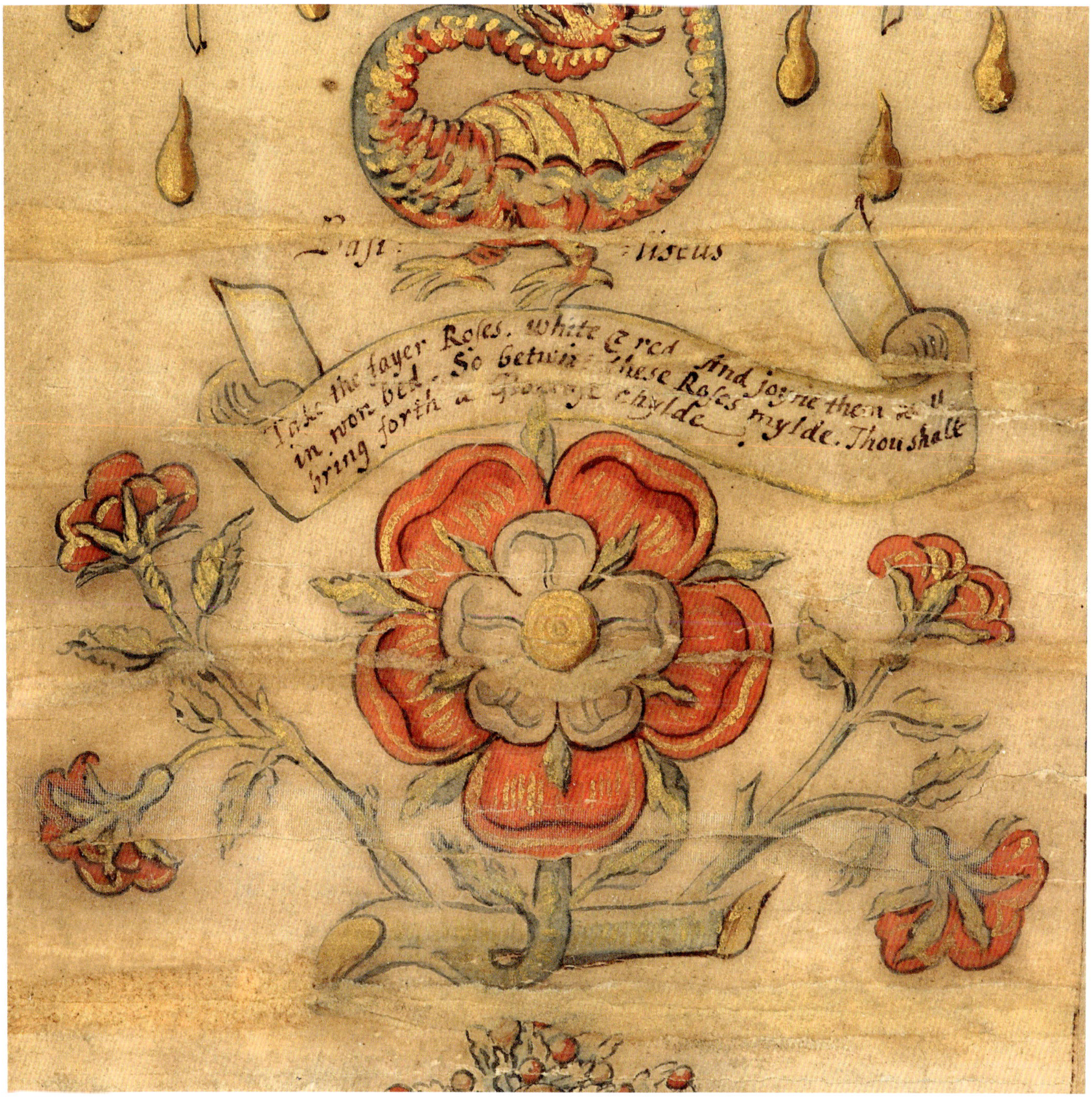

ANONYMOUS

Ripley Scroll (detail), 16th century
Manuscript, approx. 48 × 39⅜ in. / 120 × 100 cm
British Library, London

A gigantic red-and-white rose sprouts from a bush in full bloom. Above it, a floating scroll gives peculiar planting instructions, and the bottom of a dragon can be spied, alongside raindrops of gold. These mysterious symbols are not from an average garden manual but details from a sixteenth-century alchemical manuscript known as a Ripley Scroll, named after the fifteenth-century English priest George Ripley, who wrote a treatise on alchemy called *The Compound of Alchymy* in 1471. In obscure rhymes, he revealed how to manufacture the Philosopher's Stone, a mythical object that could turn base metals, such as lead, into silver and gold, and even produce an elixir of life. This rose illustrates part of the process for creating the red, white, and black stones that, once united in a trinitarian compound, would become the Philosopher's Stone. Ripley's poem was not published until 1591 and the original scroll is now lost, although subsequent copies—of which twenty-three survive—contain material from various other sources. Despite the esotericism of their field, alchemists did achieve genuine chemical and astronomical investigations, and the Ripley Scrolls intrigued scientists of the seventeenth century, such as John Dee, Robert Boyle, and Isaac Newton. The largest Ripley Scroll is nearly 20 feet (6 m) long, illustrating in detail the mystifying alchemical process. Toads, dragons, eagles, and lions enliven the instructions, and Ripley himself appears at the end of the scroll, setting off as a pilgrim with an outsize quill pen over his shoulder.

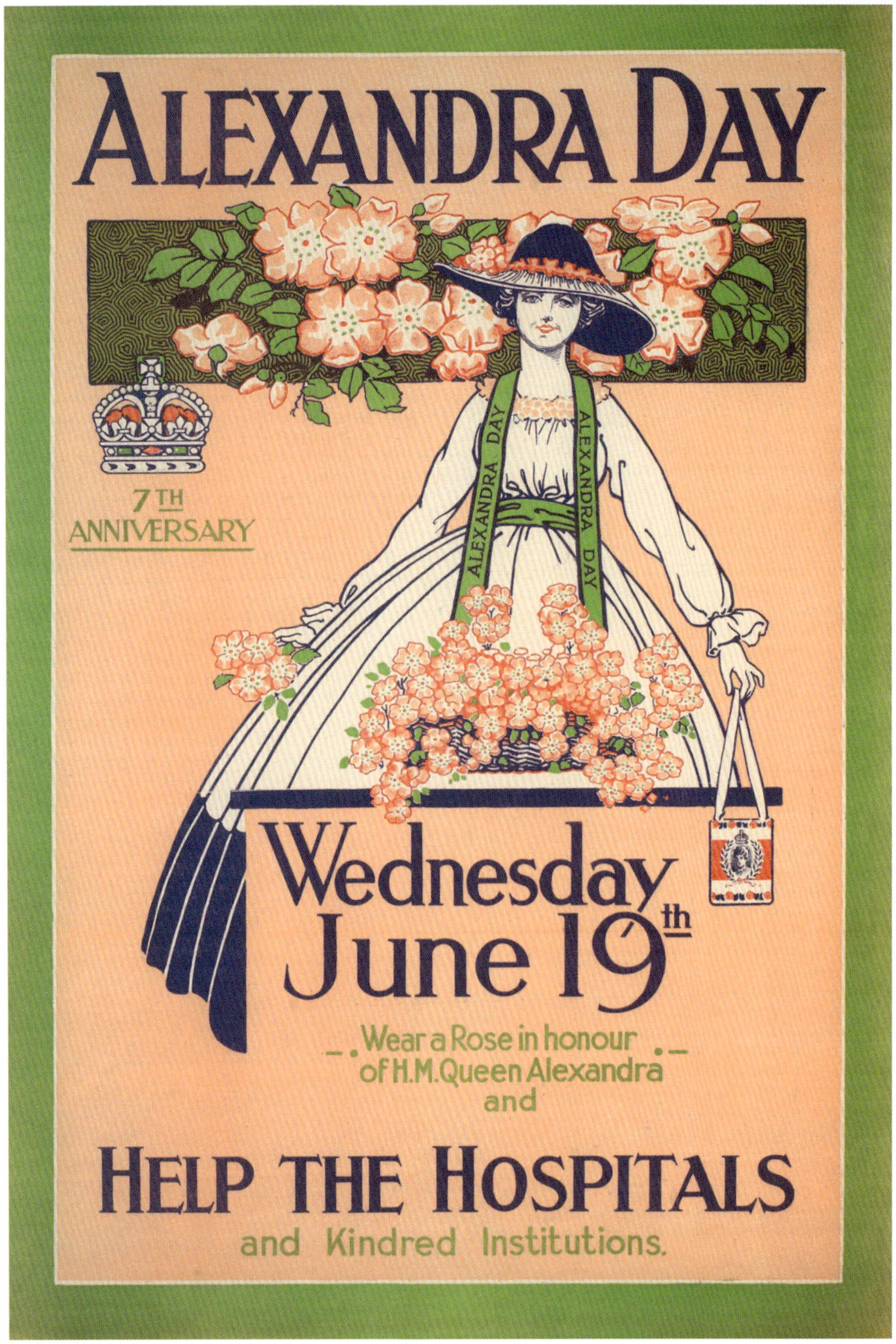

ANONYMOUS

Alexandra Day poster, 1918
Lithograph, 28¾ × 19⅝ in. / 73.1 × 50 cm
Imperial War Museums, UK

A young woman in a long white dress and a purple hat holds pink wild roses in her right hand, while around her neck a green ribbon supports a wicker basket full of the blooms in this British poster from 1918. Behind her head are yet more roses. In her left hand is a collecting tin with a small portrait of Princess Alexandra of Denmark, who arrived in Great Britain for her marriage to the Prince of Wales (later King Edward VII) in 1863. In 1912, as the fiftieth anniversary of her arrival approached, the queen requested that a charity be established in her name to raise money for London's hospitals and related charities, called "kindred institutions" on this poster. After hearing about a priest in her native country who was selling real roses to raise money for those in need, Queen Alexandra—by then the Queen Mother—applied the same idea to the United Kingdom, asking volunteers to create artificial roses out of silk and sell them for these charitable funds. Further, she stipulated that the roses be made by people with disabilities, providing them with employment, reviving a trade that had flourished in the mid- to late nineteenth century (see p.176). Thus was born Alexandra Rose Day, which was celebrated every June, even during World War I, from when this poster dates. It continues to this day, although now it raises money for food charities. Traditionally, the British prime minister buys the first rose every year.

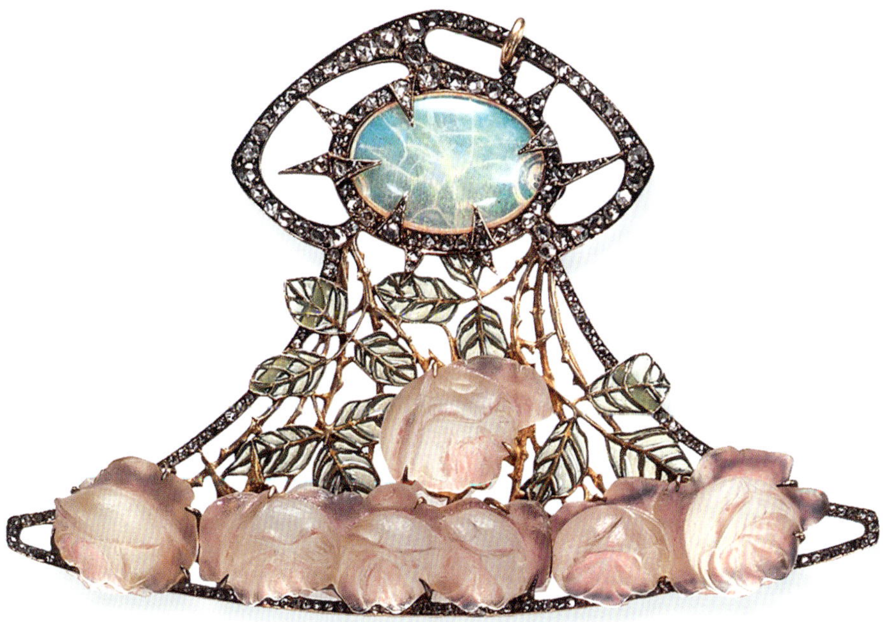

RENÉ LALIQUE

Opal and glass rose pendant, c. 1900
Enamel, opal, diamond, gold, and glass, 1¼ × 2 in. / 3.2 × 5 cm
Private collection

Opaque pink glass roses bloom in this strikingly eccentric pendant created at the beginning of the twentieth century by French artist jeweler René Lalique (1860–1945). A large opal—a stone that was a favorite of Lalique and other jewelers in the Art Nouveau period—is set in an openwork surround, studded with rose cut diamonds. A cascade of sculpted rose blooms with gold stems and translucent enameled leaves seems to tumble out from the opal center. As were many of Lalique's creations, this piece was inspired by his passion for nature. The fragility of the delicate thorny stems, plique-à-jour enamelwork, and glass roses give the flowers a stained-glass-like quality that contrasts powerfully with the solidity of the opal. Lalique, whom many consider to be the inventor of modern jewelry, launched his own studio in the mid-1880s. There, he concentrated his style on the Art Nouveau, which rejected previously popular historical aesthetics to focus on nature. Lalique's designs were often whimsical in form and fabulously eclectic, and he included glass in his jewelry to produce works of wearable art the like of which had never been seen before. For the Paris Exposition Universelle in 1900, Lalique exhibited about one hundred pieces and, alongside Emile Gallé, who crafted hand-blown glass objects, was one of the major stars of the Exposition. Despite the widespread admiration for Lalique's jewelry, by the early years of the twentieth century, he concentrated increasingly on the glassworks for which he is best remembered.

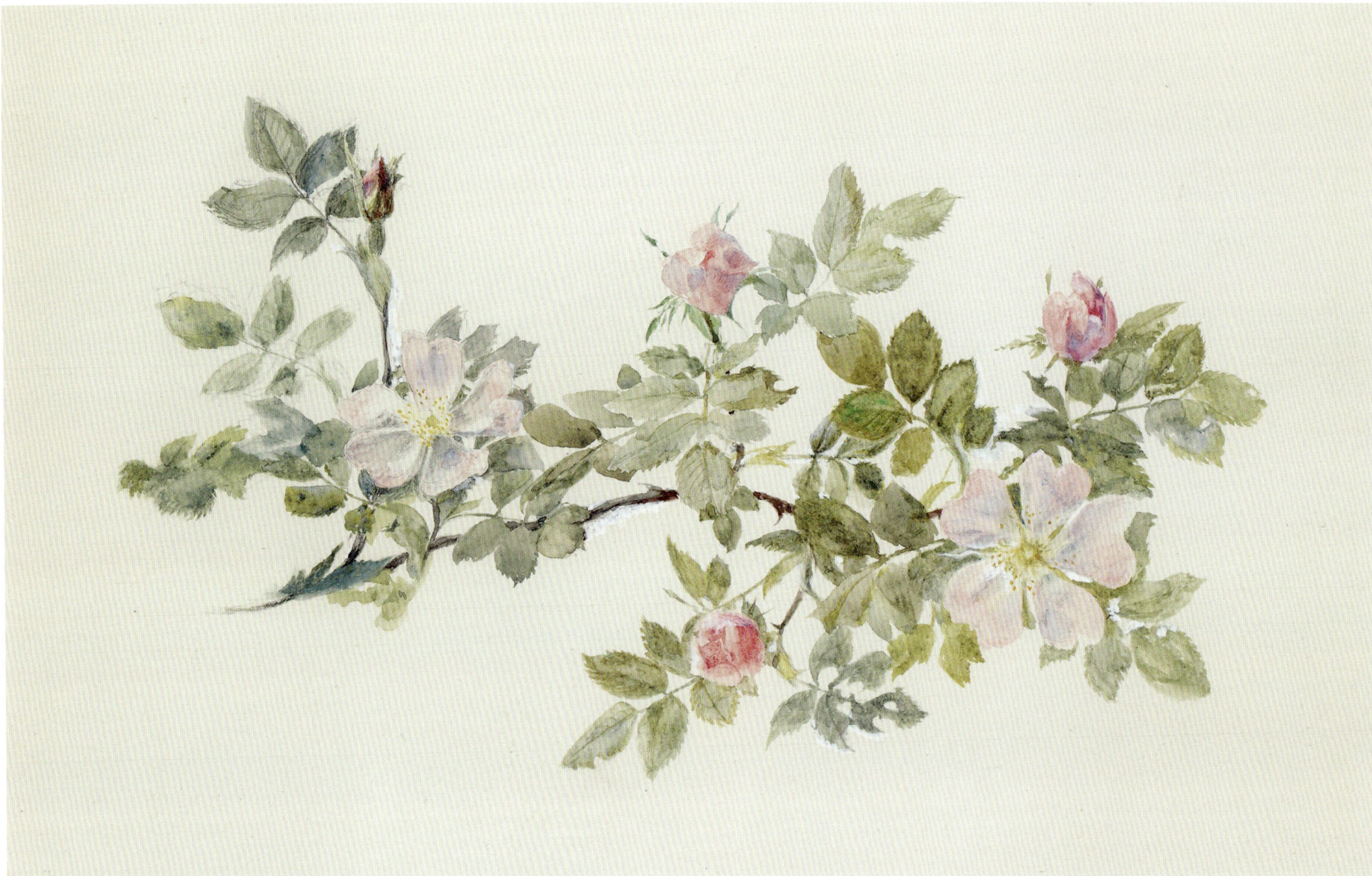

JOHN RUSKIN

Study of a Wild Rose, c. 1875
Watercolor and bodycolor over graphite on paper, 10½ × 16⅝ in. / 26.8 × 42.2 cm
Ashmolean Museum, University of Oxford, UK

In this delicate study of a wild rose, English philosopher, art critic, and artist John Ruskin (1819–1900) captures the fragility of the blush-colored open flowers and the warm pink of the buds against the soft gray-green of the leaves. It is one of the hundreds of drawings and watercolor sketches he made almost daily, focusing on aspects of the natural world: flowers, trees, birds and feathers, curled dead leaves, strands of seaweed, and close-ups of rock formations and geological features. One of the nineteenth century's most influential writers on art and architecture, Ruskin believed that drawing encouraged a close observation essential to a proper understanding of the structure of a subject, and he was thus a vital underpinning in the training of artists and architects. During his time as Slade Professor of Fine Art at the University of Oxford, he encouraged his students to draw as a way of teaching them how to look at both art and the world around them, declaring that "to see clearly is poetry, prophecy, and religion—all in one." He made no claims for his drawings as art; instead, he said that when he made a sketch it was as "a written note of certain facts," and where another artist might aim for an impression, he drew for "information." This depiction of the native British wild rose includes several leaves that have been damaged by insects, confirming that it is a study from life and not merely a decorative sketch.

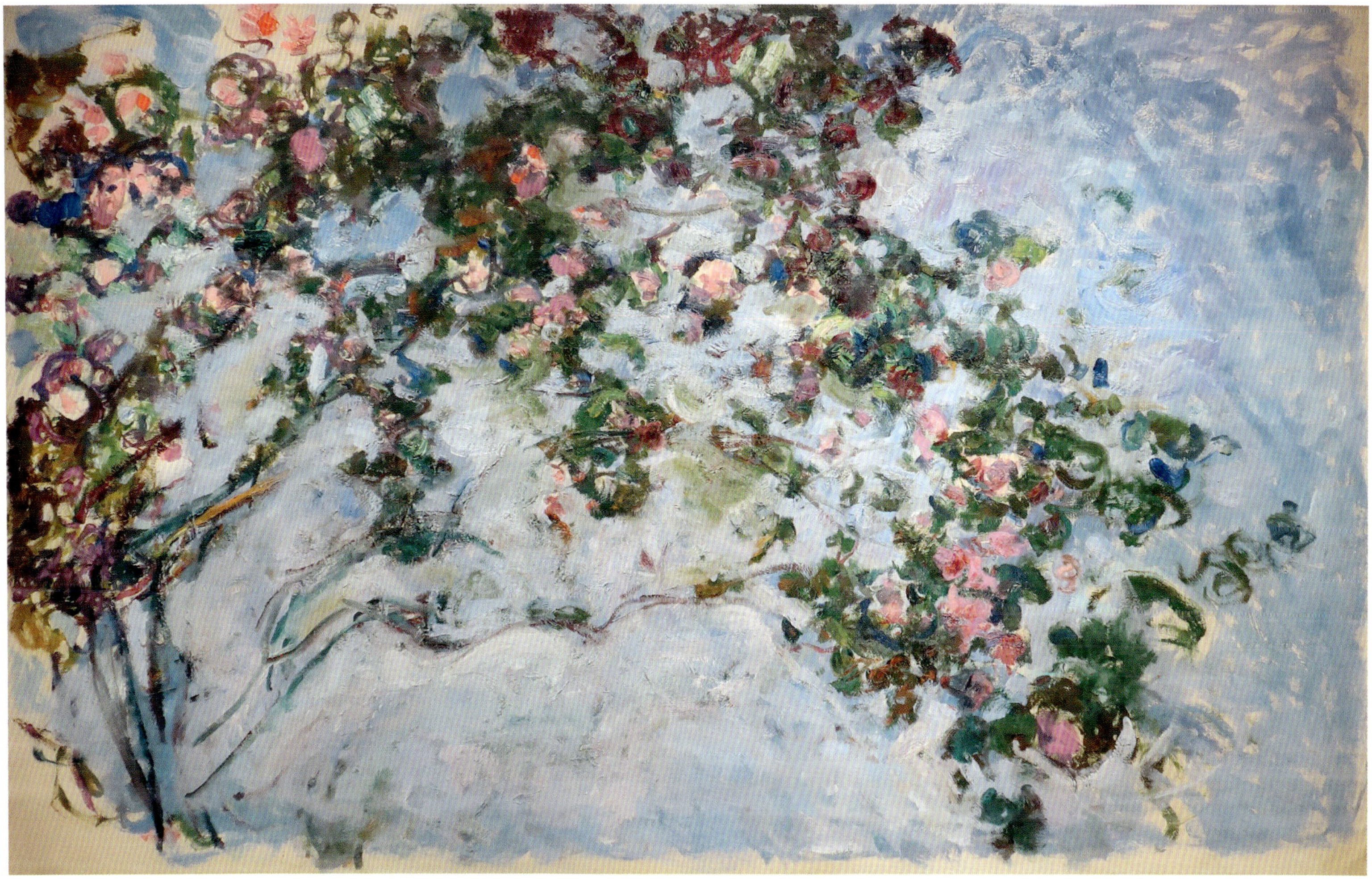

CLAUDE MONET

Les Roses, 1925–26
Oil on canvas, $51\frac{1}{8} \times 78\frac{3}{4}$ in. / 130×200 cm
Musée Marmottan Monet, Paris

Against a pale blue sky, a sketchlike rosebush laden with deep pink and pale pink flowers in full bloom fills the canvas, capturing the changing effects as the petals and leaves move in the breeze. French artist Claude Monet (1840–1926), a leading member of the Impressionist movement, is more commonly associated with the water lilies he famously painted in the aquatic garden of his home at Giverny. Toward the end of his life, however, as his vision failed, he also made several paintings of the roses growing in other parts of his garden. In 1883 Monet and his family moved to Giverny, a small country village northwest of Paris, and he lived there for the rest of his life, painting *en plein air*. His home and garden provided continual inspiration, particularly his water garden with its iconic lily pond, Japanese bridge, and weeping willows. Over the years, Monet painted thousands of different flowers and scenes from Giverny, including a trellised avenue of roses in 1920. A cataract operation in 1923 left him with a changed perception of color that was corrected only a few years later, and his style changed at the end of his life from the quick brushstrokes of his earlier paintings to markings that more resembled short commas, as seen here in the leaves of the rosebush. In his final years, Monet embarked on a frenetic painting spree, creating such works as this one in his palette of blues, greens, pinks, and reds to conjure up the beauty of his roses on a summer's day.

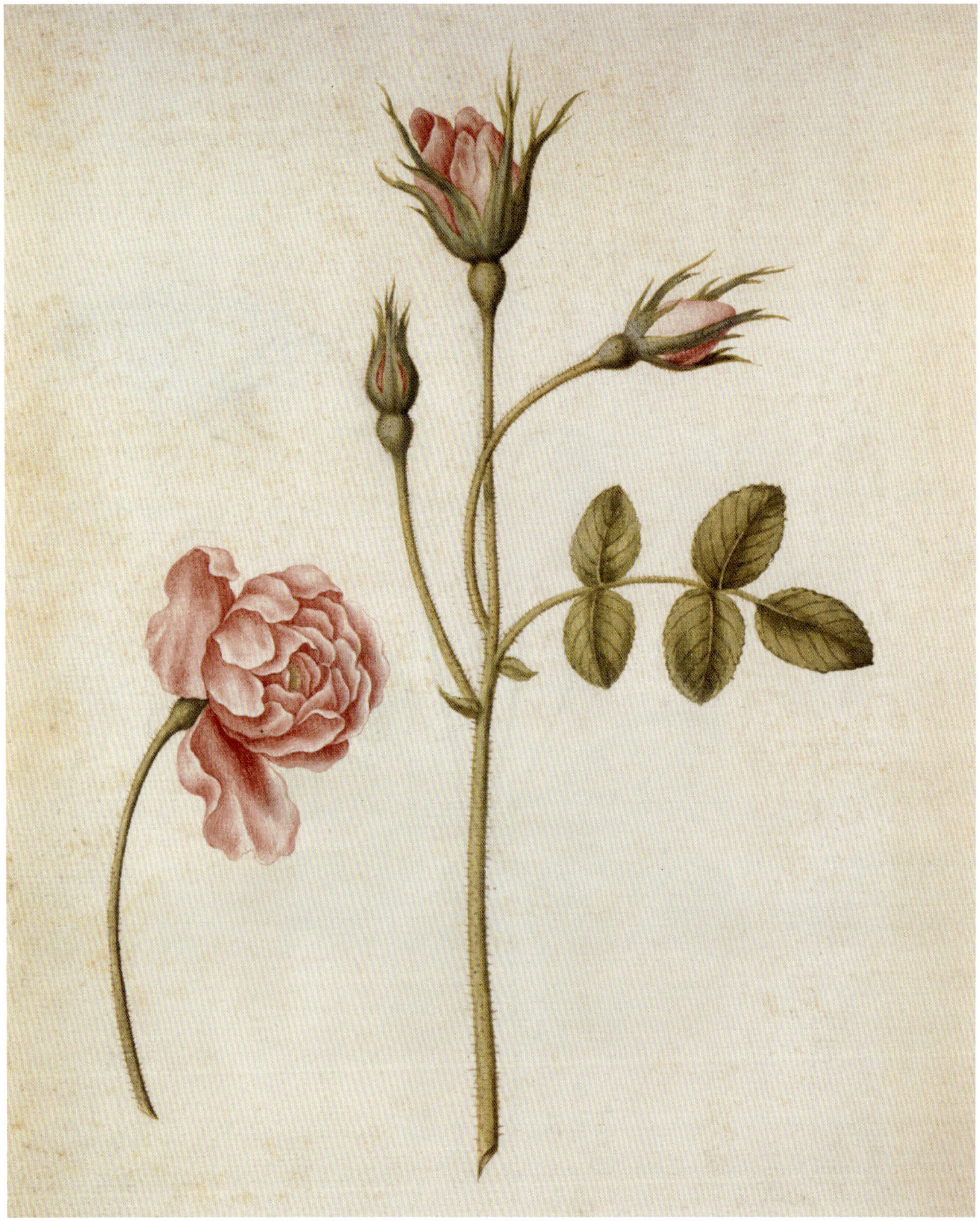

JACQUES LE MOYNE DE MORGUES

Pink Rose, c. 1568
Watercolor on paper, 10¾ × 7⅛ in. / 27.4 × 18.3 cm
Victoria and Albert Museum, London

This delicate pink rose, with its full flower set beside a spray of mossy buds, has long been attributed to French artist Jacques Le Moyne de Morgues (c. 1533–1588). The rose is one of fifty-nine studies of flowers, fruit, vegetables, and nuts, all originally contained in a lavishly bound album of a similar date, but later removed when it was acquired by the Victoria and Albert Museum in 1856. The watercolors were not actually attributed to Le Moyne until the early twentieth century, and until that point, he had been known more for his depictions of the southeastern United States, after taking part in a colonial expedition to Florida from 1564 to 1566. A Huguenot, Le Moyne later fled to England, seeking asylum shortly after the St. Bartholomew's Day massacre of Protestants in 1572, and it is assumed that he brought the album with him to London. The plant studies vary considerably in style and competence. Some appear to have been drawn from life and show insect-bitten leaves and wilting flowers alongside exquisitely painted butterflies and moths. Others, such as this rose, are awkwardly drawn and unnaturally stiff, features that suggest they were not based on the observation of living specimens but copied from earlier sources, such as herbals or printed pattern books. Recent research has cast doubt on the attribution to Le Moyne, since his authenticated works are more polished and sophisticated. Until further evidence emerges, this rose and the other studies from the album might be better considered the work of a skillful but anonymous French artist.

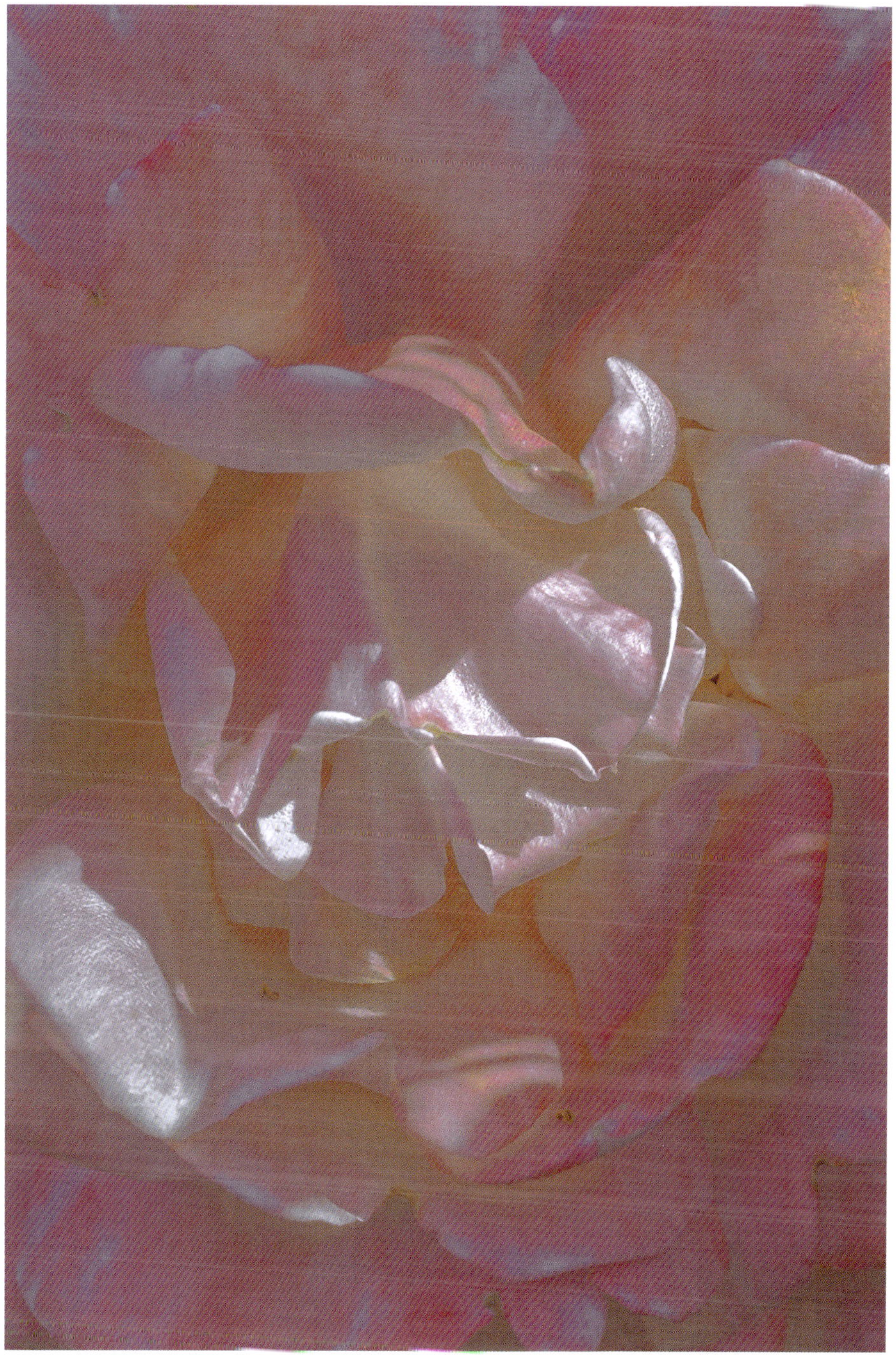

XUEBING DU

Mother of Pearl, 2022
Digital photograph, dimensions variable

Through the wonders of photography, the earthly rose is transmuted into something otherworldly, a radiant jewel-like object of shimmering textures and exquisite forms. Incorporating a wide range of pinks and yellows, and with hot spots of the purest white, Chinese photographer Xuebing Du's (b. 1990) extreme close-up of a blossoming rose captures the preternatural beauty of the flower. Du associates flowers with love and beauty. She aims for her images to convey their feminine qualities, rendering them in an almost surreal way to provide pleasure to the viewer. Inspired by the use of light in both Renaissance and Impressionist paintings, her photography centers on elements of the natural world, from deserts and gardens to individual plants and flowers. Having previously found flowers a challenging subject to photograph, Du experienced a difficult period in her personal life in 2017 that helped her to appreciate the kindness and gentleness of flowers, and to learn from them about self-love and healing. Du now lives in the South Bay area of California, where the weather supports an abundance of flowers, and her images adopt a Pictorialist approach, which draws attention to the beauty of the subject matter through texture, tone, light, and composition instead of attempting to document reality. Alternating between iridescent, metallic, translucent, and nacreous qualities, her images of flowers call to mind other organic forms, such as mother-of-pearl, the crystallized mineral compound found in some seashells, which provides the title for this work.

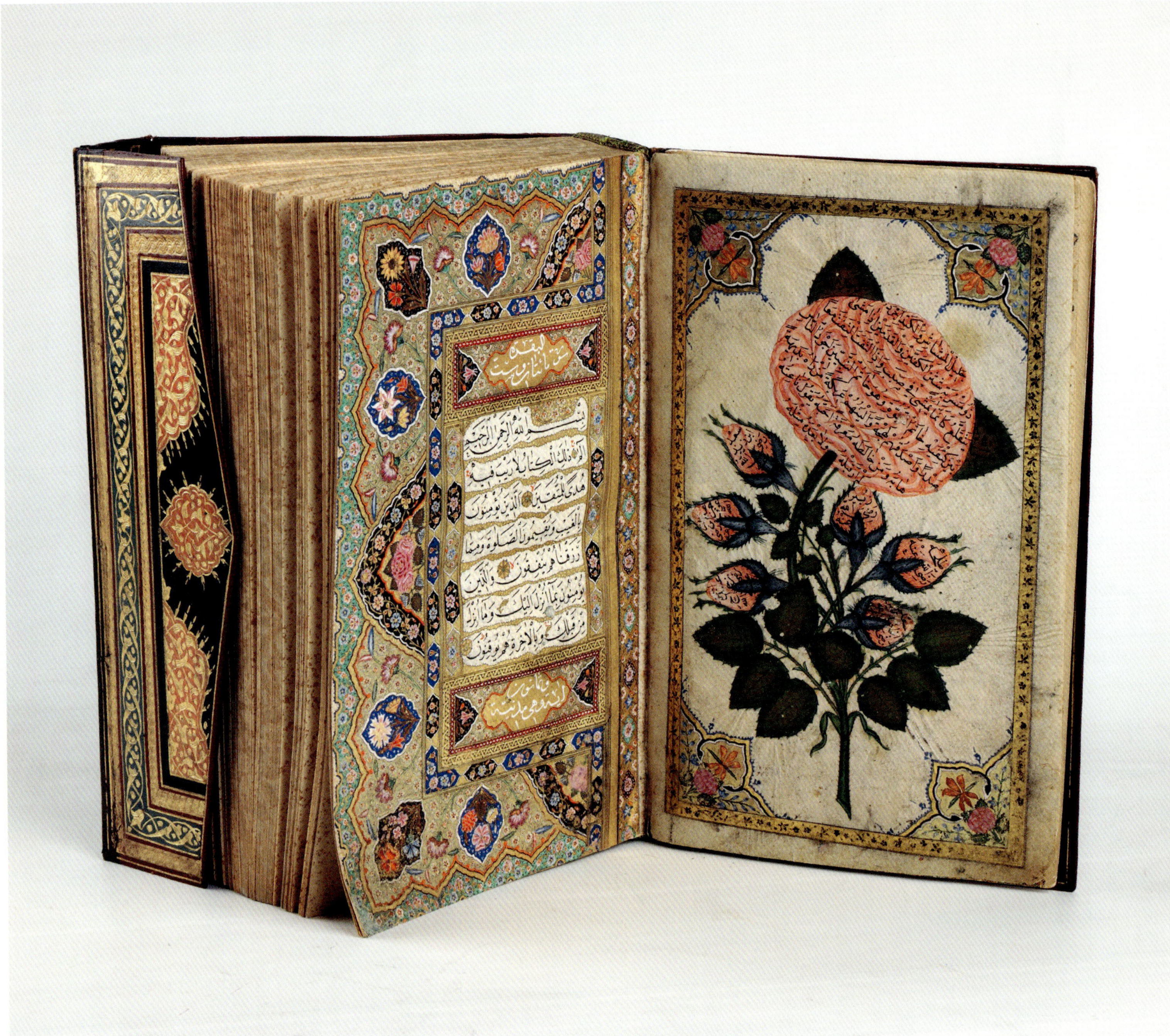

ZEHHEBEHU MUSTAFA

Folio from a manuscript of the Koran, 1744
Gold-flecked sized paper, black ink, colored paint, and gold, 7¼ × 4¾ × 1½ in. / 18.3 × 12 × 3.5 cm
Sabancı University, Sakıp Sabancı Museum, Istanbul

This exquisitely illuminated folio contains the *Al-Fatihah*, the opening chapter (*surah*) of the Koran, and *Al-Baqarah*, its second chapter, on separate pages. On each page of the folio between the two *surahs* is a pink rose, on which is written a description of the Prophet Muhammad. Symbolizing the Prophet, the rose is used often in Islamic imagery and Koranic illustrations to represent the divine without resorting to figuration. Important people are instead represented by name within sections of the rose. One bud contains the names of the four caliphs, or successors of Muhammad—Ebubekir, Ömer, Osman, and Ali—and of his two grandsons, Hasan and Hüseyi. A leaf contains the phrase *Allah celle celâlühu* (God is great), while another reads *Muhammed sallallahu aleyhi ve sellem* (Muhammad whom God confirmed as Prophet). *Surah Al-Fatihah*, often referred to as the Opening, is not merely the first chapter of the Koran but also contains the essence of the entire holy book. Its seven verses are a concise summary of the faith that was bestowed on the Prophet during his early years of prophecy in Mecca. The verses of the *Baqarah surah* outline the foundational values that form the bedrock of a righteous society, providing principles that promote justice, honesty, and kindness. The illuminator who decorated the manuscript is identified by the signature Zehhebehu Mustafa, which appears in white on a gold ground inside a large *hatayi* flower (an Ottoman decorative motif) following the last line of the book's colophon.

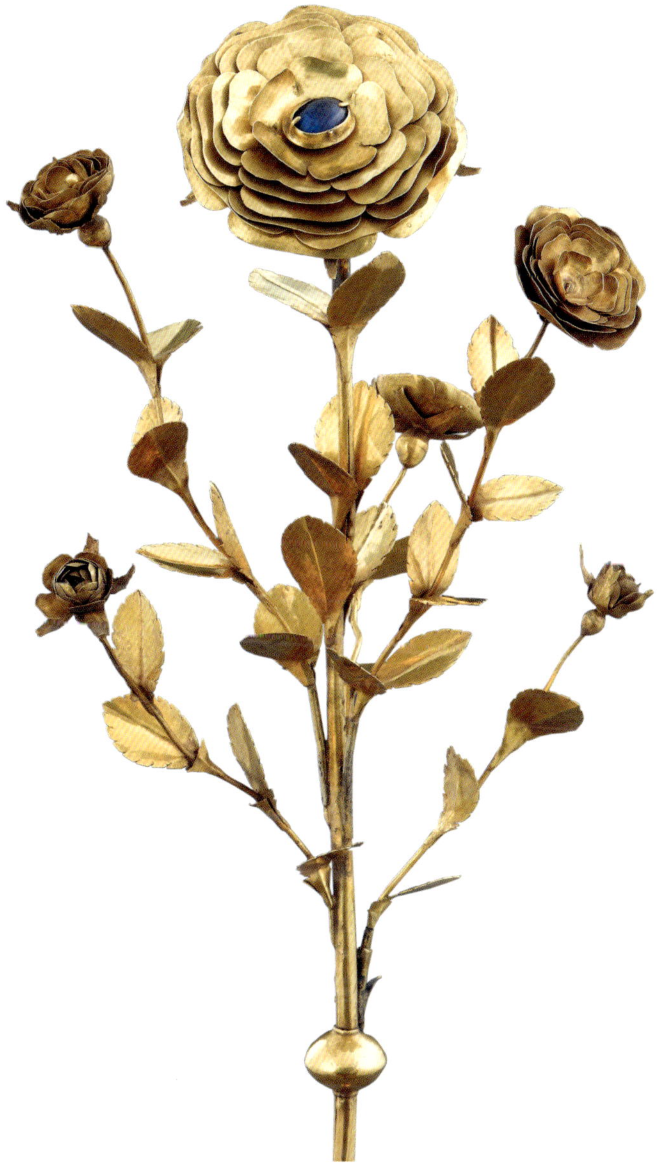

MINUCCHIO DA SIENA

Golden Rose, 1330
Gold and colored glass, H. 23⅞ in. / 60.5 cm
Musée de Cluny, Paris

This stunningly detailed golden rose is the oldest of its kind still in existence. Crafted in 1330 by the artisan known as Minucchio da Siena, it exemplifies the intricate craftsmanship and symbolic artistry characteristic of the medieval period. The Golden Rose served as a significant ecclesiastical gift, bestowed by the Pope to monarchs or monasteries as a token of favor and recognition. This award was given to the Count of Neuchâtel by Pope John XXII, the longest-reigning Avignon pope during Catholicism's Great Schism, when one pope ruled in France and another in Rome. The French identities of both the Pope and the rose's recipient indicate the political alliances of the time; while the "antipopes" in Rome were supported by the Holy Roman Empire, their rivals in Avignon were backed politically and financially by the French monarchy and its aristocrats. Although the blue stone at the center of this rose is only colored glass, other examples incorporated luxurious materials and precious gems, such as sapphires, to add brilliance and enhance their value. Their use of gold and elaborate designs reflect the medieval emphasis on creating objects that were not only visually stunning but also imbued with spiritual and political meaning. Gold represented the light of God, and the Golden Rose is thornless like the roses that grow in Heaven, symbolizing the passion of Christ and his Resurrection. The use of these gemstones not only increased the visual appeal of the artifact but also underscored its status as a prestigious and sacred gift.

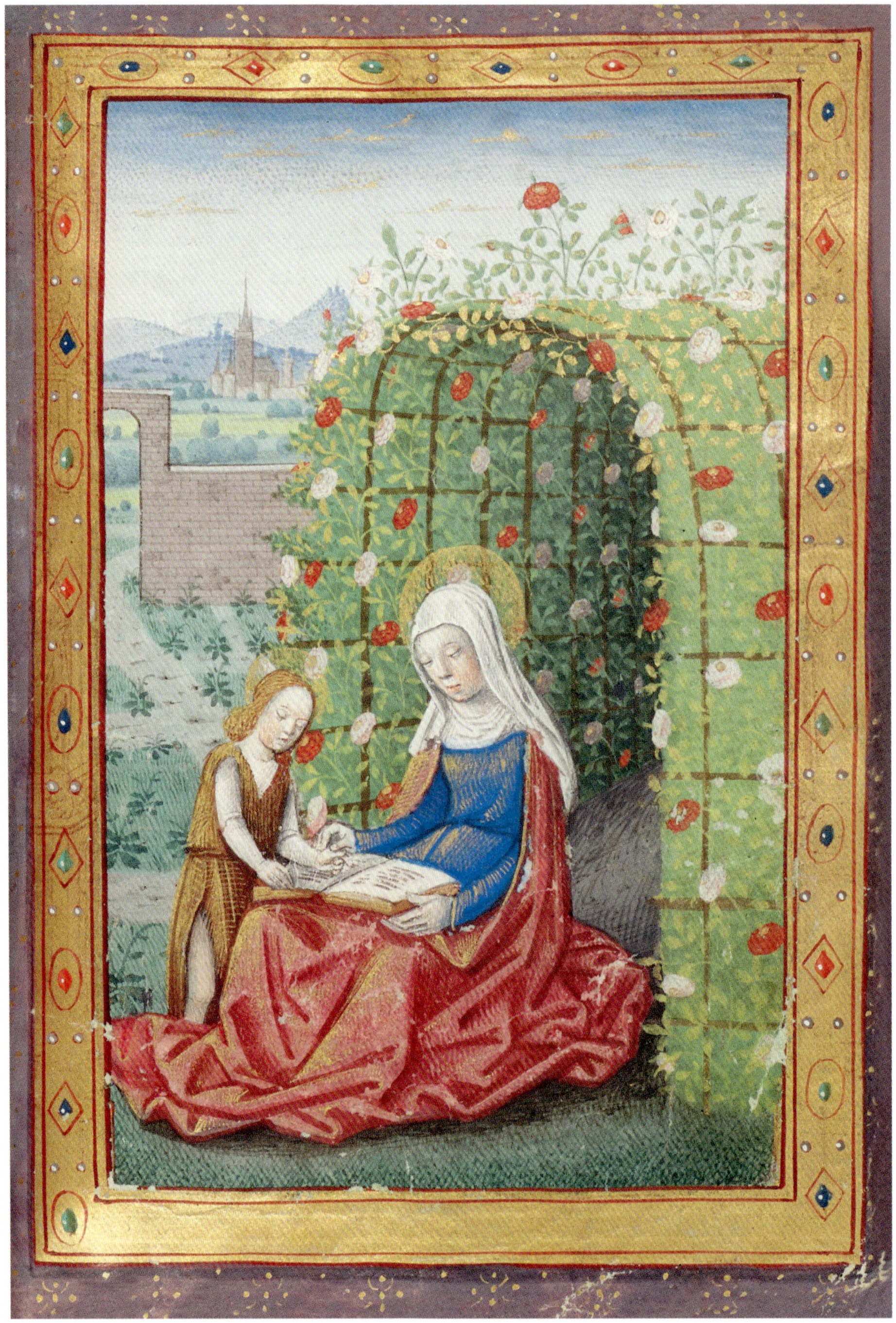

ANONYMOUS

Saint Elizabeth teaching John the Baptist to read in a rose arbor, c. 1490
Illuminated manuscript on parchment, 5⅜ × 4 in. / 13.6 × 10.2 cm
Bodleian Library, University of Oxford, UK

This illuminated manuscript from a medieval French prayer book puzzles garden historians because the arbor is covered in climbing roses. The climbing gene comes from Asiatic roses, and the first climber to arrive in Europe was the Musk rose (*Rosa moschata*), which has small white flowers. The roses that were cultivated in medieval gardens were shrub roses, either the Gallica rose, which is red or pink (see p.122), or the white *R. alba* (see p.72). Both are semidouble, meaning they have yellow stamens in the center of the flower, as this illustration shows. But these medieval cultivars are not climbers, although the wild roses of hedgerows might be trained in this way. Possibly the red and white roses pictured here were grafted onto wild climbing stock, as fruit trees certainly were, in which case this is a unique record of medieval skills in rose cultivation. Given the lack of written records about climbing roses in the fifteenth century, this illustration may not be evidence of horticultural ingenuity, but only a figment of the artist's imagination. In a lovely garden setting, Saint Elizabeth, mother of John the Baptist, teaches her son to read. This is not a flight of fancy but instead suggests that in literate families it was natural for mothers to be responsible for teaching their children this skill. Early reading books, known as primers, were derived from prayer books such as this, which opened with "Prime," the first prayers of the day. The unknown artist raises one more gentle question: did they normally retreat to the rose garden for reading lessons?

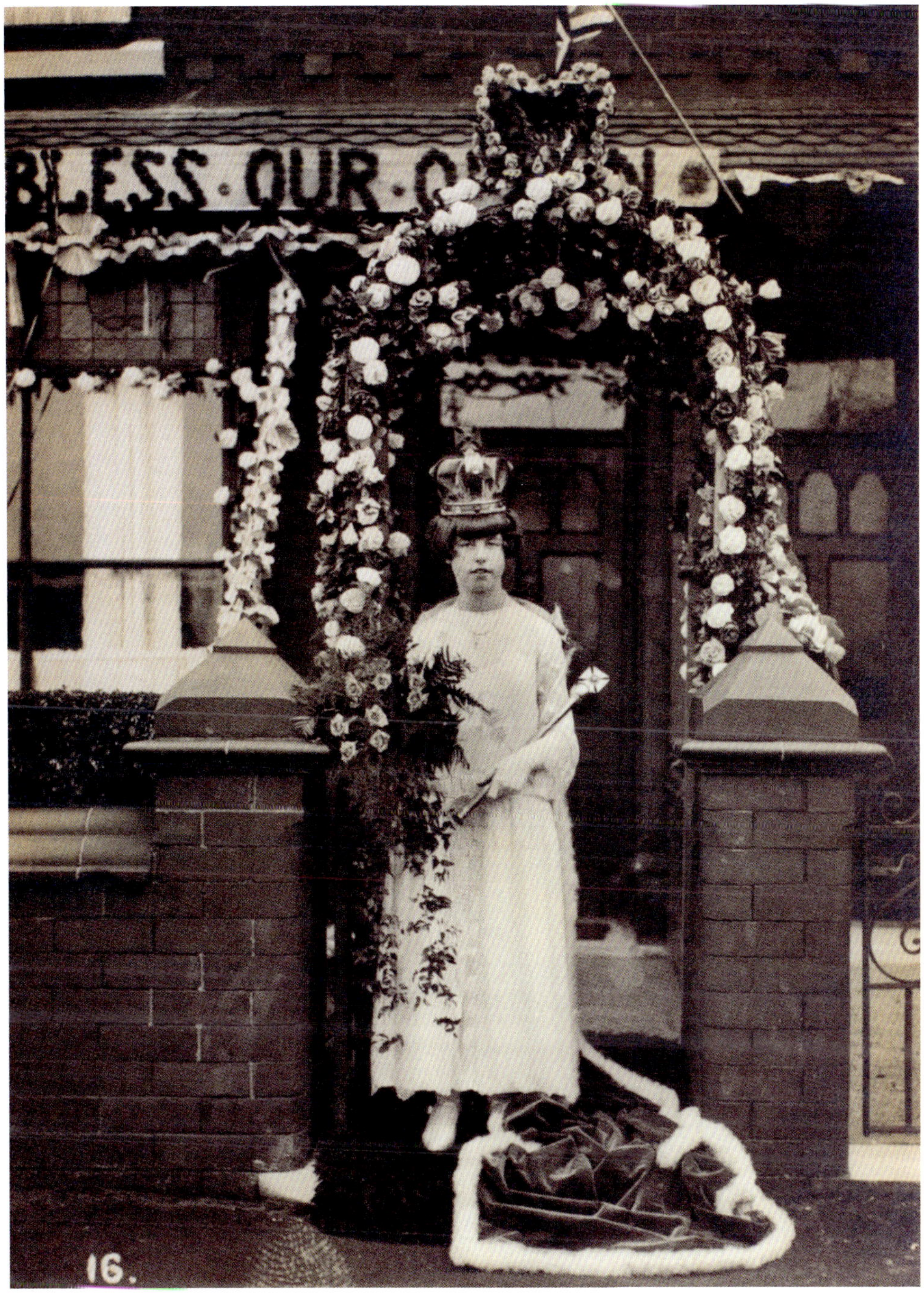

ANONYMOUS

Walkden Rose Queen, June 17th, 1922, 1922
Postcard, 5½ × 3½ in. / 14 × 9 cm
Private collection

Posing beneath an arch of fragrant garden roses, the Rose Queen is appropriately costumed with crown, fur-trimmed robe, and scepter. It is June 1922, and she will lead the annual Rose Festival in Walkden, near Manchester, England. The house behind her—presumably the family home—is also dressed for this special day with a banner that reads "Bless Our Queen," which may be a sign of respect for Queen Mary of Teck or for the rosy royal highness herself. Since ancient Roman times, rose festivals have taken place around the world, and many continue today in such places as Pasadena, California, and Tyler, Texas (see pp.187 and 211). From the 1880s, they would become major annual charitable events across the United Kingdom and would be especially popular in Lancashire (of which Walkden is historically a township), known as "the red rose county" following the fifteenth-century Wars of the Roses (see pp.33 and 164). Rose Queens would be chosen for various qualities, such as their scholastic achievements, religious observance, or perceived beauty. Often conveyed within a rose-decorated pony cart and accompanied by female attendants and, occasionally, a King, they would lead a community procession traveling to a local field for a day of dancing, games, and refreshments. In the 1920s, it would be commonplace and inexpensive to have personal photographs marking special occasions made into multiple postcards. Once a popular British pastime, few rose festivals take place in the United Kingdom today.

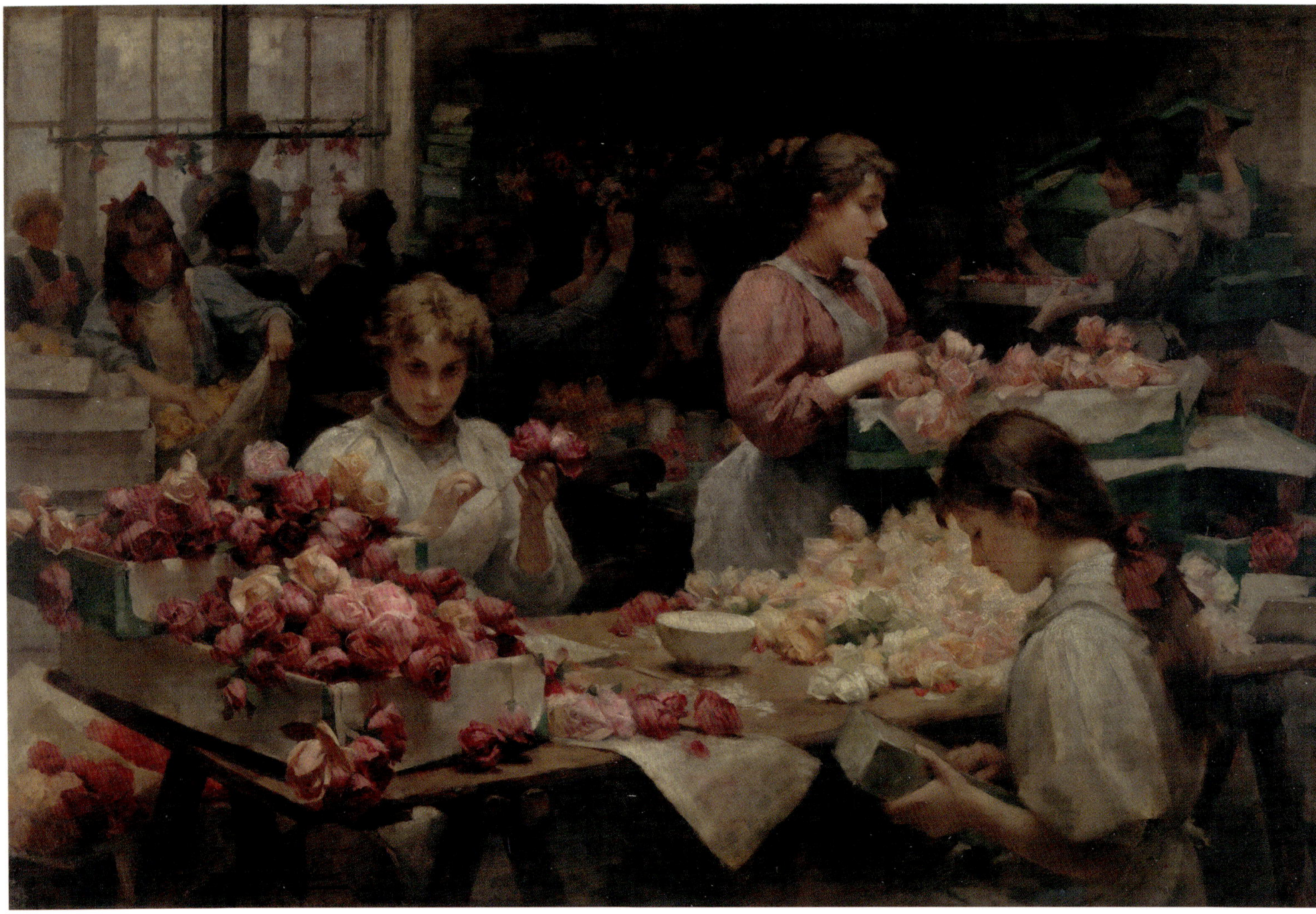

SAMUEL MELTON FISHER

The Clerkenwell Flower Makers, 1896
Oil on canvas, 42¼ × 62½ in. / 107.3 × 158.7 cm
Museum of New Zealand Te Papa Tongarewa, Wellington

Amid a bustling Victorian floristry factory overflowing with artificial roses, a young woman looks up from her work to meet the viewer's gaze. Her fellow workers are absorbed in their tasks as they create a product that was in great demand for decorating dresses, bonnets, and hats, as well as interiors. This sentimentalized scene was painted by the British artist Samuel Melton Fisher (1859–1939) and typifies the kind of social subject that was popular with nineteenth-century audiences. The women depicted here were employed by John Groom's Flower Girls' Christian Mission, founded in 1866 to help some of London's poorest and most vulnerable street vendors. The mission, based in the neighborhood Clerkenwell, became well known in the 1880s for its artificial flowers made by disabled girls and women and sold for charity. Making the astonishingly realistic roses by hand was a complex and physically demanding job. Petals were cut from fabric before being dyed, molded, and shaped with special tools that added veins and texture. They were then assembled and held in place with wire before being arranged into sprays or wreaths. The process required great skill and took several years to master. In 1891, at the height of the trade, the national census reported 4,011 flower makers in London. Fisher's painting, which celebrates the virtue of charity, reflected the values of many in Victorian society who believed it was morally right to help less fortunate individuals to improve themselves through honest hard work.

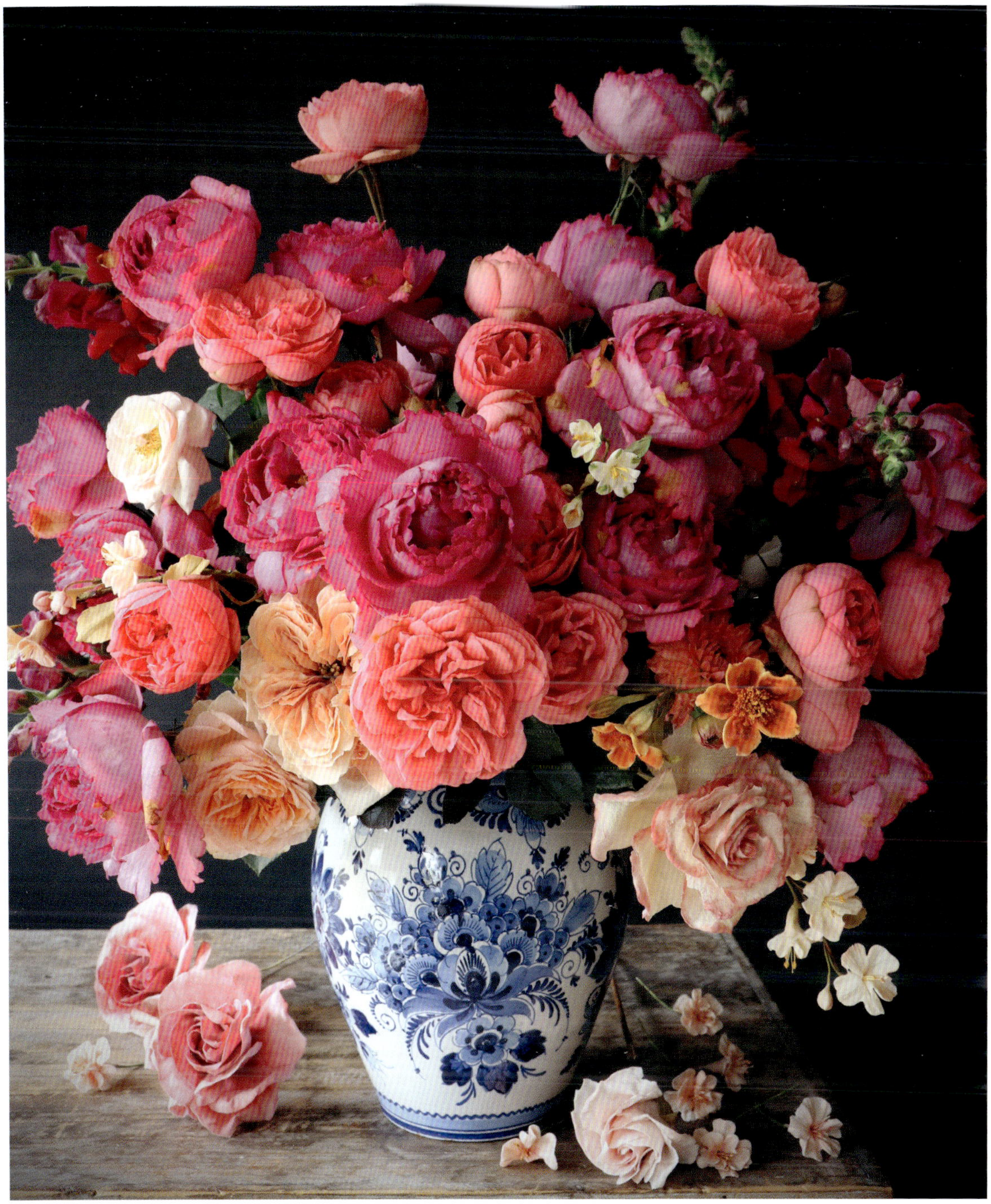

NATASJA SADI

Arrangement, 2023
Sugar flowers and fresh roses, dimensions variable

Pink and peach Centifolia roses mingle with deep pink Hybrid Teas and Floribundas in this stunning arrangement in a delftware vase, as rich and dramatic as any Dutch still-life painting. Yet each bloom in this photograph is edible, made entirely from sugar. Suriname-born, Amsterdam-based artist Natasja Sadi (b. 1975) discovered her talent for making lifelike floral confectionery while decorating her daughter's fifth birthday cake. She has since mastered the sculpting and coloring of cultivars at all stages; from buds and blossoms to wilting and decaying petals, each is rendered with remarkable verisimilitude. Her opulent crescendos of color have gained a global audience on Instagram, but her art runs deeper than simply celebrating the beauty of flowers. For Sadi, her sugar sculptures honor her African and Indonesian ancestors, who labored as slaves and contract workers on the sugarcane plantations of Suriname, a former Dutch colony in South America. In the seventeenth century, while her family members toiled in the fields, Dutch flower painters were creating Golden Age masterpieces. Sadi developed a great appreciation for such artists as Maria Sibylla Merian (see p.24) and her pupil Rachel Ruysch (see p.10), inspiring many of her own sugar compositions. By bringing together these histories, Sadi calls attention to historical injustice. As she writes in her book *A Sweet Floral Life* (2023), "Creating sugar flowers and floral arrangements in the Dutch style has been one way of breaking uncomfortable habits of silence."

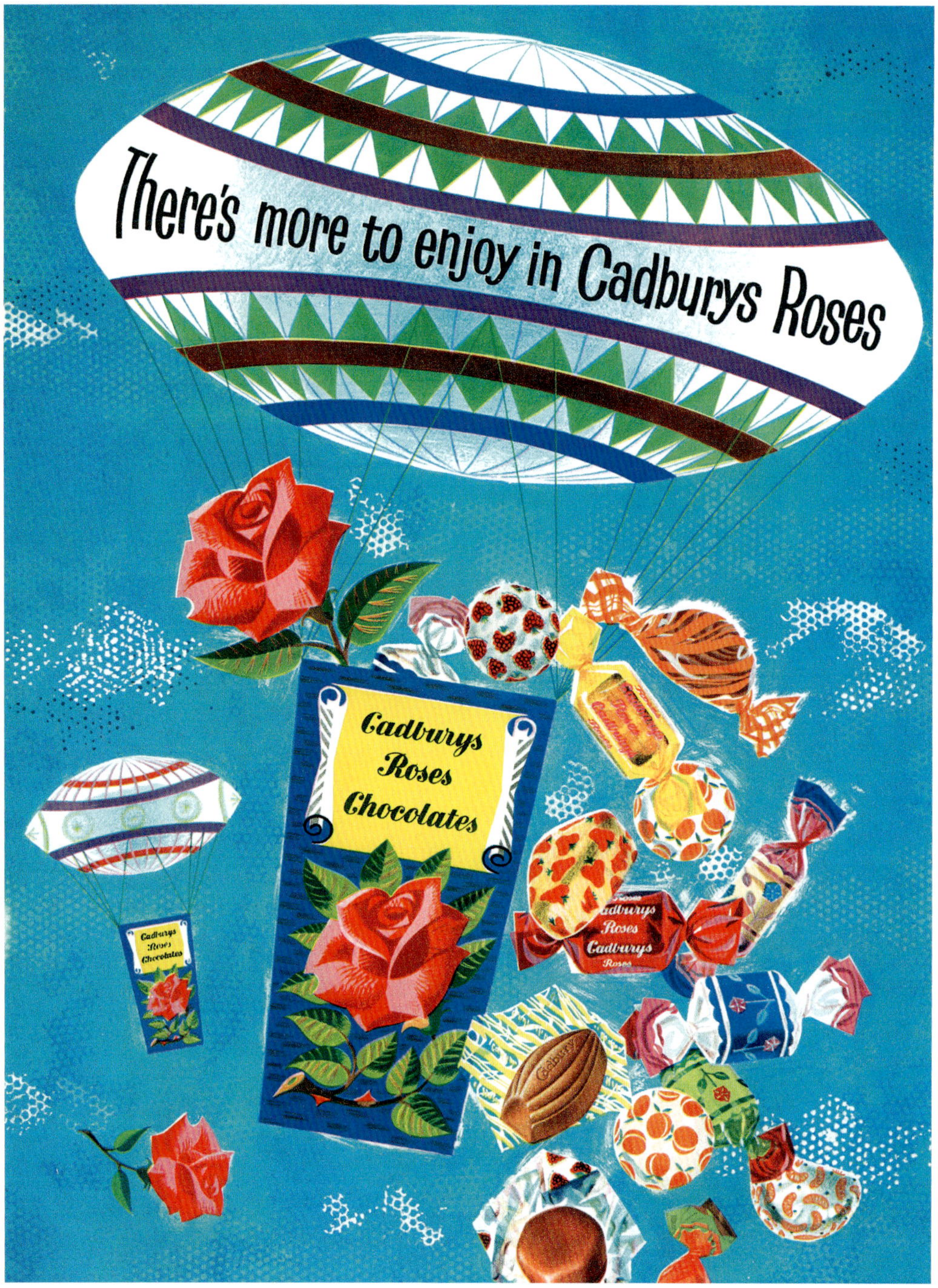

CADBURY

There's More to Enjoy in Cadbury's Roses, 1958
Lithograph, 13½ × 10 in. / 34.5 × 25.5 cm
Private collection

A colorful array of individually wrapped chocolates cascade from the sky in this 1950s advertisement for Cadbury's Roses, a favorite British treat at Christmastime. Red roses also tumble from the iconic blue carton, which dangles from a patterned parachute. The classic chocolate assortment was introduced by renowned British confectionery company Cadbury in 1938 to compete with Mackintosh's Quality Street, launched two years earlier. At first it was said that the sweets were named after English company Rose Brothers, which manufactured and supplied the machinery that wrapped the chocolates and, indeed, invented the first machine that could successfully wrap differently shaped sweets. In 2020 Cadbury offered an alternative origin of the brand name, claiming that it referred to the favorite flowers of Dorothy Cadbury—a director of the confectionery company and a skilled botanist—that grew in the gardens surrounding the first factory at Bournville near Birmingham, England. The famous blue packaging has seen many revisions since 1938, but red roses have always featured in the design. The contents have also been updated several times over the years. Today, the boxes, tubs, and tins contain nine varieties, including the popular Golden Barrel filled with caramel, the soft-centered Strawberry Dream, and the Hazel Whirl, a roasted hazelnut embedded in chocolate. Each wrapper bears the Roses logo, except for the Cadbury Dairy Milk, a pure block of the firm's quintessential milk chocolate.

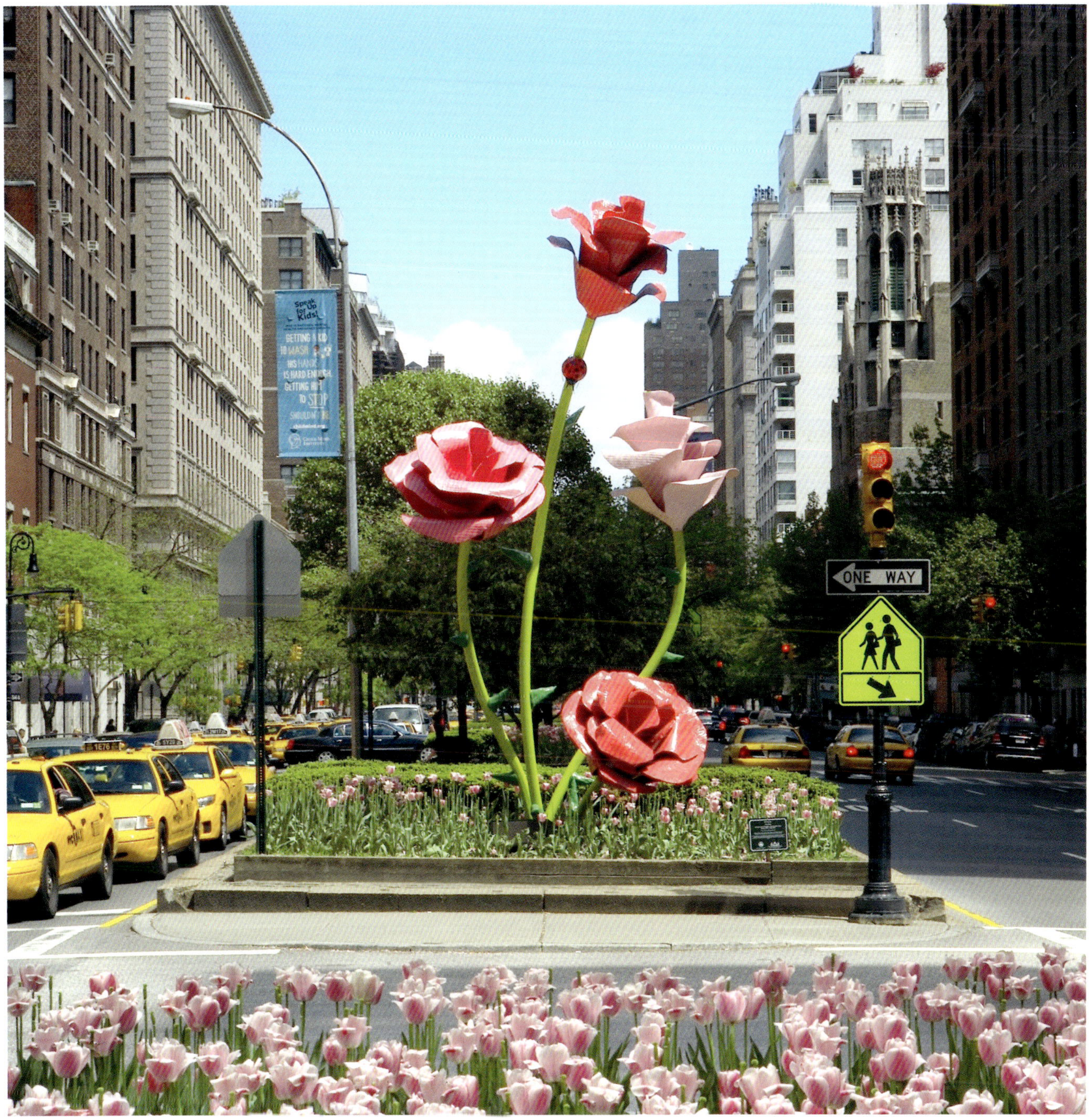

WILL RYMAN

The Roses on Park Avenue, 2010
Fiberglass resin, stainless steel, and paint, dimensions variable
Installation view, Park Avenue, New York

Gigantic roses brighten the traffic median of busy Park Avenue in New York. A site-specific work of thirty-eight whimsical fiberglass-and-steel sculptures by American artist Will Ryman (b. 1969), *The Roses on Park Avenue* transforms the mundane into the extraordinary. Four roses in shades of pink tower above thorny green stems, with a massive ladybug crawling down the tallest one. By selecting such a ubiquitous flower, Ryman aimed to elevate the everyday experience of pedestrians, encouraging them to see beauty in their daily surroundings. Exaggerating the size of the roses to a colossal extent—about 25 feet (8 m)—Ryman makes their presence unavoidable and impactful. He disrupts the viewer's sense of perspective and prompts them to reconsider their relationship with reality and nature, as well as their expectations of the urban landscape. In this way, the viewer must collaborate with the work to complete it, an element that connects to Ryman's background in theater and playwriting. These roses are characters within a larger story, and his installations, many of which feature oversize flora, often have a strong narrative. With this theatrical approach, he uses the roses to convey deeper meanings and emotions. The artist has mentioned that roses remind him of his mother, who was an enthusiastic gardener. This personal association adds a layer of intimacy and nostalgia to the installation. More than just a public artwork, these roses are a tribute to his own memories and past experiences that only roses can completely evoke and unlock.

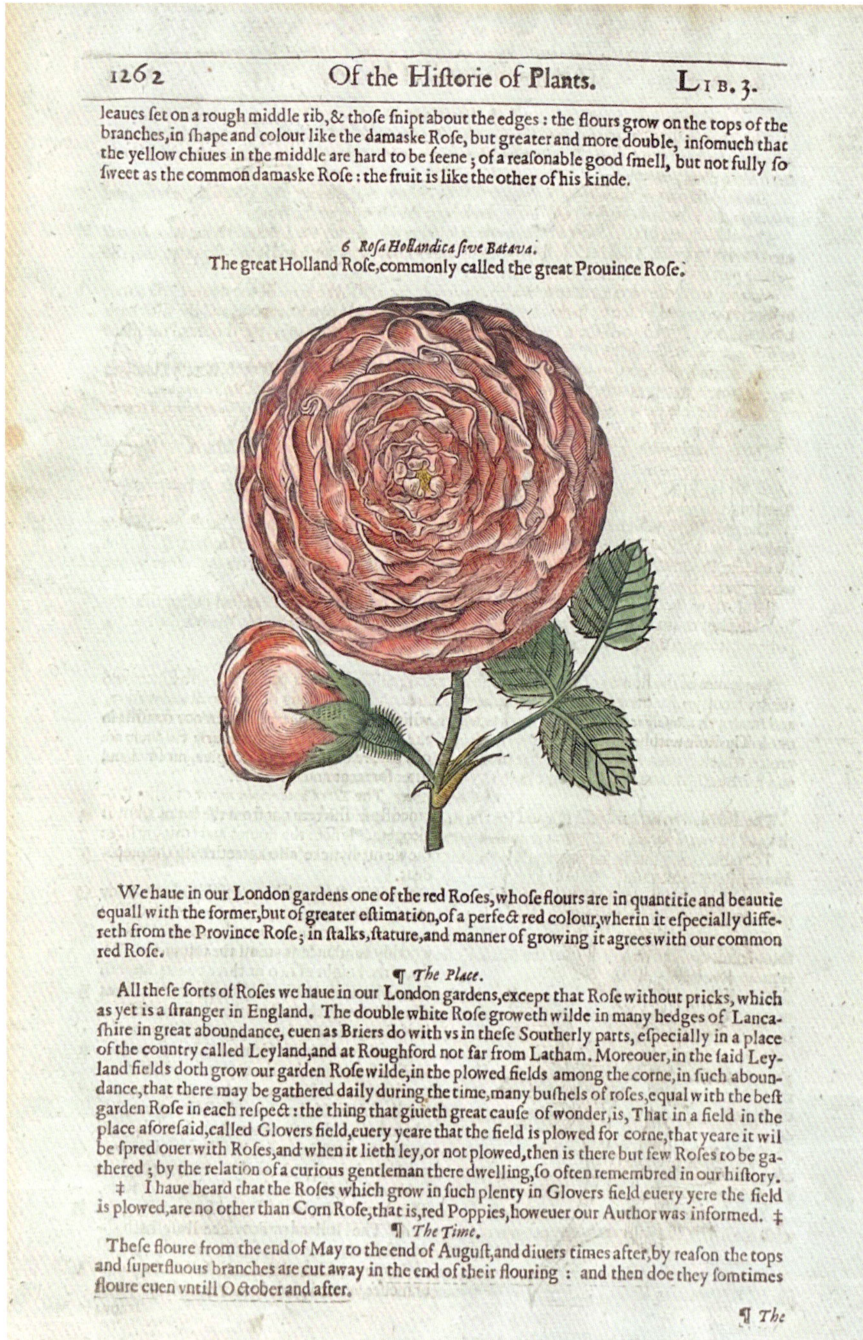

1262 Of the Hiſtorie of Plants. Lib. 3.

leaues ſet on a rough middle rib,& thoſe ſnipt about the edges: the flours grow on the tops of the branches,in ſhape and colour like the damaske Roſe, but greater and more double, inſomuch that the yellow chiues in the middle are hard to be ſeene; of a reaſonable good ſmell, but not fully ſo ſweet as the common damaske Roſe: the fruit is like the other of his kinde.

6 Roſa Hollandica ſiue Batava.
The great Holland Roſe,commonly called the great Prouince Roſe.

We haue in our London gardens one of the red Roſes,whoſe flours are in quantitie and beautie equall with the former,but of greater eſtimation,of a perfect red colour,wherin it eſpecially differeth from the Prouince Roſe; in ſtalks,ſtature,and manner of growing it agrees with our common red Roſe.

¶ *The Place.*

All theſe ſorts of Roſes we haue in our London gardens,except that Roſe without pricks, which as yet is a ſtranger in England. The double white Roſe groweth wilde in many hedges of Lancaſhire in great aboundance, euen as Briers do with vs in theſe Southerly parts, eſpecially in a place of the country called Leyland,and at Roughford not far from Latham. Moreouer,in the ſaid Leyland fields doth grow our garden Roſe wilde,in the plowed fields among the corne,in ſuch aboundance,that there may be gathered daily during the time,many buſhels of roſes,equal with the beſt garden Roſe in each reſpect: the thing that giueth great cauſe of wonder,is, That in a field in the place aforeſaid,called Glovers field,euery yeare that the field is plowed for corne,that yeare it wil be ſpred ouer with Roſes,and when it lieth ley,or not plowed,then is there but few Roſes to be gathered; by the relation of a curious gentleman there dwelling,ſo often remembred in our hiſtory.

‡ I haue heard that the Roſes which grow in ſuch plenty in Glovers field euery yere the field is plowed,are no other than Corn Roſe,that is,red Poppies,howeuer our Author was informed. ‡

¶ *The Time.*

Theſe floure from the end of May to the end of Auguſt,and diuers times after,by reaſon the tops and ſuperfluous branches are cut away in the end of their flouring: and then doe they ſomtimes floure euen vntill October and after.

¶ *The*

JOHN GERARD

The Great Holland Rose, from *The Herball*, 1597
Woodcut engraving, 12¾ × 8½ in. / 32.4 × 21.6 cm
Private collection

This hand-colored woodcut engraving of the great Holland rose reveals the principal characteristic reflected in one of its alternative names, *Rosa × centifolia*, "rose with a hundred petals." This early depiction of a pink rose is from the first edition of English herbalist John Gerard's (1545–1612) *The Herball*, or *Generall Historie of Plantes*, the most famous of all the English herbals, published in 1597. Gerard's Holland rose, now most commonly known as the cabbage rose, has a history of many names. In the late sixteenth century, Dutch nurserymen developed a flower that looked like this one, using a complex method of hybridization, hence the "Holland" epithet. In the same period, French botanist Carolus Clusius described the flower as *R. × centifolia batavica* (see p.190). Others knew it as the Provence rose (see p.212), after the region in France where *R. gallica* 'Officinalis' was supposedly introduced in 1240. Gerard's *Herball* includes 800 species of plant and more than 2,500 woodcuts; it is unsurprisingly London-centric, as Gerard's garden was in the city's neighborhood of Holborn. An entire chapter is dedicated to roses, which Gerard conceded were thorny but fragrant and beautiful. This association continued through the centuries, and the Holland rose, in particular, became a favorite icon for seventeenth-century artists to include in their portraits of women. The rose's pendulous, double, dark pink blooms have a wonderful perfume, which must have surely entranced sitters and painters alike.

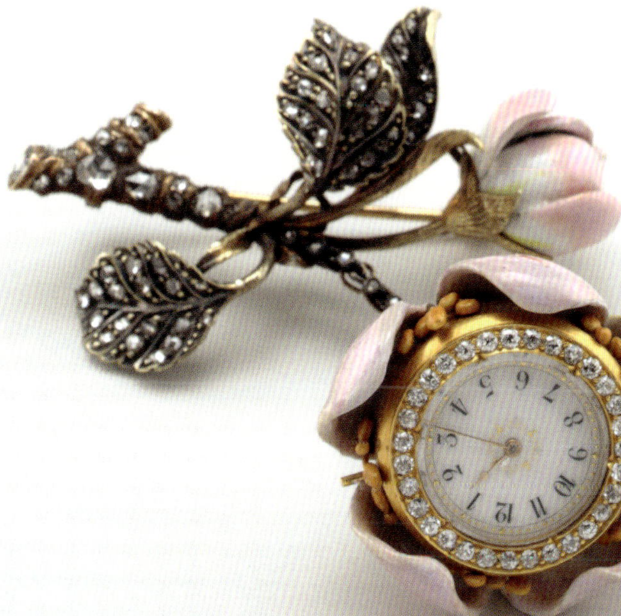

PAULDING FARNHAM

Rose lapel watch, c. 1889
Gold, enamel, and diamonds, 3 × 2½ × 1⅜ in. / 7.6 × 6.3 × 3.6 cm
Cooper Hewitt, Smithsonian Design Museum, New York

For such a small watch, this timepiece, which could be worn on the lapel or as a chatelaine pendant hooked to a decorative chain hung around the waist, is outstanding in its detail. The watch face emerges from the center of a wild rose crafted from pink enameled petals, its three leaves and stem made of gold and studded with one carat of rose diamonds. The attention to detail, precious jewels, and natural romanticism mark this piece as the intricate work of a genius. This rose watch was crafted by renowned American artist-jeweler Paulding Farnham (1859–1927) for the celebrated luxury design house and specialty jeweler Tiffany & Co. (see p.59). Farnham joined the company's design department in 1885, when he was only twenty-six years old, and rapidly established himself as a talented and innovative artist, becoming head designer in 1891. After winning a gold medal for Tiffany's at the Paris Exposition Universelle in 1889 with a series of elaborate floral brooches—including several exceptional orchids reflecting his close study of the plants themselves—Farnham was universally recognized as the genius behind the company's success. In 1900 he achieved further glory when the design house won the Grand Prize at the same event. Despite his achievements, Farnham was deposed in 1902 when Louis Comfort Tiffany assumed the role of design director, while Farnham spent the rest of his life drawing and painting, often inspired by nature.

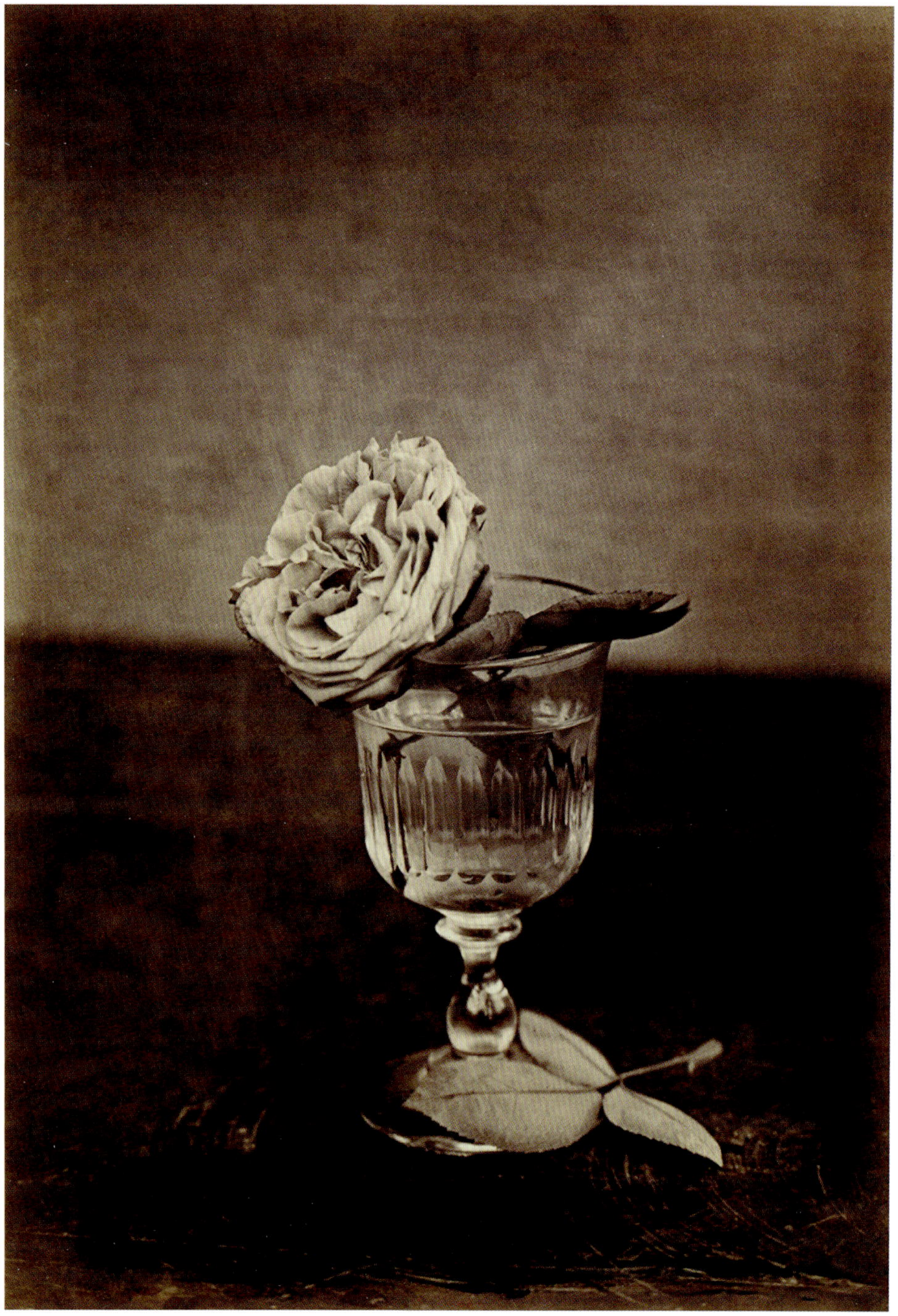

CHARLES AUBRY

Rose in a Crystal Glass, 1864
Print on albumen paper, 13½ × 9⅜ in. / 34.3 × 23.8 cm
Bibliothèque nationale de France, Paris

Placed in a French wineglass, a lone rose is the focus of a masterful composition of light and space, with its slightly asymmetrical composition and stark aesthetic. For Charles Aubry (1811–1877), a pioneer of photography, this was more than a pretty picture of a cut flower brought indoors. Having spent more than thirty years as a designer of patterns for textiles and wallpaper manufacturers, Aubry embarked on a new venture in 1864, at the age of fifty-three. Spurred on by advances and experiments in the new medium of photography, he set out to create an archive of photographs and plaster casts that artists and designers could use as models to reproduce botanical subjects. In his first year, he created about 150 negatives—including this rose—and he eventually found clients for his templates in drawing schools in France and abroad, and with such companies as the jeweler Tiffany & Co. (see p.59) and the textile manufacturers Mulhouse and Gobelins. Aubry's images of flowers, fruit, and leaves vividly capture the details and textures of the specimens, in some cases thanks to his technique of first dipping the plants in plaster to keep them looking fresh throughout the forty-five-minute exposure time required to make an image. His business proved short-lived, however, mainly due to a general lack of acceptance of photography as a medium at the time, and Aubry was forced to declare bankruptcy within a year. Despite his commercial failure, he is today considered a master of early photographic still lifes.

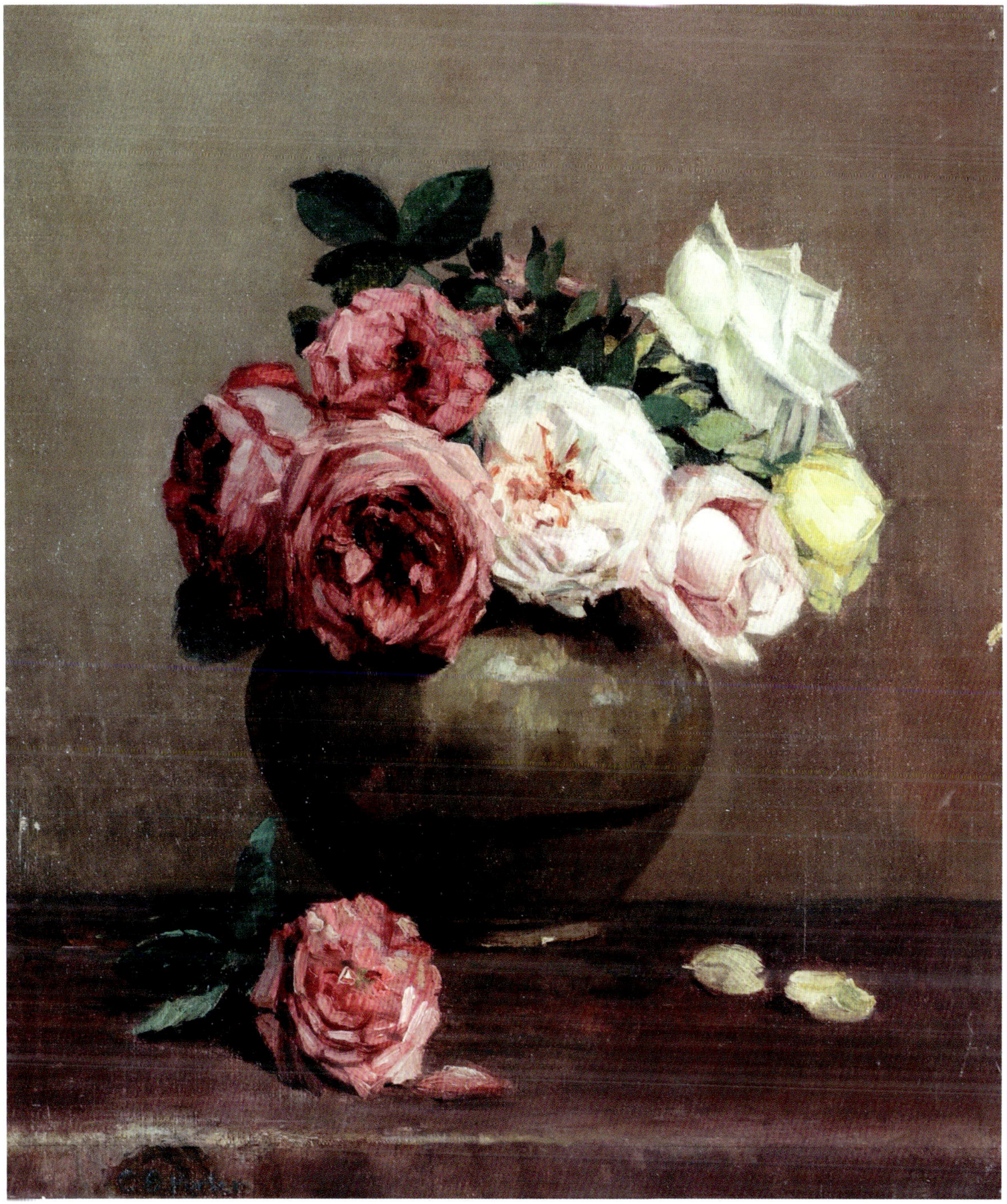

CHARLES ETHAN PORTER

Roses, c. 1881
Oil on canvas, 19¾ × 16½ in. / 50.2 × 41.9 cm
Norton Museum of Art, West Palm Beach, Florida

A bouquet of roses crowns a stoneware vessel in this delicate painting by Charles Ethan Porter (1847–1923), a Connecticut-based African American artist who excelled in still lifes and is thought to be the first Black professional artist to work primarily in the genre. The blooms here—caught at that exquisite moment when the petals just begin to fall—are so heavy that they lean for support on the neck of the jar. The range of coloration and shading is superb. Reflected in the ceramic glaze, the roses vary from dusty mauve to pale pink to almost white, with ivory-yellow for contrast. A single blossom in the left foreground rests near the table's edge, adding depth to the composition. Porter probably created this painting during his two years of study in Paris. Art historians see the influence of Henri Fantin-Latour (see p.75), the foremost floral still life painter of the time, in his work. Raised in rural poverty in Connecticut, Porter had a talent that first earned him a place at the National Academy of Design in New York, where he was one of the first African Americans to exhibit work; he established a studio in Hartford, Connecticut, in 1873. His patrons included landscape painter Frederic Edwin Church and writer Samuel Clemens—better known as Mark Twain. In a letter to Clemens from Paris in 1883, Porter wrote, "The colored people—my people—as a race I am interested in, and my success will only add to others who have already shown wherein they are capable, the same as other men."

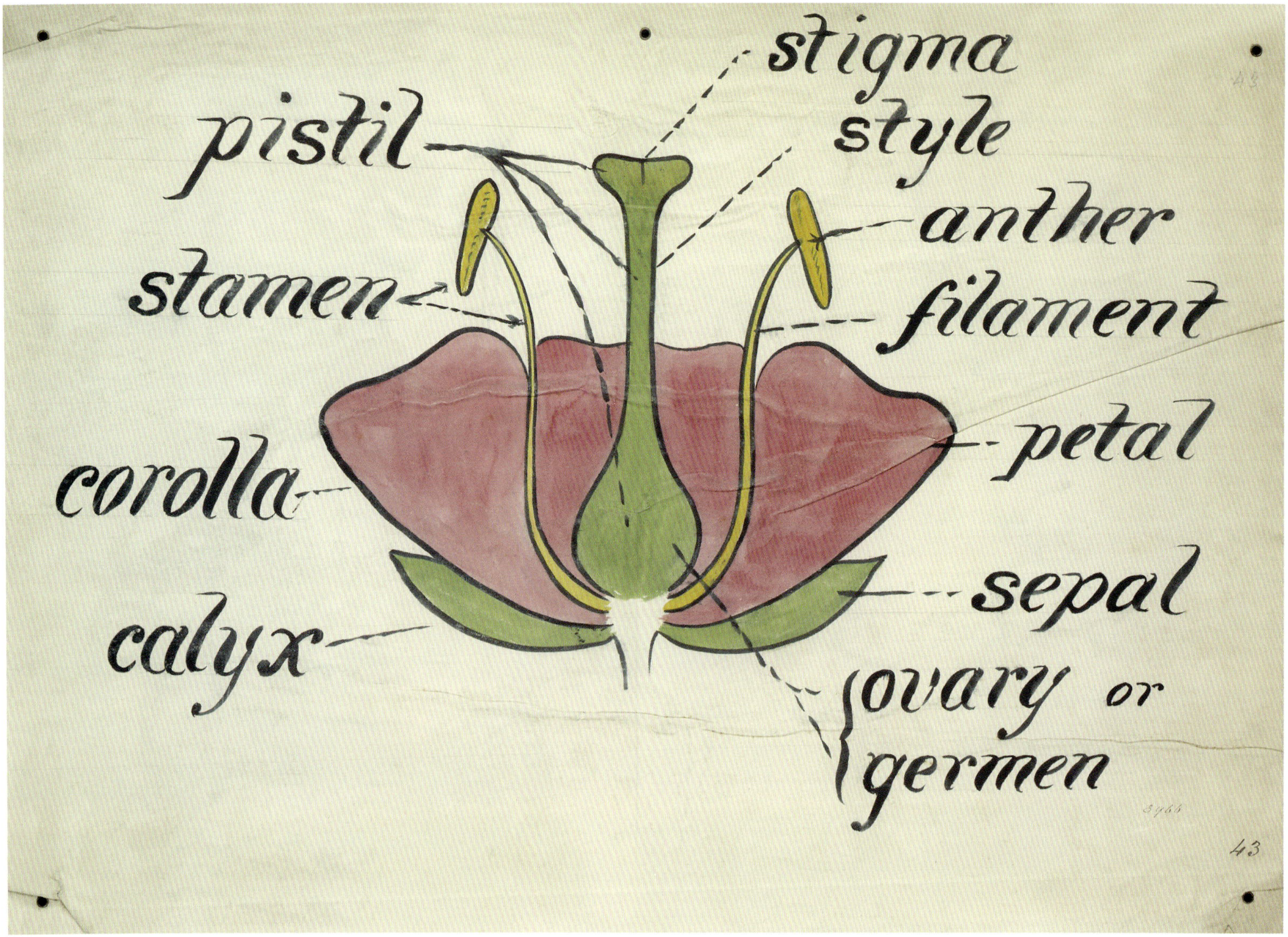

CHRISTOPHER DRESSER

Cross section through a rose, 1854–56
Pen and ink on buff paper, 21½ × 29½ in. / 54.5 × 75 cm
Victoria and Albert Museum, London

A pioneer of the late nineteenth-century design reform movement, Scottish artist and theorist Christopher Dresser (1834–1904) is widely considered a forefather of industrial design, introducing Japanese and botanical aesthetics into art objects and handicrafts. With floral motifs becoming popular in decorative arts at the time (see pp.32 and 126), Dresser recognized the importance of the scientific study of the natural world to modern design. He took a symbiotic approach to botany and art education, believing that the former depended on the use of visual images, while creative practices could pull from aesthetics already found in nature. In his role as professor of artistic botany in the department of science and art at the Government School of Design (now the Royal College of Art) in London, he integrated botanical studies within the curriculum of art education, with the Royal Botanic Gardens, Kew, regularly supplying his students with plant material. This diagram of a cross section of a rose, depicting its physiology and technical parts, is one of seventy-two drawings he produced to illustrate a lecture series on botanical morphology. Simple yet powerful diagrams like this one served as easy methods for teaching botany, as well as tools to develop his students' drawing, observation skills, and understanding of color theory. The symmetrical structure of this rose would have been especially appealing to Dresser's aesthetic since he looked for artistic inspiration in functional, natural, and beautiful objects.

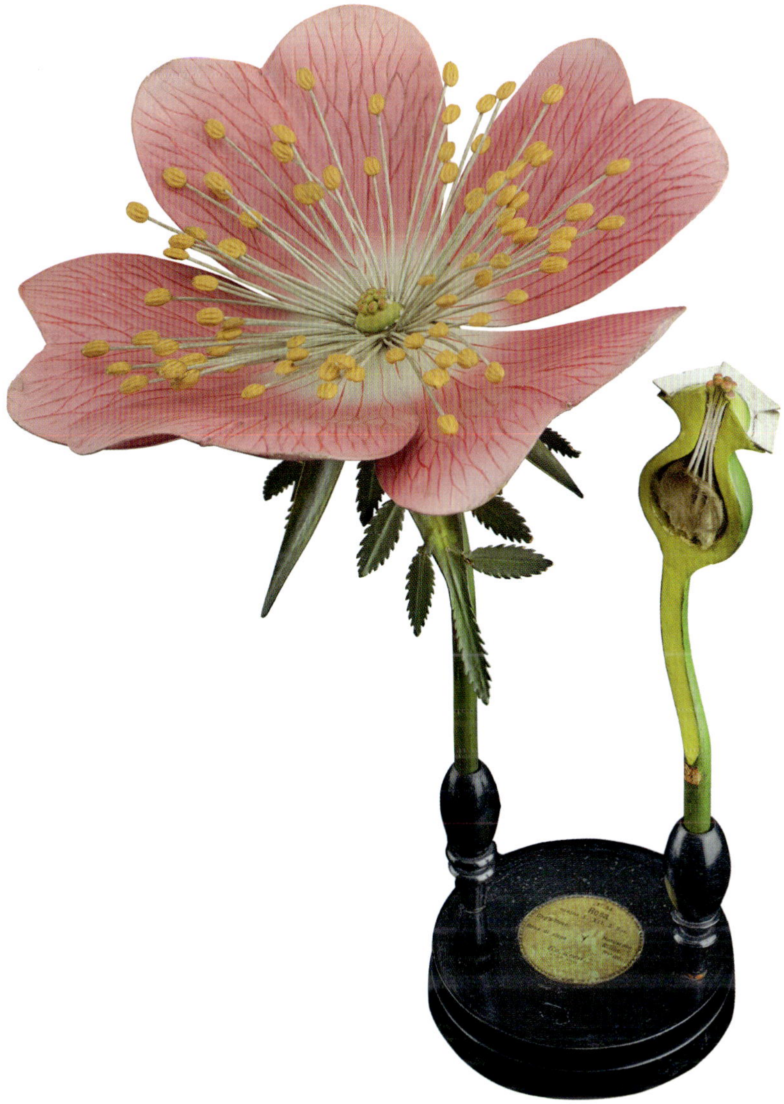

R. BRENDEL COMPANY

Rosa canina (Dog's Rose), No. 34, c. 1900
Papier mâché, wood, paper, hair, and plaster, overall 17⅝ × 15 × 13½ in. / 44.8 × 38.1 × 34.3 cm
National Museum of American History, Smithsonian, Washington, DC

Standing slightly over 17 inches (nearly 45 cm) tall, this highly detailed papier-mâché replica portrays *Rosa canina*, the dog rose, a climbing rose known for its sweet-smelling pink flowers and the sharp, hooked thorns that enable it to hold onto vertical surfaces (see pp.98 and 103). This botanically accurate model comes from the R. Brendel Company, which was started by German modelmaker Robert Brendel (c. 1821–1898) in 1866 and taken over by his son Reinhold (c. 1861–1927) in 1898. The Brendels worked closely with scientists and craftspeople to make accurate botanical, anatomical, and mineralogical models to be used as study aids in universities and medical schools at a time when the study of practical sciences was booming. Such was the beauty of the models produced by their company that they found their way out of the laboratory and into mail-order catalogs, becoming sought-after collectibles in their own right and winning prizes at art and craft exhibitions. Here, the single dog rose—some say its name comes from the similarity between its hooked prickles and a dog's canine teeth—is shown magnified so that its five petals and multiple stamens and pistils are clearly visible. On the other side of the stand, a cross section of the rose hip shows its seed pods. Native to Europe and West Asia, the dog rose is prized for its large ovoid rose hips, which are turned into syrup, tea, wine, and preserves and are valued for their high vitamin C content.

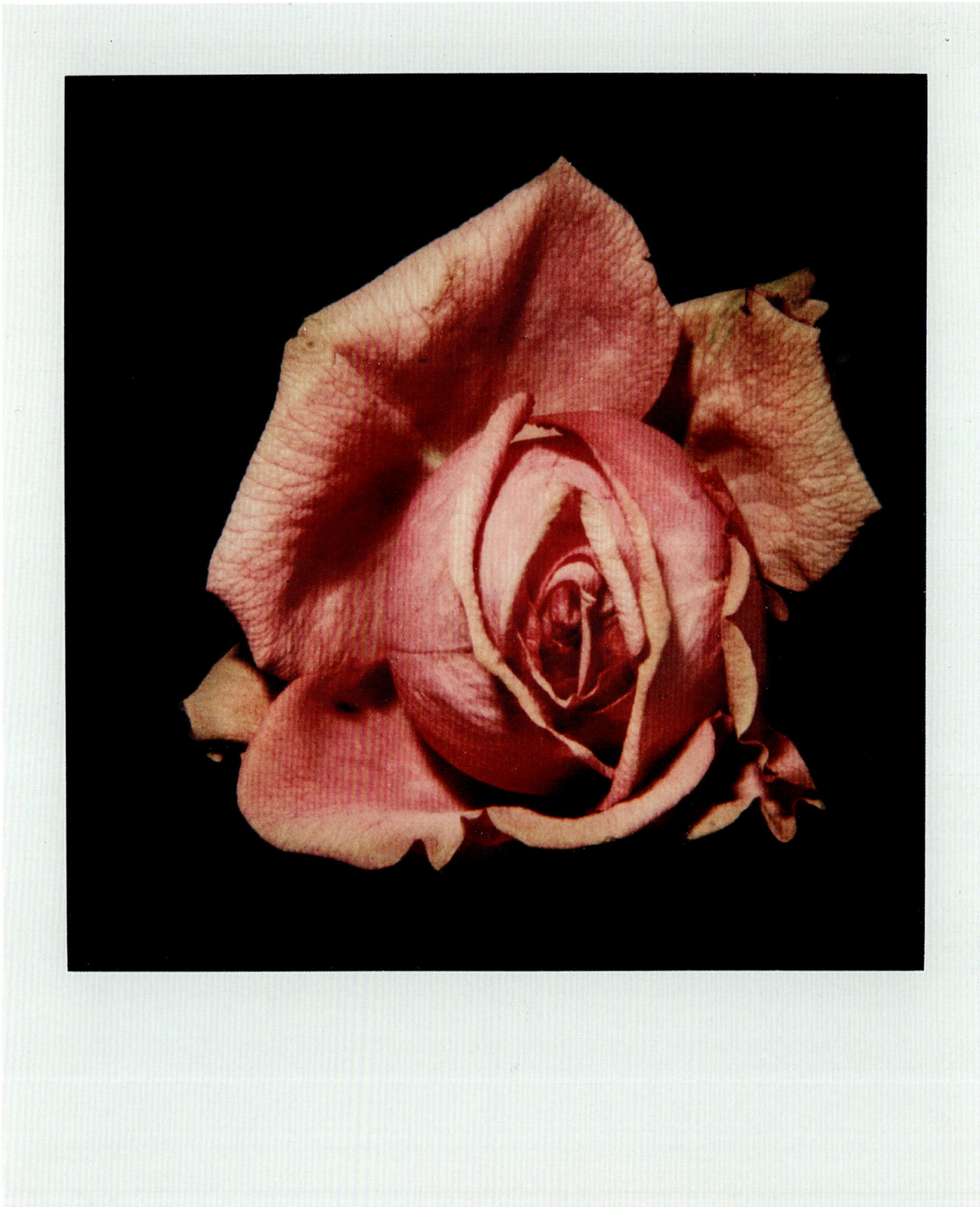

NOBUYOSHI ARAKI

Flower, date unknown
Unique Polaroid-type 600 print, 3⅛ × 2⅞ in. / 7.8 × 7.6 cm
Private collection

A glossy Polaroid picture presenting a life-size image of a rose against a stark black background, this photograph carries an air of intimacy that is reinforced by the absence of natural light. Fascinated with the spontaneous nature of photography, Japanese photographer Nobuyoshi Araki (b. 1940) follows his libido instead of his thinking mind when he creates images. During his studies, Araki focused on film and photography, after which he worked for several years as a commercial photographer in advertising. His interest in experimental approaches led him to film, collage, and, later, the Polaroid instant images with which he keeps a diaristic record of his life and experiences, sometimes also applying paint to the prints. His first photographic project, published in 1971, was a record of his honeymoon, and since then, love and death—the instincts represented by the ancient concepts of Eros and Thanatos—have been twin themes running through Araki's oeuvre. He has gained acclaim and come into the spotlight with an output featuring nudes, portraits of women in Japanese erotic bondage, female genitalia, and flowers in varying stages of decay, as well as images of his cat and everyday domestic scenes. The frontal angle from which this rose is photographed hones in on the still-closed center of the rosebud, while giving space to the unfurling, fading outer petals. Their fleshy pink, recalling the cavities of the body, lends a sensuality to the bloom that conveys an impression of both vulnerability and power.

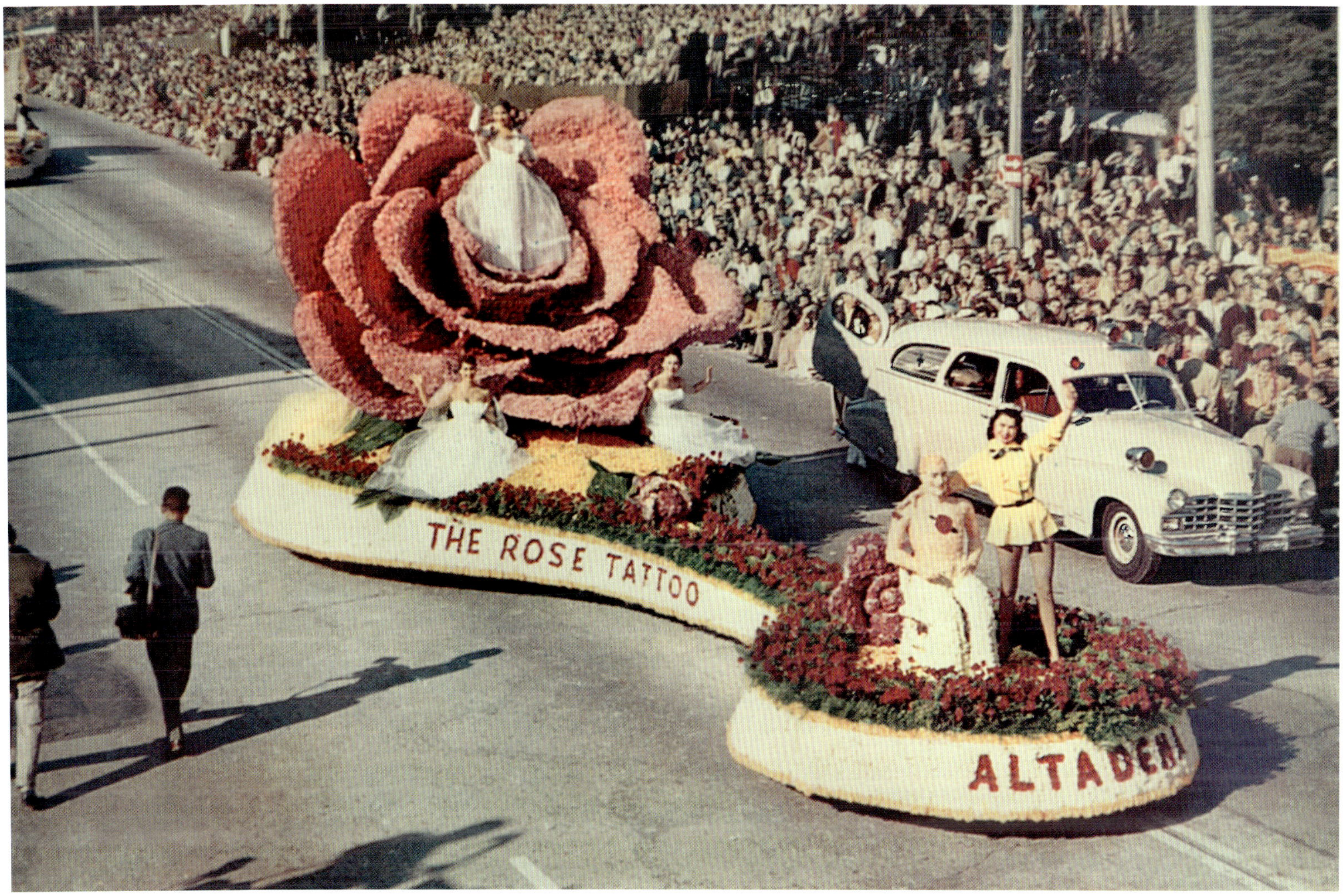

ANONYMOUS

Altadena Rose Parade Float, 1957
Promotional brochure, dimensions variable
Altadena Historical Society, California

Drawing huge crowds to witness spectacular processions of flower-bedecked floats, marching bands, and cavalcades of horse riders, the annual Tournament of Roses Parade in Pasadena, California, has been a cherished New Year's Day tradition since 1890. The parade was first organized by members of Pasadena's Valley Hunt Club, many of whom were former residents of the northeastern and midwestern United States, where the winters were harsh. Their intention was to highlight California's mild winter weather by showcasing the many flowers, especially roses, that bloomed around Christmas time. The earliest parades featured beautiful horse-drawn carriages covered in blooms. The float in this photograph, entered by the City of Altadena in 1957, is motorized and found its inspiration in the Tennessee Williams play *The Rose Tattoo* (1949–50). At the head of the float, which is completely covered with roses and rose petals, is a seated figure decorated with white and pink blossoms and a red rose tattoo on his chest. At the rear, a huge pink bloom provides a backdrop for three women wearing white wedding dresses, who appear like fairies alongside the giant flower. Altadena entered a float every year until the late 1960s, when escalating costs forced them to stop. The parade, which takes place on Colorado Boulevard and is watched by hundreds of thousands of spectators, is followed in the afternoon by the Rose Bowl, one of the major bowl games in American college football.

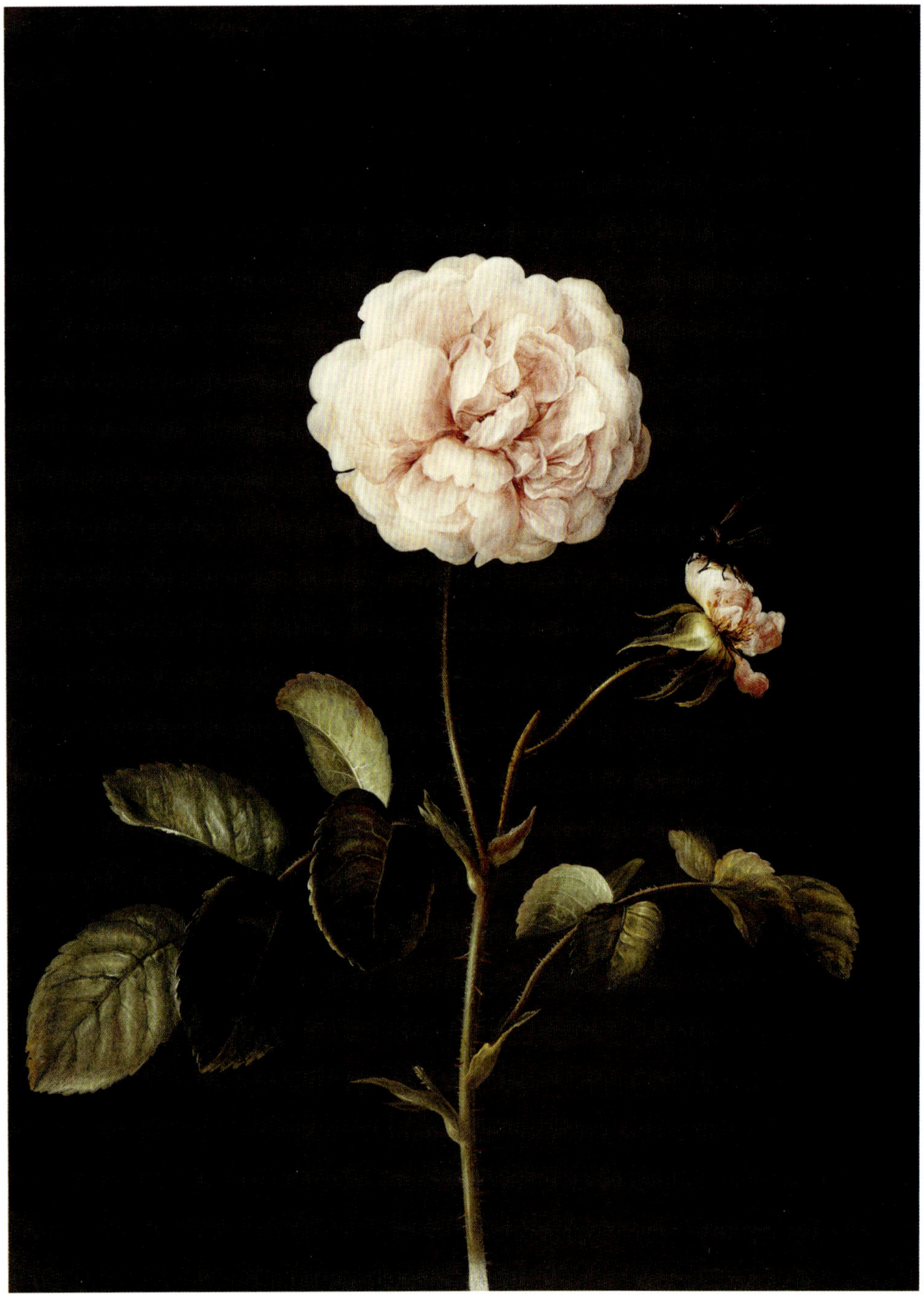

BARBARA REGINA DIETZSCH

A Gallica Rose with a Bumblebee, c. 1760
Bodycolor, gold, and gold ink on vellum, 11⅜ × 8¼ in. / 29 × 21 cm
Private collection

A single large Gallica rose shines against a plain, dark background that highlights its frothy petals, while a bumblebee sits on an adjacent, much smaller, withered bloom. Realized in highly realistic detail—some of the leaves on the thorny stem show signs of decay—this image is the work of German painter and engraver Barbara Regina Dietzsch (1706–1783), who followed in the footsteps of such celebrated artists of the German Renaissance as Albrecht Dürer and Hans Hoffmann in producing botanically accurate paintings of plants and animals. Dietzsch lived and worked in Nuremberg, which in the eighteenth century became a major center for botanical illustration in a tradition that grew directly from the work of Dutch still-life painters of the previous century. Painting on vellum or parchment, she typically used a monochromatic background so that she could paint a single plant with a precision that helped her turn many of her images into engravings. She came from a family of artists, but her gender prevented her from joining a guild to learn her trade formally. Instead, she worked from home for her entire career, declining an invitation to join the Bavarian archducal court as a painter. The Gallica is one of the oldest of all Garden roses, grown in antiquity by the Greeks and Romans, and bred into new varieties by the Dutch in the seventeenth century. Known for its variety of pink, purple, and red shades, the short, bushy shrub grows well in poor soil and is prized for its hardiness.

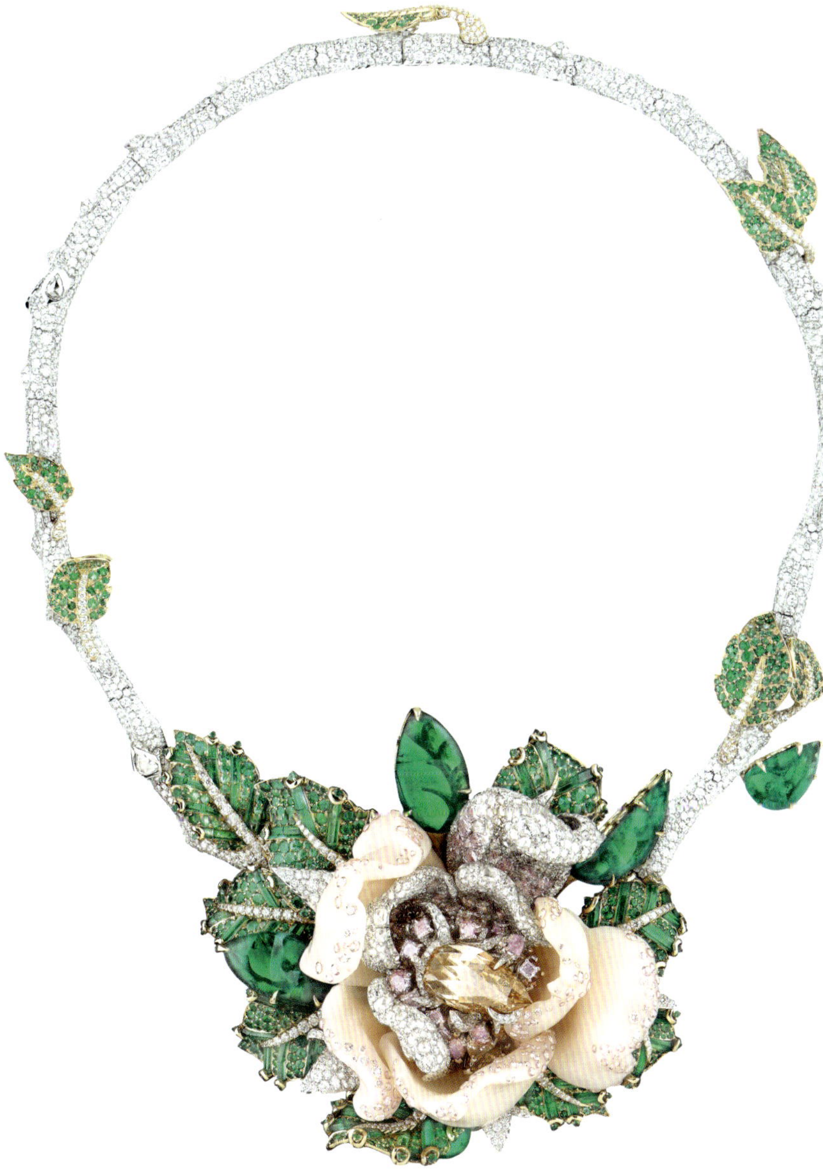

VICTOIRE DE CASTELLANE

Bal de Mai necklace, from *Le Bal des Roses* collection, 2011
White and yellow gold, diamonds (fancy brown, pink, lilac, and mauve), pink opal, and emeralds

Pink, lilac, brown, and mauve diamonds, along with pink opals, are some of the precious stones used to create this remarkable necklace, known as *Bal de Mai* (*The Rose of May*), which is finished with white and yellow gold. This exquisite piece was crafted by French jewelry designer Victoire de Castellane (b. 1962), who has, since 1998, been creative director of the French fashion house Dior's high jewelry division, Dior Joaillerie. The rose necklace pays homage to the house's storied founder, Christian Dior, whose was famed for his passion for roses, which featured in many of his designs and as the foundational note in his debut fragrance, Miss Dior (see p.37). Forming part of de Castellane's 2011 collection *Le Bal des Roses* (*The Roses' Ball*), which celebrated twelve different roses that were, in de Castellane's words, "preparing to go to a ball," this necklace is a unique, one-of-a-kind piece. De Castellane cites a number of influences for her bold designs, including nature, fairy tales, and pop culture, as well as the house of Dior itself. Her genius lies in her ability to reinterpret the familiar flower in a startlingly new form, here as if the central rose were wearing a ball gown. The House of Dior moved into jewelry in the late 1990s, when de Castellane joined the company to build a high jewelry line from the ground up. Previously, she had been working in the jewelry division at Dior's rival Chanel, where she had started her career alongside Karl Lagerfeld.

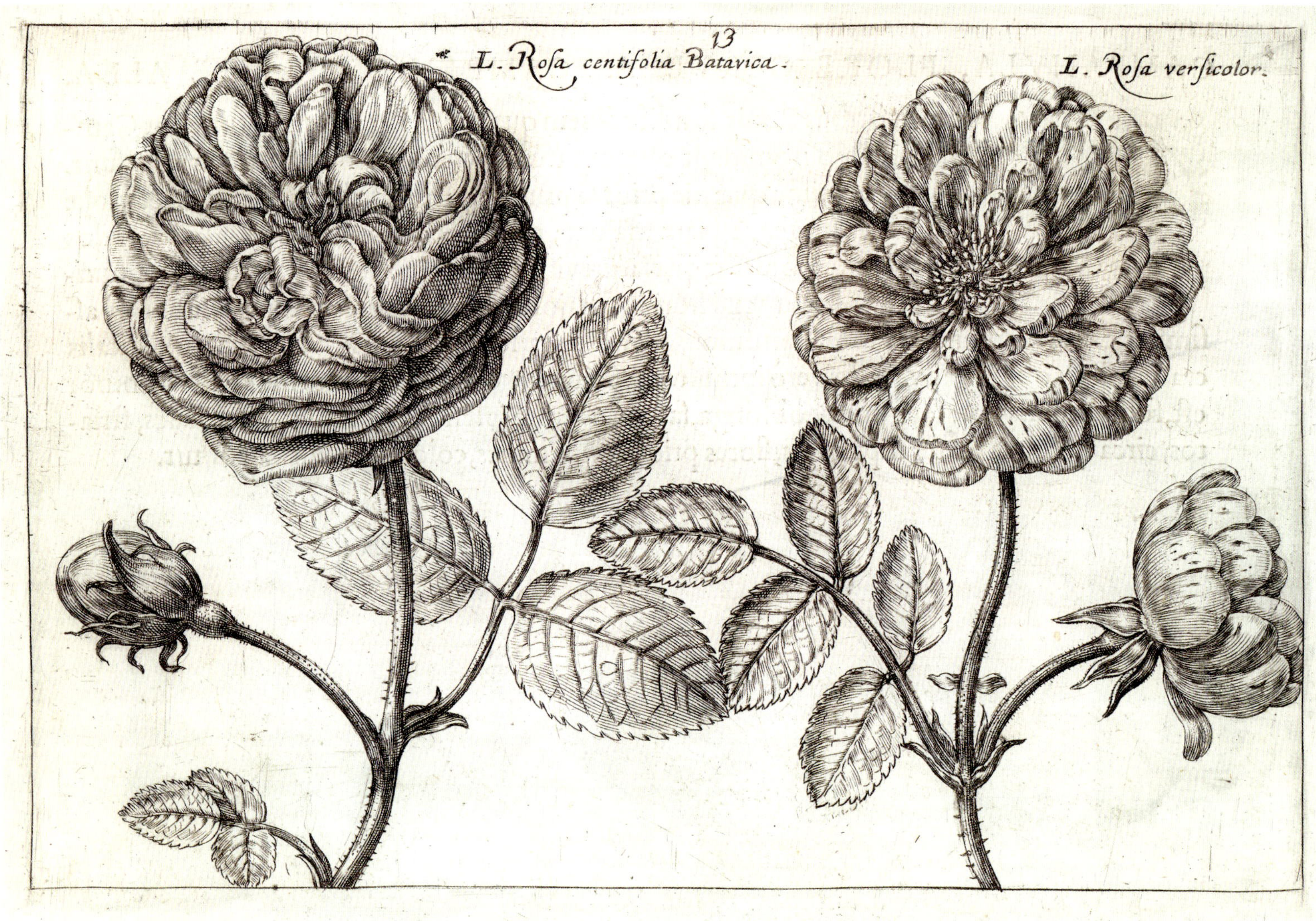

CRISPIJN DE PASSE THE YOUNGER

Rosa Centifolia Batavica and Rosa Versicolor, from *Hortus Floridus*, 1614
Metal plate engraving, 8 × 11½ in. / 20 × 29 cm
Royal Horticultural Society, Lindley Library, London

This engraved plate showing two roses, *Rosa centifolia Batavica and Rosa versicolor*, comes from one of the most influential botanical books ever made. It appeared in *Hortus Floridus* (*A Garden of Flowers*), first published in 1614 by Dutch artist Crispijn de Passe the Younger (c. 1594–1670), who spent four years preparing more than 160 copper plates of flowers to include in the florilegium. Both roses have an interesting history that merits their inclusion in this anthology of flowers, which presents the varieties that could be found in a seventeenth-century Dutch garden at a time when roses were highly valued for their scent but the number of rose varieties was limited. Centifolia roses were notable for their "hundred" petals (see pp.24 and 180) and strong perfume. A descendant of an ancient Roman rose, this one was probably a hybrid (hence the inclusion of "*Batavica*" in the name). The versicolor rose is notable for its striped petals and was first named in 1601, although a similar striped rose was discovered painted on the walls of the first-century house of Venus Marina in Pompeii. The *Hortus Floridus* was created to meet a growing public interest in plants by depicting them in an accurate scientific manner, which ultimately led to the emergence of the Dutch still-life genre (see pp.10 and 140). De Passe came from a family of illustrious engravers and publishers, and he used the greater precision of copper engravings over earlier woodcuts to include intricate detail in his works.

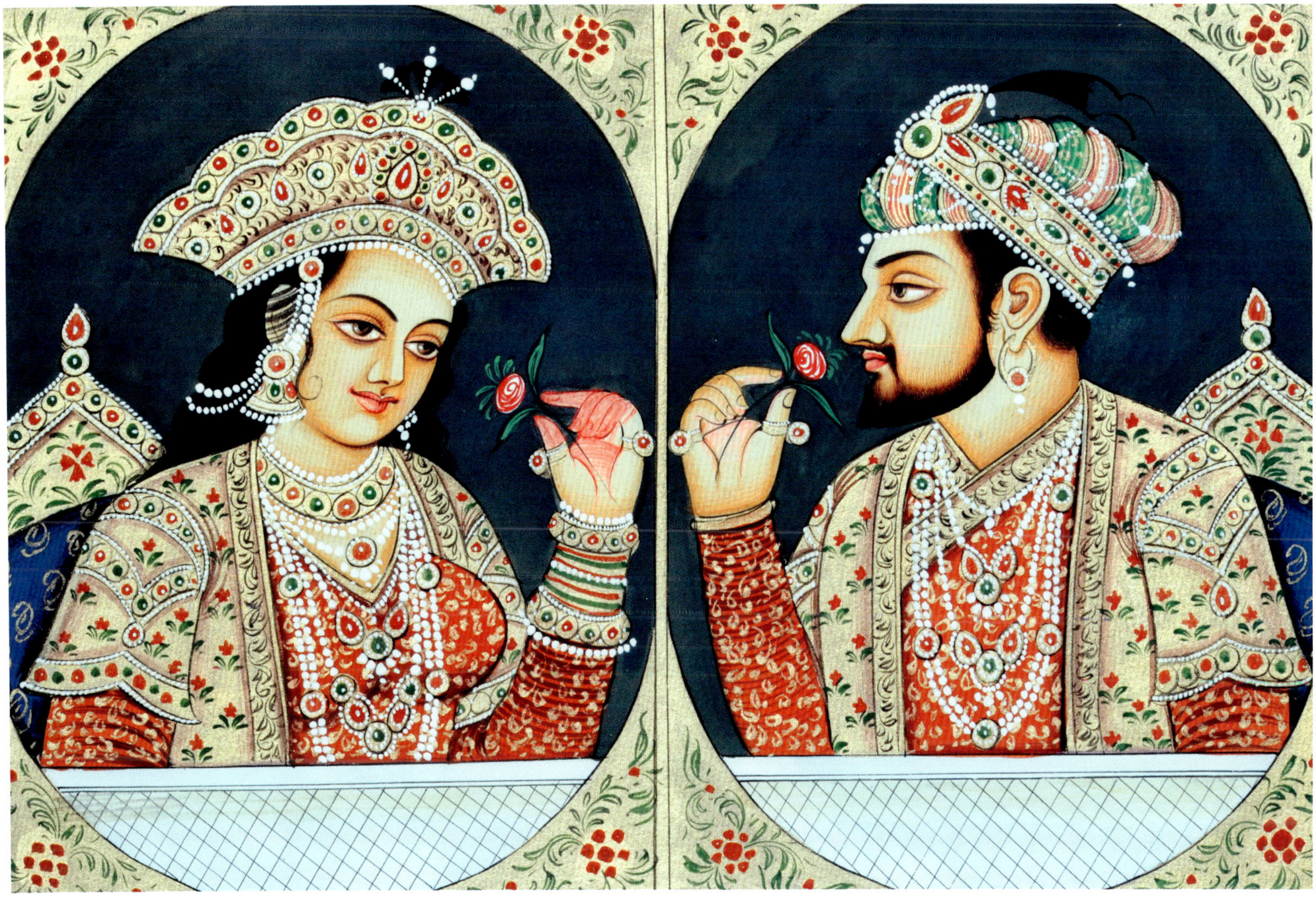

ANONYMOUS

Shah Jahan and Mumtaz Mahal smelling a pink rose, c. 17th century
Private collection

In Indian culture, roses signify both worldly and divine adoration, and they are frequently used at weddings and religious ceremonies to highlight love and devotion. In Hinduism, when the gods were said to live on Earth, the creator god Brahma and Vishnu, the god of preservation, disputed which flower was the most beautiful. Brahma preferred the lotus while Vishnu favored the rose. Upon being shown the arbor laden with lovely, fragrant roses in Vishnu's celestial garden, Brahma conceded the argument. This celebration of peaceful beauty continued into the sixteenth century, when the Muslim Mughal emperors arrived from Persia and Afghanistan to reign over India. As legend has it, the first Mughal emperor Bābur carried with him caravans of the Damask rose from his native country in present-day Uzbekistan into the region. A century or so later, another Mughal emperor, Shah Jahan, who ruled from 1628 to 1658, wanted to display in his likenesses the image of a humanitarian king, adding the rose as a symbol of beauty, patience, and love. This double portrait shows him with his favorite wife, Mumtaz Mahal. From 1632 to 1648, Shah Jahan had the Taj Mahal built to honor her after her death in childbirth. The mausoleum is covered with white marble—white being the color of mourning, harmony, and purity in India—and decorated with precious stones, such as lapis lazuli from Afghanistan. Like the icon of the rose, it is a wonder of symmetry and an emblem of collaboration between Hindu and Muslim cultures.

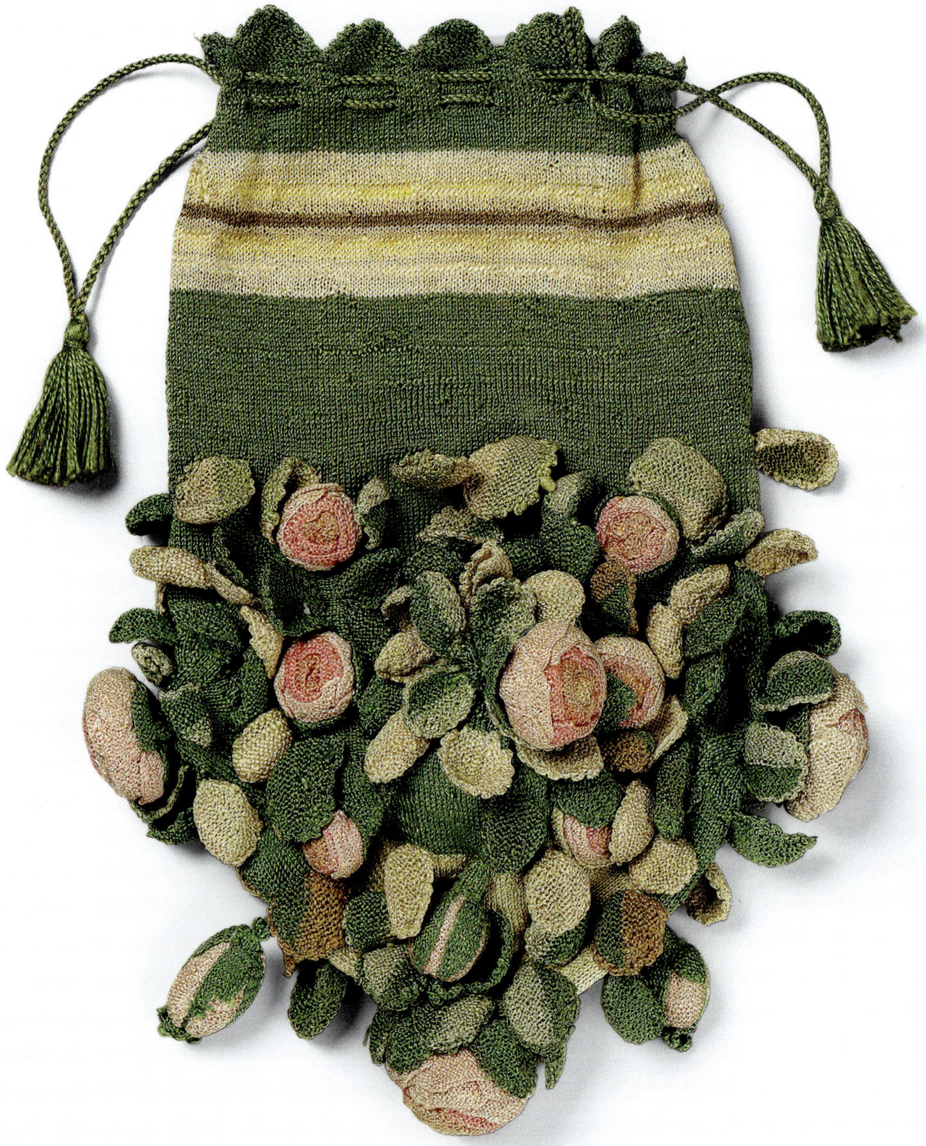

ANONYMOUS

Bag, late 18th century
Knitted silk, 10⅜ × 6⅞ in. / 26.5 × 17.5 cm
Museum of Fine Arts, Boston

Rose aficionados will find this lovely rosebud bag from the late eighteenth century highly covetable. The accessory, known as a reticule, is of green silk knit with contrasting ecru, yellow, and brown bands near its drawstring top. The pink and salmon roses and green leaves are appliquéd onto the bag, perhaps demonstrating the unknown maker's patience and apparent love for the aesthetics of the craft. For much of the eighteenth century, women kept the essential objects they wanted to carry in large pockets tied around their waists underneath their wide, gathered skirts. As fashions changed toward the end of the century from full-skirted dresses to more figure-hugging neoclassical or Empire-line dresses of lighter fabric, the previously hidden pockets became unsightly bulges under the dresses. Therefore, easily made drawstring bags, or reticules, became popular. These new accessories were made of silk, silk satin, lace, or knitted fabric, and were often embroidered or ornamented with beading, tassels, or—as in this case—beautifully crafted roses. They were also a means for refined ladies to display their needlework skills. Due to their light fabric, the bags could be used to carry a small amount of coins, a handkerchief, a fan, and other small items, but nothing too weighty. From the 1790s until the 1830s, the reticule was considered an indispensable accessory, thus giving the bag its alternative name—an "indispensable."

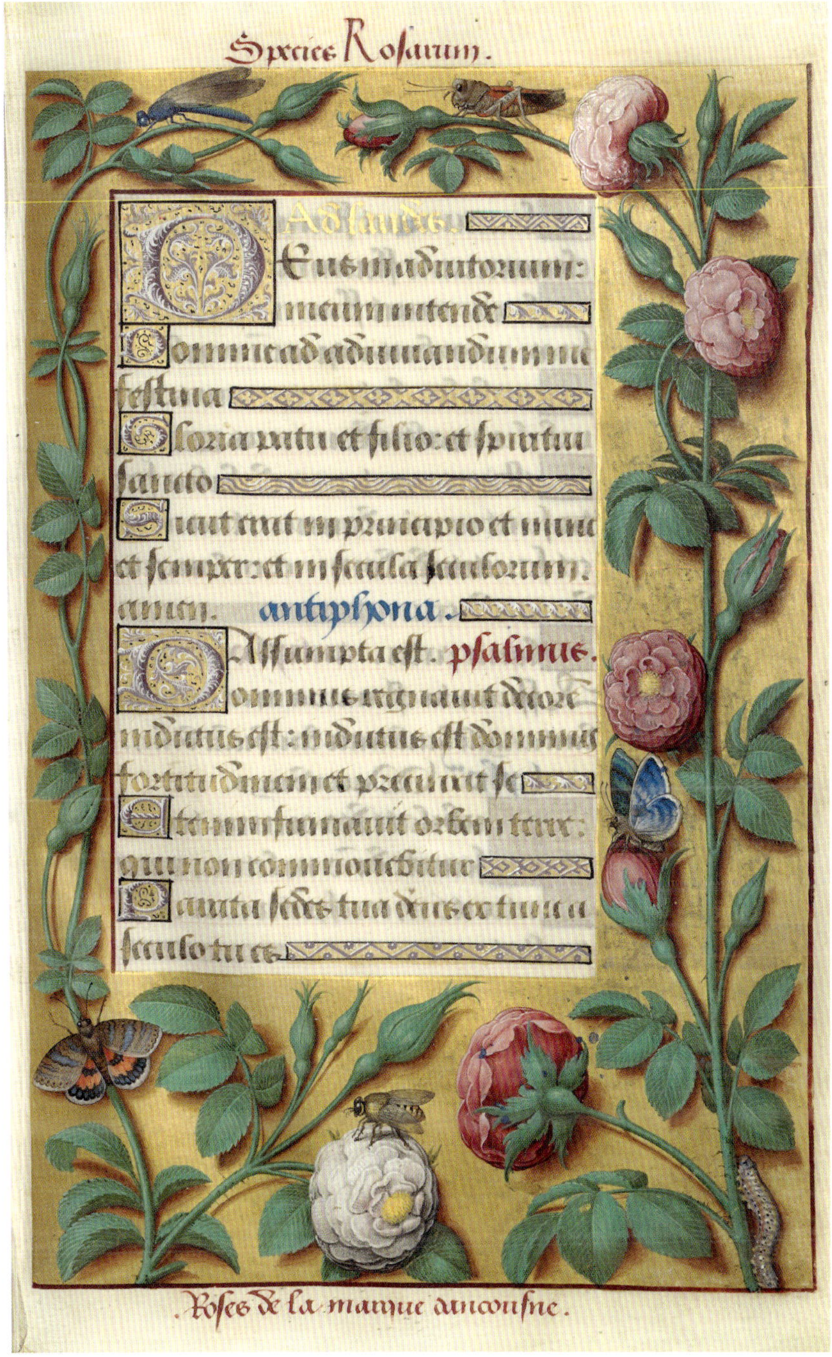

JEAN BOURDICHON

Folio 37r: Species Rosarum, from *Grandes Heures of Anne de Bretagne*, 1503–8
Parchment manuscript, 11¾ × 7½ in. / 30 × 19 cm
Bibliothèque nationale de France, Paris

Set against a gold background, dusty pink roses in various stages of full bloom and bud climb up the sides of a written prayer while a single white rose, with a fly sitting on it, decorates the bottom of the frame. This page comes from a book of hours—a volume containing devotional prayers for different days of the year—commissioned by Anne of Brittany, then queen of France, and illuminated by the leading French court artist of the time, Jean Bourdichon (c. 1457–1521), between 1503 and 1508. Bourdichon painted accurate depictions of single flower species on each page of the prayer book, including the plant's scientific and Latin names. Anne, who was the last independent ruler of Brittany before becoming queen, lived at the Chateau de Blois in the Loire Valley, where Bourdichon painted from life more than three hundred flowers and other plants growing in the garden and surrounding countryside. He often included insects and small animals—here, butterflies, a dragonfly, and a cicada—as reminders of the cycle of life and death, beauty and decay, adding to the impression of verisimilitude. The preeminent painter of richly decorated books of hours, Bourdichon worked for four different kings at the French court. Unusually for a prayer book intended for private devotion, his painstakingly detailed paintings of plants and flowers provide a valuable botanical record, while also making this one of the most magnificent of all books of hours of the early French Renaissance.

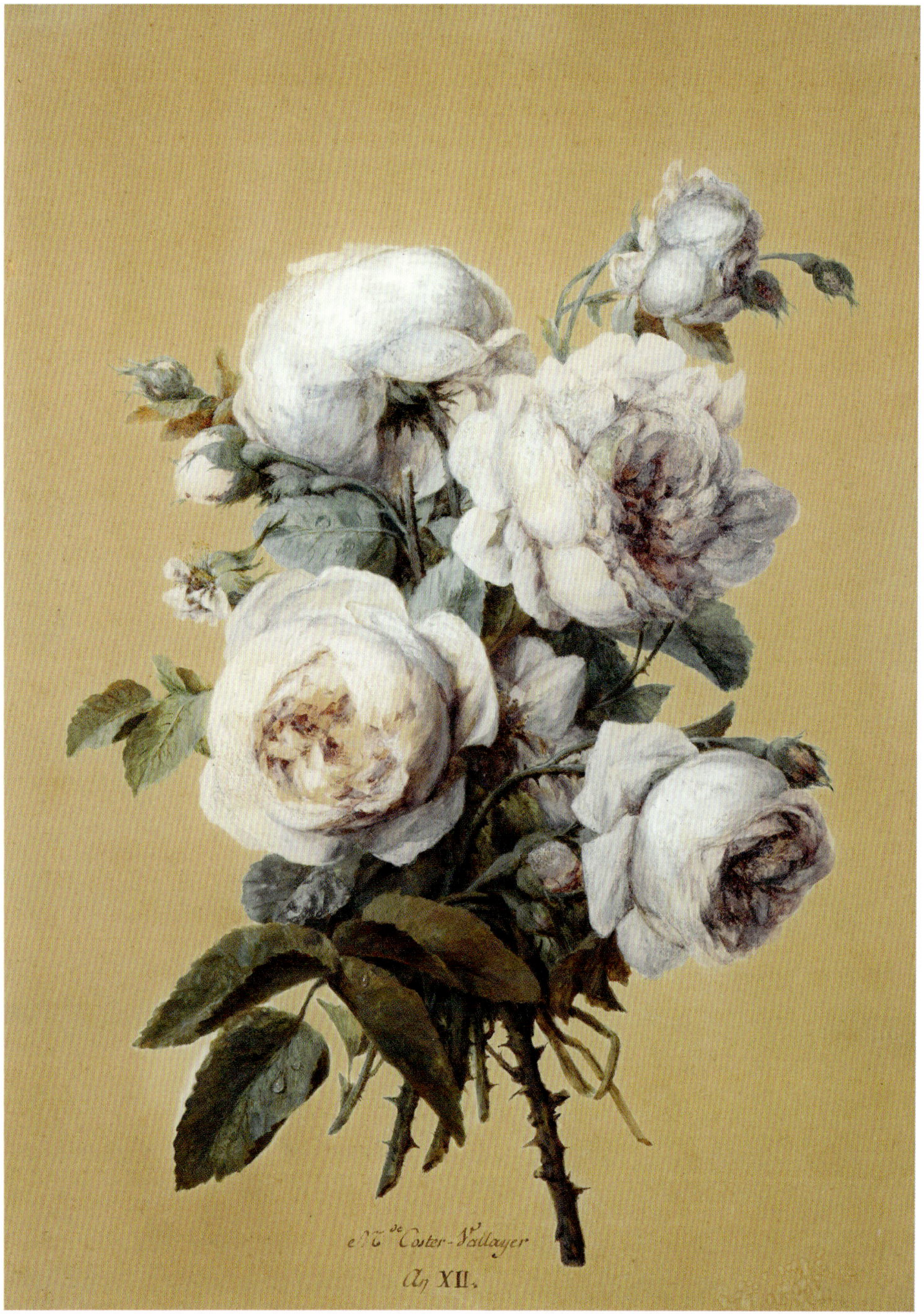

ANNE VALLAYER-COSTER

Bouquet of Roses, 1804
Watercolor, 12⅝ × 10 in. / 32 × 25.5 cm
Bois-Préau National Museum, Rueil-Malmaison, France

A bouquet of pale white roses, many in full bloom but a few still in bud, appears suspended against a pale beige background. This delicate watercolor is a later work of the French artist Anne Vallayer-Coster (1744–1818), who was renowned in her lifetime for her still-life paintings of fruit, game, seashells, and, in particular, flowers. Her close attention to detail—each petal, thorn, and leaf vein is clearly delineated here—was honed early. Vallayer-Coster's father was goldsmith to the king and taught her how to pay attention to a work's intricate elements, while her painting teacher, the watercolorist Madeleine Françoise Basseporte (see p.124), gave her instruction in the scientific art of observing and painting flowers. After exhibiting her first floral still life in 1775, Vallayer-Coster began to paint opulent floral arrangements. Her ability to capture the transient beauty of flowers gained her admirers, including the most influential woman of the period, Queen Marie Antoinette, who appointed her painter to the queen in 1780. Her paintings captured the extravagance of the French court at the time, and she remained loyal to the monarchy throughout the French Revolution, which began in 1787 and would overthrow her patron in the process. In 1770, at only twenty-six years old, Vallayer-Coster was unanimously admitted into France's leading art institution, the Académie Royale—a remarkable achievement given there were then only four female artists in the academy, three of whom were themselves the wives of painters.

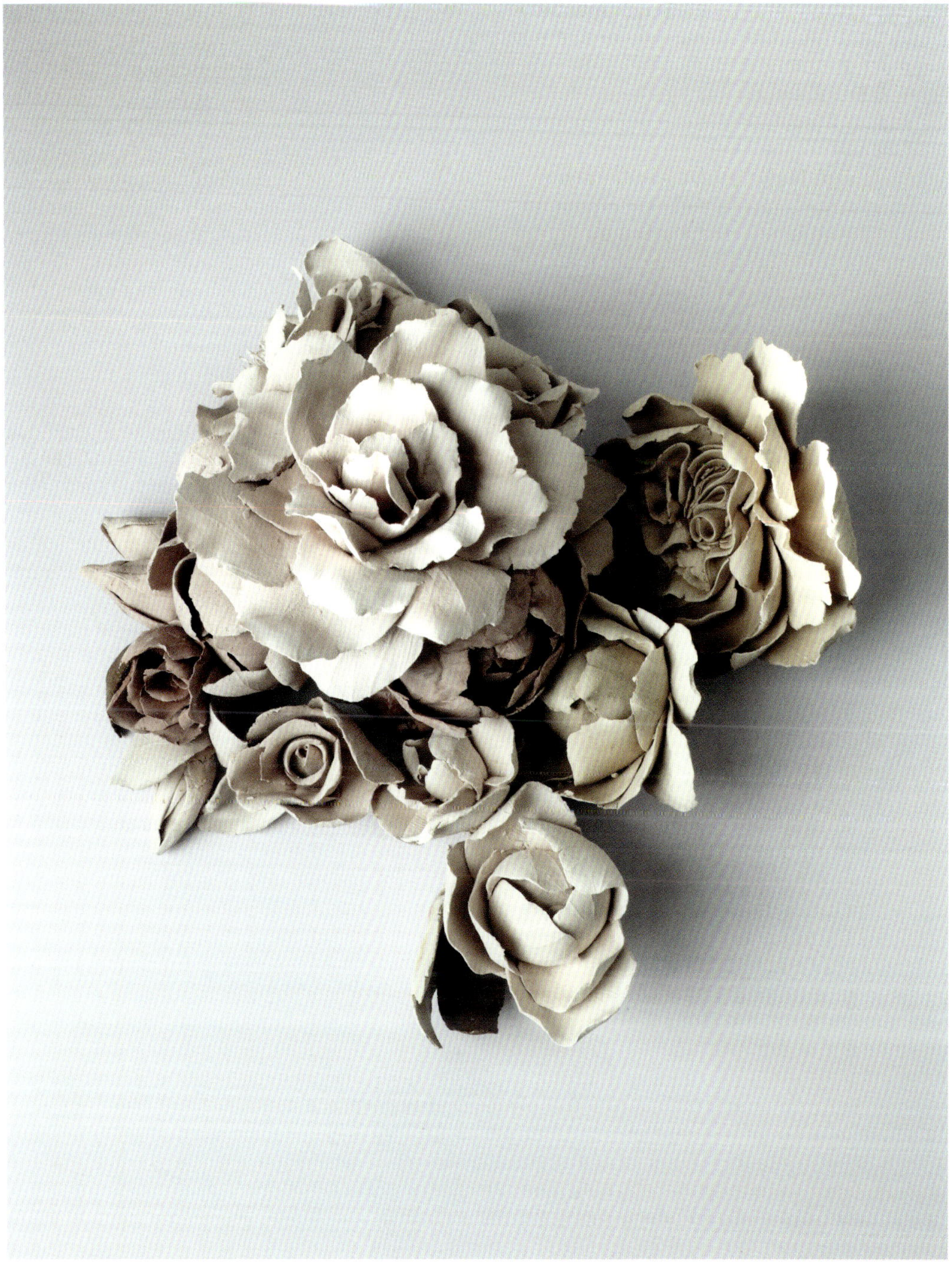

PHOEBE CUMMINGS

Rose Study, 2022
Raw clay, dimensions variable

The fulsome rose sculptures of British ceramicist Phoebe Cummings (b. 1981) are almost rococo in style, while her use of raw clay, with its shadowlike, muted palette, communicates a modernist vision. Instead of working directly from nature, she chooses to hold the memory of the objects she crafts, hybridizing botanical accuracy with an imagined gesture or essence of flowering. Cummings also has an interest in the way roses exist as patterns, symbols, fiction, and myth, and she investigates through her art how they are retold in visual culture as much as how they really exist in nature. Her ceramics technique was originally born of necessity. Lacking studio space and a kiln, she started to work with raw clay that was not fired. The decision to retain fingerprint traces that evidence the hands that cut, coax, and curl the material into bud and flower forms makes her art intensely human. In the same way that a rose can fall apart in your hand, these exquisitely fragile roses will crumble if not cradled with the utmost care. It is thus partly of necessity that her craft practice is nomadic and performative. Her installations are site-specific and made on location, often created in public view. This composition, for instance, was crafted for the exhibition *Wild & Cultivated: Fashioning the Rose* at London's Garden Museum in 2022. Cummings is also committed to sustainable working practices, and reuses the clay from each installation. Like that of the rose itself, her art is transient, its beauty fleeting.

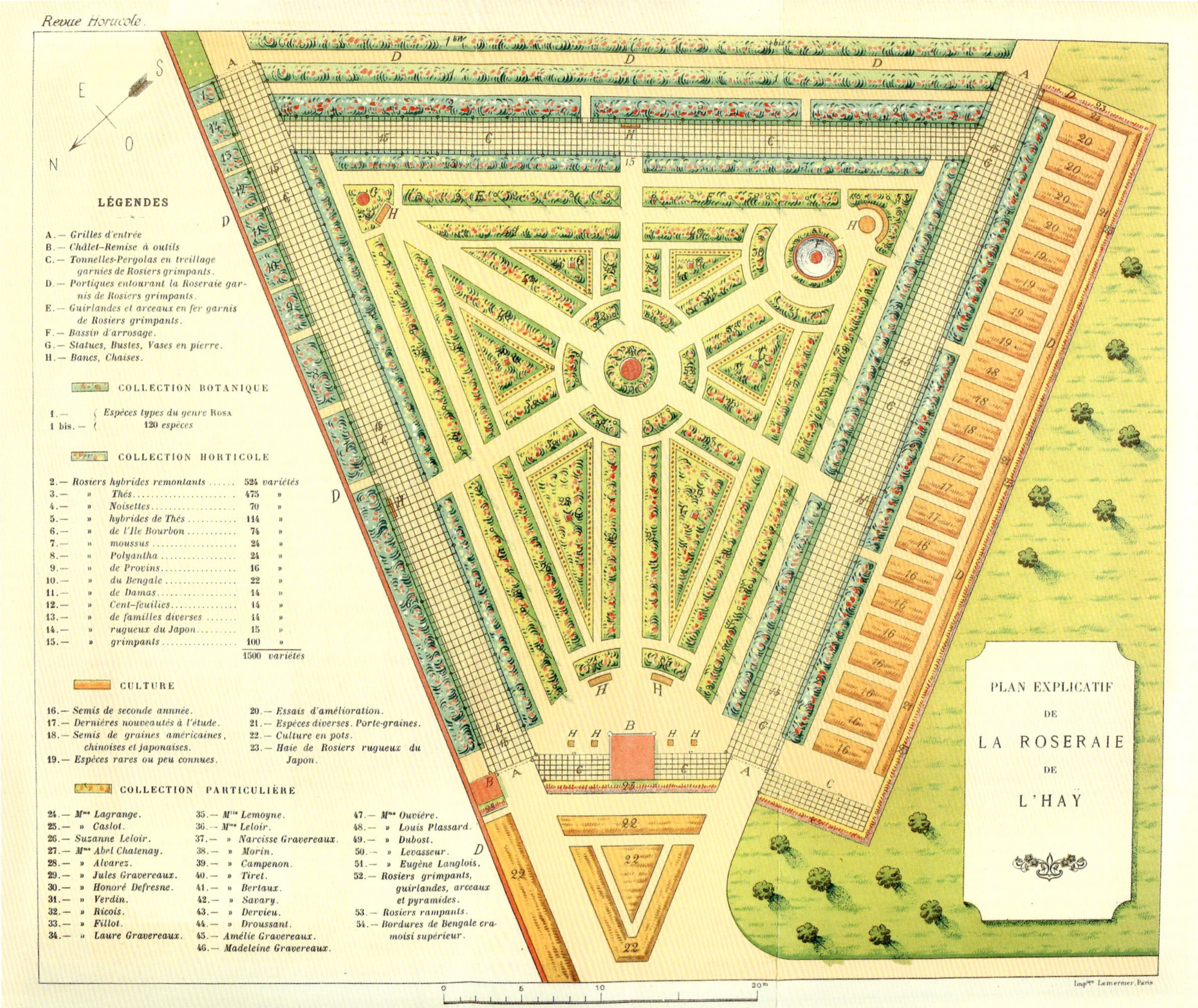

ÉDOUARD ANDRÉ

Plan for the Roseraie de l'Haÿ, from *Revue Horticole*, 1899
Lithograph, 6½ × 10⅜ in. / 16.5 × 26.3 cm
Harvard University Botany Libraries, Cambridge, Massachusetts

The Roseraie de l'Haÿ was the brainchild of entrepreneur Jules Gravereaux, a director of the Bon Marché department store in Paris and an enthusiastic rose grower. In 1893 he bought a house at L'Haÿ, a southern suburb of the city, and by the end of the century, he had a collection of three thousand rose species and varieties. In 1899, determined to make a great public garden of his collection, he hired renowned French horticulturist and landscape architect Édouard André (1840–1911) to design it. André, who had begun his career working on the layout of the great Parisian parks, was later appointed *Jardinier Principal* (head gardener) of Paris before becoming an independent landscape designer with an international practice. So famous did Gravereaux and André's garden become that the town officially changed its name to L'Haÿ-les-Roses in 1914. Gravereaux wanted his garden to be more than a decorative display of roses, famously saying, "The purely decorative rose garden is like a pretty woman without brains. She may attract attention for a time, but does not retain it." The Roseraie was above all an educational display, with a museum and separate sections for different categories of rose cultivars. It also had a long walk flanked by borders, seen in this plan, one of which shows the development of the genus *Rosa* in evolutionary terms, and the other the history of roses under cultivation. After Gravereaux's death, the garden was managed (less successfully) by his family, but it was bought by the government in 1937 and is still in operation today.

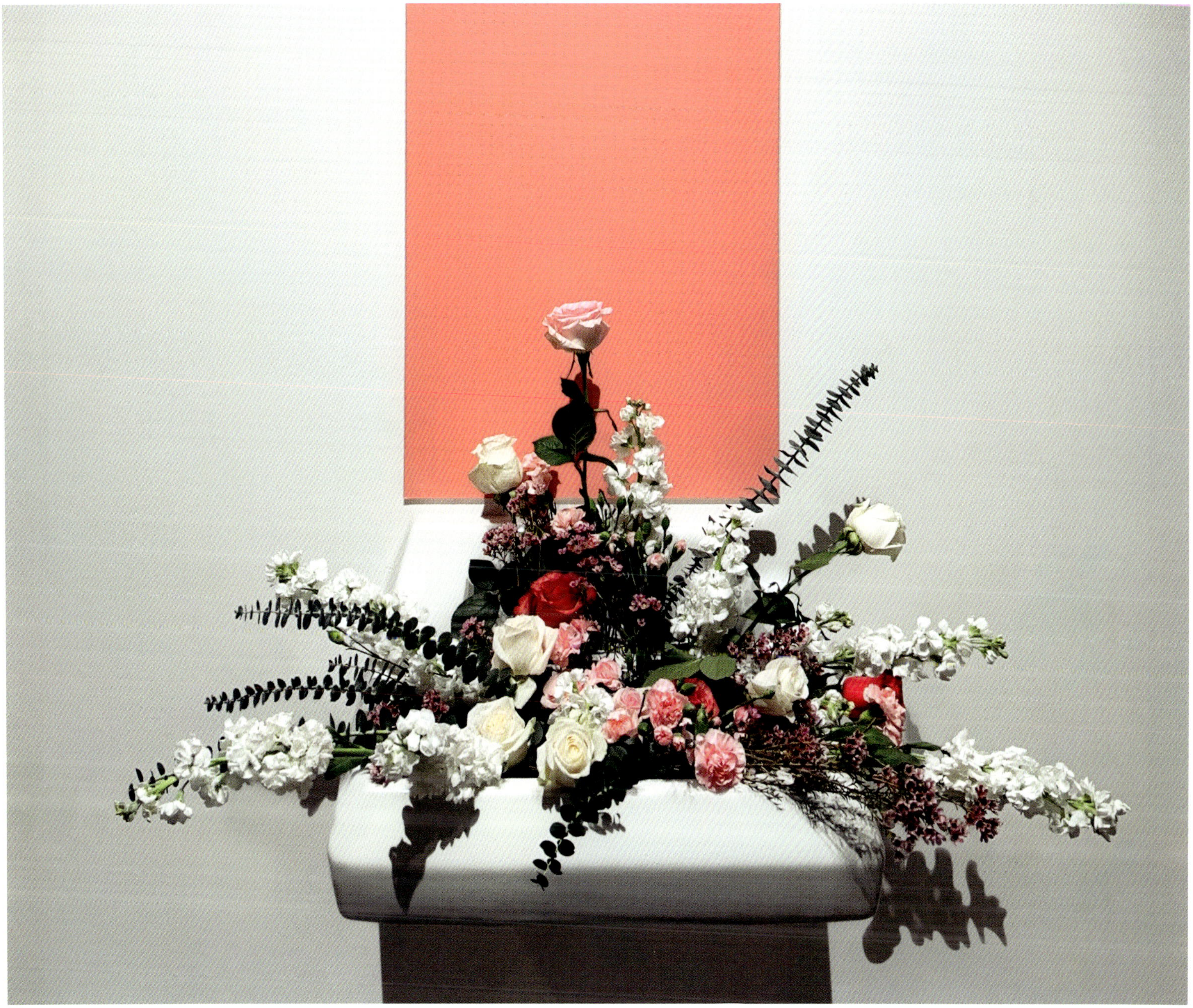

EJ HILL AND AUGUST GRAHN

Vase with flowers, 2020–22
Porcelain sink and flowers with acrylic on wood, 51 × 20 × 18 in. / 129.5 × 50.8 × 45.7 cm
Estate of August Grahn, Piedmont, California

A white ceramic sink becomes a vase for an unusual flower arrangement. American multimedia artists EJ Hill (b. 1985) and August Grahn (1994–2020) use the bathroom object as the receptacle for a selection of store-bought flowers, installed below a square piece of painted wood. Among its flowers, the makeshift bouquet includes white and deep pink roses and carnations, which Hill had purchased to cheer himself up: "I was sad and tired, so I decided to buy myself flowers." The artists lay out the roses horizontally instead of using the traditional vertical stance in a vase. Known for endurance-based performance works, Hill started to create a series of floral pieces he called *joy studies* in 2022, exploring flowers' common cultural role as a shorthand for joy and collaborating on some of them with conceptual artist Grahn. Both artists' works intersect with issues of gender, race, class, and finding one's place in the world. In *Vase with flowers*, Hill and Grahn chose to take small pleasures where they could; the act of arranging the flowers in the sink brings joy, as does simply admiring them for their beauty. In a performance at the Hammer Museum, Los Angeles, in 2018, Hill appeared with the following question written behind him in neon: "Where on earth, in which soils and under what conditions, will we bloom brilliantly and violently?" He seemingly answered that inquiry with the title of the 2022 exhibition in which *Vase with flowers* was first shown: *Wherever we will to root*.

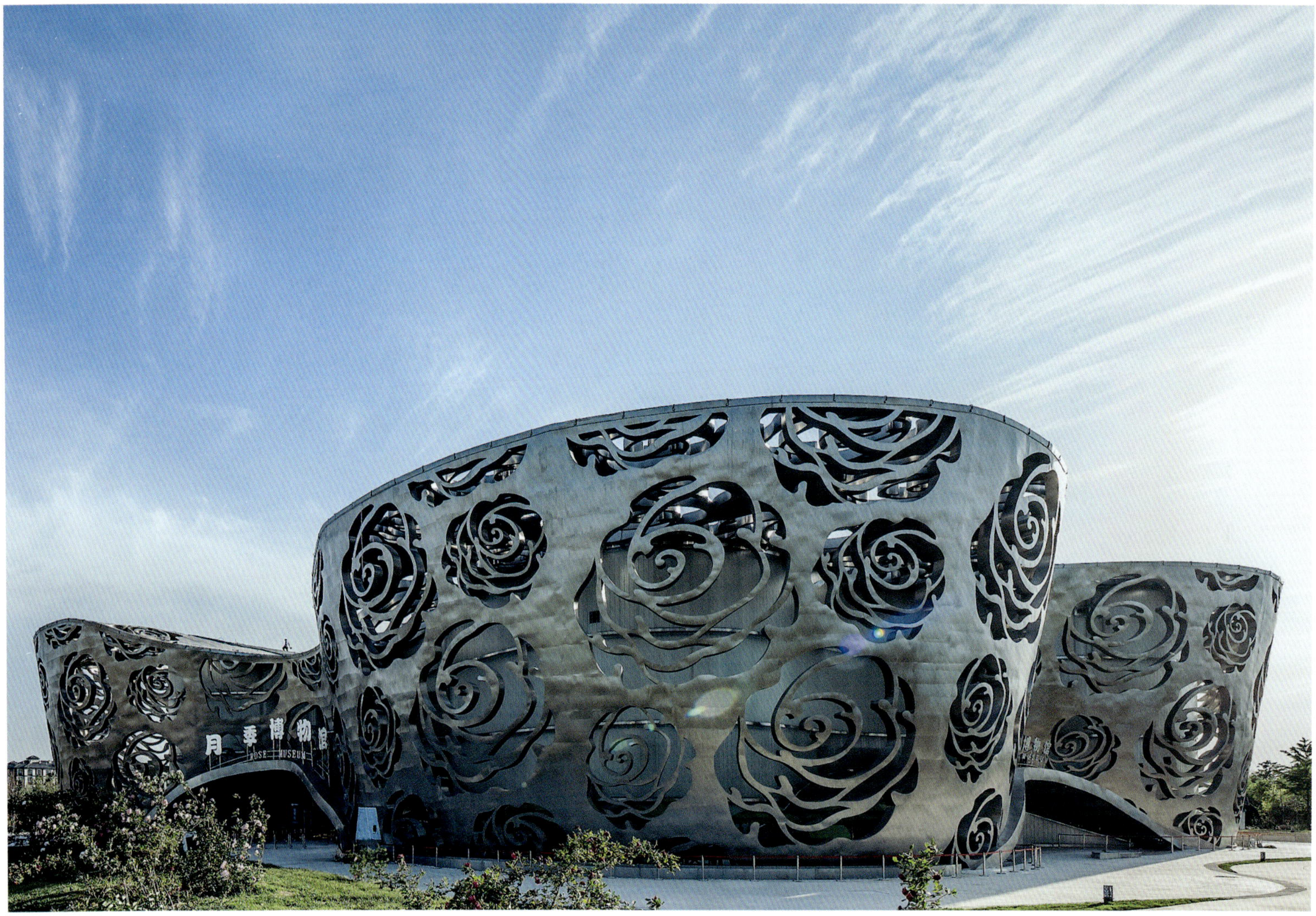

NEXT ARCHITECTS

Beijing Rose Museum, 2016
Building, dimensions variable

A pattern of Chinese roses punctures the flowing stainless steel facade that wraps the Beijing Rose Museum, the world's first museum dedicated to roses. Taking inspiration from the walled courtyards of traditional Chinese architecture, the museum's perforated facade creates four half-open courtyards. During the day, the light that enters through the perforated rose pattern creates an ever-changing play of shadow and illumination, while in the evening, the facade is lit from within, projecting rose-shaped patterns into the night. China has an ancient love of roses and floriculture, dating back to the eleventh century BCE, and the museum is in Weishanzhuang Town, in Beijing's Daxing District, which is famous for its roses. Designed by John van de Water and Jiang Xiaofei of NEXT Architects, a firm based between Amsterdam and Beijing, the Rose Museum covers 17,000 square feet (1,579 sq m) and is divided into eight themed halls covering the history, science, and culture of roses. Alongside traditional exhibits, it employs cutting-edge technology, including human motion capture, holographic animation, and virtual reality, to engage audiences with a multisensory experience. Interactive games test visitors' knowledge, and there is even an opportunity to breed their own virtual rose. Outside, a 100-acre (40.5-ha) park showcases 2,000 varieties of rose alongside art installations. The museum's opening in 2016 coincided with the World Rose Convention, held at the new institution, in which more than thirty countries participated.

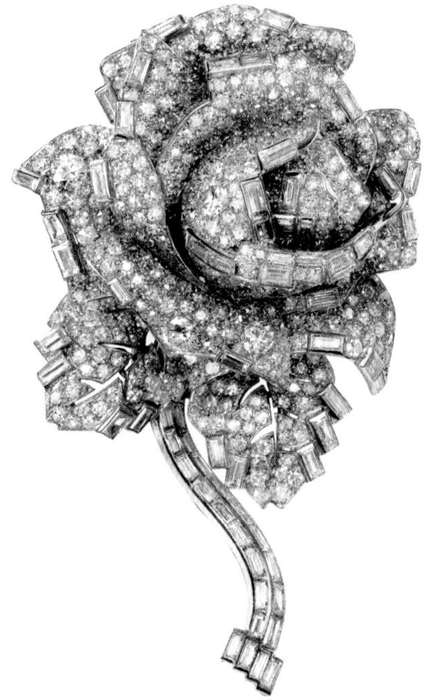

CARTIER

'Rose' clip brooch, Cartier London, special order, 1938
Platinum and diamonds, 2⅞ × 1¾ in. / 7.3 × 4.3 cm
Provenance: Her Royal Highness Princess Margaret, Countess of Snowdon
Cartier Collection

This three-dimensional brooch of a rose made entirely from platinum and diamonds is truly a flower fit for royalty. Baguette diamonds are used to build the stem and the outline of each petal, while old, single-cut diamonds fill out each petal to create a realistic-looking rose, albeit one that is dazzlingly glittering. Produced by celebrated French jewelry house Cartier, it was made during a period when the firm was combining its deep understanding and knowledge of diamonds with a return to portraying highly refined elements of nature. The flower is clearly in the shape of a rose, but its solidity takes it far from the ephemeral fragility of a garden rose. Little is known about the first fifteen years of the brooch's existence, but, by 1953, it was owned by Princess Margaret of Great Britain (whose middle name was Rose), younger sister of the British queen Elizabeth II. It is thought she was presented with the brooch after launching the ship *Maori* in Newcastle upon Tyne for the New Zealand Union Steamship company in 1952. During the 1950s, the princess was often photographed wearing the brooch, most notably to her sister's coronation in 1953 in Westminster Abbey and two years later for a series of photographs by Cecil Beaton (see p.74) to mark her twenty-fifth birthday. Cartier, founded in 1847, has a long history of making jewelry inspired by natural flora, as well as of creating unique pieces for European royalty.

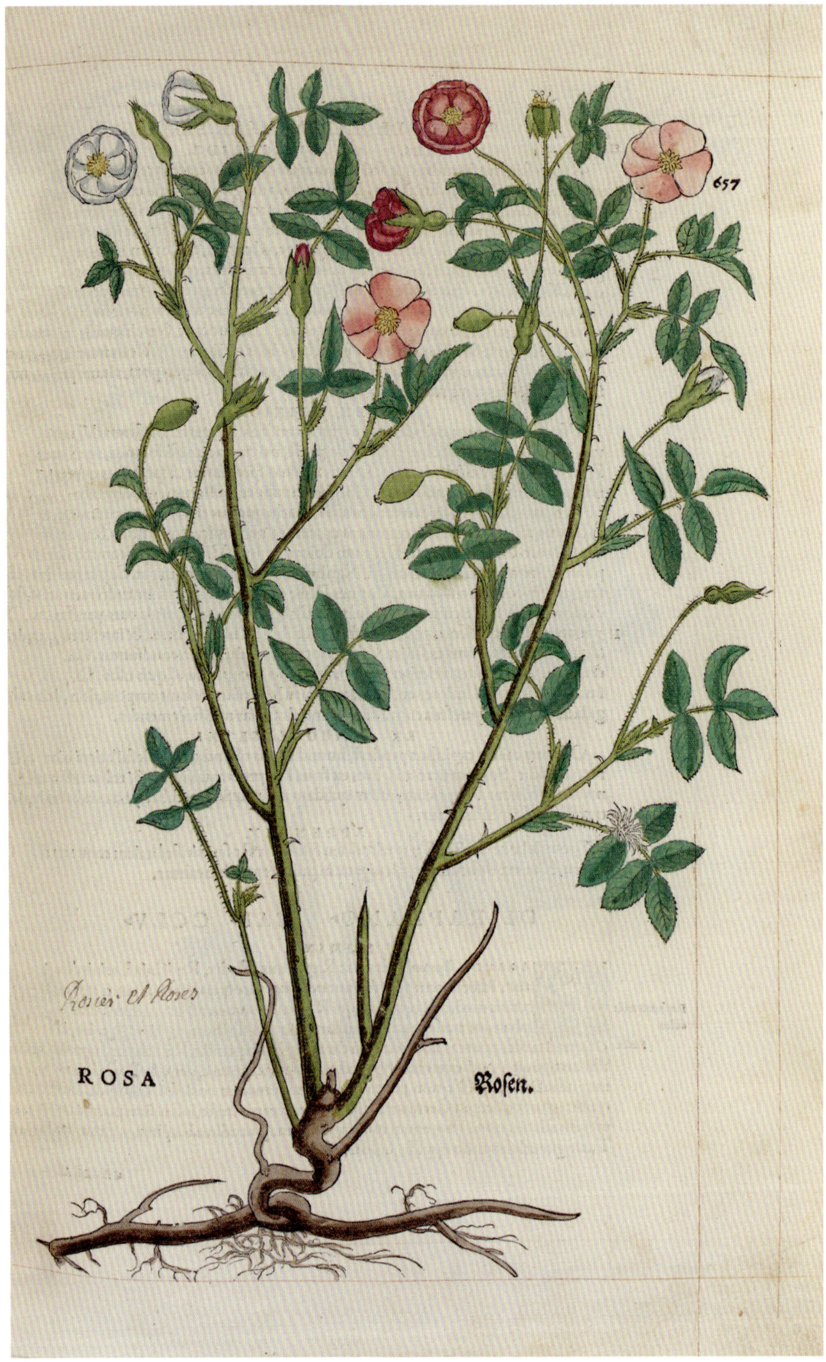

LEONHART FUCHS

Rosa, from *De historia stirpium*, 1542
Hand-colored woodcut, 15¼ × 10¾ in. / 39 × 27 cm
Bibliothèque nationale de France, Paris

In 1542 and 1543, three books were published that challenged the accepted world order. In *On the Revolutions of Celestial Spheres*, Nicolaus Copernicus suggested that Earth was not the center of the universe. In *On the Fabric of the Human Body*, Andreas Vesalius published the first evidence-based book of human anatomy. And in *On the History of Plants* (*De historia stirpium*), German botanist Leonhart Fuchs (1501–1566) published the first comprehensive botanical study. Fuchs's book included four hundred botanically accurate woodcut images of local plants, many of which grew in his own garden in Tübingen, and an additional one hundred foreign plants. This illustration of an evergreen rose shows the bush's shallow root system, flowers in bud and bloom, and spindly stems filled with green leaves. Fuchs asked the book's illustrators, Heinrich Füllmaurer and Albrecht Meyer, to draw from life to produce as accurate an illustration as possible for easy identification by sixteenth-century herbalists. Each image, which was prepared for printing by the woodcutter Veyt Rudolff Speckle, was accompanied by an essay that described the plant's features, history, and medicinal powers. Since the first herbals produced by the ancient Greeks, plant illustrations had often been misleading or omitted in botanical studies, so Fuchs specified that each plant should be drawn with its flowers in various stages of bloom. As well as the rose, Fuchs's book introduced previously unknown plants from the Americas, such as pumpkins and chili peppers.

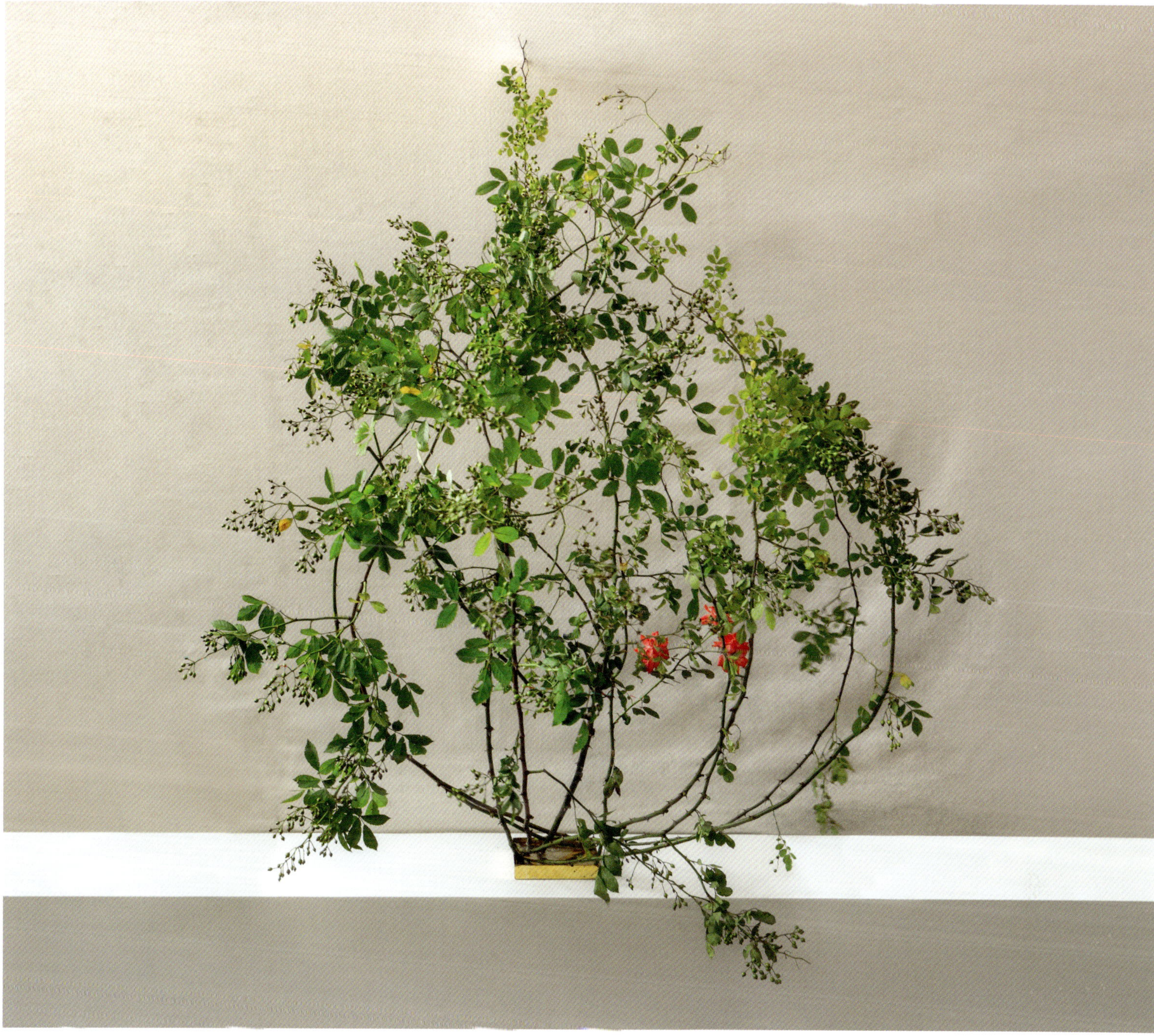

EMILY THOMPSON

Untitled, Rose Hip Floribunda and Rosa 'Knock Out', 2024
Floral installation, dimensions variable
Photograph by Julianne Nash

A splash of color from three blooms of a 'Knock Out' rose emerges from a leafy mass of green stems and hips, which have been teased into an airy, onion shaped dome, creating a living sculpture so delicate as to barely count as an object. This installation is the work of innovative American sculptor-turned-florist Emily Thompson (b. 1973). Thompson deliberately seeks out unlikely and rarely used blooms, seed pods, and branches to create wildly original architectural floral sculptures. Here, she places the fruit of the rose, the rose hip, center stage. She celebrates the ability of rose stems to creep and bend, their thorns acting as joints to form what is both a bramble and a natural scaffold. The resulting frothiness of the leaves and hips helps to accentuate the vivid pink roses, which remain partly concealed. A disciple of eighteenth-century English garden theorist William Gilpin, who championed the concept of "magnificent style," Thompson is also inspired by English florist Constance Spry (see p.113), who caused a sensation in 1928 when she used hedgerow flowers for a floral window display in London's exclusive West End. Following in Spry's footsteps, Thompson has built a highly successful flower business by reimagining the natural world for her clients, using unexpected combinations and juxtapositions not only of flowers, but also of overlooked undergrowth and foraged produce, to create gracefully disordered arrangements that express a sensitive understanding of nature's seasons and their unsung nuances.

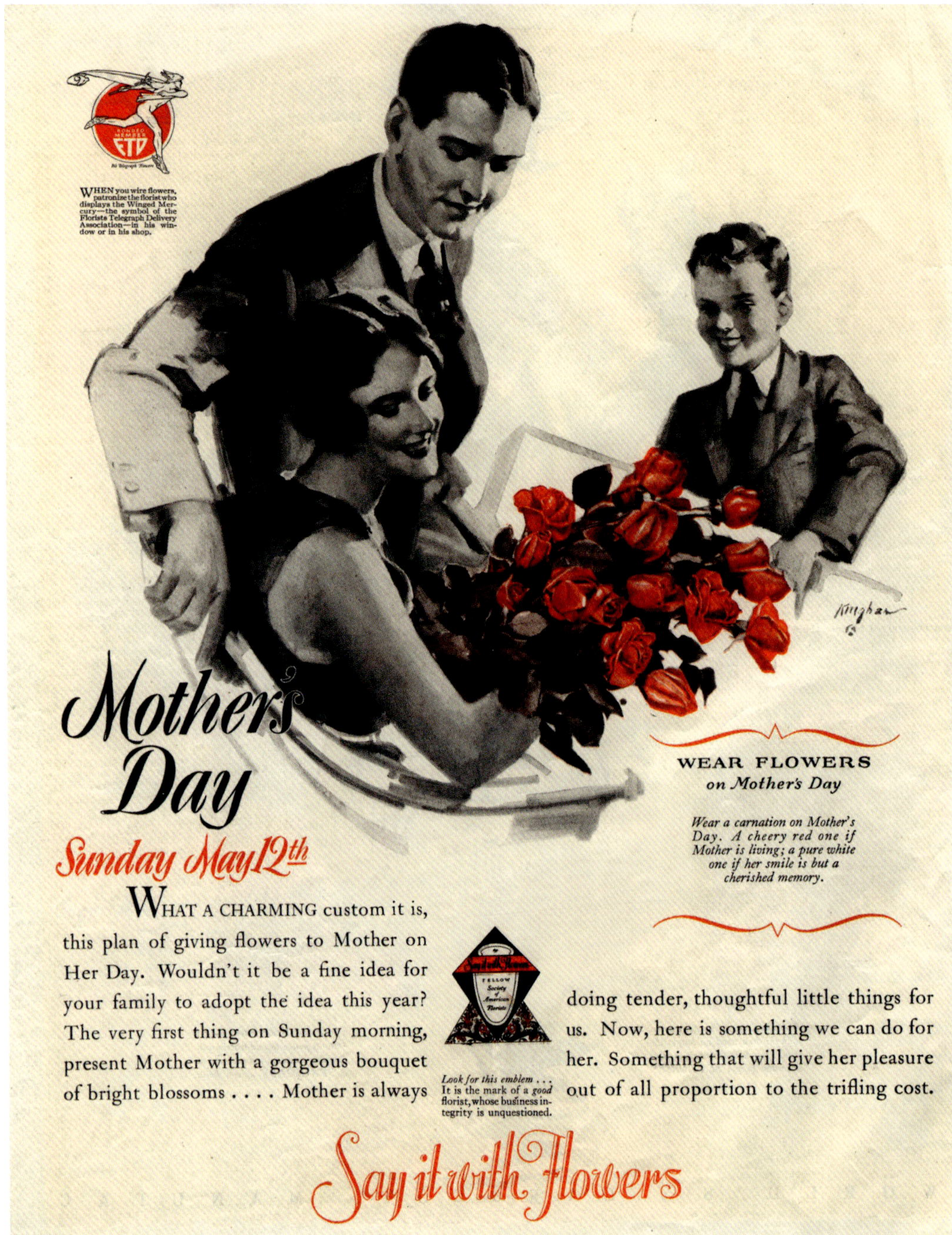

FLORISTS' TRANSWORLD DELIVERY

Mother's Day: Say it with Flowers, 1930s
Lithograph, 13½ × 10 in. / 34.5 × 25.5 cm
Private collection

"Say it with Flowers" advises this hand-drawn advertisement from the 1930s, playfully urging families across the United States to celebrate the upcoming Mother's Day on May 12 by sending their mothers flowers. A mother beams as she clutches her bouquet of red roses while her husband and young son look on. To highlight the specialness of the occasion, and to make the advertisement stand out in a magazine, only the roses, the date, and the slogan are colored in red. Most important, the ad makes it clear that the roses have been delivered to this family's home—the young boy is even holding the box the flowers arrived in—for this is an advertisement for Florists' Transworld Delivery (FTD), a national flower-delivery service that was founded in 1910. The concept was simple. In the past, flowers were sent by train across the country to the recipient, which, depending on the journey time, meant that the blooms often arrived past their best. The aptly named John Valentine of Rochester, New York, reasoned that if florists in different parts of the country could telegraph their orders to the florist nearest to the recipient, the flowers would arrive as fresh as possible, and an idea was born. Originally established as a collective of ten florists, FTD used the telegraph to transform the practice of giving flowers. Their floral revolution continues to this day, and local florists still deliver fresh flowers ordered from hundreds or even thousands of miles away.

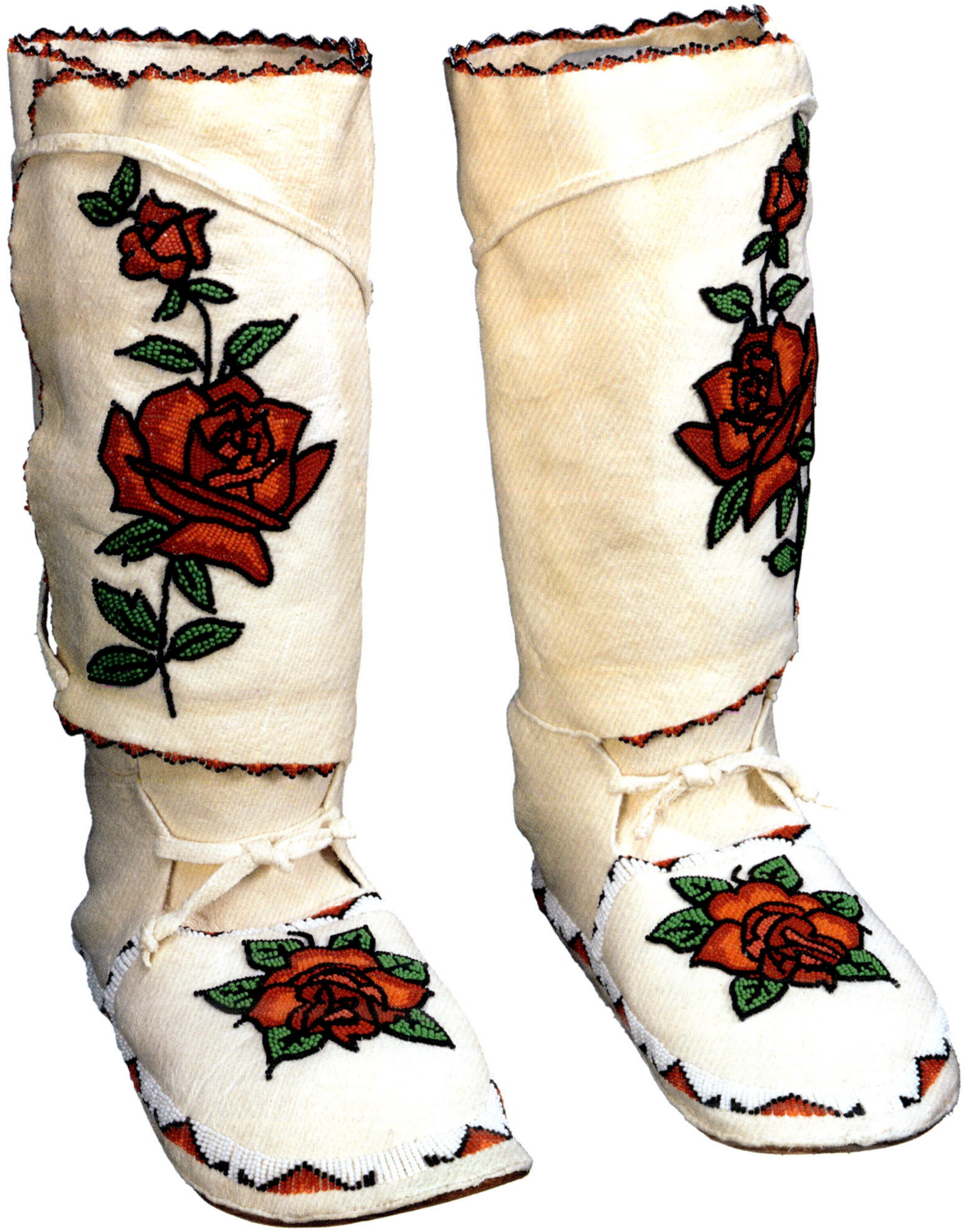

SANDRA SWALLOW-HUNTER, VIVIAN SWALLOW, AND WILMA JEAN MCADAMS-SWALLOW

Woman's Moccasins, Eastern Shoshone, Fort Washakie, Wyoming, 2000
Tanned elk hide and glass beads
Buffalo Bill Center for the West, Cody, Wyoming

Contemporary beadwork artists Wilma Jean McAdams-Swallow, Sandra Swallow-Hunter, and Vivian Swallow are Eastern Shoshone (Gweechoon Deka) tribal members from the Wind River Reservation in central-west Wyoming. With their skilled and creative handiwork, Wilma and daughters Sandra and Vivian are dedicated to preserving Eastern Shoshone art and traditions. The adornment of garments has long been an important part of Indigenous American culture, and beadwork is an art form expressed and practiced by many tribes, with each nation having distinct designs, colors, patterns, and techniques. The skills and designs are often learned from other family members and frequently applied to traditional Shoshone clothing and regalia worn to powwows and other ceremonies. Shoshone artwork is known for its beautiful floral designs, and the red rose is frequently depicted in its beadwork and painting. The rose has cultural significance and is an important plant resource. Highly valued for its medicinal properties, it is used in herbal medicine, and its petals and vitamin C–rich hips are dried for infusing in teas and use in cooking. Several rose species are used, but the wild *Rosa woodsii* is one of the most common and is native to the Wind River area. In Shoshone artwork, rose motifs are often used as a symbol of life and to represent survival and vitality. Although glass beads are a more recent material, as in this pair of Moccasins embellished with vibrant red roses, the plant's hips were once used as beads themselves.

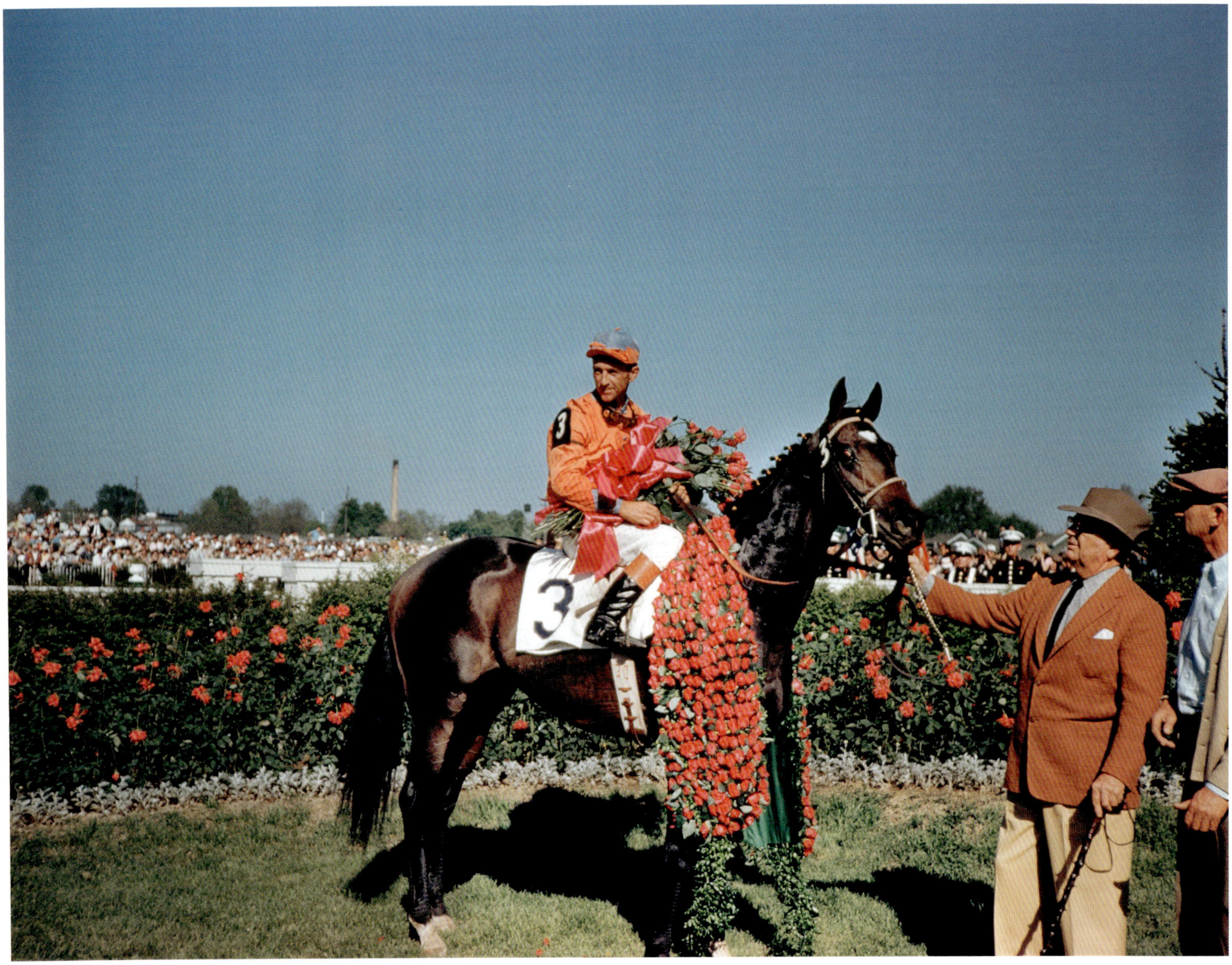

JERRY COOKE

Winner of the 1956 Kentucky Derby, 1956
Photograph, dimensions variable

The jockey David Erb sits atop Needles shortly after the pair's victory in the 1956 Kentucky Derby in this color photograph by Ukraine-born American photographer Jerry Cooke (1921–2005). Needles is swathed in a spectacular shawl of red roses, a Kentucky Derby tradition with origins that are hotly debated. What is not up for debate, however, is the fact that the rose has been the official flower of this prestigious horse race since 1884. The first published account of the winning horse being draped in roses dates from 1896, when Ben Brush was presented with a collar of white and pink roses. This became a custom early in the twentieth century, and a photograph of the 1906 Derby shows the winning horse with a garland. However, it was not until 1932 that a Garland of Roses was officially commissioned. It was made by Louisville florist Grace Walker using five hundred deep red roses and greenery, stitched onto a cloth-backed blanket. Over the years, the Garland of Roses has come to symbolize the annual race, which is the highlight of the most watched horse-race meet in the United States, when three-year-old thoroughbreds race over a distance of about 1¼ miles (2 km). The 1956 Derby was just one of forty-two Kentucky Derby races that Cooke attended over a prolific career, during which his images appeared frequently on the covers of *LIFE* and *Sports Illustrated* magazines.

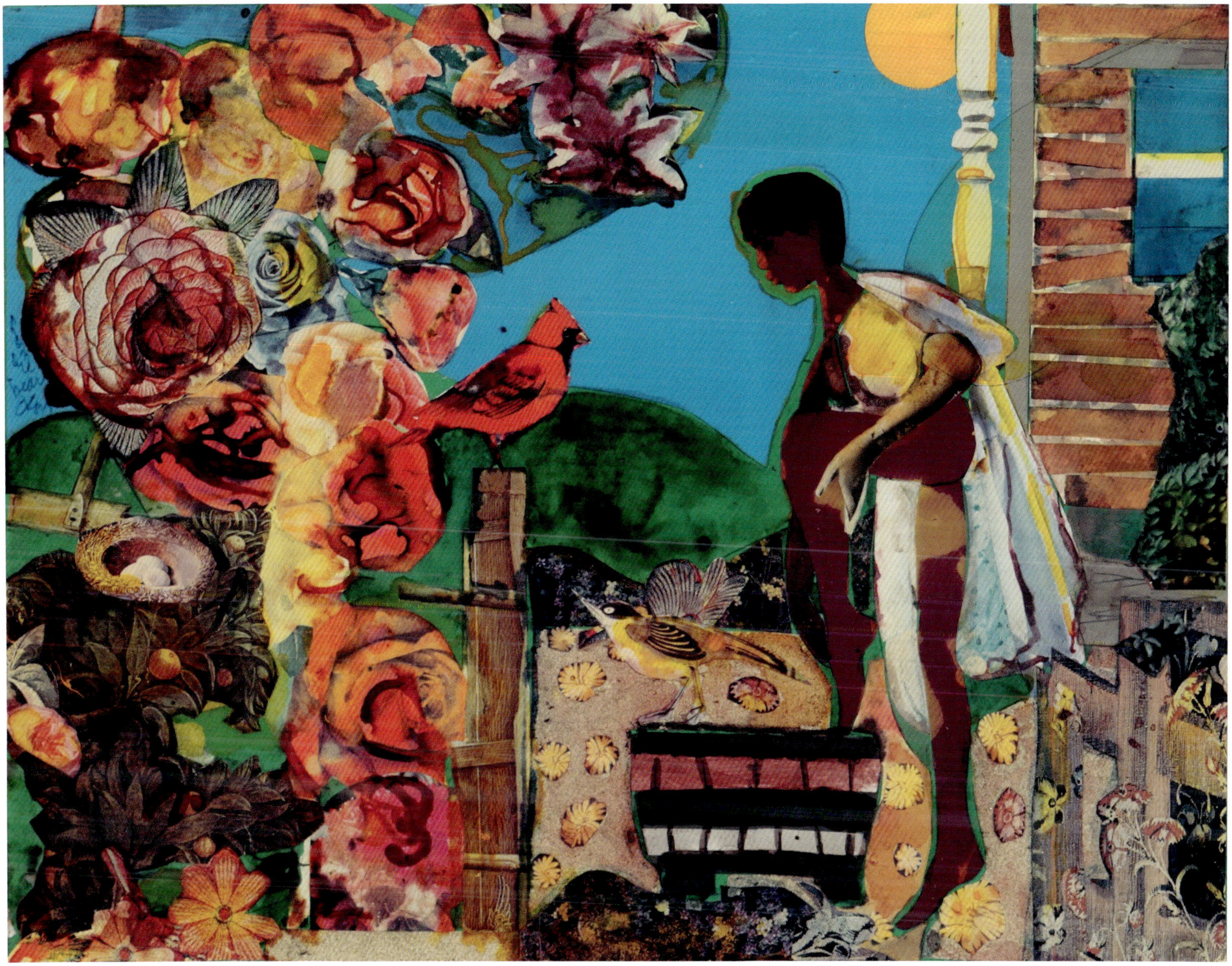

ROMARE BEARDEN

Morning of a Red Bird, 1975
Collage with acrylic wash on board, 24 × 28 in. / 61 × 71.1 cm
Private collection

In this boldly colored collage by American artist Romare Bearden (1911–1988), a profusion of lush roses transports an outdoor bather into a brief spell of natural harmony. Created during a period of significant social and cultural change in the United States—marked by the aftermath of the Civil Rights Movement, the rise of the Black Power Movement, and a growing emphasis on Black identity and cultural heritage—Bearden's work maps an emotionally charged personal trajectory. Born in Charlotte, North Carolina, Bearden moved north during the Great Migration, later becoming an artist integral to the Harlem Renaissance. His art often incorporated elements of Black folklore, music, and everyday life, blending them with modernist techniques and a sharp sense of social commentary. Bearden began experimenting with collage in the 1960s and it soon became his signature technique. He used the medium to freely juxtapose different textures, colors, and images in layered, multidimensional pieces, of which *Morning of a Red Bird* is a prime example. As with much of his work during this period, the collage explores themes of memory, history, and identity, using symbolic elements to convey deeper meanings related to Black culture and experience. Representing beauty, resilience, and the complexity of life, the climbing roses also held a deep personal significance for Bearden. The flowers reflect his connection to the rich tapestry of Southern life and flora, as well as the cultural heritage of Black Americans in the South and beyond.

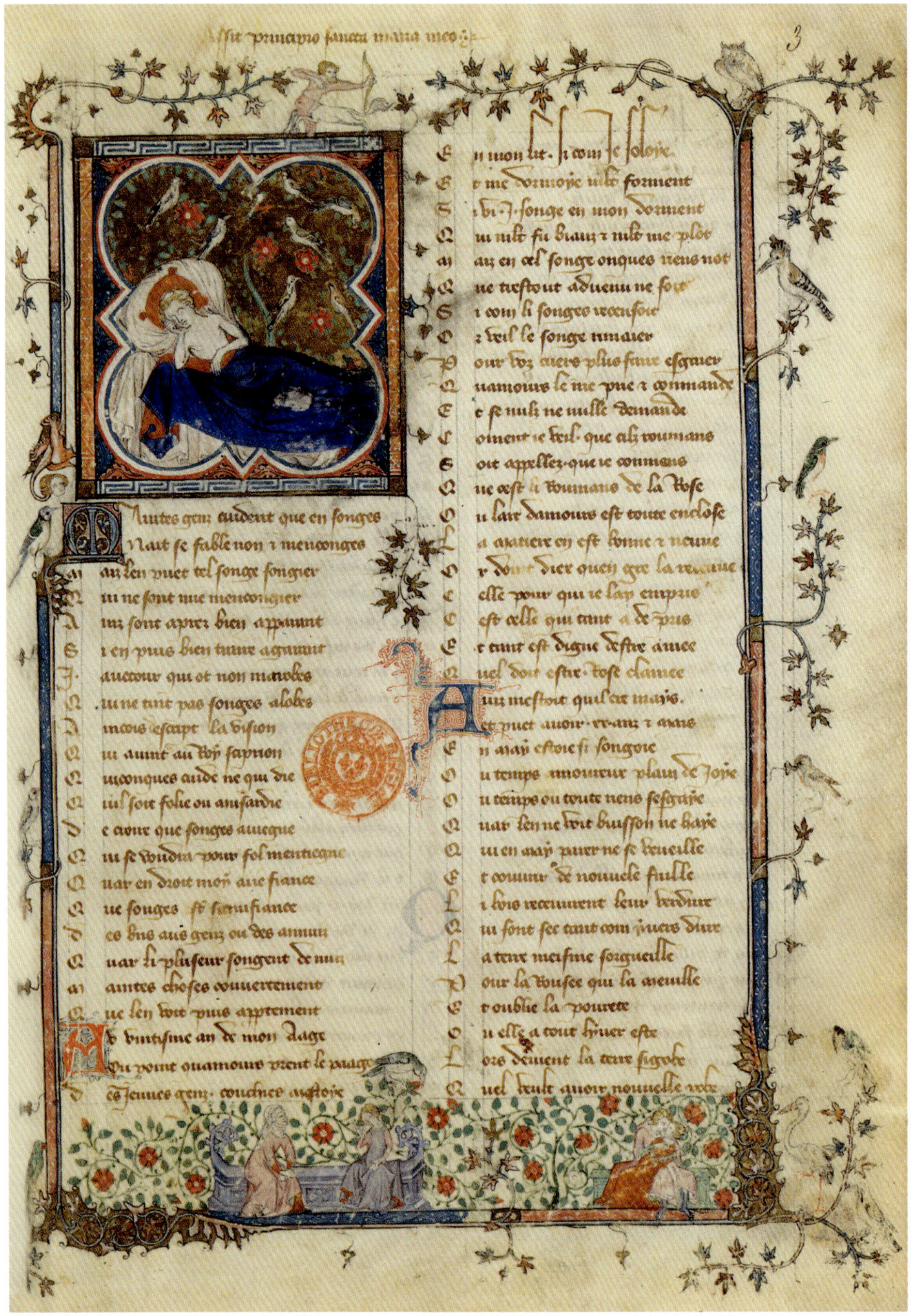

GUILLAUME DE LORRIS AND JEAN DE MEUN

Folio 3r, from *Roman de la Rose*, c. 1326–50
Illuminated manuscript, 11¼ × 7¾ in. / 28.5 × 20 cm
Bibliothèque nationale de France, Paris

Le Roman de la Rose (*The Romance of the Rose*), a medieval poem of epic proportions, starts with a dream, here illustrated in the first initial of an illuminated manuscript at top left, where the lover sleeps in a rose garden full of songbirds. In his quest for the special rose symbolizing his love, the poem's hero enters a walled garden peopled with allegorical figures, some helpful and some hostile, where the rose is heavily guarded. The ambivalence of the poem is further explored in the lower margin of this page, depicting again a garden of roses, where a demure lady in blue is under watch while a scarlet lady is passionately embraced. Significantly, the surrounding birds, including a parrot, a stork, and an owl, are not all symbols of love. The text was written in Old French by two poets; Guillaume de Lorris (c. 1200–1260) began the poem in 1225–30 and Jean de Meun (c. 1240–1305) completed it about 1280. Guillaume's original poem came from the charming world of troubadours and courtly love, but Jean's verses shifted it to an encyclopedic and cynical morality tale, which divided medieval opinion. Some said it was about lust, while others read it as a search for wisdom or a pilgrimage toward redemption. One of the most read poems during the Middle Ages, the *Romance* reflects the era's fixation on the symbolism of the rose. Almost three hundred manuscript versions still exist in various languages, including this fourteenth-century one, and some even date from after 1500, when printed versions with woodcut illustrations proliferated.

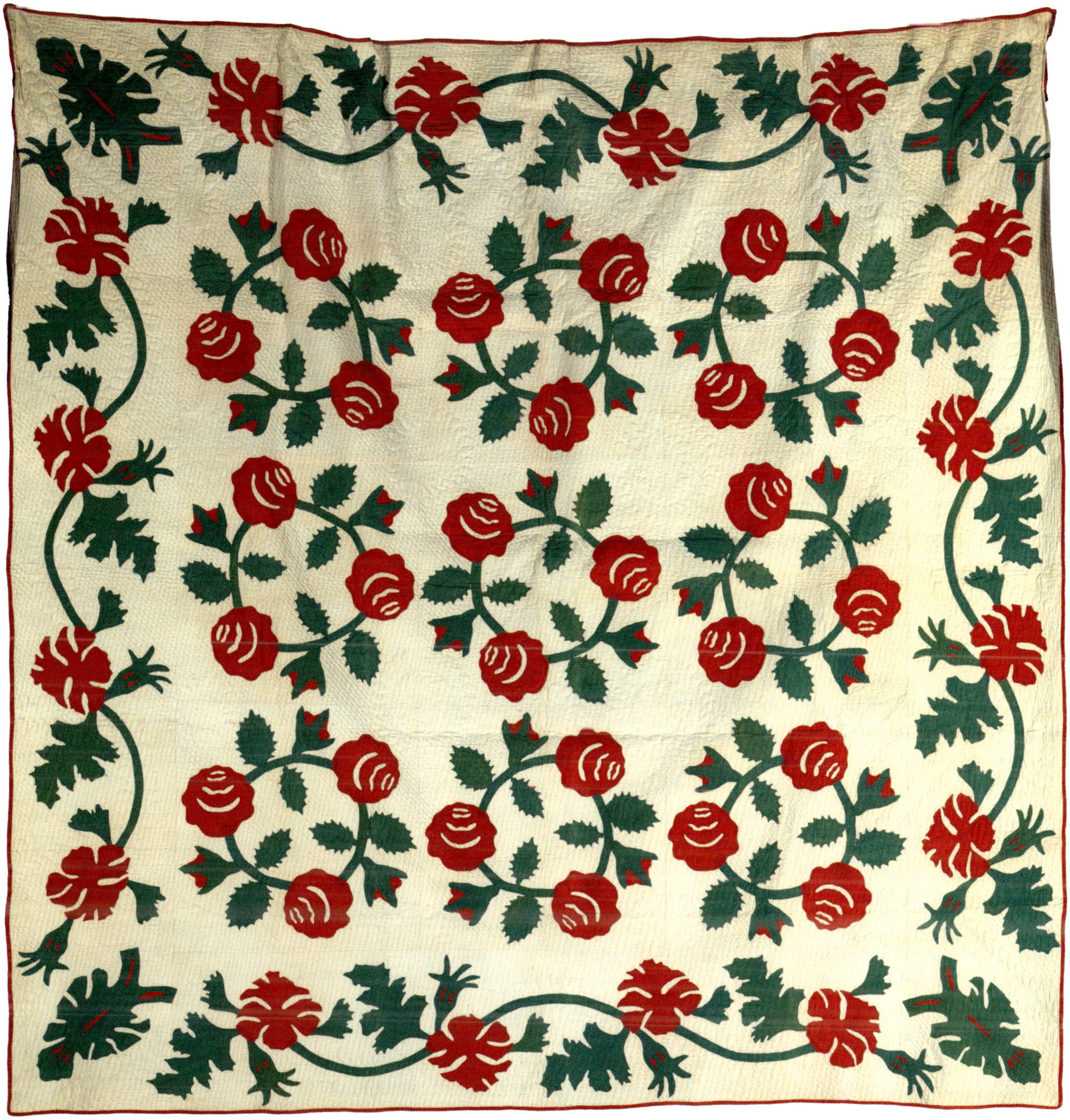

CHRISTIANN DRICK SMITH

Rose of Sharon, 1866
Cotton quilt, 80 × 80 in. / 203.2 × 203.2 cm
Smithsonian American Art Museum, Washington, DC

A bold arrangement of roses, buds, and leaves fills the square of this handsewn quilt, appliquéd in red and green on a cream background. The design repeats with nine identical rose circlets in the center, vines along each edge, and abstract figures angled in the corners. This traditional pattern has garnered dozens of variations in form and an equal number of names based on shape (Wreath of Roses), place (Kentucky Rose), politics (Whig Rose), and the Bible (Rose of Sharon), to give a few examples. The biblical reference is from the Song of Solomon (2:1), a celebration of romantic love. The plains of Sharon are a fertile region in Palestine. In the passage, the beloved calls herself "the rose of Sharon, the rose of the valley." The rose of Sharon symbolizes beauty and fruitfulness. In gardeners' parlance, "rose of Sharon" is a common name for several different plants, including the common hibiscus (*Hibiscus syriacus*), although for quilters, the genus *Rosa* is typically represented in the Rose of Sharon pattern. Rose of Sharon quilts were linked to courtship and marriage, often made for a girl's hope (or dowry) chest. Christiann Drick Smith made this quilt in north central Pennsylvania in 1866, the year after the American Civil War ended. Perhaps she stitched with upcoming nuptials in mind. While her work is dated, other facts about Smith's life, including her birth and death dates, are unknown—but her prowess with the needle is certain.

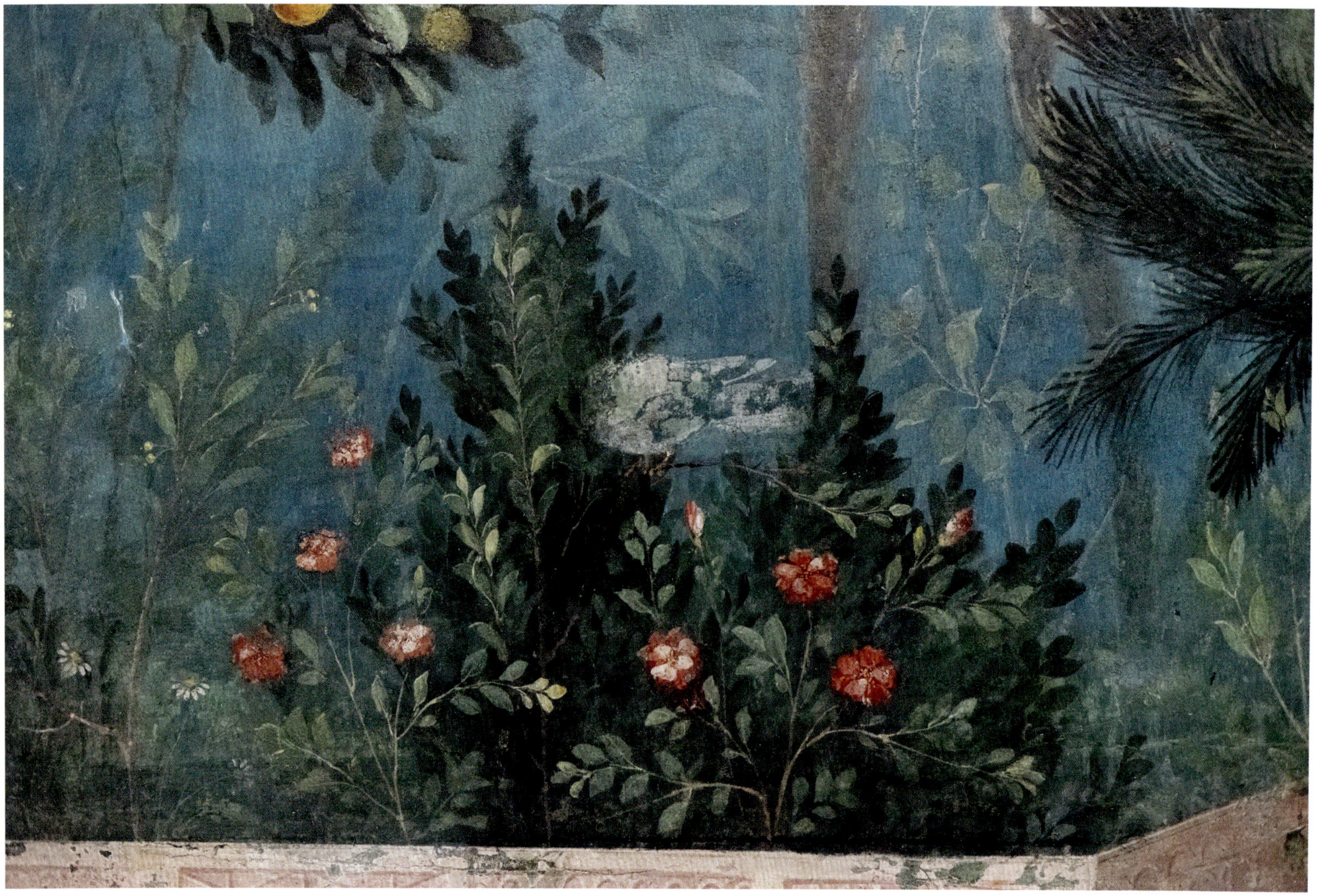

ANONYMOUS

Garden Room fresco (detail) in the Villa of Livia, c. 30–20 BCE
Pigment on plaster, L. 38½ ft. / 11.7 m
National Roman Museum—Palazzo Massimo, Rome

In this paradisiacal garden, all fruit are ripe at once, and spring, summer, and fall flowers bloom at the same time. The fall-ripening quince and fruit of the strawberry tree shade summer-flowering daisies and spring- and early summer-blooming roses. For the ancient Romans, roses represented rebirth and rejuvenation. This small room in the Villa of Livia—a country residence built in c. 30–20 BCE that belonged to Livia Drusilla, wife of the emperor Augustus—was a semi-subterranean summer *triclinium* (dining room) decorated on all four interior walls with realistic garden scenes showing plants, flowers, trees, and birds. Even in the enclosed underground space, the frescoes and the barrel-vaulted ceiling transport the viewer to a peaceful garden outside. The painting is divided into several levels, and there are two barriers, one a bamboo fence and another a stone wall featuring latticework. The site was discovered in 1596 but was not acknowledged as the Villa of Livia until the nineteenth century. In 1951, because the site was deteriorating, the frescoes were removed from the walls, and in 1998, they were installed in the Palazzo Massimo, in a room built to the dimensions of the original *triclinium*. The frescoes are also politically symbolic. As Rome's first emperor, Augustus wanted his reign to be regarded as a Golden Age following decades of civil strife, with a return to customary Roman values and morals. This illusionistic garden makes such an impression, its roses offering a picture of serenity, prosperity, and abundance.

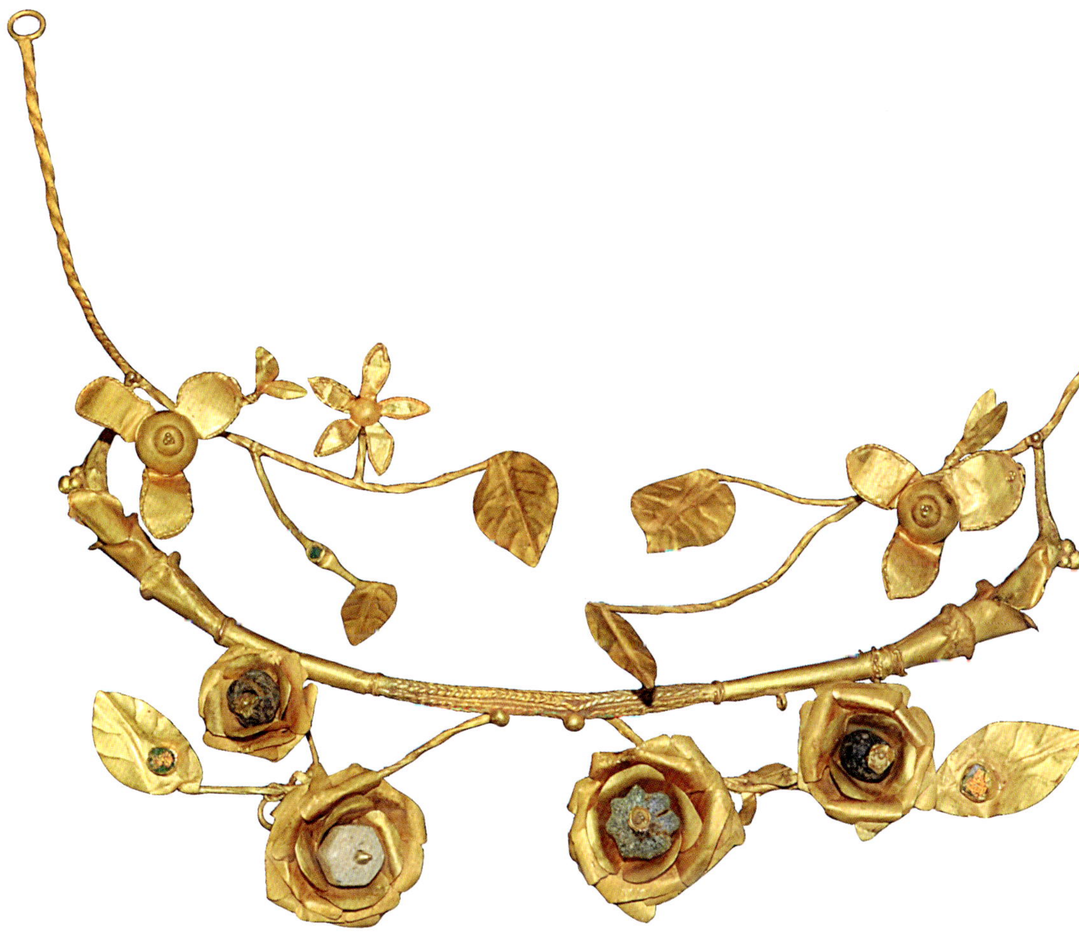

ANONYMOUS

Floral wreath, c. 3rd–2nd century BCE
Gold and glass beads, dimensions variable
Private collection

This Greek Hellenistic (323–31 BCE) diadem features golden rose flowers, each decorated with a central glass bead, as well as leaves and buds on a twisted and braided wire. Even to the ancient Greeks, roses had major associations with deep emotions, such as victory or love; they were the sacred flower of the Greek goddess Aphrodite (known to the Romans as Venus; see p.55), and Cleopatra had her pillows stuffed with fresh rose petals every night. In ancient Greece, winners of Panhellenic athletic events were given wreaths made from the branches of real sacred plants, such as the olive tree, and garlands of roses, violets, or myrtle leaves were also worn at aristocratic dinner parties known as *symposia*. In classical Greece, gold crowns were commonly sent as tribute to Delos and Athens by members of the Delian League, a confederation of Greek city-states in the mid fifth century BCE. Sacrificial rituals, as well as festival rites of marriage, birth, and death, all involved wreaths that synthesized the organic with the metal crown, such as this one. Gold wreaths were made to resemble their natural counterparts but were too fragile to wear except on rare or special occasions. They were exclusive to the aristocracy and often given as wedding presents or as rewards for favored officials. While not much is known about this specific example, Hellenistic gold funerary crowns are the main type of golden wreath to survive, and most have been found in tombs.

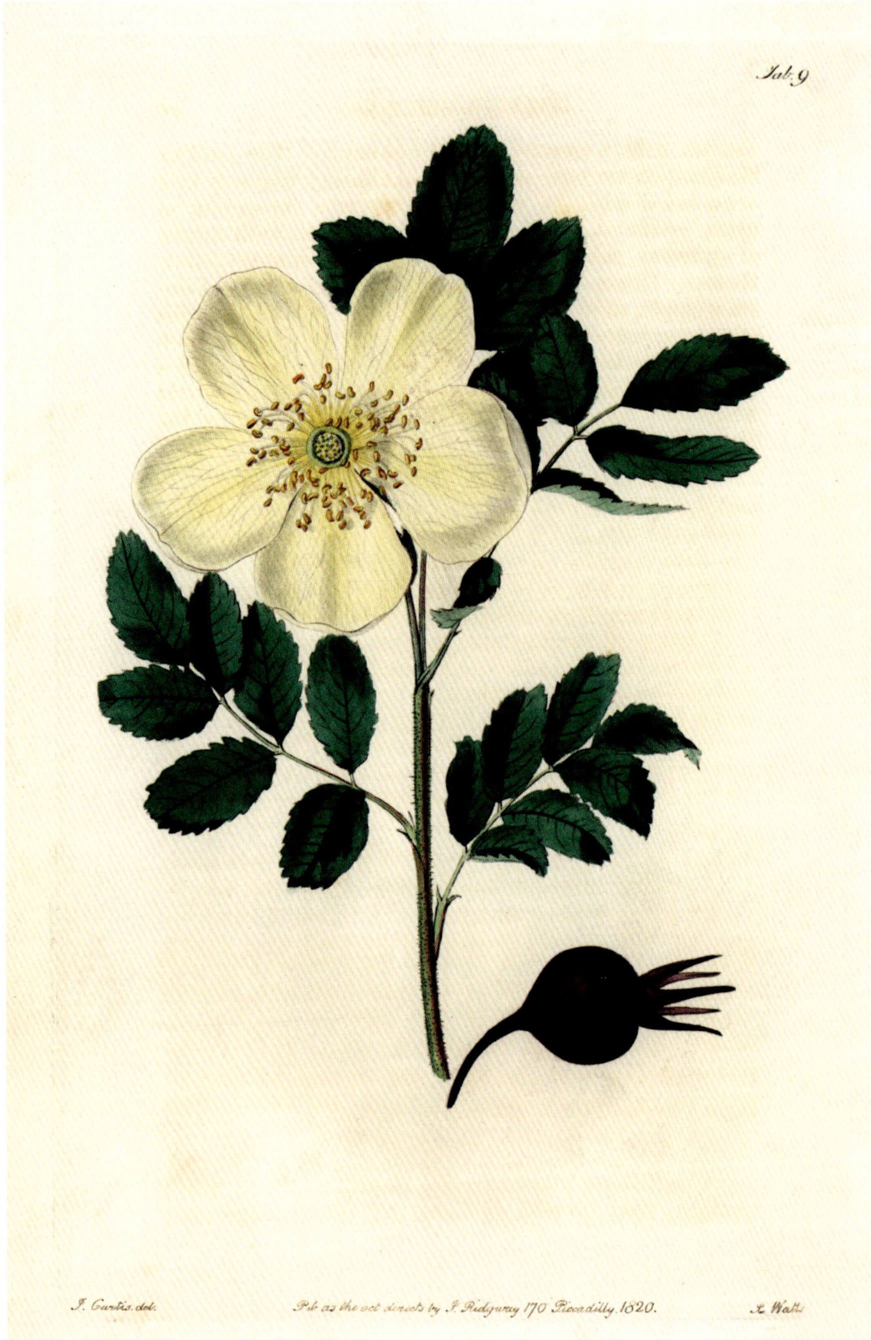

JOHN LINDLEY

Yellow American Rose (Rosa lutescens), from *Rosarum Monographia*, 1820
Hand-colored copperplate engraving, 12½ × 8⅞ in. / 31.8 × 22.5 cm
Peter H. Raven Library, Missouri Botanical Garden, St. Louis

This engraving by English botanist John Lindley (1799–1865) depicts a cutting of a buttery yellow rose with a large, turniplike black rose hip. Confusingly, there is no single rose that goes by the Latin name written at the bottom of the illustration, *Rosa lutescens*, and there are three possible identifications. The first belongs to the Banksia group, *R. banksiae* var. *lutescens*, a native of China whose flowers look vaguely like this illustration, but whose leaves, bristly stem, and hips are completely different, ruling it out. The second potential flower is 'Harison's Yellow' (*R. lutea* var. *hogii*, more commonly known as *R.* × *harisonii*), otherwise known as the Yellow Rose of Texas, but, while it does have an American connection, it has semidouble flowers and is a much brighter yellow. The third and correct rose is *R. hispida*, which comes true from seed, suggesting that it is a species, or native, rose. Then again, roses are notorious for hybridizing, and the black hip of this particular flower reveals that it may, in fact, have a common parent with 'Harison's Yellow' in the Scotch rose (*R. spinosissima*). Why Lindley titled this illustration *Yellow American Rose* is a mystery, since *R. hispida* most probably originated in Siberia, of all places. Lindley was renowned for his knowledge of orchids, but he began his career studying roses, publishing his *Rosarum Monographia* (*The Botanical History of Roses*) when he was only twenty-two. A landmark botanical publication, it distinguished seventy-six species and described thirteen new roses in nineteen plates painted by Lindley himself.

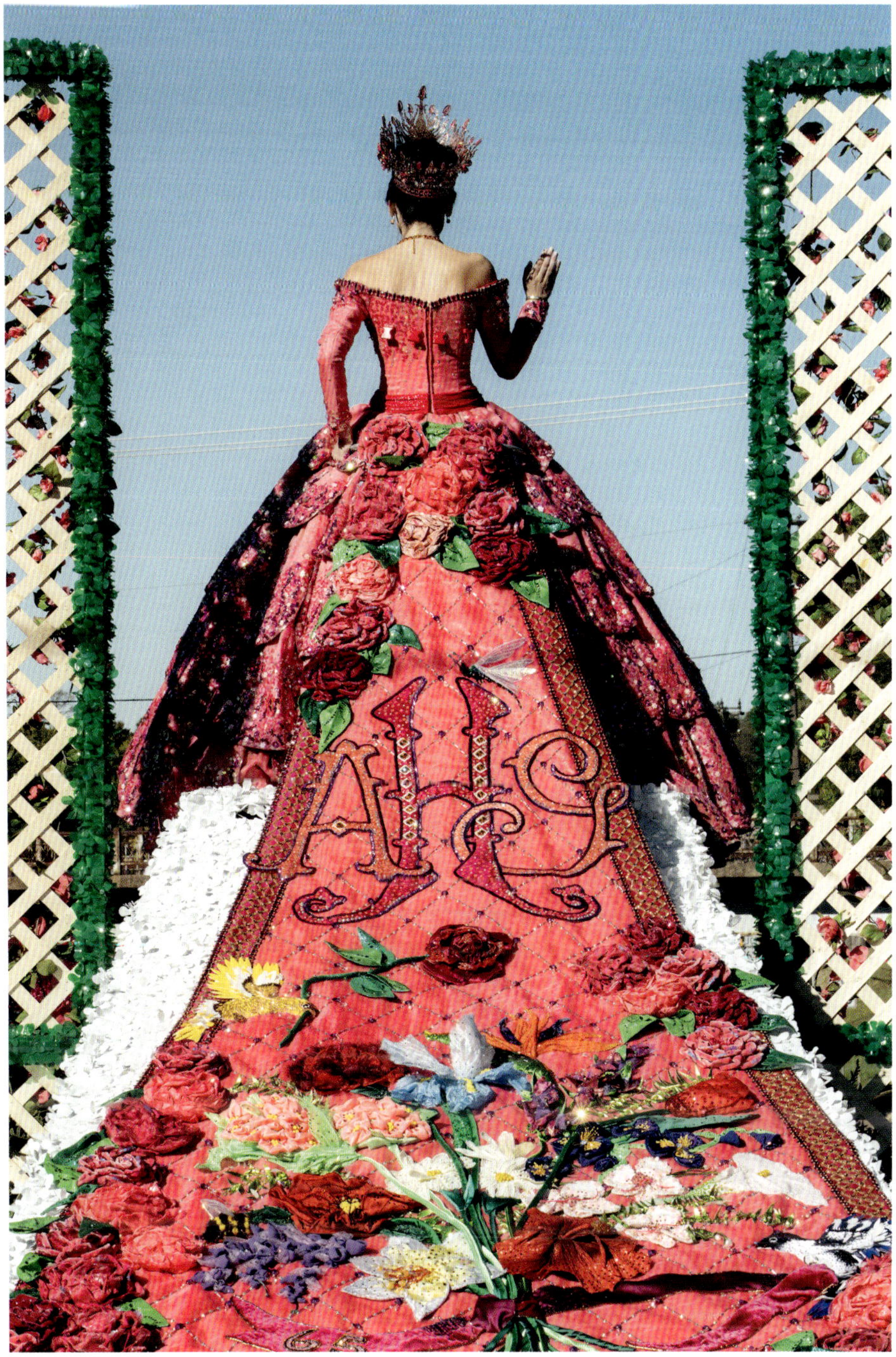

ELI DURST

Rose Queen of Texas, 2021
Photograph, dimensions variable

The city of Tyler in Texas has a good claim to be the rose city of the United States. In a state better known for oil and cattle, Tyler is an anomaly; its wealth comes from the roses that thrive in its sandy soil and temperate climate. Tyler's devotion to the rose is visible everywhere—not just in the rows of blooming plants, but also in the city's crest, which appears on Tyler's water tower, trash cans, and trail markers. But nothing celebrates the importance of the rose to the city as much as the annual Texas Rose Festival and its climax, the crowning of the Rose Queen, photographed here by Texas-based photographer Eli Durst (b. 1989). The 2021 queen, Anna Grace Hallmark, is shown in an extravagant pink organza and lamé dress with a long silk-trimmed train adorned with giant roses and her embroidered initials. The outfit—which weighed an impressive 42 pounds (19 kg)—was topped by a crown studded with Swarovski crystals and enamel roses. Since 1933, the Texas Rose Festival has been held on the third weekend in October every year, a tradition broken only by World War II and the COVID-19 pandemic. Highlights of the festival include the Rose Parade, which boasts 120 floats; the coronation of the Rose Queen; and tea with the Queen and her court. For those who want to see more roses, the Tyler Rose Garden, which opened in 1938, has 12 acres (4.86 ha) filled with 38,000 stunning rosebushes.

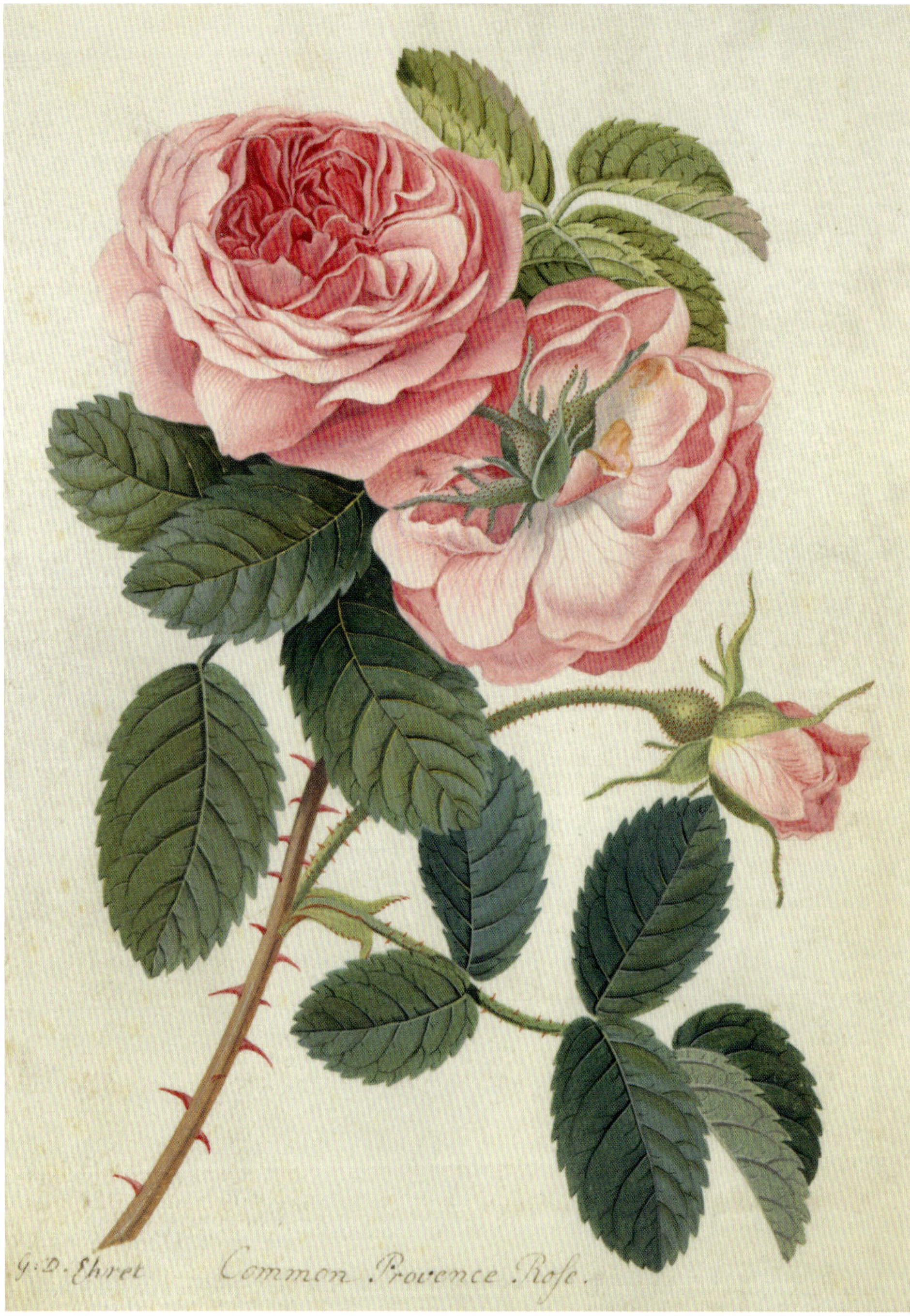

GEORG DIONYSIUS EHRET

Common Provence Rose, c. 18th century
Graphite, watercolor, and bodycolor on vellum,
7 × 4¾ in. / 17.8 × 12.2 cm
Fitzwilliam Museum, Cambridge, UK

Common Provence Rose is the title of this work by German painter Georg Dionysius Ehret (1708–1770), but few rose names have been more productive of confusion. First, it was confused with the *Provins* rose, named after the French city where it was supposedly introduced, but that was a Gallica rose, which this one is not. "Provence" suggests an origin in the south of France, for which there is also no evidence. This flower is *Rosa* × *centifolia,* the "hundred-leaved" (or hundred-petaled) rose. While it was described by Carl Linnaeus as a species, the general consensus now is that it had a hybrid origin, hence the multiplication sign in its scientific name. In his *Rariorum Plantarum Historia* (1601), French botanist Carolus Clusius said that he was given a specimen in 1589 by a Dutch official, and he helped to propagate and distribute it around Europe. As a result, it was also called a Holland rose (see p.180), which is more accurate than Provence, and more flattering than its later common name, the cabbage rose. Ehret was born in Heidelberg, and, after training as an artist, made his name by drawing the botanical illustrations for Linnaeus's *Hortus Cliffortianus* (1737). He then illustrated various works by German botanist Christoph Jacob Trew before moving to England, where his work proved to be in great demand. This undated drawing was bequeathed to the Fitzwilliam Museum by Lord Fairhaven, the art collector and gardener who created the gardens at Anglesey Abbey in Cambridgeshire.

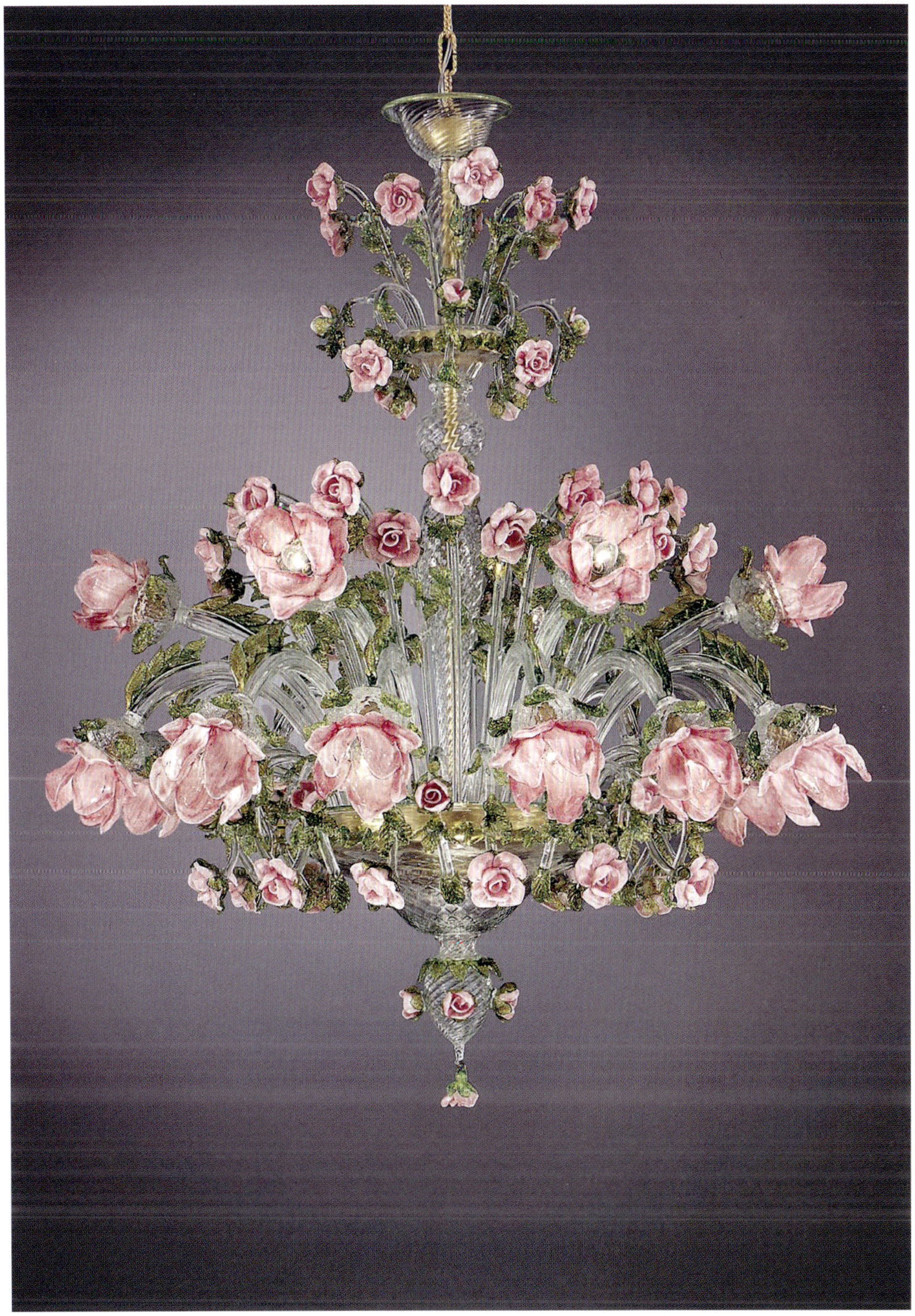

CARRARA GLASSMAKERS

Rosai Chandelier, c. 2020
Glass, 59 × 49½ in. / 150 × 125 cm
Private collection

A large salon on the *piano nobile* (first floor) of a Venetian palace overlooking the Grand Canal is perhaps the most suitable setting for this spectacular rose chandelier. It would be a fitting choice because this floral light fixture was handmade on the neighboring island of Murano by the factory Carrara Glassmakers, embodying the exuberance of the world-famous Murano glass, which has been made there since the thirteenth century. The chandelier's glass body was created using traditional tools, such as pliers, chisels, and blowers, to shape the glass, which was heated to extreme temperatures to allow it to become soft and workable before cooling and hardening into the shape of a hanging rosebush. The pink roses are shown in varying degrees of flowering from tight buds to full blooms, and each was handcrafted individually before being added to the chandelier; the large flowers at the ends of the clear stems face downward to house the light bulbs. The glass masters added the metal erbium to the molten glass to make the roses pink, and chromium to turn the glass green for the leaves. Created in the twenty-first century, this piece was made using a process that has remained largely unchanged since 1709, when the first Murano chandelier was made for King Frederick IV of Denmark, who had a passion for the Italian glass. Continual innovation has helped Murano glassmakers to maintain their reputation for original design, which includes developing a lighter-weight glass specifically for use in ornate chandeliers, such as this one.

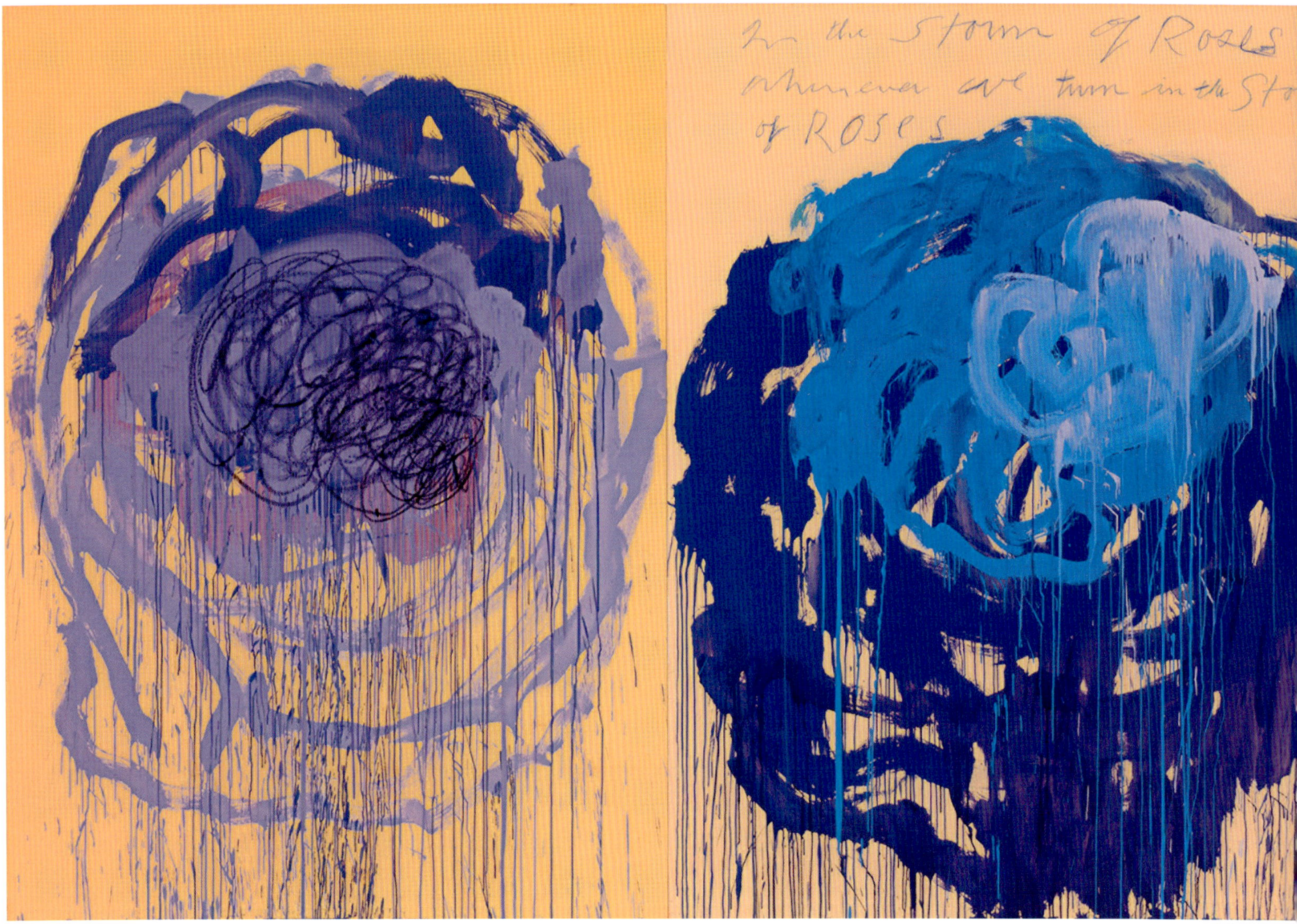
In the Storm of ROSES
whenever we turn in the Sto
of ROSES

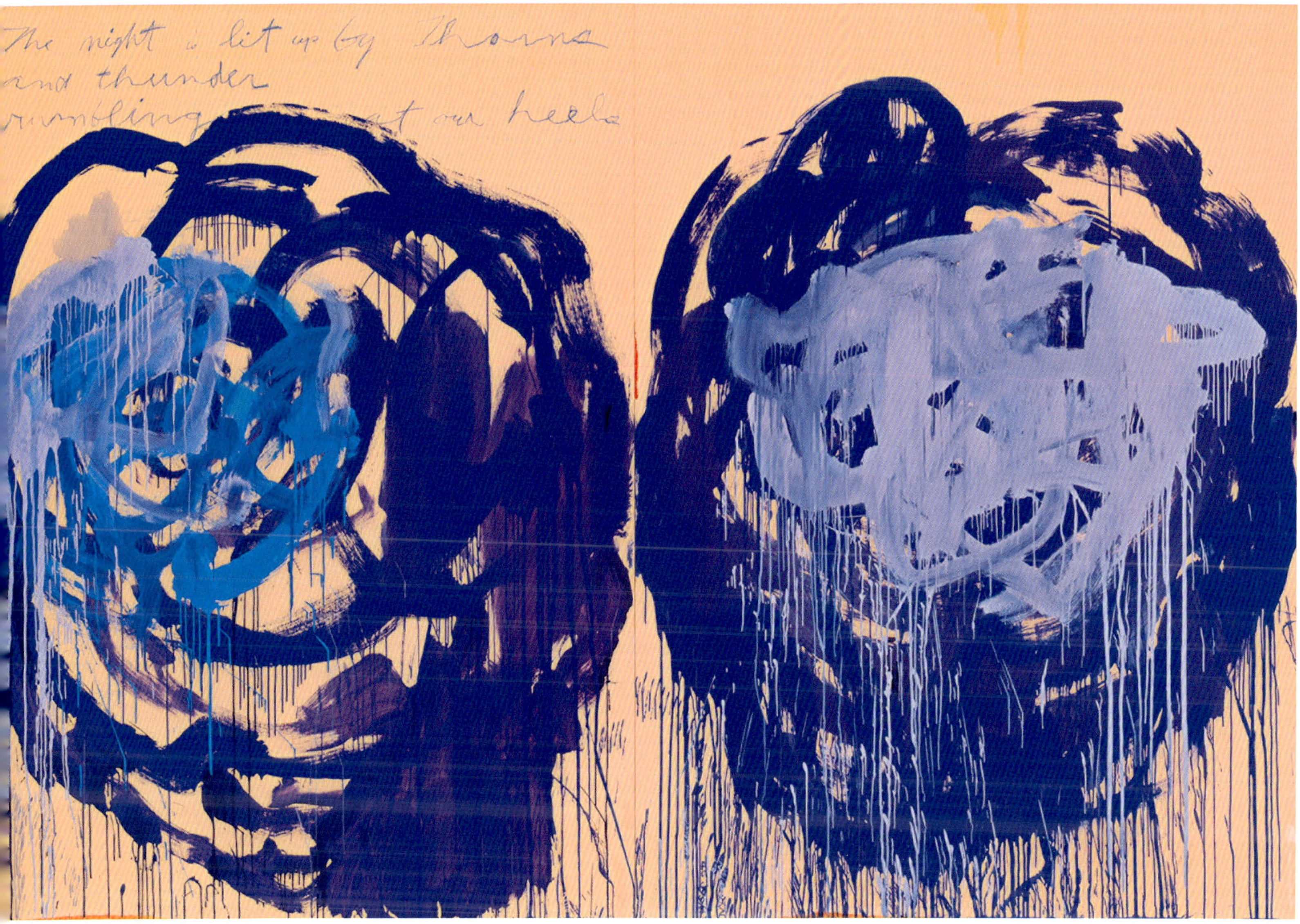

CY TWOMBLY

Untitled (Roses), 2008
Acrylic and crayon on wood, 8 ft. 3½ in. × 24 ft. 4 in. / 2.5 × 7.4 m
Bavarian State Museum Collections, Museum Brandhorst

Four seemingly casually painted representations of rose heads float in repeated fashion across a canvas, each constructed from a series of concentric circles in what could be one continuous brushstroke. The looseness of the painting, including drips of paint, builds up an idea of the flower heads instead of offering a strict representation of how they appear in reality. The lines at the center of each head, where they are at their tightest, could almost be considered a scrawl. However, each brushstroke, shape, and drip is executed intentionally by American artist Cy Twombly (1928–2011), who built his entire practice around an exploration of how gesture and the physical act of painting can create their own distinct artistic language. The mauve flower head Twombly paints is rare in real life, while blue roses—because of genetic limitations—do not exist in nature at all, so these flowers are largely made up or painted from his imagination. And although the title of the "untitled" picture parenthetically acknowledges the inspiration of roses, it seems as though Twombly may not be interested in the rose as such but is instead using its form to explore a range of painterly gestures. He explores the materiality of paint as he moves his brush across the canvas, describing the flowers and the act of painting them at the same time.

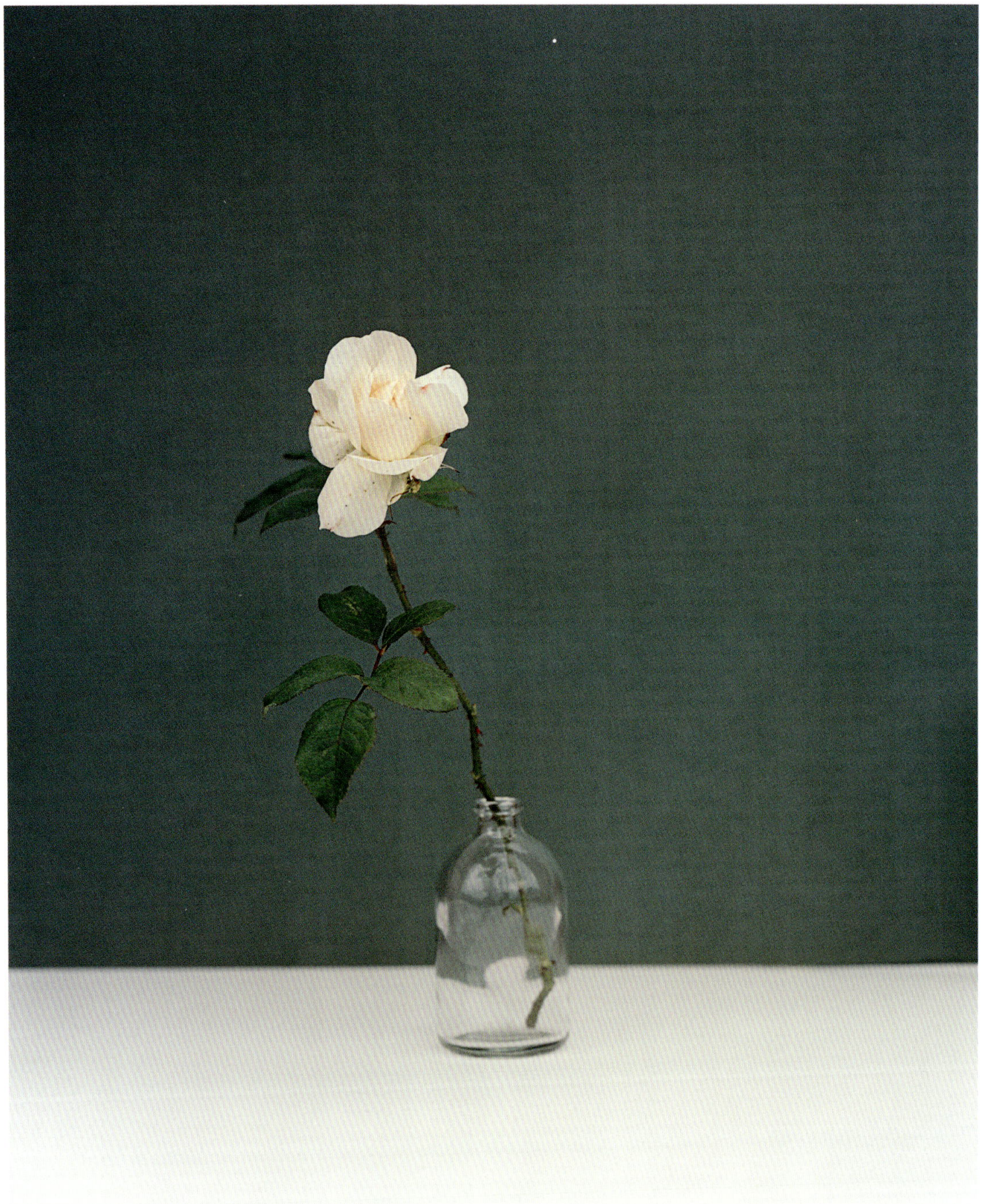

KATE FRIEND

John Pawson, Garden Rose, Cotswolds, 2020
C-type color print, 60¼ × 48 in. / 153 × 122 cm

A single stem rises out of a simple bottle against a background the color of a stormy sea. This unlikely depiction of architect John Pawson is part of British photographer Kate Friend's (b. 1983) series of botanical portraits *As Chosen By . . .*, which fuses two typically separate genres of photography: botanical still life and portraiture. From the age of eighteen, Friend traveled in Mongolia, China, Japan, and Indonesia, and her creative influences include the cyanotypes of pioneering English botanist and photographer Anna Atkins; the colored collotypes of Japanese photographer, printer, and publisher Ogawa Kazumasa; and the wooden camera made by German photographer and sculptor Karl Blossfeldt. To realize one of her images, Friend visits her subjects at home, asking them to select a favorite flower from their garden and to provide a container chosen from their home. Together, the flower—grown for pleasure—and the chosen vessel convey something of the person's private life. Once she has the flower and vessel, Friend sets up an informal studio using colored boards and available light, and photographs the flower onsite. The finished images invite speculation about the connections, resemblances, and parallels between the identity and physical characteristics of the flower and of the person. In this image, a pale rose, its petals tinged with pink and marked with a trail of tiny specks, turns its blossom up to the sky as it stands upright in its austere glass bottle: a vision of authenticity and pared-back aesthetics.

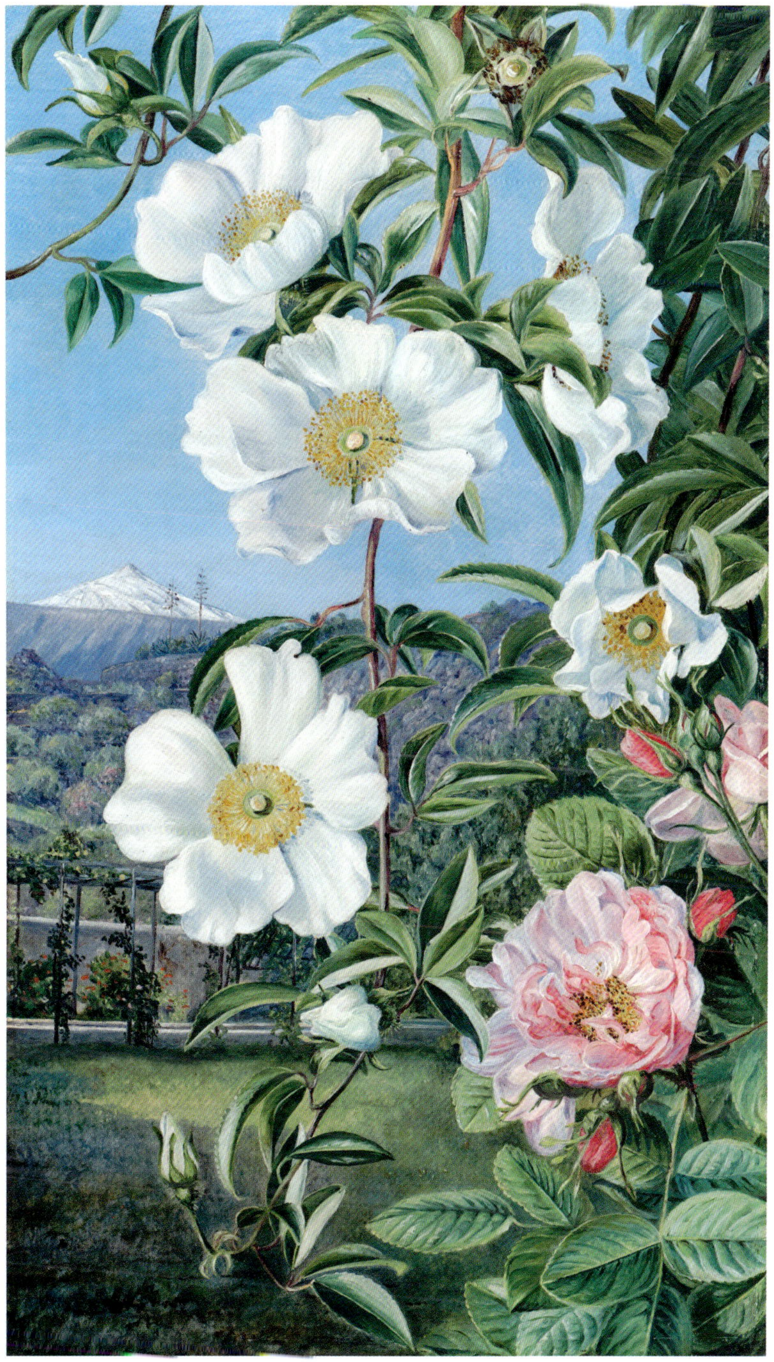

MARIANNE NORTH

Cherokee Rose with the Peak of Tenerife in the Distance, 1875
Oil on board, 18⅞ × 11⅜ in. / 50.4 × 28.9 cm
Royal Botanic Gardens, Kew, London

The large white single-flowered Cherokee rose (*Rosa laevigata*) and an unnamed pink rose almost leap from the canvas of this painting by English biologist and botanical artist Marianne North (1830–1890). The Peak of Tenerife in the background of this striking work signals the rose's exotic location of La Orotava in the northern part of Tenerife, one of the Canary Islands of Spain. An intrepid explorer, North eschewed the conventional mode of Victorian botanical painting and set forth across the globe to record vegetation as she saw it in the landscape. In the garden of her host in Tenerife, humongous sweetly scented Cherokee roses cover a great arbor and tool shed. Beyond the garden plants, aloes, euphorbias, and many eucalyptus were growing on the lava-enriched hills depicted here in the middle ground of the painting. An invasive plant in the Canary Islands, the Cherokee rose originated in China, Laos, and Vietnam; and it is not known when it was introduced to the Macaronesian islands. The rose was probably imported to the southeastern United States in the mid-eighteenth century, when it took on erroneous connotations with the Indigenous Cherokee culture of the Southeastern Woodlands, hence its common name (see pp.120 and 127). On her return to Great Britain in the late 1870s, after traveling as far afield as Borneo, Jamaica, Java, and Japan, North donated her spectacular collection of paintings—as well as a building to hold them—to the Royal Botanic Gardens, Kew.

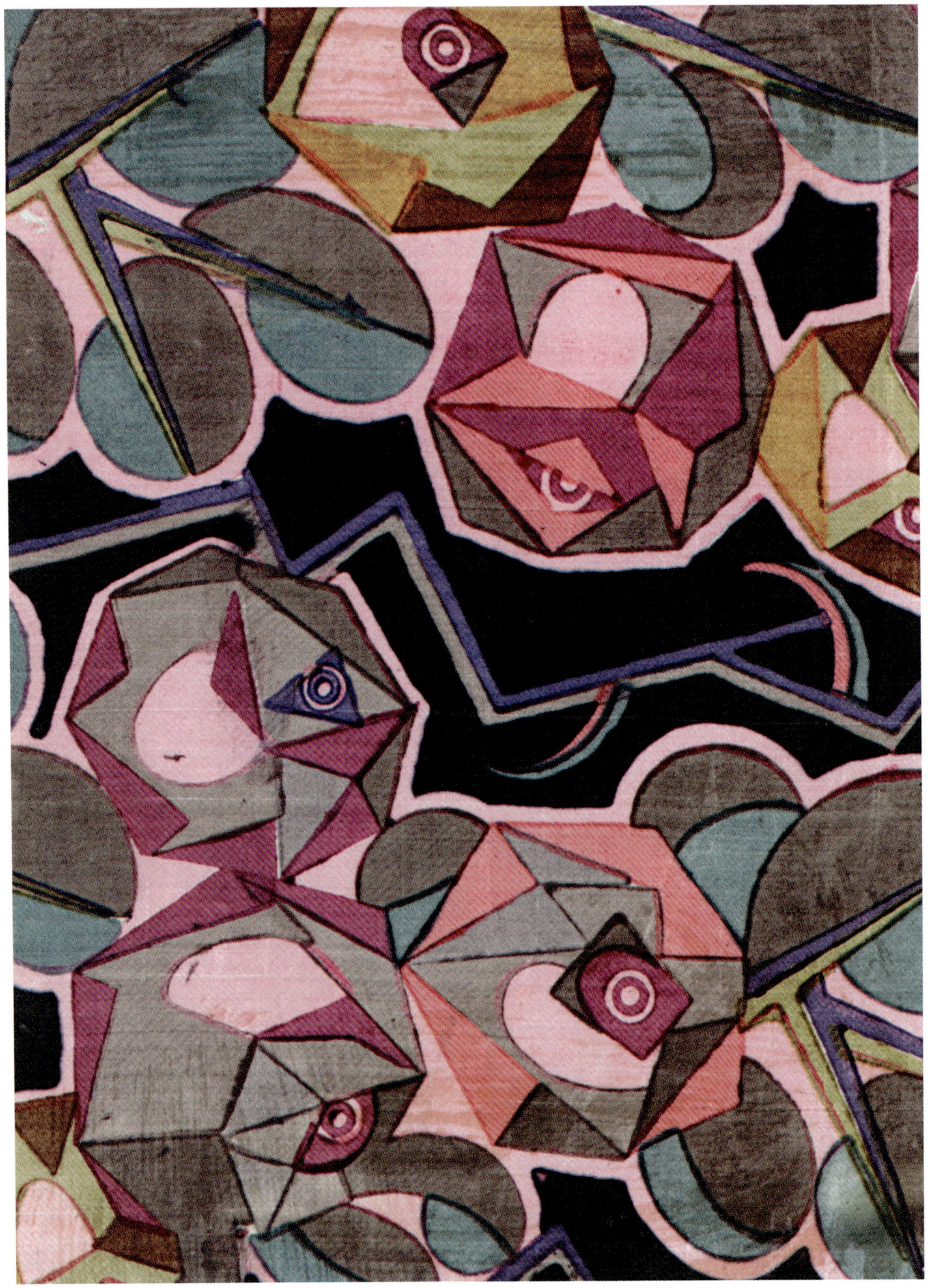

WIENER WERKSTÄTTE

Textile sample, 1910–28
Silk, 11 × 7¼ in. / 27.9 x 18.4 cm
Metropolitan Museum of Art, New York

Established in 1903, the Wiener Werkstätte (Vienna Workshop) was an enduring design movement of the early twentieth century that was fundamental to the development of European modernism. Founded by Josef Hoffmann and Koloman Moser, both of whom had been central figures in the Vienna Secession art movement, the Wiener Werkstätte aimed to bring good design and craft into all areas of life. The house style was one of geometric symmetry that eschewed ornamental excess. The abstracted, flattened flowers printed on this silk are recognizable as roses. The designer is unknown, but the textile recalls the distinctive flowers depicted in the work of Scottish architect and designer Charles Rennie Mackintosh (see p.38) and artist Margaret Macdonald Mackintosh, both of whom were an influence on Hoffmann. The Wiener Werkstätte worked with an alliance of progressive artists and craftspeople marrying traditional methods of manufacture with an avant-garde aesthetic. Their output was ambitious and included complete architectural projects through smaller-scale forms of the decorative arts that were commercially available, including jewelry, ceramics, silverware, furniture, graphic arts, and textiles. A standalone textiles department was established by the Wiener Werkstätte in 1910, working with many designers before ceasing production in 1932. Its textile output embraced a broad range of styles, from the angular to the more rounded naturalistic motifs that are evident in this example.

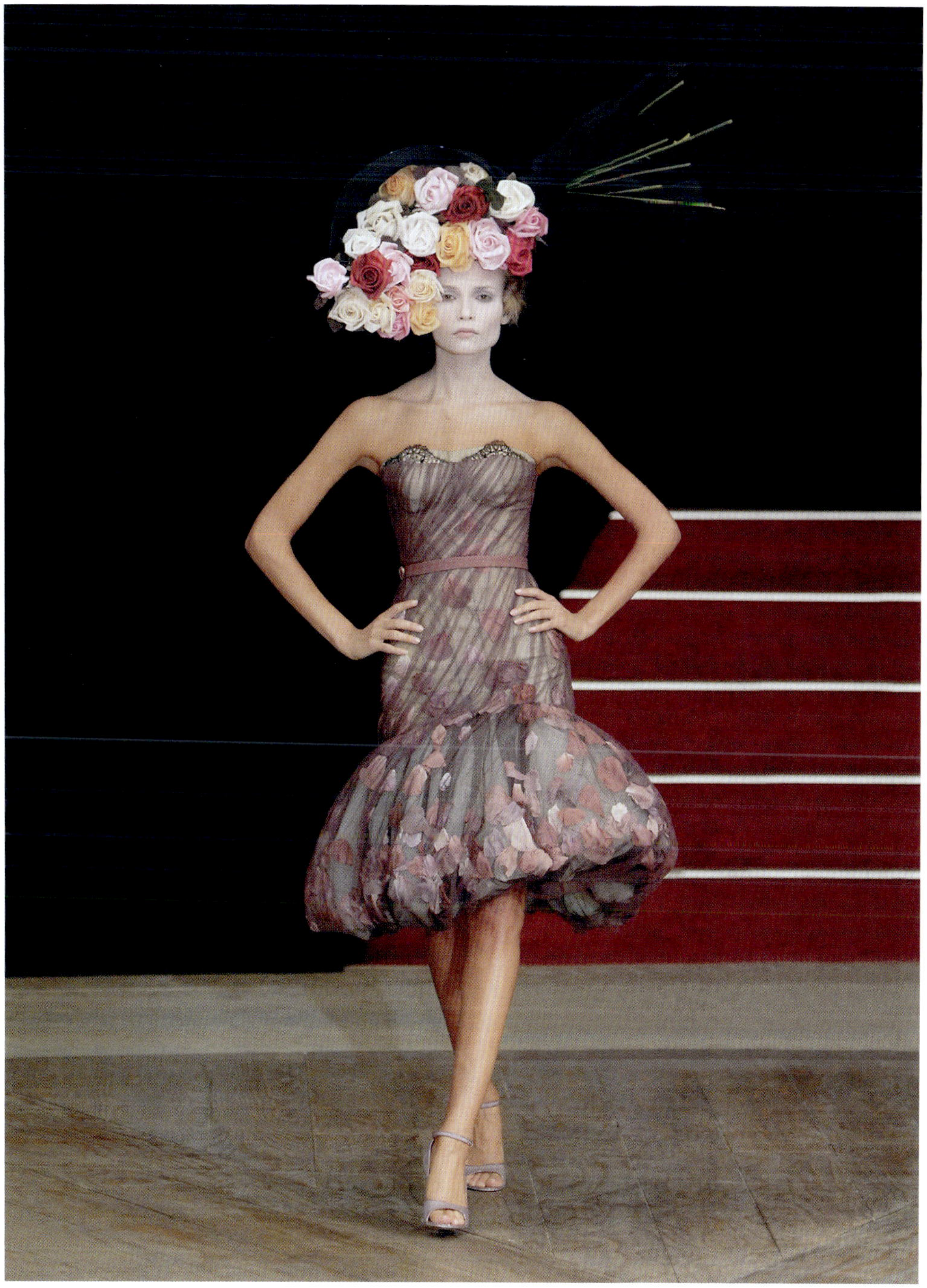

ALEXANDER MCQUEEN

Dress, from *Sarabande* collection, 2007

British fashion designer Alexander McQueen (1969–2010) was modern fashion's rosarian, and *Sarabande* his most rose-filled collection. The palette of this evening dress from McQueen's Spring/Summer 2007 runway show includes shades of mauve, rose-pink, peach, ivory, cream, tea, gray, and black. The construction comprises hundreds of silk rose petals that appear to cascade down the body, trapped within organza, while the skirt flounce takes the form of a bloom in reverse. The magnificent hat by leading London milliner Philip Treacy is a bouquet of long-stemmed, full-bloom, cultivated roses in bold polychrome silks, tied with an oversize black silk bow. Although this dress was made with artificial flowers, McQueen also made use of real ones; the collection's finale dress was covered in fresh roses and hydrangeas, which were sewn into place an hour before the show. McQueen envisioned *Sarabande* as a display of decadent, decaying grandeur, a theme that was further evoked by the runway in the form of a deserted theater. He derived inspiration from a wide array of sources: from Stanley Kubrick's use of George Frideric Handel's eighteenth century composition "Sarabande" in his film *Barry Lyndon* (1975) to Romantic Spanish painter Francisco Goya's haunting portraits. This was the first collection that Sarah Burton, who took the helm of the brand following McQueen's untimely death, worked on with him. She too is a rose devotee, who recalled the rose festivals of her childhood in Lancashire (see p.175) for this collection.

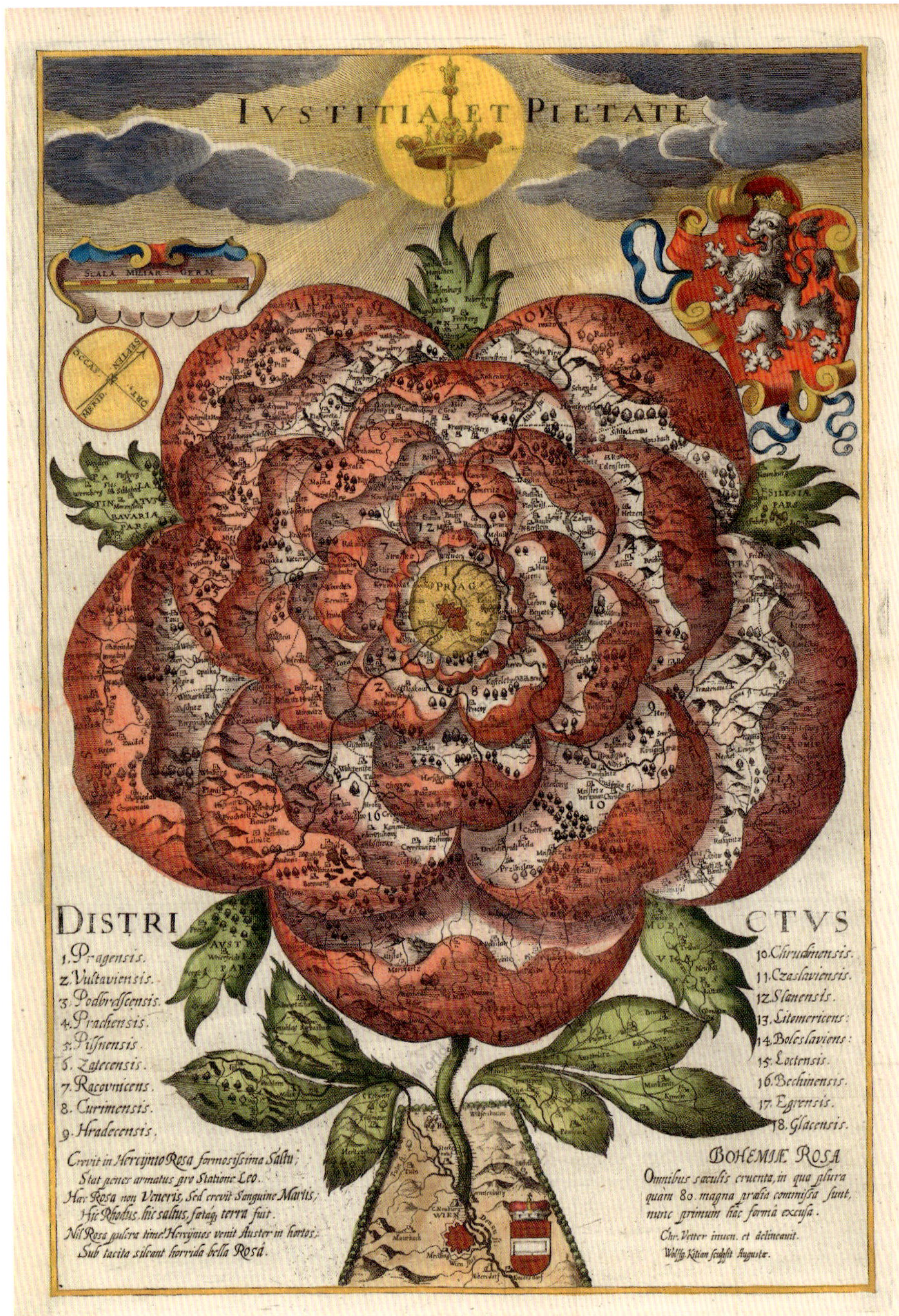

CHRISTIAN VETTER

The Rose of Bohemia, c. 1677
Hand-colored copper engraving, 15⅜ × 10⅜ in. / 38.9 × 26.2 cm
Private collection

This red rose in full bloom is actually a novelty map of Bohemia, the central European country that was part of the Hapsburg empire between 1526 and 1848 with borders that roughly correspond with the present-day Czech Republic. Prague, the capital of Bohemia, is the pistil at the center of this rose map, while Vienna, the seat of the empire, is nestled in the roots at the bottom of the stem. The rose's leaves represent the regions that immediately surrounded Bohemia: Austria, Bavaria, Moravia, Silesia, and Meissen. At the top of the map is the motto of Holy Roman Emperor and king of Bohemia, Leopold I, "*Iustitia et Pietate*" (Justice and Pity). Bohemia itself is the treasure of the empire, a flower of beauty drawing its energy from the Hapsburgs' seat of power. Drawn by Bohemian cartographer Christian Vetter (1575–1650) and engraved in copper by Wolfgang Kilian, the hand-colored map was included in a book about the early history and geography of Bohemia, *Epitome Historica Rerum Bohemicarum* (*Historical Epitome of Bohemian Affairs*), by Jesuit priest Bohuslav Balbinus. Often referred to as the Czech Pliny, Balbinus dedicated his life to promoting the Czech language and culture in the face of the growing cultural dominance of the German-speaking Hapsburgs. The rose was a common symbol in southern Bohemia. Two of the country's most important families featured roses in their family crests; the red rose was the symbol of the Rožmberks, while a black rose was used by the lords of Hradec.

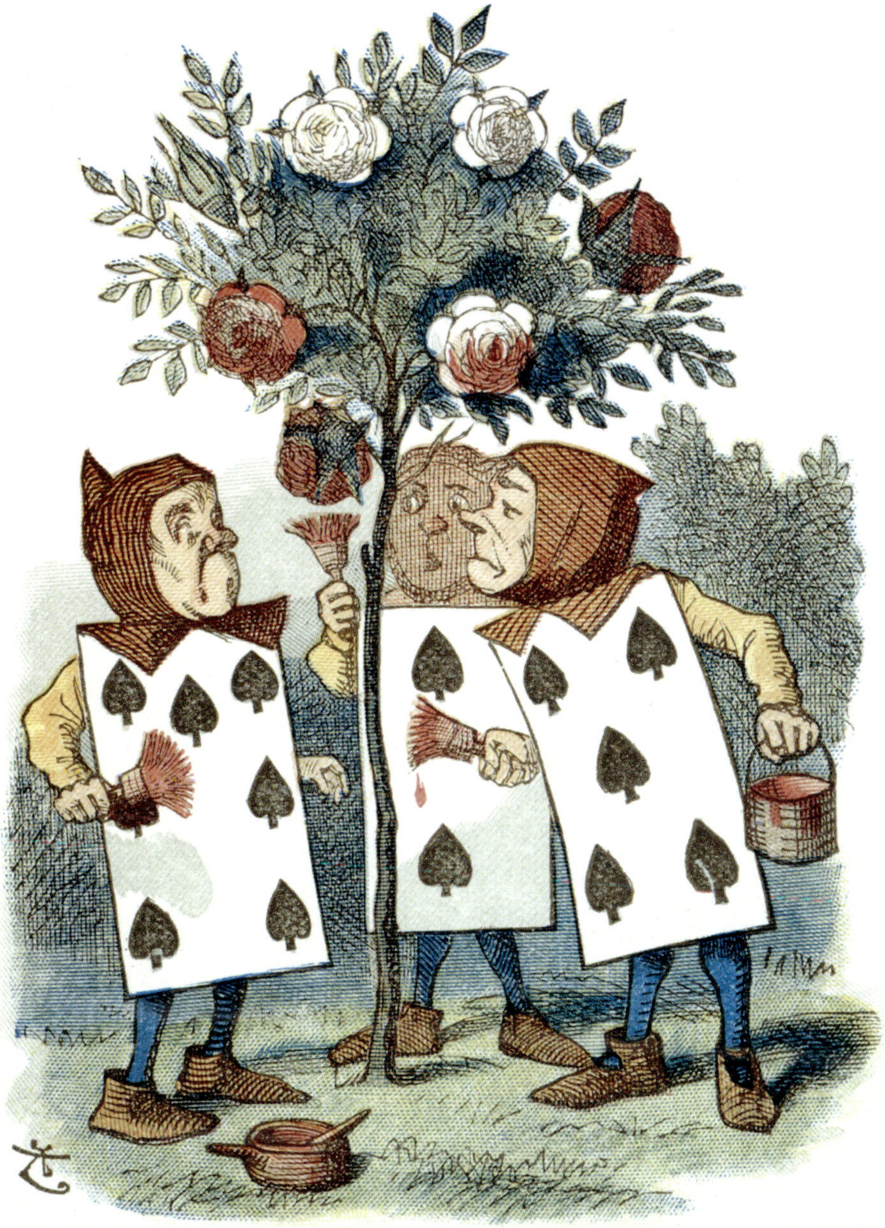

JOHN TENNIEL

The Playing Cards, from *The Nursery Alice* by Lewis Carroll, 1890
Illustration, dimensions variable
British Library, London

It is every rosarian's dream: instead of replacing a rose with a color that is wrong for the spot, find a quick way to change its hue in situ. British cartoonist John Tenniel (1820–1914) had an opportunity to do just that while illustrating Lewis Carroll's classic *Alice's Adventures in Wonderland* (1865). In accordance with the text, he simply gave the gardeners paintbrushes. The three playing cards shown here are on the horticulture staff for the Queen of Hearts. A play on the vocabulary of gardening, they are the two, five, and seven of spades. When Alice encounters them outside of the royal croquet grounds, they are busy splashing red paint on the blooms of a white rose to suit the preference of Her Majesty. An academically trained artist, Tenniel was already well known as a political cartoonist for the satirical magazine *Punch* by the time Charles Dodgson—better known by the pen name Lewis Carroll—asked him to illustrate the first edition of *Alice* in 1865. The publisher later commissioned an abridged nursery edition with Tenniel's illustrations in color to appeal to a younger age group. When Carroll wrote the manuscript of what he first called *Alice's Adventures Underground*, he illustrated it himself. His playing cards had rudimentary features, and the plant they paint is unrecognizable as the genus *Rosa*. Tenniel's signature talents were depicting movement and expression, seen here in the true-to-life stances and stubborn, stunned, horrified looks of the gardeners. His realistic standard rose, whether red or white, is in full flush.

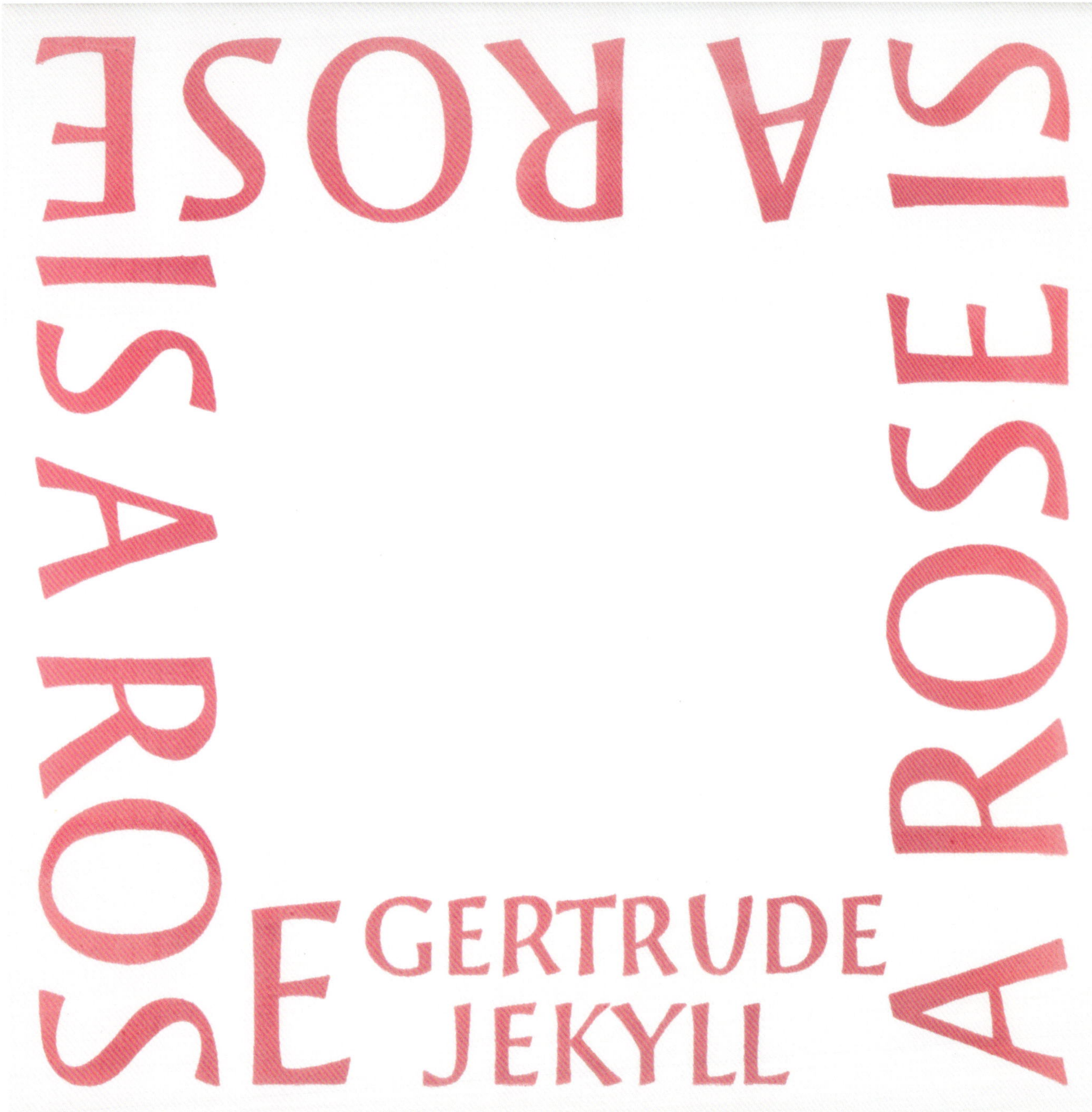

IAN HAMILTON FINLAY

A Rose Is a Rose, 1975–76
Screenprint on ceramic tile, 6 × 6 × 2⅜ in. / 15.3 × 15.3 × 6 cm
Tate, London

A square white ceramic tile is framed by a deceptively simple, repetitive phrase, written in a cheerful pink. To the uninformed, this textual work may be nonsensical, but its simplicity belies deeper and rosier meanings. It is the work of Scottish artist and poet Ian Hamilton Finlay (1925–2006), who is known not only for the wry sense of humor and imagination of his work, but also for Little Sparta, the iconic garden he created with his wife, Sue, in the Pentland Hills near Edinburgh. Much of Finlay's art, including stoneworks scattered throughout the garden, celebrates the power of words. "Rose" may conjure a multitude of different blooms, along with the color or a woman's name. Here, Finlay is, in fact, referencing Gertrude Stein's poem of 1913 about a young woman, "Rose is a rose is a rose is a rose." He guides the viewer by referring to another Gertrude, influential English gardener Gertrude Jekyll (see p.73), who in the late nineteenth and early twentieth century created a new style of gardening that used plants as paints on a palette. Finlay challenges the onlooker to imagine their own rose. There is, in fact, a pink rose named after Jekyll. An English Old Rose hybrid, it is notable for its strong perfume and tight buds that grow into large, deep pink, rosette-shaped heads. However, it was created long after Finlay made his tile. David Austin (see pp.14, 77, and 105), the famous English rose breeder, created this cultivar in 1986, since when it has twice been voted Britain's favorite rose.

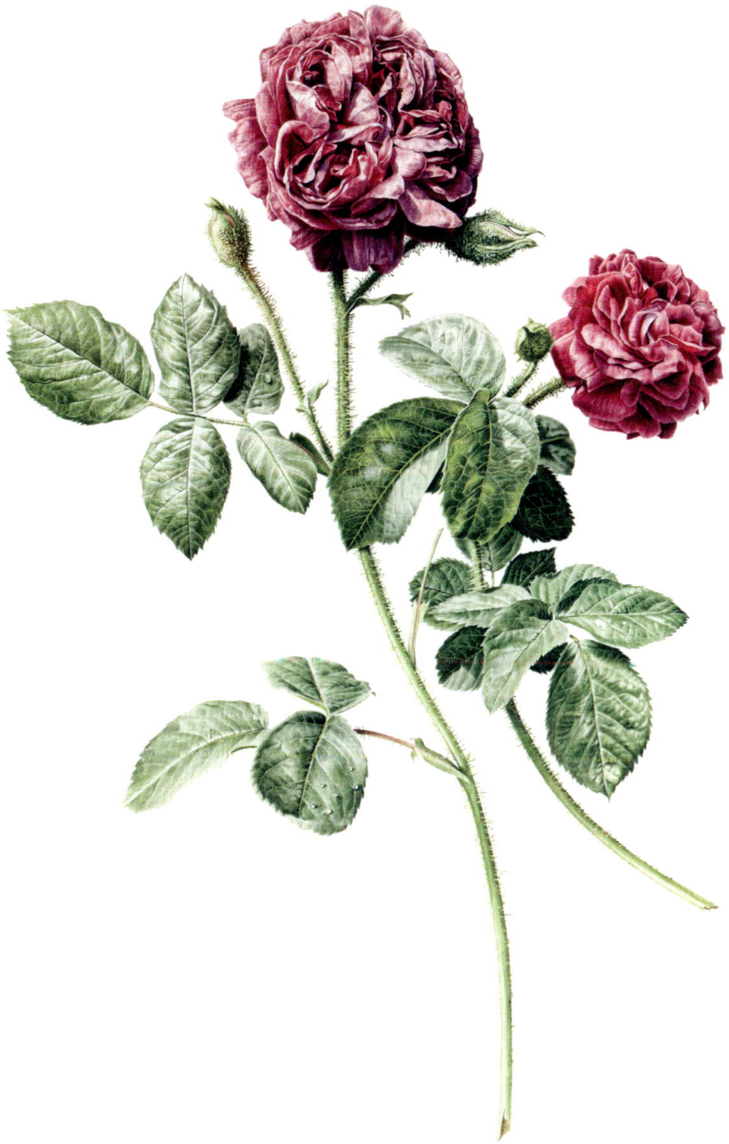

RORY MCEWEN

Rose 'William Lobb', 1976–78
Watercolor on vellum, 22½ × 18¾ in. / 57.2 × 47.7 cm

In this vivacious portrait of an old-fashioned Moss rose, Scottish botanical artist Rory McEwen (1932–1982) has skillfully captured the satiny texture of the densely packed deep mauve petals, the glossy sheen of the leaves as they catch the light, and even some sparkling drops of dew. From the 1960s onward, McEwen chose to paint his plant studies on vellum instead of paper because the material's smooth, polished surface was so well suited to recording the characteristic textures of flowers and foliage, as well as fine details. Here, for example, he has captured the tiny furlike thorns, known as "moss," that cover the sepals, the outer green parts enclosing the buds. Aside from some early lessons in drawing from his French governess, McEwen was largely self-taught, but he derived this technique of painting with watercolor on vellum from the "old masters" of botanical art—Claude Aubriet, Georg Ehret (see p.212), and Pierre-Joseph Redouté (see p.13)—whose work he had studied at the behest of Wilfrid Blunt, author of a definitive account of the subject, *The Art of Botanical Illustration* (1950). Like his predecessors, McEwen invested his chosen specimens with individuality and character and presented them in elegant arrangements isolated on blank white grounds. Although the emphasis is on the decorative qualities of his subjects, he always paired them with a precise scientific accuracy. In his finest studies, such as this one, McEwen perfected a flawless naturalism, creating a trompe l'oeil effect in which the plant appears to lift from the page.

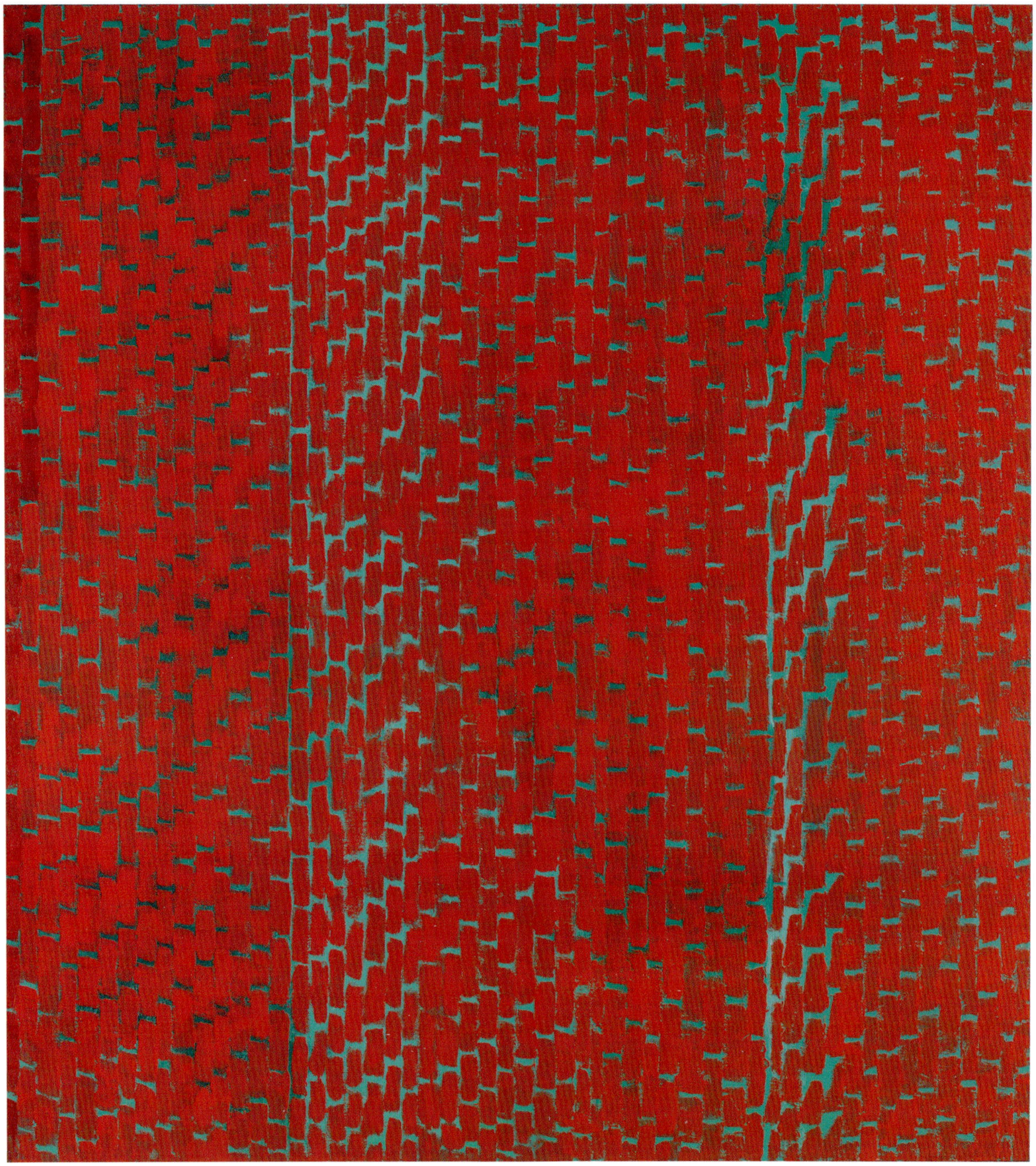

ALMA THOMAS

Red Roses Sonata, 1972
Acrylic on canvas, 59⅞ × 54 in. / 152.1 × 137.3 cm
Metropolitan Museum of Art, New York

In this dazzling abstract painting, a modulated teal, blue-green ground has been overpainted with a field of vertical scarlet dashes. The overall effect evokes the vibrancy of red roses in bloom, reflecting the boundless energy of nature that American artist Alma Thomas (1891–1978) sought to convey through many of her horticulture-inspired works. Her love of gardening stemmed from the beautiful flower beds that she remembered her parents planting at the family home in Columbus, Georgia. Her observations of flowers, foliage, and other plants fed into the mosaic-like patterns and careful balance of vibrant complementary colors that characterize the artist's dynamic paintings. Thomas preferred abstraction over figuration because it liberated her from having to represent her subjects literally, giving her the freedom to explore the possibilities of pattern and color while also conveying movement and rhythm. She believed that nature has its own musicality, and, as is the case with many of her works, the title of this piece makes a connection between music and roses (titles of other paintings include *Red Rose Cantata*, *White Roses Sing and Sing*, and *Red Rambling Rose Spring Song*). In the same year that Thomas made this painting, she became the first African American woman to receive a solo exhibition at the Whitney Museum of American Art in New York. By then, she was nearly eighty years old and had spent decades working as an art teacher before she was able to pursue her own art during her retirement.

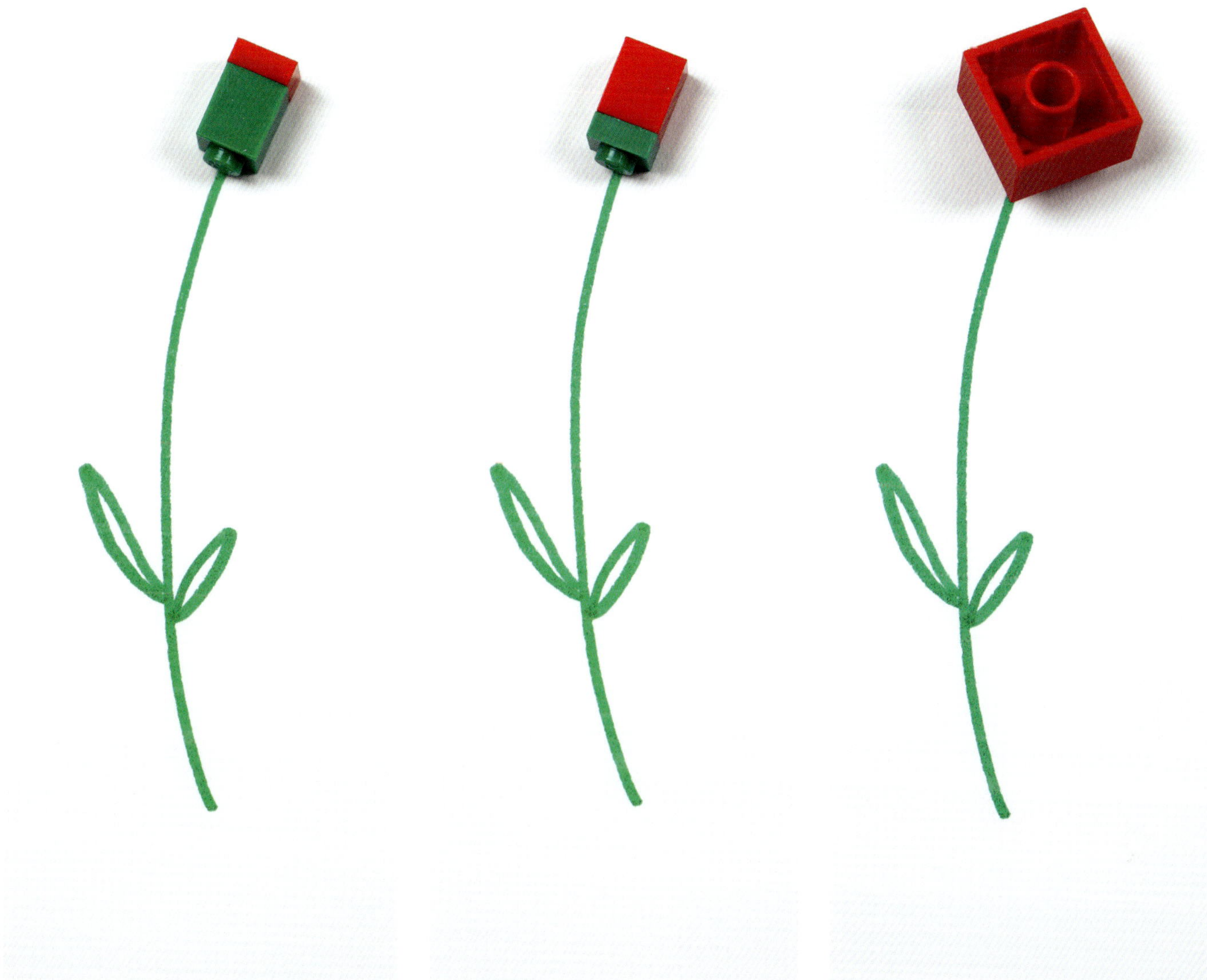

CHRISTOPH NIEMANN

Sunday Sketch (Spring), 2021
Offset print, 19⅝ × 19⅝ in. / 50 × 50 cm

Against a stark white background, three red roses lie in a row, each in a different stage of maturity. If the sequence of a flower-bud opening is not unusual in art, the depiction is. These are Lego roses in which each flower is formed from red and green toy building blocks laid on white paper. The work of German-born, New York–based illustrator and graphic designer Christoph Niemann (b. 1970)—renowned for his many covers for *New Yorker* magazine—the roses belong to an exercise he calls his "Sunday Sketch." On Sunday mornings, Niemann takes a random object, stares at it, and moves it around, waiting for inspiration to strike. He places the week's chosen object at different angles to see if it connects with other objects nearby. Here, with the deft addition of drawn green stems, his five Lego bricks (three red, two green) are transformed into roses. On the left, the flower is still in bud, in the middle it is starting to open, and, on the right, it is in full bloom . . . as far as an upturned Lego piece can become a rose. Lego—which is an abbreviation of two Danish words "leg godt" (play well)—is more typically associated with children than with art or roses. The Lego company, founded in 1932 by Danish carpenter Ole Kirk Kristiansen, was originally a maker of wooden toys, but it was the invention of the interlocking plastic bricks in 1958 that helped to unleash the creative potential of children—and artists—across the world.

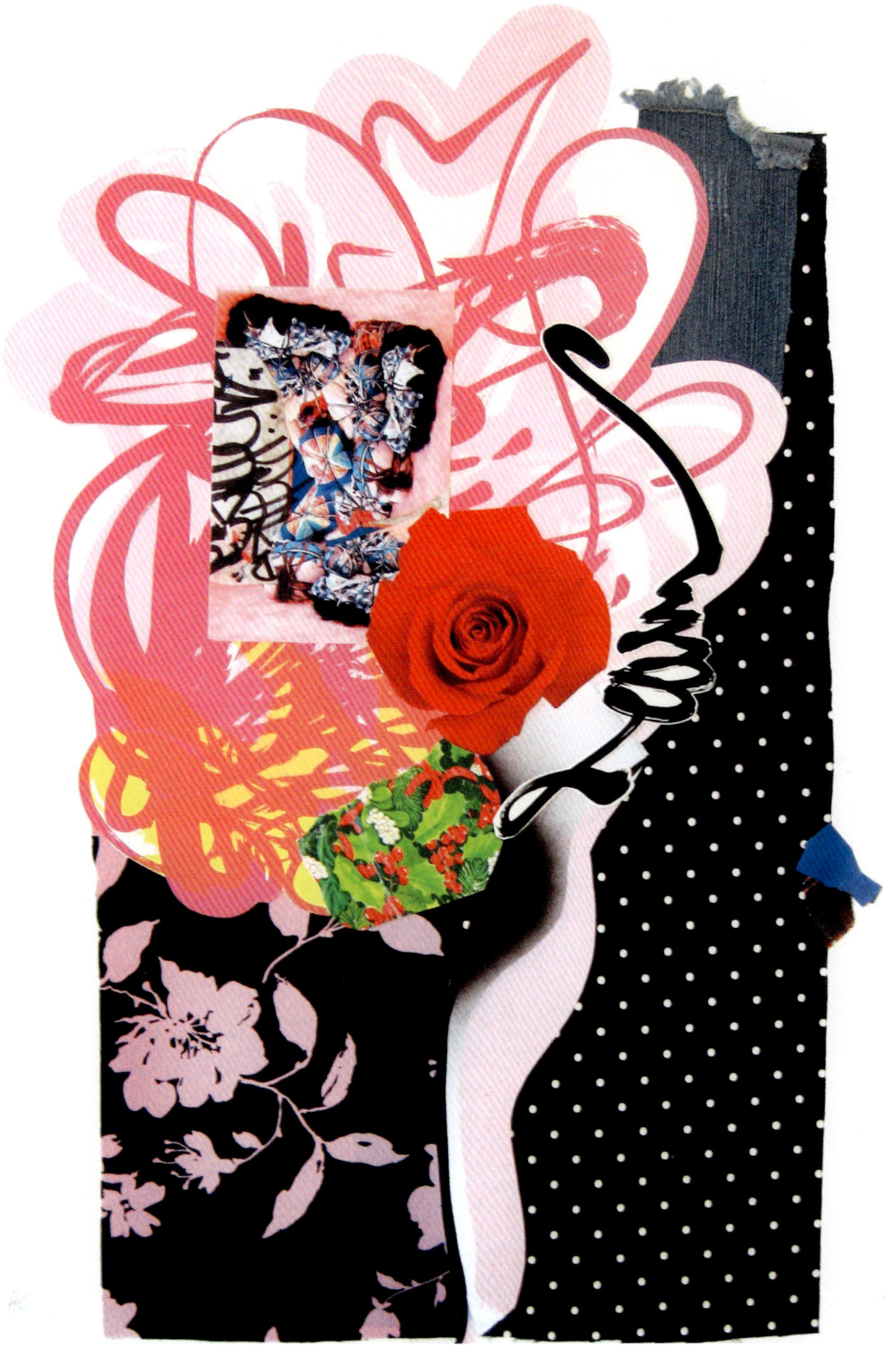

SHINIQUE SMITH

Red Rose, 2008
Silkscreen, fabric, and collage, 30¼ × 22½ in. / 76.8 × 57.2 cm
Private collection

At the heart of this dynamic configuration of expressive marks and graphic patterns is a single red rose, bringing a point of calm and focus to the busy composition. Resembling an abstracted bouquet, it is the creation of American artist Shinique Smith (b. 1971) and began life as a collage created from commercial fabrics, drawn marks, printed images, and heterogeneous found materials of the type Smith typically uses in her monumental sculptural works. Each element was then reproduced and reconstructed using digital techniques and screenprinting. For Smith, it serves as a kind of scrapbook, a collection of elements representing some of her favorite visual imagery. Roses are a recurring motif in her semi-abstract paintings and collages, a fact that she attributes to memories of her grandmother's garden in Maryland, where, as a child, she observed colorful roses blooming against the urban backdrop of Baltimore. Such visual contrasts permeate her works, manifested in the juxtaposition of hard and soft forms, smooth and rough textures, and free-flowing and rigid lines. Smith's distinctive painted gestures are reminiscent of Islamic or Japanese calligraphy, although, in fact, they derive from her time spent tagging with a local graffiti crew while she was studying painting and drawing at Baltimore School for the Arts. Mixing these hand-painted elements with pasted fabrics and colorful found images torn from posters and magazines, her lively compositions are filled with suggestions of freedom and abundance.

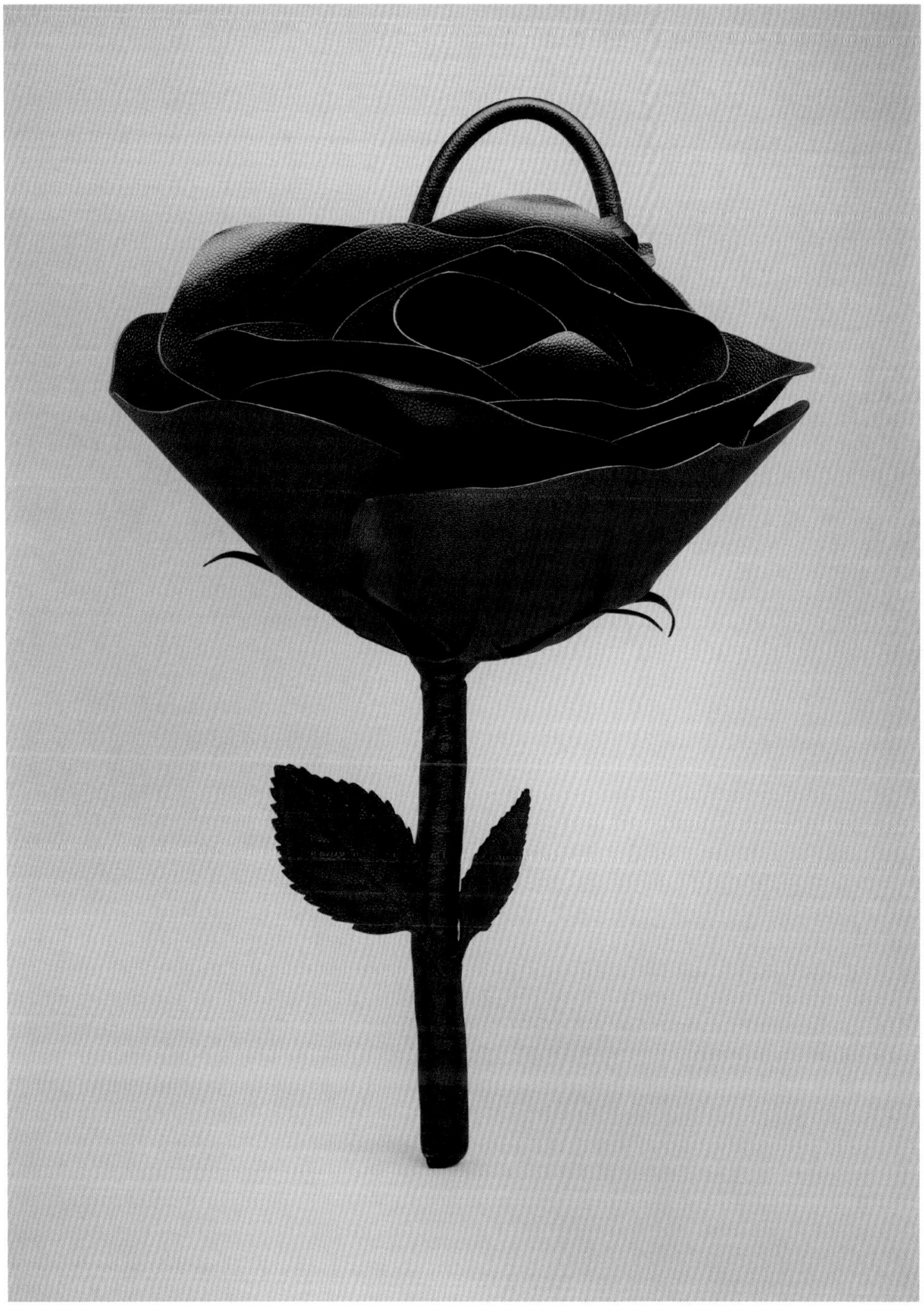

THOM BROWNE

Pebble Grain Rose Bag, 2024
Leather and brass

This large black pebble-grain leather rose might resemble some kind of Gothic sculpture, but it is actually a modern-day handbag. A leather stem with two attached leaves supports the flower's head from below. When carried by the short handle concealed in the petals, the rose appears to float by its owner's side, its folds of petals facing out toward the world and its stem trailing behind in a zany reversal of the traditional gesture of holding a cut flower by its stem. Making a statement, this rose bag is from the 2024 collection of American fashion designer Thom Browne (b. 1965). The New York–based designer has come a long way since he began his men's fashion business with just five suits in a West Village shop. Widely credited with having modernized men's tailoring by questioning the traditional cut of the suit, Browne launched his debut women's collection in 2011 and dressed Michelle Obama for the presidential inauguration in 2013. Roses have often featured in his clothes, and Browne uses them to embellish everything from women's sweaters to men's ties. His 2013 runway show, entitled *Rose Garden*, featured an abundance of dresses embellished with red roses, as well as, most memorably, stockings bearing protruding cut flowers with crimson petals and verdant green stems that ran up the models' legs. Continuing his passion for the flowers, Browne's bag takes his interpretation of the rose in a new direction, using a consciously luxurious material to create a fashion accessory unlike any produced before.

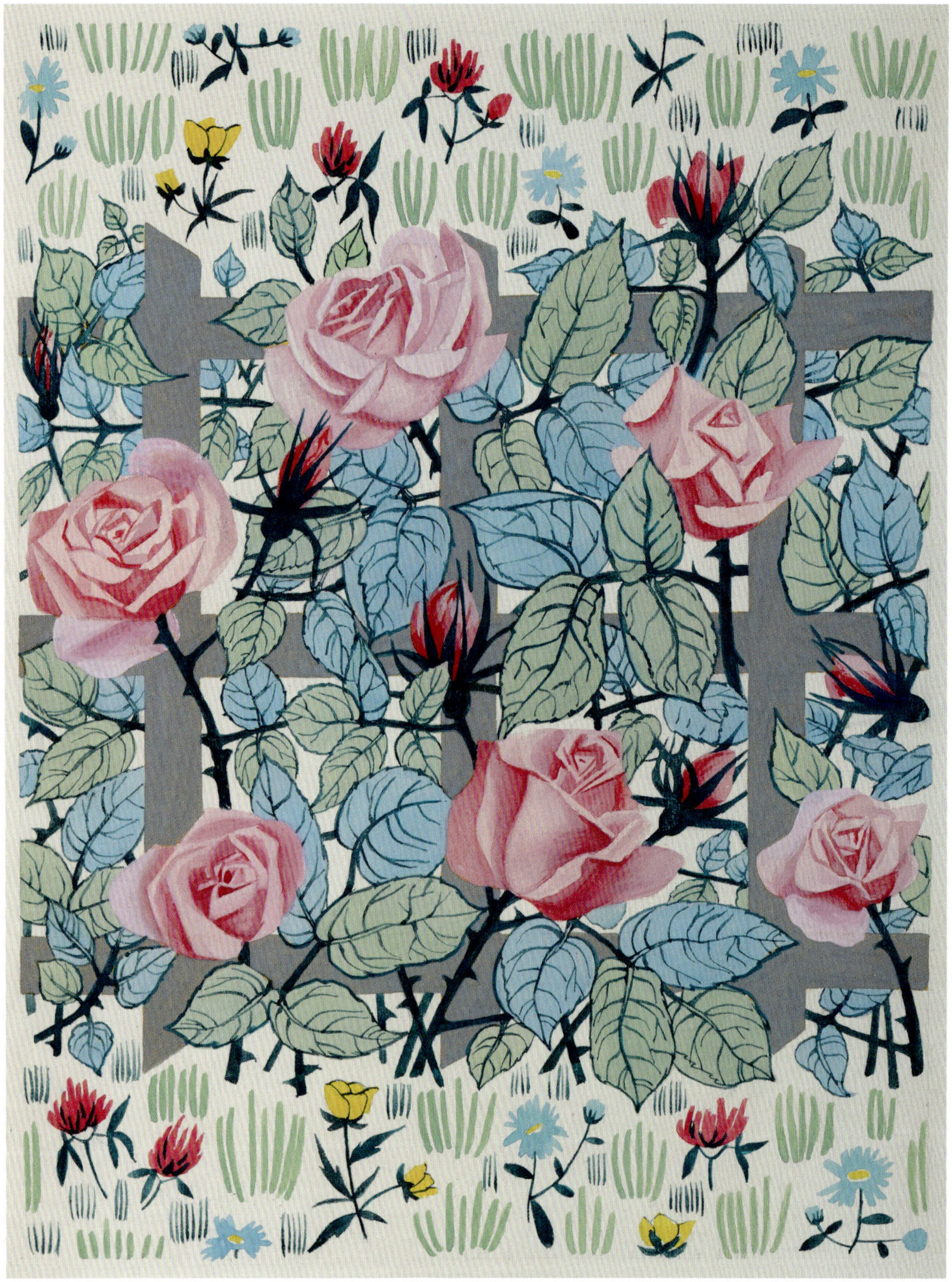

SHEILA BOWNAS

Pattern design, c. 1950s
Gouache on paper, dimensions variable
Leeds Museums and Galleries, UK

Sheila Bownas (1925–2007) was a little-known but talented British designer and botanical illustrator whose work was sold for use in wallpapers and textiles. Clients included popular retailers Liberty of London, Marks and Spencer, and Crown Wallpapers, so her designs found their way into the homes of many British people throughout the 1950s, 60s, and 70s. This hand-painted pattern illustrates large pink roses in a trellis design interspersed with smaller depictions of pink clover, yellow buttercups, green grass, and small blue flowers. Colorful, maximalist, and stylized, it is representative of her botanical illustrations, as well as the larger vibrant floral aesthetic of decorative arts during the mid-twentieth century. After World War II, Bownas's bright and bold designs chimed with the public appetite for cheerful optimism as the United Kingdom rebuilt itself, and in 1962, she was commissioned by the Natural History Museum in London to produce a set of botanical studies for an exhibition. Designers' names were rarely featured on the products they created for large manufacturers, so—despite her abundant creativity—her name slipped into obscurity. However, in 2008, gallerist Chelsea Cefai bought a collection of about 200 hand-painted Bownas designs at auction and began to research the life and work of this intrepid artist. The collection now resides in the Leeds Museums and Galleries, near where Bownas grew up in the Yorkshire Dales, and she is finally recognized as an important part of British design history.

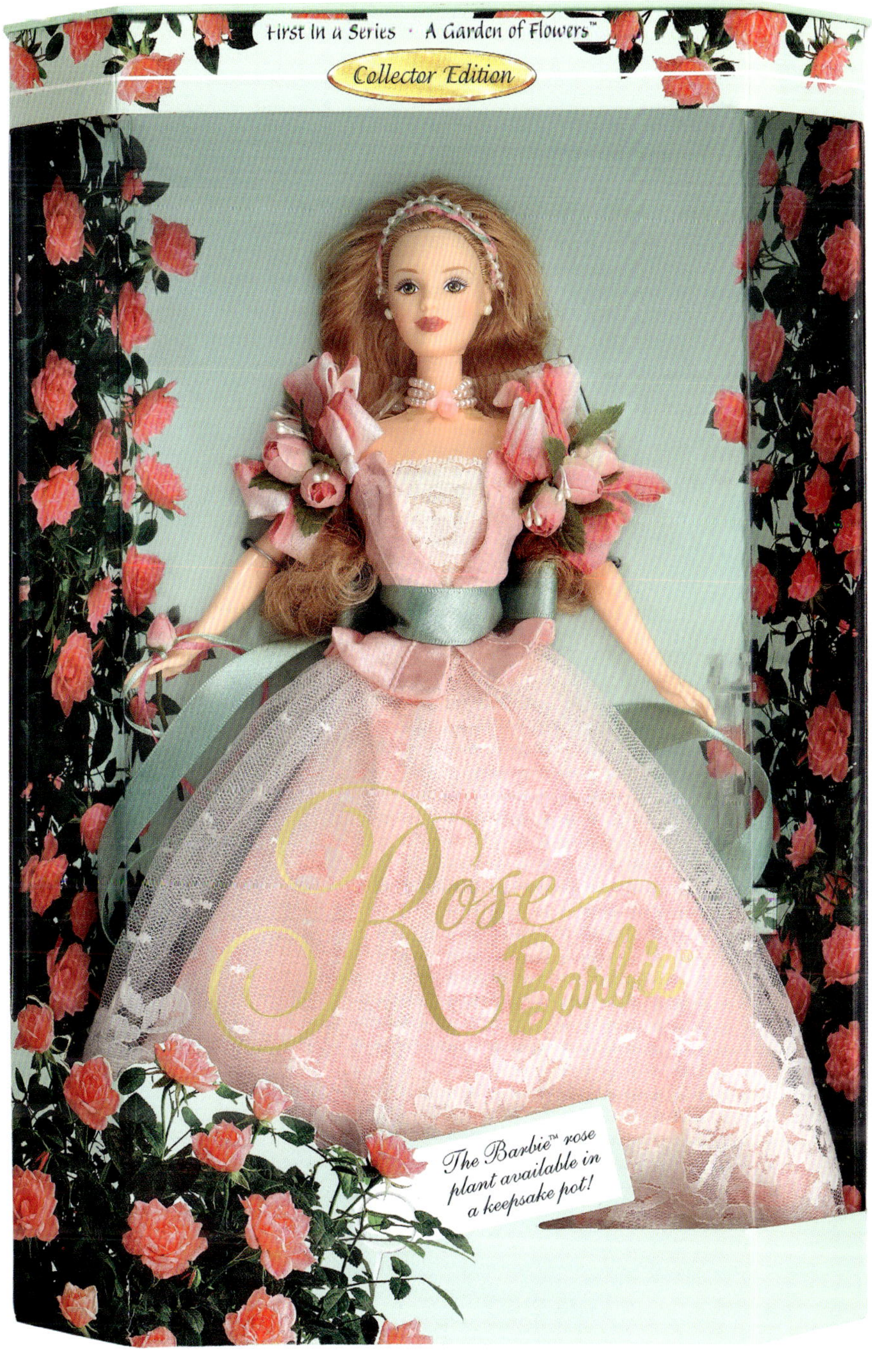

MATTEL

Rose Barbie, 1999
Fabric, vinyl, plastic, synthetic hair, and paint, dimensions variable

Ruth Handler, cofounder of American toymaker Mattel, introduced Barbie in 1959. With its grown-up body, striped black-and-white swimsuit, and high heels, the doll was an instant hit. Over the following decades, Mattel launched hundreds of themed versions, giving Barbie dozens of careers: among them astronaut, surgeon, fashion model, veterinarian. She won Olympic Gold, ran for President of the United States, and Andy Warhol painted her portrait. A sweetheart named Ken, friends like Midge, and sister Skipper soon came along. She drove a sports car, and her closet bulged with outfits, not the least of which was a sequined ensemble by Christian Dior (see p.37). By the time of her fortieth anniversary in 1999, Barbie was the undisputed queen of the fashion-doll world. A new collector's edition would connect her to an icon of the floral world: the rose. The first in the Garden of Flowers series, Rose Barbie sports a striking damask gown in her signature pink, decorated with roses; her strawberry-blonde hair is banded with pink and green ribbons, and she carries a single long-stemmed rosebud. Barbie's box is garlanded with images of blooming roses, a miniature pink rose selection named (what else) 'Barbie'. Eager collectors who purchased one of their own from rose cultivator Jackson & Perkins received it in a specially designed ceramic keepsake pot. While the 'Barbie' rose is no longer in the trade, with her recent box-office success, the Barbie phenomenon—like the celebration of roses—is still strong.

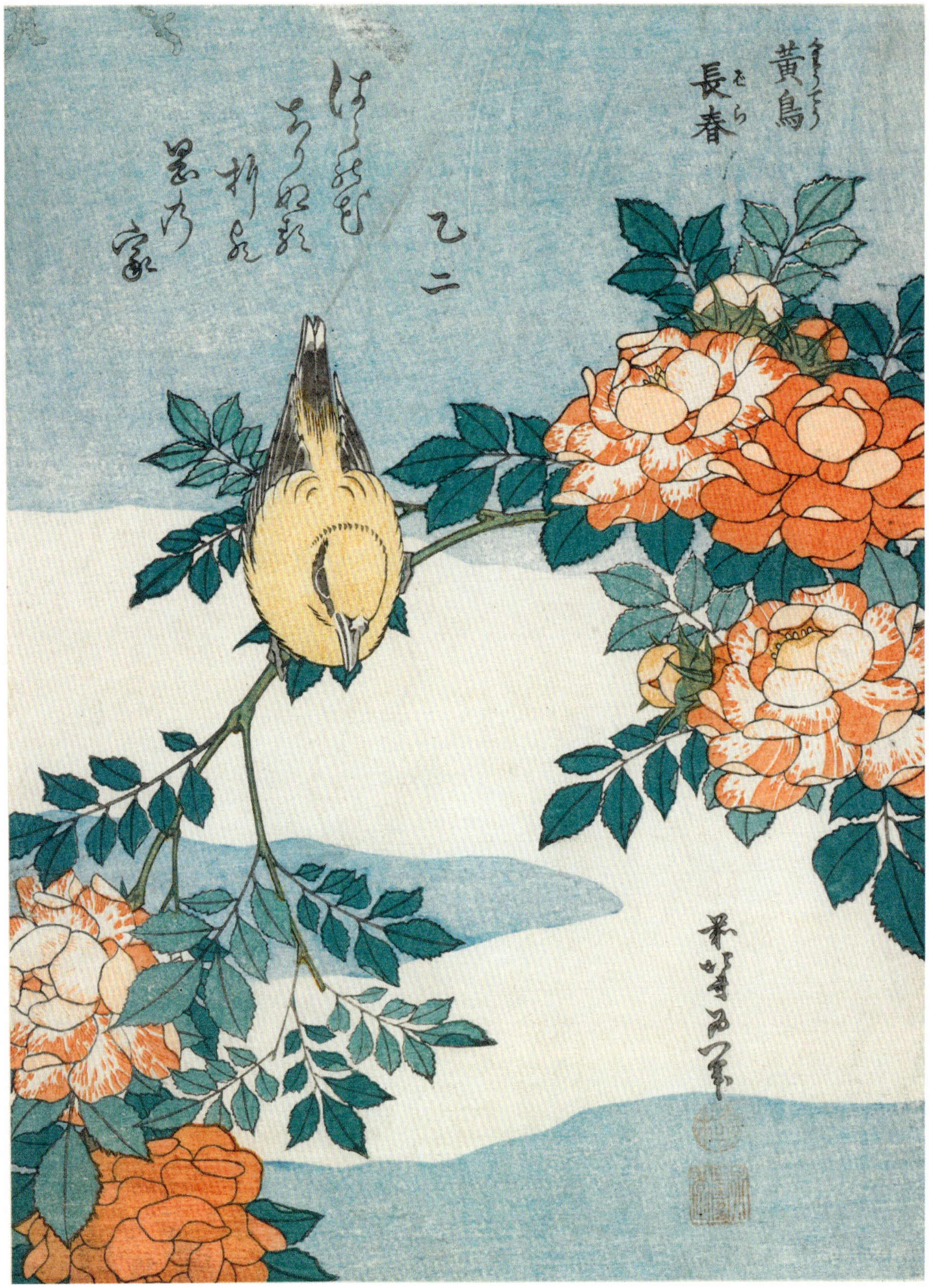

KATSUSHIKA HOKUSAI

Warbler and Roses, c. 1834
Ink and color woodblock print, 9$\frac{5}{8}$ × 7$\frac{1}{8}$ in. / 24.5 × 18 cm
Minneapolis Institute of Art

High on a hill above a lake, a pale yellow bush warbler perches on a briar rosebush against a stylized landscape in this print by prolific Japanese artist Katsushika Hokusai (1760–1849). The rose's reddish flowers are captured in full bloom, but the bleaching of some of the petals by the sun is an indication that this is the moment just before the petals start to fall. Accompanying the image is a short verse (*hokku/haiku*) written by poetry master Iwama Otsuni, which reads, "So now's the season/ when roses droop and fall?/ Home on a hill." Typically, warblers in Japanese woodblock prints appear with plum trees; however, Hokusai chose a rose for his image to suggest that moment when the summer draws to an end. One of the most well-known of all Japanese artists, Hokusai had a career spanning eight decades during Japan's Edo period (1603–1867); as a maker of ukiyo-e, or prints considered "pictures of the floating world," he sought to capture the transient nature of daily life, such as the flowering and dying of the wild roses that grow on mountainsides and beaches in Japan. He is perhaps best known for his woodblock print series *Thirty-Six Views of Mount Fuji*, which includes the iconic image *The Great Wave off Kanagawa* (1831). Despite never leaving Japan, Hokusai incorporated European painting styles into his work. This image comes from later in his career, when he turned to birds and flowers, attempting to capture their transitory but unique beauty.

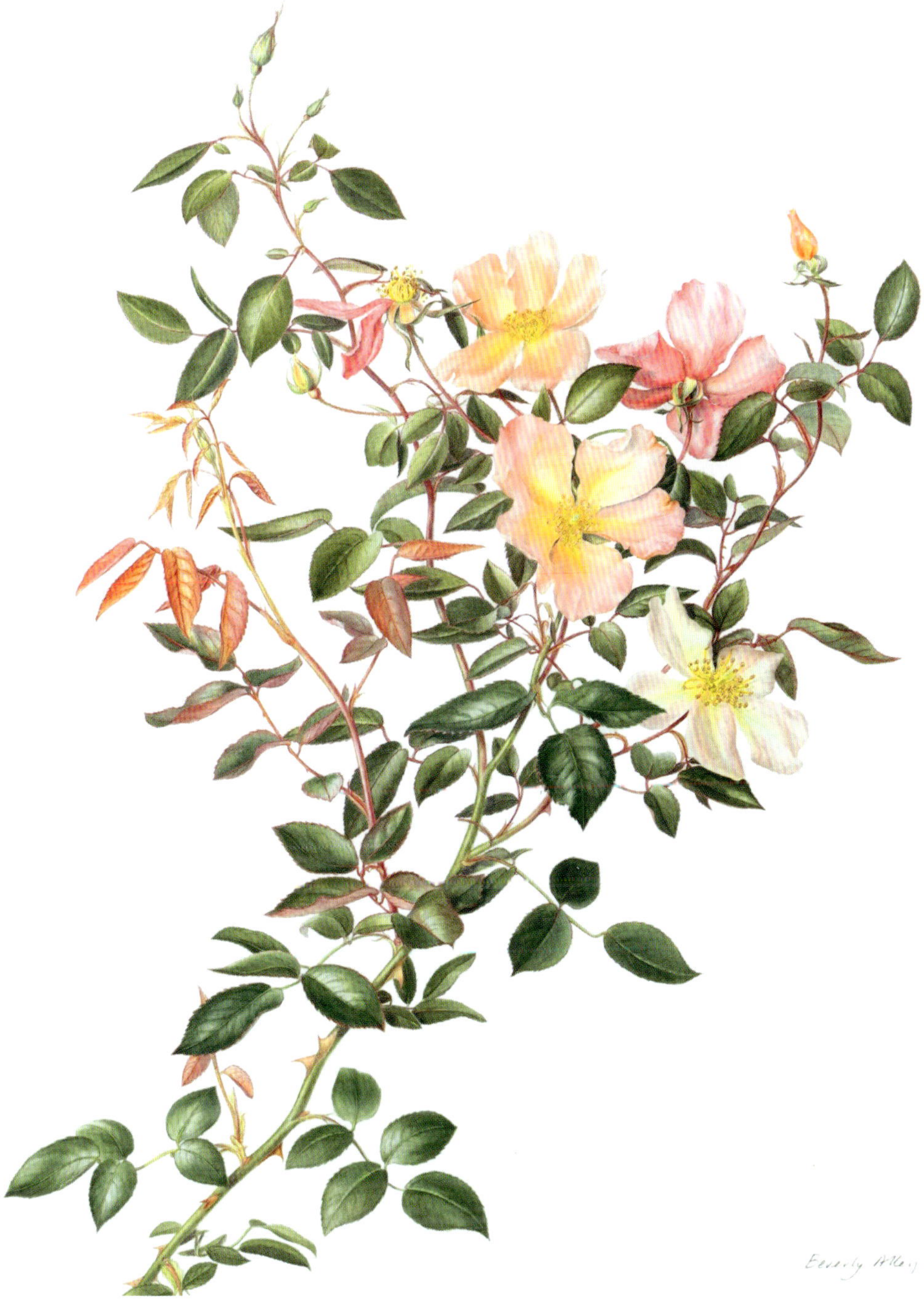

BEVERLY ALLEN

Rosa chinensis 'Mutabilis', 2023
Watercolor, 19⅝ × 15¾ in. / 50 × 40 cm
Private collection

China roses grow particularly well in Australia, and *Rosa chinensis* 'Mutabilis', known commonly as the butterfly rose, is beloved among rose aficionados in Sydney, where Australian botanical artist Beverly Allen (b. 1945) lives. 'Mutabilis' is a large and vigorous shrub rose that flowers year-round in warm climates and is rarely affected by pests and diseases. Its new growth is a handsome red-bronze, and its blooms change from pale yellow through pink to magenta as they age, as can be seen in Allen's watercolor, which was painted from life. The pale multicolored, single-petaled flowers resemble butterflies alighting on a bush, providing the inspiration for its common name. Although its origins are unknown, 'Mutabilis' is thought to be a hybrid that was introduced into Europe before 1894. It is said to have been given as a gift to Swiss botanist and hybridizer Henri Correvon of Geneva by Prince Ghilberto Borromeo, from his famous garden on Isola Bella, an Italian island in Lake Maggiore. Naming it 'Mutabilis' because its blooms are so liable to change in color, Correvon described it in 1934, when it first came to the attention of botanists. Allen, an internationally recognized botanical artist, cofounded the Florilegium Society of the Botanic Gardens of Sydney. She is known for her large or life-size works depicting plants that are native to Australia or otherwise thrive in her garden, perhaps including this vibrant and dynamic butterfly rose.

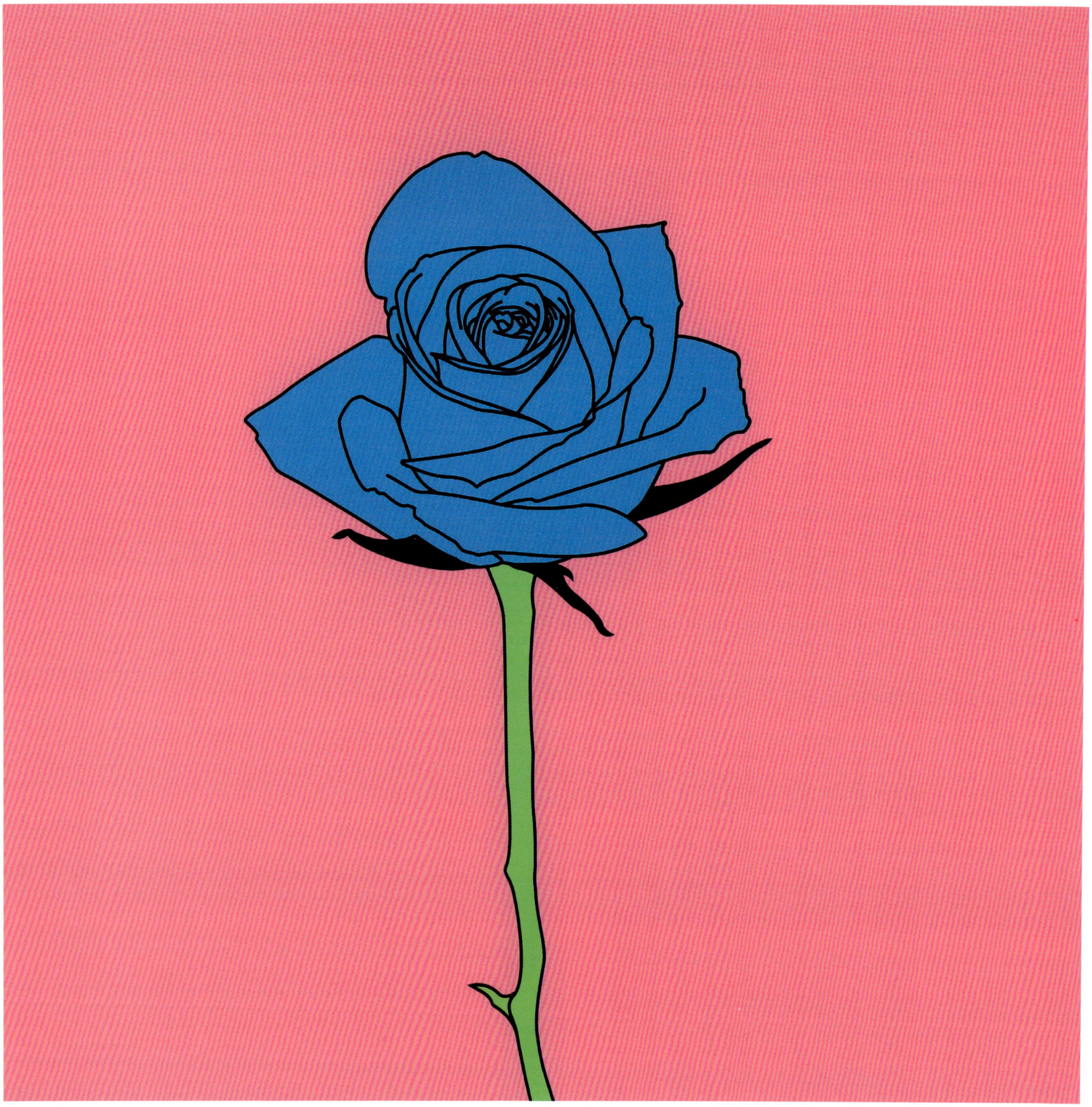

MICHAEL CRAIG-MARTIN

Rose, from *Domesticated Flowers*, 2021
Screenprint on paper, edition of 40, paper: 27⅝ × 27⅝ in. / 70 × 70 cm, image: 22 × 22 in. / 55.9 × 55.9 cm

This perfect blue rose comes from a series by Irish artist Michael Craig-Martin (b. 1941) titled *Domesticated Flowers*, in which various stylized images of plants, such as tulips and gerberas, are set against a single-color ground. A truly blue rose does not exist in nature, as roses lack the gene to produce this hue, so Craig-Martin's choice of color is intended to underline his contention that many of the flowers and fruit we consider part of the natural world are, in fact, created by humans. He has said, "We domesticated and bred [plants] to become what we want them to be. They are in fact as much mass-produced objects as iPhones." His work previously focused on manufactured objects—everyday things ranging from light bulbs to laptops, safety pins to sneakers, chairs to coffee cups—all represented as pared-down generic forms with black outlines and saturated non-naturalistic colors. His depictions of flowers, fruit, and vegetables, all presented in vivid, unnatural hues, were prompted by his visits to the supermarket during the COVID-19 pandemic lockdown in 2020, when he realized that unromanticized objects had much in common with his established graphic lexicon of commonplace objects, which are generally so familiar that people take them for granted. For Craig-Martin, the only essential difference between a wristwatch and a rose is that the rose will quickly fade and die, but in his images, both have the same pristine air of permanence.

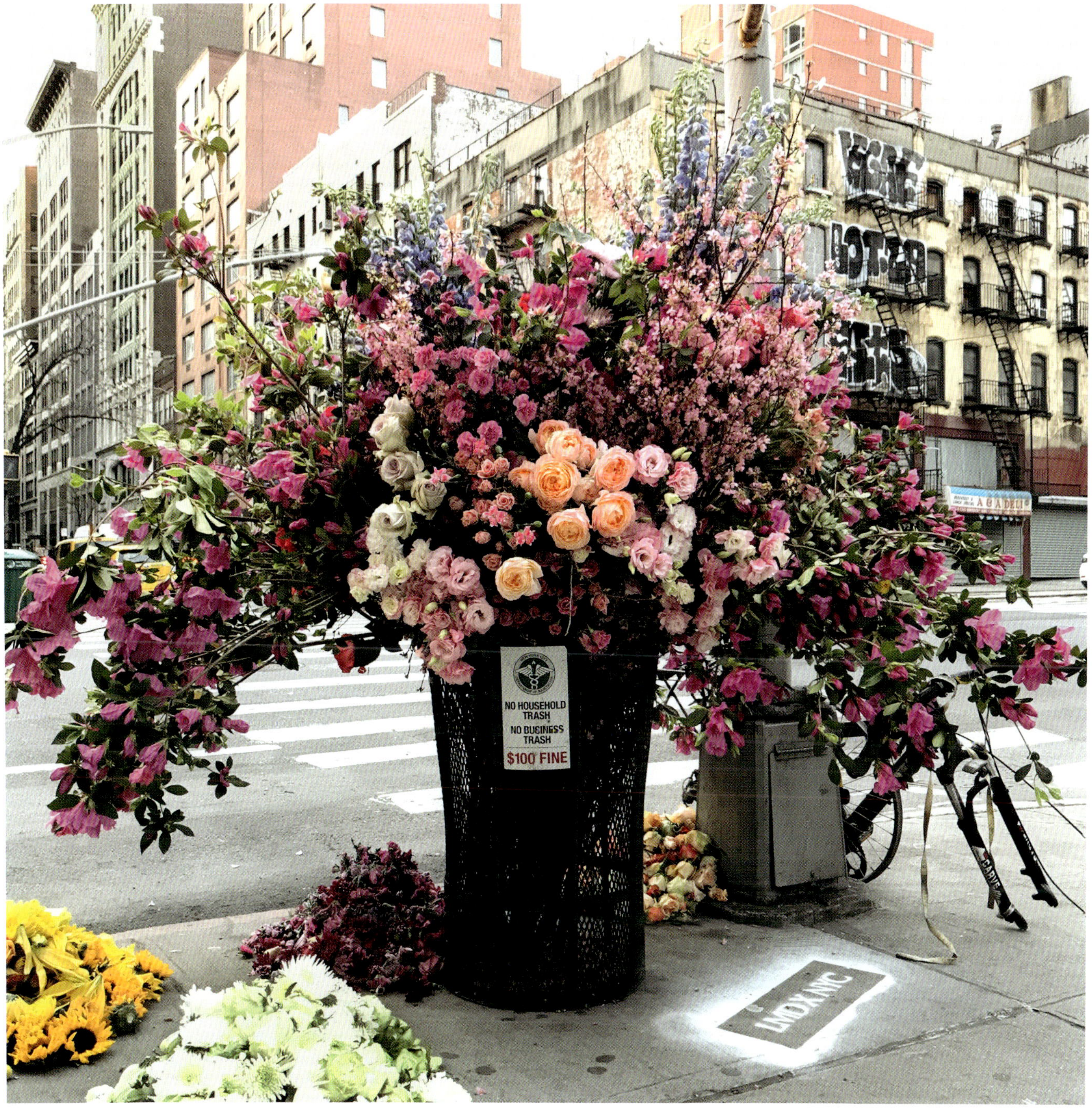

LEWIS MILLER

Flower Flash: W 22nd Street/7th Avenue,
April 11, 2019
Floral installation, dimensions variable

The grimy trash cans of New York can hardly be described as attractive—unless, that is, they happen to be overflowing with exuberant flower arrangements by American floral designer Lewis Miller (b. 1974). This photograph captures the explosion of color that greeted passersby on the corner of West 22nd Street and 7th Avenue in the neighborhood of Chelsea one bright spring morning in 2019. An abundance of pink, white, and peach-colored roses at the heart of the display are accompanied by an array of seasonal flowers, while yet more roses lie on the sidewalk. Miller, who has been hailed as the "botanical Banksy," has been making these eye-catching guerrilla street-art installations since 2016. Known as Flower Flashes, they have appeared across Manhattan on random street corners, over bus canopies, on traffic medians, in front of the Whitney Museum of American Art, and even at the John Lennon Memorial in Central Park. The arrangements, which are installed secretly and without permission from the city, are created before dawn in locations that are scouted out in advance. Miller and his team work fast, usually completing each Flower Flash in less than thirty minutes, using leftover blooms from the large events that are the mainstay of his floral-design company, Lewis Miller Design. For him, they are a gift to the city that he hopes will bring moments of delight and wonder to countless New Yorkers, brightening the day of all who encounter them.

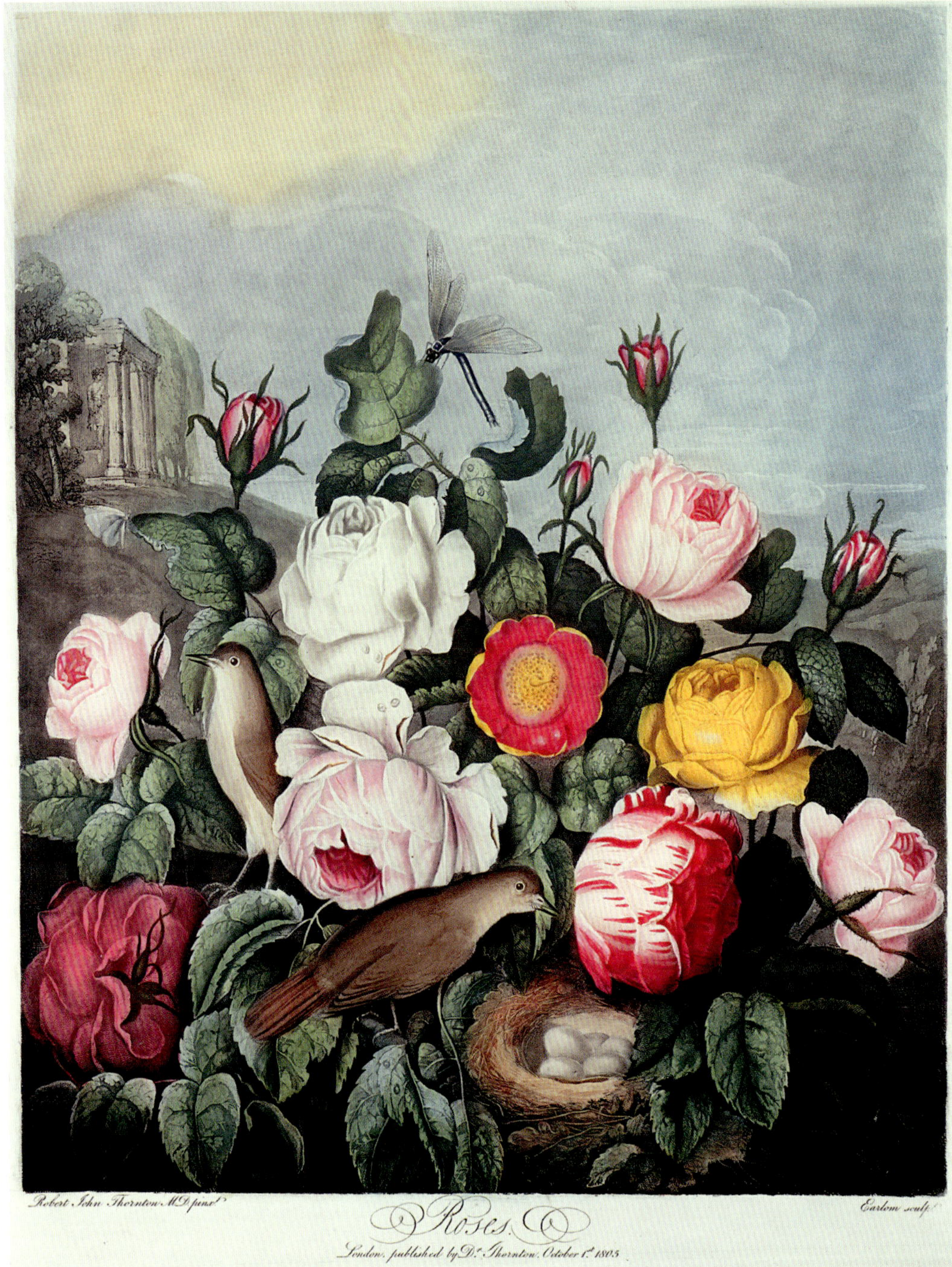

ROBERT JOHN THORNTON

A Group of Roses, from *The Temple of Flora*, 1805
Colored engraving, 22⅝ × 17⅞ in. / 57.5 × 45.5 cm

Robert John Thornton (1768–1837) was an English physician with a special enthusiasm for botany, who set out to promote Carl Linnaeus's system of classification with a great illustrated work, the *New Illustration of the Sexual System of Linnaeus* ("new" because the rest of the title had been used by botanist John Sebastian Miller twenty years earlier). Thornton spared no effort to make this a beautiful production, employing a variety of artists to illustrate it with colored mezzotint engravings depicting plants in decorative landscape settings. He nearly bankrupted himself on the publication and was eventually forced to set up a lottery to raise funds. During the eight years it was being issued in parts, from 1799 to 1807, many of the plates had to be revised, so they exist in multiple versions. Once it was completed, Thornton published a smaller version under the title *The Temple of Flora* in 1812. This illustration was published in October 1805, but it was engraved again in a smaller format in 1812. *A Group of Roses* is the only one of the major color plates to have been drawn by Thornton himself; he generally contributed illustrations of botanical details. The roses are not identified by variety, nor can the anatomy of the flower be discerned, while the insects, birds, and landscape background are irrelevant to the plant. Although a stunning reproduction, the vagueness of the illustration mirrors *The Temple of Flora*'s larger inadequacies when it came to innovations in both botanical content and illustration.

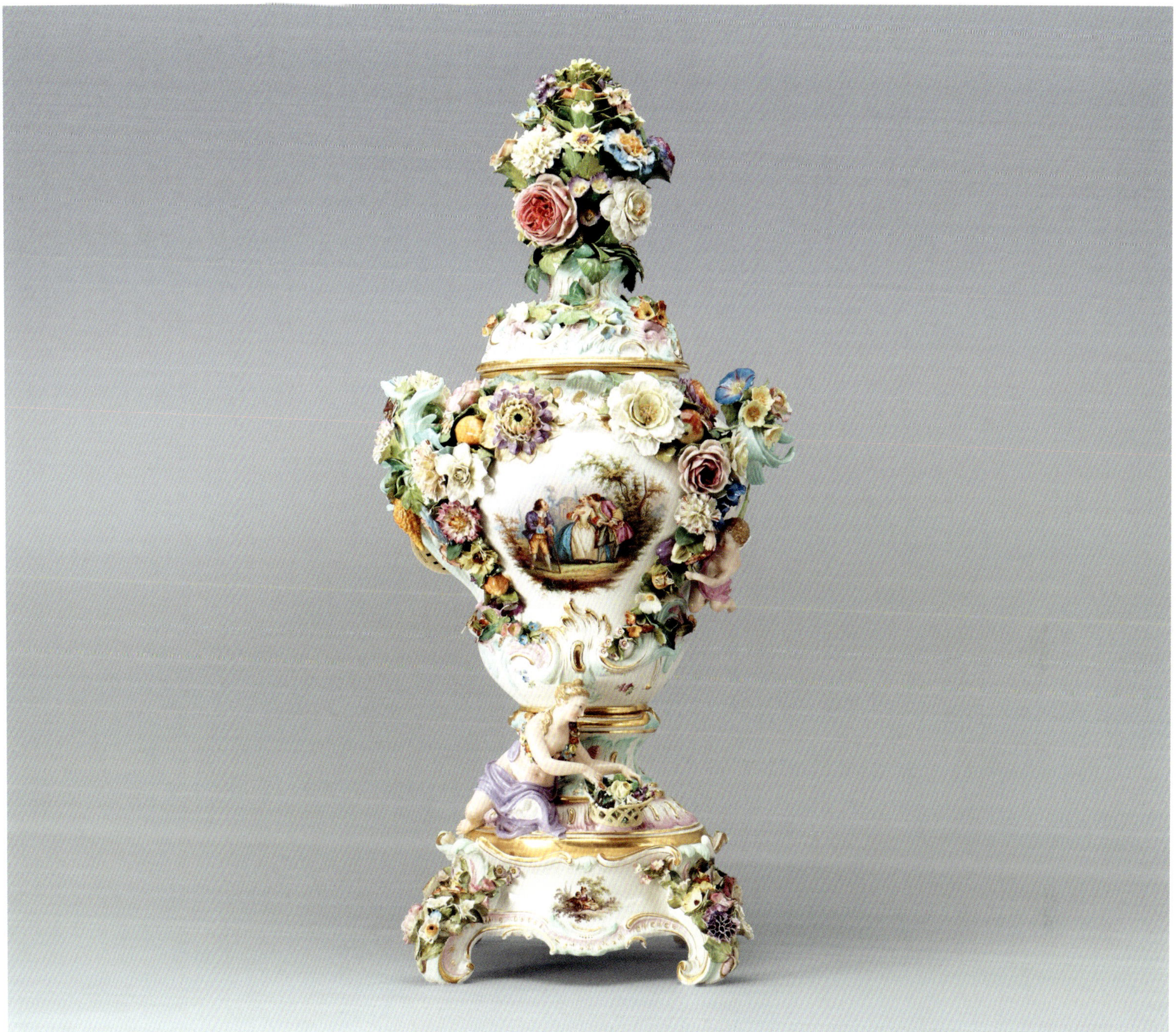

MEISSEN PORCELAIN FACTORY

Potpourri vase, late 19th century
Porcelain, H. approx. 35⅜ in. / 90 cm
Private collection

A profusion of porcelain flowers tumble down this extravagantly decorated vase, designed appropriately enough to hold potpourri, a scented mixture of dried flower petals and spices used to perfume rooms. Roses are everywhere in the design, most prominently in the encrusted cover at the top and along the sides of the urnlike container. Crafted in the late nineteenth century by the German porcelain factory at Meissen, this vase recalls the exuberant Rococo style popular over a century earlier. Beyond the pastel florals, it oozes the dramatic romance characteristic of Rococo in its anthropomorphic symbols of love (a clinging cherub and seminude woman, appropriately proffering a basket of flowers) and painted courtship scene in the cartouche. This piece belongs to a period of creative renaissance for the Meissen workshop, when Ernst August Leuteritz led his skilled artisans in a return to the company's heritage of producing opulent porcelain, kick-starting a second golden age between 1849 and 1886. Meissen's first period of success had begun in 1708, when founder Augustus the Strong, Elector of Saxony, spared no expense to assuage what he called his "porcelain sickness." An obsession with delicate porcelain from China led him to sponsor alchemists to discover the precise recipe used by the Chinese so that Saxony could make its own porcelain—a challenge that was eventually solved by Johann Friedrich Böttger, paving the way for the creation of major manufacturers, including Derby Porcelain Factory (see p.81) and Meissen, which is still in operation today.

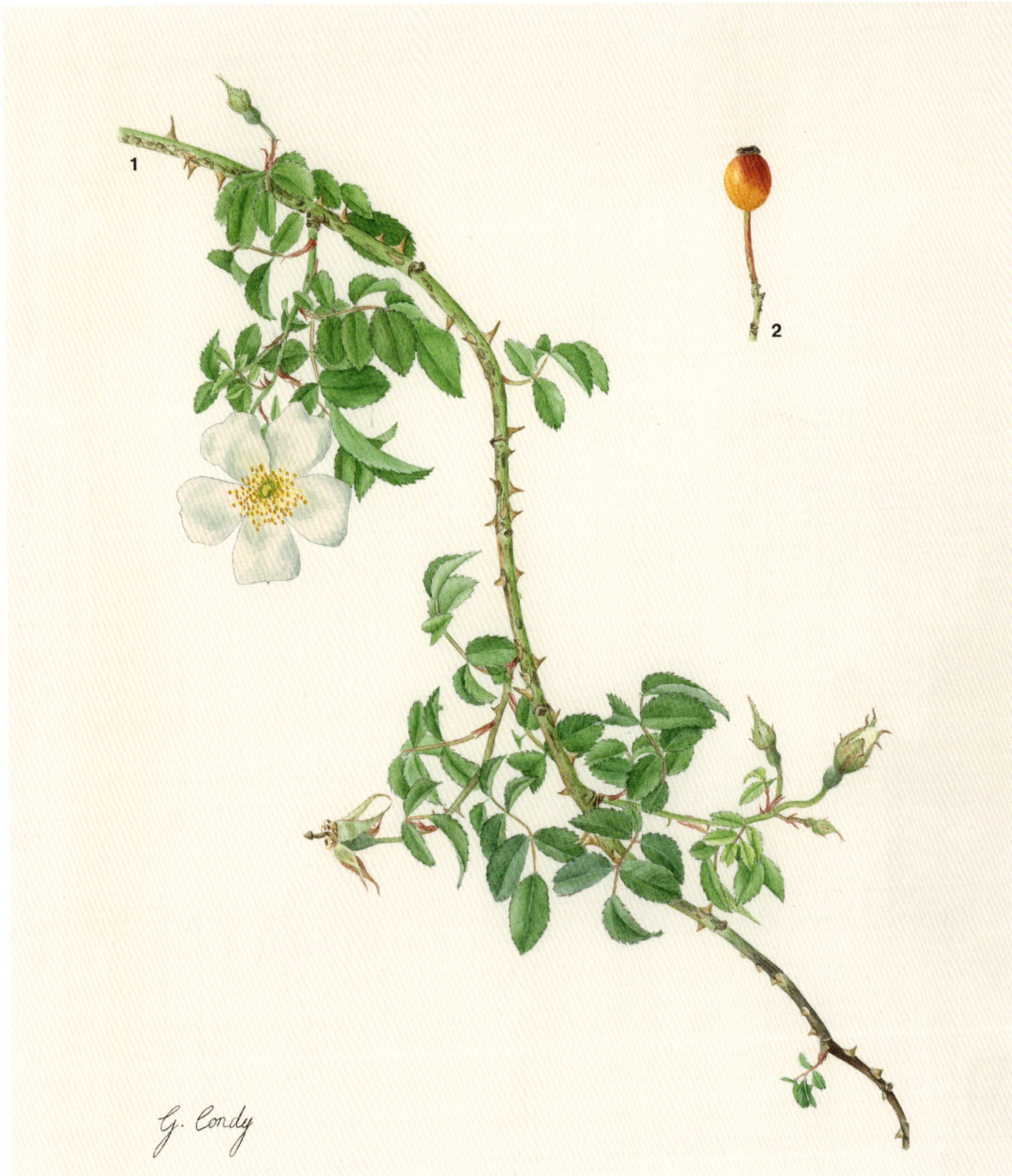

GILLIAN CONDY

Rosa abyssinica R.Br., from *Flowering Plants of Africa*, 1987
Watercolor on paper, dimensions variable
South African National Biodiversity Institute, National Herbarium, Pretoria

This detailed botanical painting shows a flowering and fruiting branch belonging to the only rose native to Africa, *Rosa abyssinica*, which is found principally in Eritrea, Ethiopia, Somalia, and Sudan, as well as Saudi Arabia and Yemen. Notable for its hardiness, this evergreen climbing shrub is able to survive in inhospitable conditions that include rocky outcrops, bushland, and dry forests. It can grow as tall as 23 feet (7 m) and has leathery leaves and attractive whitish yellow flowers that emit a lovely fragrance. This illustration is the work of award-winning botanical artist Gillian Condy (b. 1952), who was born in Kenya but is based in South Africa, and was the resident botanical illustrator of the South African National Biodiversity Institute for thirty-four years from 1983 until 2017. In her illustration, Condy paints the fruit separately; the green rose hip ripens to orange-red and grows to a length of 1 inch (2.5 cm). Sometimes described as a "big prickly dog rose," *R. abyssinica* was unknown outside Africa until the nineteenth century, when it was first cataloged by Scottish botanist Robert Brown, better known for his discovery of the nucleus of the cell. Condy has spent much of her career illustrating Africa's many floral varieties. She made more than 200 plates for the multivolume magazine *Flowering Plants of Africa*, which continues today more than a century after it started in 1921, with the remit to publish descriptions and illustrations of all the flowers of the continent and its neighboring islands.

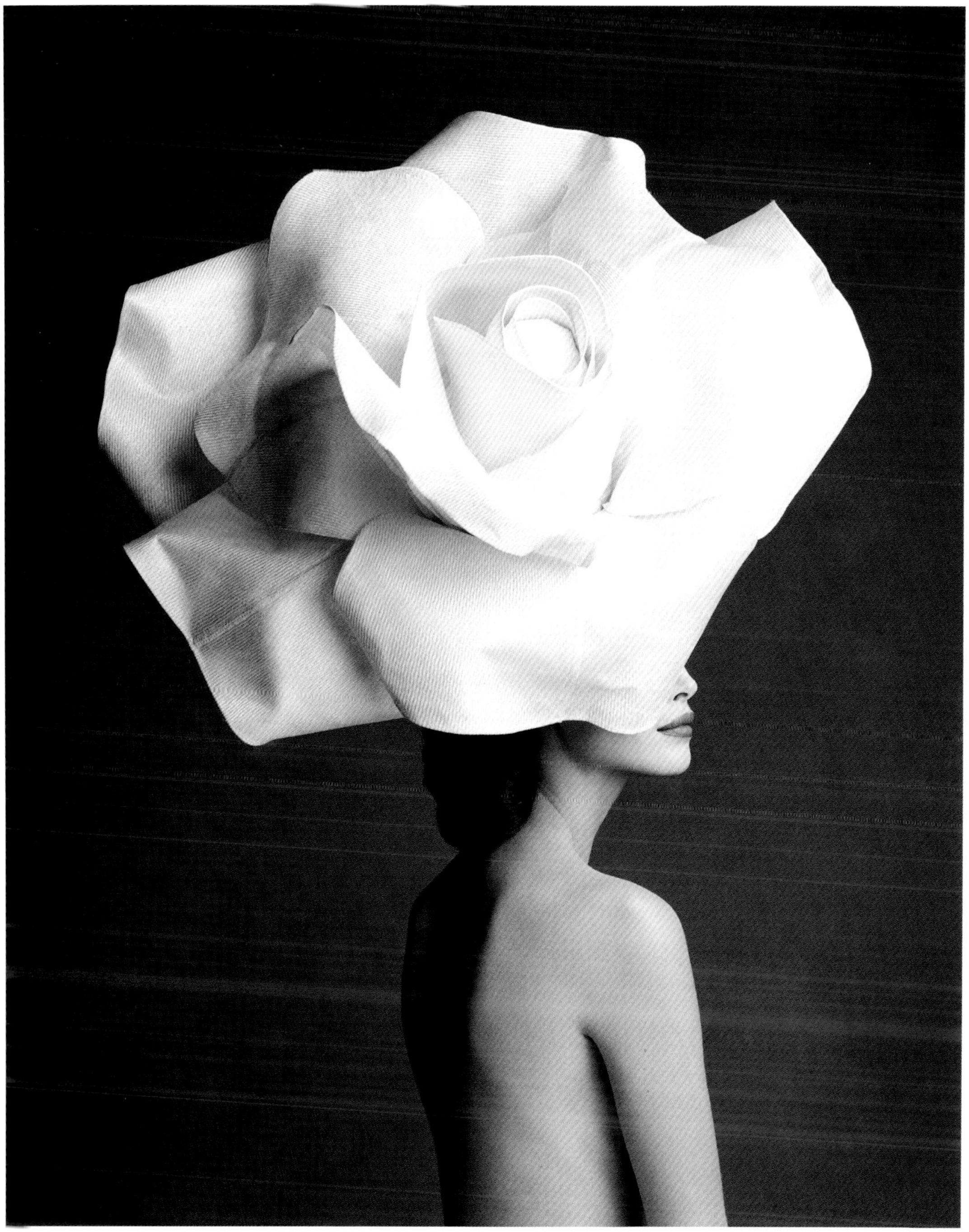

PATRICK DEMARCHELIER

Christy Turlington, New York, for British *Vogue*, 1990
Platinum palladium print, 32 × 26 in. / 81.3 × 66 cm

Within floriography, the white rose is associated with youth, chastity, innocence, and femininity. Here, French fashion photographer Patrick Demarchelier (1943–2022) portrays American supermodel Christy Turlington as a modern *femme-fleur*, wearing a huge white rose headpiece. The portrait plays with Surrealism's love of inflated scale and of entwining the human body with elements from the natural world. But, unlike Surrealism's desire to shock, Demarchelier's black-and-white portrait is quietly beautiful, exemplifying the elegant style of 1990s fashion photography. Turlington's oversize rose hat was created by British fashion designer Jasper Conran for the Broadway revival of *My Fair Lady* in 1992, and this image comes from an article in British *Vogue* detailing Conran's updated take on the costumes for the musical. Originally staged in 1956 with flamboyant fashions by Cecil Beaton (see p.74), *My Fair Lady* follows the upward trajectory and transformation of Eliza Doolittle, a Cockney flower seller. "When I got the job," Conran told *Vogue*, "all I could think about were flowers." In line with his minimal 1990s aesthetic, and partly inspired by the close-up floral portraits of Irving Penn (see p.99), Conran brought the musical's floral theme to life in the famous scene at the Ascot Racecourse. His dresses resembling green flower stems were all topped by humongous hats fashioned as flower blossoms, so the women on stage resembled tulips, pansies, lilies, and, of course, roses walking around in full bloom.

THE LANGUAGE OF THE ROSE

Shane Connolly

In early February, in almost all parts of the world, a long-forgotten language is miraculously resurrected when billions, perhaps trillions, of roses are given as symbols of love on Valentine's Day. Most of these roses are red, and most come with an increasingly unacceptable carbon footprint from vast industrial flower farms in Africa and South America. It is ironic that a rose representing love is one of only two words *vaguely* remembered from the eighteenth and nineteenth centuries' preferred medium of romantic communication: the language of flowers. (Thanks, perhaps, to Ophelia in Shakespeare's *Hamlet*, rosemary for remembrance is the other.) In its heyday, the language of flowers, or floriography, was a valid romantic vocabulary, with many lexicons of meanings published to guide the would-be romantic (see p.112). The authors of these floral dictionaries relentlessly researched the history of each flower, studying their medicinal uses, sacred links to ancient gods and goddesses, and symbolism in the arts, before linking each to a uniquely appropriate word or phrase to create a coded language like none before.

Roses have been connected to goddesses of love since time immemorial (see p.55), and early Christians happily transferred that imagery to the Virgin Mary (see p.130). The flowers' links with love, lust, and devotion were well established by the time this new language seeped into Victorian popular culture. So naturally, these floral dictionaries seized on roses, in all their glorious and increasingly rich variety, with wild enthusiasm. Roses quickly added more options to the rather limited vocabulary of love that existed then, with distinctions inspired by the many different colors and varieties available at the time. Most still recognize red roses as symbols of the ultimate declaration of passionate love, like the rush of blood to the enamored heart. But have we forgotten pink roses, which, like a warm flush on a healthy cheek, became the symbol of perfect happiness? We might prefer to forget that yellow roses—a color traditionally associated with bile and jealousy—were linked with infidelity and betrayal. This new language of love had its negatives as well as its positives.

Obvious physical characteristics gave the prickly sweetbriar rose (*Rosa rubiginosa*) the meaning "I wound to heal," as anyone who has tried to prune them will understand. Dog roses (*R. canina*; see pp.98 and 103), another prickly number, symbolized pleasure and pain and were often, perhaps unkindly, associated with marriage. On the other hand, Banksia roses—a relative newcomer introduced before 1824—were instantly connected to love at first sight, since they are practically thornless. Sometimes the scent of a rose helped to decide its fate at the hands of these floral linguists. The Musk rose (*R. moschata*), with its rather sensual scent, defined a capricious beauty. Even dead roses were taken on board the etymological ship, with wilting and withering white roses standing for a loss of innocence, and a bunch of dead roses clearly telling a would-be admirer that their love had been completely rejected. If, on the other hand, the object of affection immediately returned a fresh leaf torn from the roses they had received, the message was a very encouraging, "You may hope." The endless symbolic possibilities were embraced wholeheartedly by artists and authors of the time to bring extra meaning to their work, which they peppered with relevant flowers to entice the viewer into their hidden depths.

The language of flowers thrived until World War I, when life lost its innocence and the language of love became more urgent and direct. Yet somehow the red rose, the one flower that ironically could not be more expensive or less seasonal in February, survived the roll call of extinction. Although the historians and romantics who embroidered the symbolism of flowers into the romantic idiom might not recognize the rose that is actually on offer in February today, the red Valentine's rose is a remarkable example of the way the language of flowers lives on. For millennia, we have grown them, tended them, drawn them and, most of all, given them to those we love. We continue to turn to roses to communicate with each other in ways that go far beyond human language.

GLOSSARY OF ROSE TYPES

Michael Marriott

Plants in the genus *Rosa* first appeared approximately 30 million years ago, and today about 150 wild species are recognized, mostly native to Asia, although some appear in other regions north of the equator, from North America across to Europe and North Africa. Woody, perennial plants, all the species roses have five petals, but forms with extra petals were described as early as 371 BCE by Greek philosopher Theophrastus, and would have probably been chosen preferentially for easier harvesting, extra scent, and general attractiveness. Two antique roses that are still grown today, *Rosa gallica* 'Officinalis' and 'Quatre Saisons', may well have been two of those early varieties, the latter being unique among European roses in its ability to repeat-flower, or bloom several times in a growing season. Thus, humanity has a long history of rose cultivation, whereby new varieties and cultivars have been created by hybridization or sporting (small mutations), leading to the tens of thousands of rose cultivars that exist today. In the West, five main groups developed over the centuries—Gallicas, Damasks, Albas, Mosses, and Centifolias—all flowering once a year (with a few exceptions) and only in shades of white, pink, and purple. In China, meanwhile, philosopher Confucius told of roses grown in imperial gardens over 2,500 years ago that could flower repeatedly over a single season. It was the combination of European varieties with the so-called China roses in the early eighteenth century that heralded the great majority of roses that are grown in gardens today.

During the nineteenth century, Portlands, Bourbons, Chinas, Teas, and Hybrid Perpetuals were developed by rosarians through various methods of hybridization, the majority repeat-flowering and looking more or less like the Old Garden roses. Toward the century's end, breeders combined Teas and Hybrid Perpetuals to create the famous Hybrid Tea roses, and the development of the Floribundas came later, in the 1950s. These new varieties were very free flowering, which made them fantastically popular with amateur and professional gardeners alike, peaking roughly in the forty-year period after World War II, when nearly every garden in the West boasted at least several rose plants. In the 1980s, the popularity of roses dipped, exacerbated by the chemical sprays that were often required to keep the roses healthy. There was also a reaction against the formal shape of bloom and bush of the Hybrid Tea, which led to a renewed appreciation of the historic roses, and so the great popularity of varieties bred by English rosarian David Austin. Most recently, roses from another group, the Persica Hybrids, have gained traction owing to their novelty value, low maintenance, high level of disease resistance, and diversity of use in the garden. Thanks to these new trends in cultivation, roses have once again become a popular—yet still supremely beautiful—modern garden plant.

ALBAS

An ancient group of roses, some dating back to Roman times, when they were cultivated for their fragrant blooms and hips. With distinctive gray-green foliage and white or pale pink blooms, they are particularly tough, coping well in shade and surviving with little care.

BOURBONS

The result of a chance crossing in the early nineteenth century of two old but repeat-flowering varieties, *R.* 'Quatre Saisons' and the China rose 'Old Blush', which were grown together in hedges on the Island of Bourbon (now Réunion) in the Indian Ocean. Nearly all repeat-flower, have pink blooms (with some white, red, and striped ones), are fragrant, and look much like the historic roses.

CENTIFOLIAS

The youngest of the historic roses, with the parent species called *R.* × *centifolia* (literally "hundred-leaved" thanks to its proliferation of petals, and other names including the cabbage rose, *Rose de Mai*, or the Provence rose), dating back to Holland in the late sixteenth century. Nearly all are various shades of pink, and none repeat-flowers. The most commonly used type in rose-scented fragrance, oil, and perfume, as well as still lifes by the Dutch masters.

CHINAS

Developed from the four varieties—'Old Blush', 'Slater's Crimson', 'Hume's Blush Tea-scented China', and 'Parks's Yellow China'—that originally came to Europe from China in the late eighteenth and early nineteenth centuries. The genes for repeat-flowering that we see in our roses today, as well as those for yellow blooms, originated from these four varieties.

CLIMBERS

Distinct from ramblers in generally not being so vigorous, having larger flowers and stiffer growth, and repeat-flowering, although they are extremely variable in all characters. Perfect for training against walls, up pillars, or across arches.

DAMASKS

The second-oldest group (after the Gallicas), comprising some of the most garden-worthy of all roses, with beautiful, strongly fragrant blooms. Most are vigorous and have thorny stems. With one intriguing exception ('Quatre Saisons'), all are once-flowering with pink or white flowers. It is from the Bulgarian variety 'Kazanlik' that *attar* of roses is extracted.

ENGLISH (DAVID AUSTIN)

Bred by David Austin beginning in the 1960s to combine the charm and fragrance of Old Garden roses with the repeat-flowering and wide color range of modern roses. Notable for their diverse fragrances, and much admired both as shrubs in mixed borders and as climbers.

FLORIBUNDA

Bred to develop varieties that are particularly tough and flower freely from summer into fall. Blooms are often brightly colored but are not known for being particularly shapely or fragrant.

GALLICAS

The oldest of the Garden roses, some dating back to Roman times and so named for its native region of Gaul. All once-flowering, they are identifiable by their dark leaves, bristly stems, and generally dark pink or purple flowers. The oldest, *R. gallica* 'Officinalis', was used for many centuries throughout Europe for its culinary, cosmetic, and pharmaceutical properties, hence its other common name, the apothecary's rose.

GROUNDCOVER
Best regarded as low-growing, mounding shrubs, most of which repeat-flower. Blooms are mostly small and come in a range of colors. With a few notable exceptions, they have little or no fragrance.

HYBRID MUSKS
First bred by British rosarian Joseph Pemberton in the early twentieth century to develop varieties that were tough and easy to grow. Today, they include excellent varieties that are repeat-flowering, fragrant, and colorful, and blend perfectly into mixed borders.

HYBRID PERPETUALS
Developed in the late nineteenth century, these come in a great variety of shrub size and all shades except yellow. Despite their name, not all repeat-flower, and they often do not make the most attractive shrubs, because breeders concentrated on flower shape instead. Very popular with exhibitors trying to attain the perfectly shaped bloom.

HYBRID TEAS
The best-known of all the groups of roses, introduced in 1867 with the cultivar 'La France' and characterized by flowers with high pointed centers and often bright colors at the top of rather stiff, upright stems. Hugely popular for most of the twentieth century, when they were planted en masse in gardens and widely grown for exhibiting at rose shows.

MINIATURES
The smallest roses both in height and in size of flower. In the right conditions, they can flower continuously, although few are fragrant. Blooms on some are rosette-shaped, exactly like miniature versions of historic roses.

MOSSES
First appeared in the mid-seventeenth century and easily recognizable for the mosslike growth around the bud and sometimes down the stem with a delicious resinous smell. A mixed group with colors ranging from white and pink through striped, dark purple, and even soft yellow. The great majority do not repeat-flower.

NOISETTES
The first variety was 'Blush Noisette', grown in South Carolina by a French nurseryman, Philippe Noisette. From this developed a wide range of climbers with small to large, often strongly fragrant flowers of all colors, including yellow. Most are best suited to Mediterranean climates.

OLD GARDEN (ANTIQUE/HISTORIC)
A class of rose existing before the year 1867, when the first Hybrid Tea was introduced. Historic roses can be split into two groups: those that do not repeat-flower and those that do. Belonging to the former are Gallicas, Damasks, Albas, Centifolias, and Mosses, while the latter contains the Chinas, Portlands, Teas, Bourbons, and Hybrid Perpetuals. A very mixed group in terms of size of shrub, petal count, fragrance, and garden-worthiness.

PERSICA HYBRIDS
Originating from *R. persica*, a species native to desert regions of Iran and Afghanistan, and possessing the original Persica's distinctive coloring with a dark blotch at the base of each of the yellow petals. All have either single or semidouble flowers, and the modern varieties are disease resistant.

POLYANTHAS
A small group of roses characterized by small to medium-size flowers and bushy, compact growth, making them particularly suitable for borders and containers. The best-known variety is 'Cecile Brunner', which is available in two forms: a repeat-flowering shrub and a once-flowering climber.

PORTLANDS
A small but garden-worthy group of roses that are antique in character but can repeat-flower. Many are strongly fragrant as well as tough and reliable. The first, introduced in the 1780s, was named 'Duchess of Portland' after the enthusiastic plant-collector of the same name.

RAMBLERS
Distinct from climbers in being more vigorous, with small flowers and rather lax growth. Most do not repeat-flower, but they do come in the whole range of colors, and many have a musk fragrance that wafts on the air. Many are very well suited to growing up into trees.

RUGOSAS
An excellent group of tough, disease-resistant roses that have the advantage of repeat-flowering, often being strongly fragrant and setting a crop of large orange or red hips. They originate from *R. rugosa*, which is native to the coasts of northern Japan, North Korea, and northern China.

SHRUB
Any rose that does not have the ability to climb falls into this class. Hugely variable in size and habit of plant as well as style of flower. Size is dependent on pruning and climate, so the difference between shrubs and climbers is not fixed.

WILD (SPECIES)
The original roses that first appeared about 30 million years ago nearly everywhere north of the equator. They have simple, five-petaled flowers ranging in color from white through all shades of pink to red as well as yellow. Apart from their flowers, their hips are often quite striking.

INDEX

Page numbers in *italics* refer to illustrations

A
Aerial View of Elizabeth Park Rose Gardens, Hartford, Conn. (anonymous) 82, *82*
Aerolux Light Corporation 123
af Klint, Hilma, *Dog Rose Bush, Cow Vetch* 98, *98*
Alba rose 5
Albers, Josef 145
Alexandra Day poster (anonymous) 166, *166*
Alighieri, Dante 130
Allen, Beverly, *Rosa chinensis* 'Mutabilis' 231, *231*
Alma-Tadema, Lawrence, *The Roses of Heliogabalus* 16, *16*
Alméras, Henri 129
Altadena Rose Parade Float (anonymous) 187, *187*
American Rose Society 5
American Society of Landscape Architects 152
André, Édouard, plan for the Roseraie de l'Haÿ 196, *196*
Anne of Brittany 193
Ann Lowe Originals 154
Aphrodite 209
Araki, Nobuyoshi, *Flower* 186, *186*
Arpège 9, 61, *61*
Art Deco 71, 159, 161
Art Nouveau 6, 7, 34, 38, 59, 167
Artois, Count of 41
Arts and Crafts 32, 94, 126
Atkins, Anna 216
Aubriet, Claude 124, 223
Aubry, Charles, *Rose in a Crystal Glass* 182, *182*
Augustus, Emperor 208
Augustus the Strong 235
Austin, David 5, 14, 38, 73, 77, 105, *105*, 222
Ayrshire rose 62, *62*

B
Bābur 191
Baker, John Gilbert 62
Balbinus, Bohuslav 220
Baqir, Muhammad, vanity box 54, *54*
Barbie 229, *229*
Barker, Cicely Mary, *The Rose Fairy* 89, *89*
Barrie, J. M. 89
Basseporte, Madeleine 194
Double and Single Roses 124, *124*
Batty, Dora M., *To Regent's Park to See the New Rose Garden* 161, *161*
Bauhaus 136
beach rose 108
Bearden, Romare, *Morning of a Red Bird* 5, 205, *205*
Beaton, Cecil 199, 237
Still Life of Roses 74, *74*
Beijing Rose Museum 198, *198*
Bernard of Clairvaux, Saint 130
Berry, Duchess de 70
Berthier, Charlotte 111
Besler, Basilius, *Hortus Eystettensis* 160, *160*
Bessa, Pancrace 112
A Rosebush 70, *70*
Beyoncé 132
Birmingham Church Bombing (1963) 149
Black Lives Matter 19
Black Power Movement 205
Blake, William, *The Sick Rose* 78, 79, *79*
Blanchett, Cate 39
Blossfeldt, Karl 216
Blunt, Wilfrid 223
Boleyn, Anne 14
Bonnet, Marc, *Fist and Rose* 159, *159*
Botanical Art & Photography Show (2023) 139
Böttger, Johann Friedrich 25, 81, 235
Botticelli, Sandro, *The Virgin Adoring the Sleeping Christ Child* 91, *91*
Boucher, François, *Madame de Pompadour* 36, *36*
Bourbon roses 5
Bourdichon, Jean, *Folio 37r: Species Rosarum* 193, *193*
Bourdin, Guy, *Charles Jourdan* 17, *17*
Bourgeois, Louise, *Rose* 117, *117*
Bownas, Sheila, pattern design 228, *228*
Brahma 191
Brendel Company, *Rosa canina (Dog's Rose)* 185, *185*
British Rose Growers' Association 153
Brooklyn Botanic Garden 106
Brown, Lancelot "Capability" 64
Brown, Robert 236
Browne, Thom 5, 7, 158
Pebble Grain Rose Bag 227, *227*
Brueghel, Jan the Elder 26, 31
Bruno Court perfumery 143, *143*
Buckley, Jonathan 105, *105*, *110*
Bulgarian rose 61
Burnet rose 51, *51*
Burns, Robert 22
Burton, Sarah 7, 219
butterfly rose 231, *231*
Byredo, Rose of No Man's Land 9

C
cabbage rose 8, 10, 48, 88, 212
Cadbury, There's More to Enjoy in Cadbury's Roses 178, *178*
Caillebotte, Gustave, *Roses in the Garden at Petit Gennevilliers* 111, *111*
calling card (anonymous) 4, 121, *121*
Cant, Benjamin R. 100
Cant, Frank, & Co., Rose Catalogue 100, *100*
Caput Vada, Tunisia 162, *162*
Carrara Glassmakers, *Rosai Chandelier* 213, *213*
Carroll, Lewis 221
Cartier, 'Rose' clip brooch 199, *199*
Castellane, Victoire de 5
Bal de Mai necklace 189, *189*
Catholic Church 50, *173*
centifolia roses 5, 8, 9, 10, 24–6, *24–6*, 47, 48, 75, 88, 130, 143, *143*, 156, 177, *177*, 190, *190*
Chanel 189
Chanel No. 5 5, 9, 47, *47*, 129, 143
Chanel, Coco 129, 143
Chelsea Flower Show 107, *107*
Cherokee roses 90, 120, *120*, 127, *127*, 217, *217*
Childs, John Lewis, *Childs' Rare Flowers, Vegetables and Fruits* 22, *22*
China roses 5, 8, 84
Chopard, floral necklace 6, 39, *39*
Christ 5, 91, *91*, 130, 131, 173
Christianity 4–5, 7, 33, 87, 91
Church, Frederic Edwin 183

Civil Rights Movement 205
Clarke, Danny 105
Claude, Queen of France, *Book of Flowers Studies* 72, *72*
Clemens, Samuel 183
Cleopatra 8, 209
Clifford, Rosamund 80
Clusius, Carolus 180, 212
Cocteau, Jean 71
Concours International de Roses Nouvelles 41
Condy, Gillian, *Rosa abyssinica R.Br.* 236, *236*
Confucius 4
Conran, Jasper 237
Cooke, Jerry, *Winner of the 1956 Kentucky Derby* 5, 204, *204*
Cooper, Fred G., *The Subway Sun–Brooklyn Botanic Garden–900,000 Rose Display* 106, *106*
Corke, Henry Essenhigh 40
Correvon, Henri 231
Courtiers in a Rose Garden: A Lady and Two Gentlemen 118, *118*
Craig-Martin, Michael, *Rose* 232, *232*
Crane, Walter, *Wild Roses* 7, 109, *109*
Cranford Rose Garden, Brooklyn Botanic Garden 106
Cummings, Phoebe, *Rose Study* 195, *195*

D
Daché, Lilly 150
Dalí, Salvador 71
The Rose 87, *87*
Woman with a Head of Roses 150
Damask roses 5, 8, 54, 60, 76, *76*, 88, 128, *128*
da Siena, Minucchio, *Golden Rose* 173, *173*
David Austin Roses 5, 14, 38, 77, 105, *105*, 107
Dee, John 165
Delany, Mary, *Rosa cluster Damask* 128, *128*
Demarchelier, Patrick, *Christy Turlington, New York* 237, *237*
De Morgan, William, *Rose and Trellis* 94, *94*
Derby Porcelain Factory 235
rose basket 81, *81*
Dietzsch, Barbara Regina, *A Gallica Rose with a Bumblebee* 188, *188*
Dilich, Wilhelm 156
Dior 158, 189
Dior, Catherine 37
Dior, Christian 5, 6, 8, 189, 229
Tableau Final 6, 37, *37*
Dior Joaillerie 189
Dioscorides, Pedanius, *Rose* 4, 122, *122*
dish with roses (anonymous) 35, *35*
dog roses 5, 43, *43*, 48, 98, *98*, 103, 185, *185*
Dong Qichang 97
Dresser, Christopher, cross section through a rose 4, 184, *184*
Driscoll, Clara 59
Du, Xuebing, *Mother of Pearl* 171, *171*
DuBois, Blanche 133
Duchamp, Marcel 18
Dukovic, Pari, *Fragrant Harvest: How Many Roses Go Into a Bottle of Chanel No. 5?* 47, *47*
Dunn, Jourdan 144
Dürer, Albrecht 188
Durst, Eli, *Rose Queen of Texas* 211, *211*

E
Eastern Shoshone 7, 203
Ehret, Georg Dionysius 223
Common Provence Rose 212, *212*
Elizabeth, Queen Mother 153
Elizabeth, Saint, teaching John the Baptist to read in a rose arbor 174, *174*
Elizabeth I, Queen 5, 33, *33*, 138
Elizabeth II, Queen 150, 199
Elizabeth Park Rose Gardens, Hartford, Connecticut 82, *82*
Elizabeth of York 164
English Old Rose 222
Enlightenment 5, 124
Erb, David 204
Erdem 7
Erizku, Awol, *Switch – Roy Woods* 132, *132*
Eros 4, 186
Eugénie, Empress 104
Evans, Cerith Wyn, *Untitled (yellow rose)* 5, 18, *18*

F
Fabergé, *Rose Trellis Egg* 65, *65*
Fantin-Latour, Henri-Théodore 5, 183
A Basket of Roses 75, *75*
Farnham, Paulding, rose lapel watch 181, *181*
Farrand, Beatrix 152
Feria de Abril 23
field rose 62, *62*
Finlay, Ian Hamilton, *A Rose Is a Rose* 222, *222*
Finlay, Sue 222
Fisher, Samuel Melton, *The Clerkenwell Flower Makers* 176, *176*
FitzGerald, Edward 34
Fjura, *Rosa #12* 11, *11*
Flavin, Dan, *Roses* 123, *123*
Flora 109, 162
floral wreath (anonymous) 209, *209*
Floribunda roses 5, 82, 106, 153, 177, *177*
floriography 4, 7, 112, 121, 132, 237
Florissant Fossil Beds National Monument, Colorado 4
Florists' Transworld Delivery, Mother's Day (anonymous) 202, *202*
Flower Girls' Christian Mission 176
Forestier, Jean-Claude Nicolas 41
Fragonard 138
Frank Cant & Co., Rose Catalogue 100, *100*
Fraysse, André 61
French Socialist Party 159
Friend, Kate, *John Pawson, Garden Rose, Cotswolds* 216, *216*
Fuchs, Leonhart, *Rosa* 200, *200*

G
Gallé, Emile 167
Galliano, John 158
Gallica roses 5, 48, 88, 156, 174, 188, *188*
Garden Museum, London, *Wild & Cultivated: Fashioning the Rose* 195
Garland of Roses 204
Gemmingen, Johann Konrad von 160
General Electric Additive and Protolabs 6, 144
Genzken, Isa, *Rose III* 58, *58*
Gerard, John, *The Great Holland Rose* 5, 180, *180*
German Rose Association 157
German Rose Society 57
Ghilberto Borromeo, Prince 231
Gilpin, William 201
Giverny, France 169, *169*
Glasgow Style 38
Glass, T. M., *White Hawthorne and White Shrub Rose in a Blue and White Chinese Vase* 63, *63*
Gobelins 182
God 44, 173
Godard, Antoine 157
Gooch, Simone, *Rosa #12* 11, *11*
Gothic 227
Goya, Francisco 219
Grahn, August, *Vase with flowers* 197, *197*
Grandiflora roses 5, 106
Grandville, J. J., *Rose* 7, 151, *151*
Grasse, France 5, 8, 9, 37, 47, 61, 138, 143
Grateful Dead 5, 34, *34*
Gravereaux, Jules 196
Gray, Alexander Hill 157
Griffith, Fleda, *Rose Garden, New York Botanical Garden* 152, *152*
Groom, John 176
Grow to Know 105
Gupta, Sourabh, *Always Romantic* 43, *43*

H
Hagemann 103
Hague Rose Trials 153
Hallmark, Anna Grace 211
Hammer Museum, Los Angeles 197
Handel, George Frideric 219
Handler, Ruth 229
Harlem Renaissance 205
Harvey, Cig, *Rose Petals* 145, *145*
Hasan 172
Hawara 4
Heade, Martin Johnson, *Cherokee Roses* 120, *120*
Hélène, Duchesse 13
Heliogabalus, Emperor 16
Henderson, Derek, *Rosa #12* 11, *11*
Henrietta Maria 76
Henry VII, King 32, 94, 164
Heritage Rose Park, Hartford, Connecticut 82
Heritage roses 5
Hétreau, Rémy, *Arpège, Scandal, Prétexte, Rumeur, My Sin* 61, *61*
Hewitt, Mattie Edwards, *Armsea Hall Rose Trellis* 85, *85*
Hill, EJ, *Vase with flowers* 197, *197*
Hilliard, Nicholas, *Portrait of Queen Elizabeth I* 33, *33*
Hinduism 191
Historische Beschreibung der fürstlichen Kindtaufe Fräulein Elisabethen zu Hessen, Folio 16v (anonymous) 156, *156*
Hitchcock, Alfred, *Vertigo* 5, 101, *101*Hoefnagel, Joris 26
Hoffman, Zelia 85
Hoffmann, Hans 188
Hoffmann, Josef 218
Hokusai, Katsushika, *Warbler and Roses* 230, *230*
Holland roses 8, 212
House of Schiaparelli 7, 71
Hudson, William 62

Huysum, Jan van 31
  Flower Still Life 140, *140*
Hybrid Musks 157
Hybrid Perpetual roses 5, 57, 82
Hybrid Persicas 5
Hybrid Tea roses 5, 6, 45, *45*, 57, 82, 106, 113, 135, 144, 177, *177*

I
ikebana 11, 119
Impressionism 75, 111, 169, 171
Iranian art 54, 60
Iribe, Paul, *La Rose de Rosine* 4, 9, 129, *129*
Islam 4, 44, 54, 60, 172

J
Jackson & Perkins 229
Jahan, Shah, smelling a pink rose with Mumtaz Mahal 191, *191*
Jakuchu, Itō, *Roses and Small Bird* 90, *90*
James, Charles 27, *27*
  La Sylphide dress 6, 49, *49*
Japanese art 38, 90, 108, 119
Jarman, Derek 145
Jeffrey & Co. 126
Jekyll, Gertrude 222
  Roses for English Gardens 73, *73*
Jiang Xiaofei 198
John, Mr. 150
John the Baptist 174, *174*
Johnston, Frances Benjamin, *Armsea Hall Rose Trellis* 85, *85*
Jones, Stephen, *Rose Royce* 7, 158, *158*
Joséphine, Empress 5, 13, 70
Jourdan, Charles 17
Juliana Anicia Codex 122
Jullien, Jean, *Untitled* 23, *23*
Jung Koch Quentell, *Rosa dumetorum / Heckenrose* 103, *103*

K
Kaman, Helen 82
Kangxi, Emperor 115
Kapp, Helen, *Rose Recipes* 142, *142*
Katz, Alex, *Roses on Blue 20–1*, 21
Kawahara Keiga, *Rosa* 84, *84*
Kazumasa, Ogawa 216Kelley, Alton, *Skull and Roses, Grateful Dead and Oxford Circle at Avalon Ballroom* 34, *34*
Kelly, Ellsworth, *Rose, Palm Springs* 15, *15*
Kennedy, Jacqueline Bouvier 154
Kentucky Derby 5, 204
Kessel, Jan van the Elder, *A Swag of Roses Suspended from Blue Ribbons* 26, *26*
Kiftsgate Court Gardens, Gloucestershire 42
Klimt, Gustav, *Orchard with Roses* 137, *137*
Knight, Nick, *Saturday 22nd October, 2016* 31, *31*
Koran 4, 44, *44*, 172, *172*
Kristiansen, Ole Kirk 225
Krøyer, Marie 69
Krøyer, Peder Severin, *Roses* 69, *69*
Kubrick, Stanley 219
Kunkel, Dennis, *Rose Petal Upper Surface* 116, *116*
Kuramata, Shiro, *Miss Blanche* 133, *133*
Kusama, Yayoi, *Rose Garden* 119, *119*

L
La Barre, Antoine de 72
Lagerfeld, Karl 189
Lakshmi 7
Lalique, René, opal and glass rose pendant 6, 167, *167*
Lambert, Peter 57, 157
Lambertianas 157
Lancaster, House of 5, 32, 33, 72, 164
Lanvin, Jeanne 9, 61
Lanvin Parfums 61
La Tour, Charlotte de, *Le Langage des Fleurs* 112, *112*
Lawrence, Des, *David Austin* 77, *77*
Lego 225, *225*
Leigh, Simone, *Birmingham* 149, *149*
Le Moyne de Morgues, Jacques, *Pink Rose* 170, *170*
Leopold I 220
Lesage 71
Leuteritz, Ernst August 235
Lewis Miller Design 233
Liberty of London 135, 228
Lincoln, Abraham 144
Lindley, John, *Yellow American Rose (Rosa lutescens)* 210, *210*
Linnaeus, Carl 4, 62, 212, 234
literati painting (*wenrenhua*) 97
Livia Drusilla 208
Li Yu 68
London Underground 161
Lorris, Guillaume de, *Folio 3r, from Roman de la Rose* 206, *206*
Louis XV, King 5, 6, 8, 25, 36, 124
Lowe, Ann, dress 6, 154, *154*
Lubin, Pierre-François 8

M
McAdams-Swallow, Wilma Jean, *Woman's Moccasins, Eastern Shoshone* 203, *203*
Macartney rose 48
McClintock, Russ, *Sexy Rexy* 153, *153*
McEwen, Rory, *Rose* 'William Lobb' 223, *223*
McGredy, Sam IV 153
Mackintosh, Charles Rennie 218
  rose and teardrop textile design 7, 38, *38*
Mackintosh, Margaret Macdonald 38, 218
McQueen, Alexander 5, 31
  dress from the *Sarabande* collection 6, 219, *219*
Mahal, Mumtaz, smelling a pink rose with Shah Jahan 191, *191*
Maison Francis Kurkdijan, À la rose 9
Manet, Édouard 75
Mann, Sally 102, 145
Mapplethorpe, Robert, *Flowers (Blue Rose)* 5, 146, *146*
Margaret, Princess 199
Maria Amelia, Duchesse d'Orléans 13
Maria Feodorovna 65
Marie Antoinette, Queen 5, 6, 8, 41, 194
Marshall, William, *Rosa damascena* 76, *76*
Mary, Queen 161, 175
Mattel, *Rose Barbie* 7, 229, *229*
Matthews Nurseries Ltd. 11
Maurice, Landgrave of Hesse-Kassel 156
Mawley, Edward, *Roses for English Gardens* 73, *73*
Maywald, Willy 37, *37*
Mehmet II, Sultan 114
Meissen Porcelain Factory, potpourri vase 235, *235*
memorial rose 108
Merian, Maria Sibylla 177
  Rose with Metamorphosis of Leaf Roller and a Glued Beetle Larva 24, *24*
Met Gala 6, 144
Meun, Jean de, *Roman de la Rose* 206, *206*
Miller, John Sebastian 234
Miller, Lewis, *Flower Flash: W 22nd Street/7th Avenue, April 11, 2019* 233, *233*
Millet, Jean-François 87
Modica, Andrea 145
Modotti, Tina, *Roses, Mexico* 28, *28*
Monet, Claude 5, 69, 75, 111
  Les Roses 169, *169*
Morandi, Giovanni Battista, rose varieties 148, *148*
Morris & Co. 32, 94
Morris, William 94, 109, 126
  Tudor Rose Pattern Printed Fabric 32, *32*
mosaic *emblema* with cupids gathering roses in a garden (anonymous) 46, *46*
Moser, Koloman 218
Moss roses 5, 48, 223, *223*
Mottisfont Abbey, Hampshire 151
Mouse, Stanley, *Skull and Roses, Grateful Dead and Oxford Circle at Avalon Ballroom* 34, *34*
Mucha, Alphonse 34
Muhammad, Prophet 4, 44, 54, 60, 76, 172
Mulhouse 182
multiflora roses 108
Murayama, Macoto, *Rose-iii-bc* 67, *67*
Museum of Modern Art, New York 58
Museum of National History, Paris 70
Musk roses 5, 8, 174
Mustafa, Zehhebehu, folio from a manuscript of the Koran 172, *172*
Mustard Seed Garden (*Jieziyuan*) 68

N
Napoleon Bonaparte 5
Nash, Arthur 59
National Collection of Old Roses 51
National Garden Scheme 105
National Horticultural Society of France 57
National Rose Society 100
National Trust 110
Natural History Museum, London 31, 228
Neel, Alice, *Roses* 52, *52*
New Museum, New York 58
New Romantic 158
New York Botanical Garden 5, 93, 152, *152*
NEXT Architects, Beijing Rose Museum 5, 198, *198*
Nicholas II, Tsar 65
Nicholas Tolentino, St. 50, *50*
Nichols, Clive, *Frosty Rose* 'Warm Welcome' 53, *53*
Nickels, Clarence, *Cultur, Benennung und Beschreibung der Rosen* 136, *136*
Nicolson, Harold 110
Niemann, Christoph, *Sunday Sketch (Spring)* 225, *225*
Nishimura Hodo, *Wild Vine Rose* 108, *108*
Noisette roses 5, 110, 157, *157*
North, Marianne, *Cherokee Rose with the Peak of Tenerife in the Distance* 217, *217*
Notre-Dame de Paris, north rose window 5, 131, *131*

Novak, Kim 101, *101*
nueva trova 92
Nugroho, Eunike, *Rose Petals* 139, *139*

O
Obama, Michelle 227
O'Brien, Esme 49
Ofuna Botanical Garden, Japan 42
O'Keeffe, Georgia, *Abstraction White Rose* 5, 147, *147*
Old Garden roses 5, 14, 32, 51, 77, 80, 88, 110, 113, 188
Ottoman Empire 44

P
Packer, Jennifer, *Yellow Roses* 19, *19*
Parc de Bagatelle, Paris 5, 41, *41*
Parfumerie Bruno Court, Grasse, Triage des Roses 143, *143*
Parfums de Rosine 4, 9, 129, *129*
Paris Exposition Universelle 167, 181
Parkinson, John, *Rosa damascena* 5, 76, *76*
Parr, Martin, *RHS Chelsea Flower Show, London, England* 107, *107*
Parsons, Alfred, *Rosa arvensis Huds.* 62, *62*
Pasadena Tournament of Roses Parade 5, 187, *187*
Passe, Crispijn de the younger, *Rosa Centifolia Batavica and Rosa Versicolor* 190, *190*
Patou, Jean 129
Paul, George 100
Paul, William 100
Pawson, John 216
Payne, Henry Albert, *Plucking the Red and White Roses in the Old Temple Gardens* 164, *164*
Pedder-Smith, Rachel, *Rose Hips 134–5*, 135
Peggy Rockefeller Rose Garden 5, 152
Pégomas 47
Penn, Irving 237
Rose: Blue Moon 99, *99*
perfume 8–9, 37, 47, 61, 129, 143, 163
Pernet-Ducher, Joseph 57
Petrauskiene, Emilija, *Roses Herbarium* 102, *102*
Philosopher's Stone 165
Planché, Andrew 81
Plant Patent Act (1930) 5
Pliny 4, 103, 156
Poiret, Paul, *La Rose de Rosine* 4, 9, 129, *129*
Polyantha 5
Pompadour, Madame de (Jean-Antoinette Poisson) 5, 6, 8, 25, 36, *36*, 124
Pop art 7, 92
Porter, Charles Ethan, *Roses* 183, *183*
Portillano, Ralph, *Rosa × odorata* 'Viridiflora' (Andrews) Sweet 93, *93*
Portlands 5
Portraits of Celebrities of the Persian Court 60, *60*
Posen, Zac, rose dress 6, 144, *144*
Post-Impressionism 29
Potin, Julien 57
Poulsen, Svend 57
Pre-Raphaelite Brotherhood 12, 32, 89
Proctor, Thea, *The Rose* 125, *125*
Provence rose 8, 180

Q
Quaker religion 93
Queen Mary's Gardens, Regent's Park 153, 161, *161*
Quétier, Pierre-Victor 13

R
Ray, John 62
Ray, Man 17
Redouté, Pierre-Joseph 70, 223
Rosa Gallica Aurelianensis 5, 13, *13*
Regent's Park, Queen Mary's Gardens 153, 161, *161*
Renaissance 5, 91, 148, 171, 188, 193
Renoir, Pierre-Auguste 5, 75
Roses and Landscape 155, *155*
Repton, Humphry, *The Rosary at Ashridge* 64, *64*
reticule (anonymous) 192, *192*
Richard of York 164, *164*
Riches, Lizzie, *The Rose Dress* 138, *138*
Rimsky-Korsakov, Nikolai 146
Ripley Scroll 165, *165*
Robert, Nicolas, *Rosa × Centifolia Variegata and Rosa × Centifolia* 88, *88*
Roberts, Hera 125
Rocha, Simone 7
Rockefeller, David 152
Rockefeller, Peggy 152
Rococo style 36, 195, 235
Rogers, Millicent 49
Rohde, Eleanour Sinclair, *Rose Recipes* 142, *142*
Roosevelt, Eleanor 150
Rosa 4, 5, 6
R. abyssinica 236, *236*
R. alba 72, *72*, 148, *148*, 174
R. 'Alba Maxima' 69, *69*, 72
R. 'Alba Semi-plena' 72
R. 'Alister Stella Gray' 157
R. 'American Beauty' 154
R. 'Angèle Pernet' 57, *57*
R. arvensis 62, *62*
R. balsamica 103
R. banksiae var. *lutescens* 210
R. 'Bosatsu' 90
R. bracteate 48
R. canina 5, 43, *43*, 48, 98, *98*, 103, 185, *185*
R. × *centifolia* 8, 9, 10, 24–6, *24–6*, 88, *88*, 122, 143, *143*, 180, *180*, 212, *212*
R. × *c. batavica* 180
R. × *c. variegata* 88, *88*
R. 'Charles Rennie Mackintosh' 38
R. chinensis 5, 35, *35*
R. c. 'Mutabilis' 231, *231*
R. c. 'Viridiflora' 5
R. 'Constance Spry' 5, 77
R. × *damascena* 9, 148, *148*
R. 'Dannahue' 105, *105*
R. 'Dorothy Perkins' 85
R. 'Duchesse d'Orléans' 13
R. 'Elizabeth of Glamis' 153
R. filipes 'Kiftsgate' 42, *42*, 51, *51*
R. foetida 122
R. f. bicolor 48
R. 'Frau Karl Druschki' *56*, 57
R. gallica 5, 8, 122
R. g. aurelianensis 13, *13*
R. g. 'Officinalis' 32, 80, *80*, 180
R. g. 'Versicolor' 32, 80
R. 'Général Jacqueminot' 100
R. 'Gertrude Jekyll' 73, 105
R. gigantea 35
R. 'Gloire de Dijon' 100
R. 'Grevillei' 90
R. 'Harison's Yellow' 210
R. hispida 210
R. 'Jude the Obscure' 105
R. 'Julien Potin' *56*, 57
R. 'Kirsten Poulsen' *56*, 57
R. laevigata 90, *90*, 120, *120*, 127, 217, *217*
R. 'La France' 5
R. 'Lena Schmidt-Michel' 157
R. lucieae 108
R. lutea multiplex 148, *148*
R. l. simplex 148, *148*
R. l. var. *hogii* 210
R. lutescens 210, *210*
R. 'Lyda' 102, *102*
R. 'Madame Alfred Carrière' 110
R. 'Madame Ferdinand Jamin' 154
R. 'Mademoiselle Jeanne Philippe' 157
R. 'Maid Marion' 14
R. moschata 5, 8, 174
R. moyesii 51, *51*
R. 'Mr. Lincoln' 144
R. mulliganii 110
R. m. bower 110, *110*
R. multiflora 90, *90*, 108
R. m. 'Platyphylla' 90
R. mundi 80
R. 'Nymphenburg' 51, *51*
R. 'Old Blush' 84, 93
R. 'Ormiston Roy' 51, *51*
R. palustris 5
R. 'Parson's Pink' 84
R. 'Penelope' 51, *51*
R. persica 53
R. 'Picasso' 153
R. 'Pretty Polly' 153
R. punicea 48
R. 'Qi Zi Mei' 90
R. 'Queen of Sweden' 102, *102*
R. 'Roald Dahl' 14
R. rugosa 5, 51, *51*, 108
R. 'Schneewittchen' 157
R. 'Sexy Rexy' 153, *153*
R. spinosissima 51, *51*, 210
R. sylvestris 62
R. 'Teasing Georgia' 153
R. 'Tuscany Superb' 48
R. versicolor 190, *190*
R. 'Village Maid' 153
R. 'Warm Welcome' 53, *53*
R. 'William Lobb' 223, *223*
R. 'William Shakespeare' 102, *102*
R. woodsii 203
Rose, Barbara 123
Roseberry, Daniel 7, 71
Rose Brothers 178
Rosebud Perfume Company, Smith's Rosebud Salve 163, *163*
Rose de Mai 8, 9, 143, *143*
Rose Festival, Walkden 175
Rose of Lima, Saint 50, *50*, 95
Rosen-Zeitung 157
rose parades 5, 187, *187*, 211
Rose Queens 6, 175, *175*, 211
Roseraie de l'Haÿ, Paris 5, 196, *196*

Roses and a Carnation (anonymous) 44, *44*
Rose of Sharon 207, *207*
Roses and Wisteria (anonymous) *96–7*, 97
Rossetti, Dante Gabriel, *Venus Verticordia* 55, *55*
Rostgaard, Alfredo, *Cancion Protesta* 92, *92*
Royal Botanic Gardens, Kew 45, 48, 135, 161, 184, 217
Royal Horticultural Society (RHS) 42, 62, 107, 139
Ruskin, John 109
Study of a Wild Rose 168, *168*
Ruysch, Rachel 177
Roses, Convolvulus, Poppies, and Other Flowers in an Urn on a Stone Ledge 10, *10*
Ryman, Will, *The Roses on Park Avenue* 179, *179*

S
Sackville-West, Vita, *Rosa mulliganii bower* 5, 110, *110*
Sadi, Natasja, arrangement 177, *177*
Saint-Exupéry, Antoine de, *The Little Prince in the Rose Garden* 83, *83*
Sanders, Rosie, *A Rose in Winter* 45, *45*
Sappho 4
Scheufele, Caroline 39
Schiaparelli, Elsa, evening coat 7, 71, *71*
Schmidt-Michel, Lena, *Mlle Jeanne Phillipe und Alister Stella Graii* 157, *157*
Schynvoet, Simon 80
Scotch roses 51, *51*, 62, 210
Scott Smith, Troy 110
Seven Sisters 90, *90*
Sèvres Porcelain Factory 25, *25*
Shakespeare, William 4, 164
Sheffield Botanical Gardens 139
Shirley Sherwood Gallery of Botanical Art 45
Shrigley, David, *Untitled* 78, *78*
Siebold, Philipp Franz von 84
Sissinghurst Castle Garden, Kent 5, 110
Smith, Christiann Drick, Rose of Sharon 207, *207*
Smith, Shinique, *Red Rose* 5, 226, *226*
Socialist International 159
Somerset, Duke of 164, *164*
Spaendonck, Gerard van 70
Speckle, Veyt Rudolff 200Spry, Constance 74, 201
Alabaster Chalice with Red Roses 113, *113*
Stein, Gertrude 7, 18, 222
Steiner, Rudolf 98
Stephenson, Luke, *The English Rose* 14, *14*
Stieglitz, Alfred 147
Sufis 44
Süleyman the Magnificent 114
Sullivan, Edmund Joseph 34
Surrealism 7, 61, 71, 78, 87, 237
Swallow, Vivian, *Woman's Moccasins, Eastern Shoshone* 203, *203*
Swallow-Hunter, Sandra, *Woman's Moccasins, Eastern Shoshone* 203, *203*
Swamp rose 55Swim, Herbert 144

T
Taglioni, Marie 49
Takashimaya 18
tapestry with roses (anonymous) 95, *95*
Tasker, Dain L., *A Rose* 66, *66*
Taylor, John 64
teapot (anonymous) 30, *30*
tea roses 35, 157
Tenniel, John, *The Playing Cards* 221, *221*
Tennyson, Alfred Lord 12
Texas Rose Festival 211Theophrastus 156
Thomas, Alma, *Red Roses Sonata* 224, *224*
Thomas, Graham Stuart, *Rose hips of 'Nymphenburg', R. moyesii, 'Penelope', R. filipes 'Kiftsgate', R. rugosa, and 'Ormiston Roy'* 51, *51*
Thomas, H. H., *The Rose Book* 40, *40*
Thomas, Owen 40
Thompson, Emily, *Untitled, Rose Hip Floribunda and Rosa 'Knock Out'* 201, *201*
Thornton, Robert John, *A Group of Roses* 234, *234*
Thory, Claude-Antoine 13
Tiffany & Co. 181, 182
Tiffany, Louis Comfort 59, 181
Tiffany Studios, rosebush table lamp 59, *59*
tile with a garden of roses, tulips, rosettes, and carnations (anonymous) 114, *114*
Tillmans, Wolfgang, *Rose* 86, *86*
Treacy, Philip 219
Trew, Christoph Jacob 212
Triumph of Neptune, figure of Spring (detail) 162, *162*
Tudor Rose 5, 32, *32*, 33, *33*, 94, 164
Turlington, Christy 237, *237*
Turner, Tiffanie, *Croquembouche* 141, *141*
Twain, Mark 183
Twining, Elizabeth, *Various Species of Rosa* 48, *48*
Twombly, Cy, *Untitled (Roses) 214–15*, 215
Tyler Rose Garden, Texas 211

U
Underground Railroad 93

V
Valentine, John 202
Valentine's Day 4, 5
Valentino 6
Vallayer-Coster, Anne, *Bouquet of Roses* 194, *194*
Van Gogh, Vincent, *Roses* 29, *29*
vanitas 10, 19
Velázquez, Diego 129
Venus 4, 7, 55, *55*, 91, 209
Venus Marina 190
Verein Deutscher Rosenfreunde 157
Vetter, Christian, *The Rose of Bohemia* 220, *220*
Victor, Sally, hat 7, 150, *150*
Victoria and Albert Museum, London 170
village (cottage) maid 88, *88*
Villa of Livia, Garden Room fresco (anonymous) 4, 208, *208*
The Virgin and Child in a White Rose 130, *130*
Virgin Mary 5, 7, 33, 50, *50*, 91, *91*, 118, 130, 131
Virgin of Pomata with St. Nicholas Tolentino and St. Rose of Lima 50, *50*
Vishnu 7, 191
Vogue 17, 61, 74, 99, 150, 237, *237*
Voysey, C. F. A., *The Standard Rose* 126, *126*
Vries, Bouke de, *Rose in Kang Xi Vase* 115, *115*

W
Walkden Rose Queen, June 17th, 1922 175, *175*
Walker, Grace 204
Wang Gai, *Mustard Seed Garden Manual of Painting* 68, *68*
Wang Hui 97
Warner, Chris 53
Wars of the Roses (1455–87) 5, 32, 33, 72, 164, 175
Watanabe, Noriko, *Rosa filipes 'Kiftsgate'* 42, *42*
Water, John van de 198
Waterhouse, John William, *The Soul of the Rose* 12, *12*
Weber, Bruce, *Charles James Dress and Roses, Kent* 27, *27*
Weeks Rose Growers 144
Weston, Edward 28
Westwood, Vivienne 6, 135, 158
Whitney Museum of American Art, New York 224, 233
Wiener Werkstätte, textile sample 7, 218, *218*
Wigström, Henrik 65
Williams, Tennessee 133, 187
Willmott, Ellen 62
Wills, W. D. & H. O., Roses cigarette cards *56–7*, 57
Wirth, Theodore 82
Withoos, Alida, *Rosa Mundi* 80, *80*
Withoos, Matthias 80
Withoos, Pieter 80
World Rose Convention (2016) 198
Worth, Charles Frederick, evening dress 6, 104, *104*
Wyatville, Jeffry 64

X
Xu Chongsi 97

Y
Yellow Rose of Texas 210
Yokohama Nursery Co., *Rosa laevigata (Cherokee Rose)* 127, *127*
York, House of 5, 32, 33, 72, 164
Yun Shouping 97

ROSE GARDENS AND FESTIVALS

UNITED KINGDOM

Borde Hill, West Sussex
Bowes-Lyon Rose Garden at Royal Horticultural Society Garden Wisley, Surrey
Castle Howard, North Yorkshire
Cliveden, Buckinghamshire
Coughton Court, Warwickshire
David Austin Rose Gardens, Shropshire
Hampton Court Palace, Greater London
Helmingham Hall Gardens, Suffolk
Kiftsgate Court Gardens, Gloucestershire
Lowther Castle, Westmorland and Furness
Mottisfont Abbey, Hampshire
Peter Beales Rose Garden, Norfolk
Queen Mary's Gardens, Regent's Park, London
Royal Botanic Gardens, Kew, London
Royal Horticultural Society Garden Rosemoor, Devon
Savill Garden, Surrey
Sissinghurst Castle Garden, Kent
Wynyard Hall, County Durham

Festivals

Rose Festival Weekend, Peter Beales Roses, Norfolk
The Rose Society Summer Rose Festival, Squire's Garden Centre, Shepperton, Middlesex
Royal Horticultural Society Garden Rosemoor Festival of Roses, Devon

FRANCE

Château du Rivau, Lémeré
Les Jardins de Roquelin, Meung-sur-Loire
La Roseraie Internationale au Parc de la Tête d'Or, Lyon
La Roseraie Les Chemins de la Rose, Doué-la-Fontaine
Roseraie Loubert, Les Rosiers-sur-Loire
La Roseraie at Parc de Bagatelle, Paris
La Roseraie de Provins, Provins
Roseraie du Val-de-Marne, L'Haÿ-les-Roses

Festivals

ExpoRose, Grasse
Festival des Roses, Chédigny
Journées de la Rose at the Domaine de Chaalis, Fontaine-Chaalis

REST OF EUROPE

Castle Hex, Heers, Belgium
Europa-Rosarium, Sangerhausen, Germany
Fredriksdals Rosarium, Helsingborg, Sweden
Princess Grace Rose Garden, Monac
Quando Fioriranno le Rose, Assisi, Italy
La Rosaleda, El Retiro Park, Madrid, Spain
Rose Garden Baden-Baden, Germany
Rose Garden, Mondorf les Bains, Luxembourg
RosenPark Dräger, Bad Nauheim, Germany
Rozentuin Coloma, Sint-Pieters-Leeuw, Belgium
Roseto Botanico Carla Fineschi, Cavriglia, Italy
Roseto Comunale, Rome, Italy
Valby Park, Copenhagen, Denmark
Volksgarten, Vienna, Austria

Festivals

Festa delle Rose, Busalla, Italy
Festival de la Rose, Le Roeulx, Belgium
Festival des Roses, Estavayer-le-lac, Switzerland
Kazanlak Rose Festival, Kazanlak, Bulgaria
Rose Festival at the Palm Garden, Frankfurt, Germany
Rozenfeest, Lottum, the Netherlands
Sant Jordi Festival, Barcelona, Spain
Steinfurth Rose Festival, Bad Nauheim, Germany

UNITED STATES

American Rose Society Rose Garden, Shreveport, Louisiana
Antique Rose Emporium, Brenham, Texas
Columbus Park of Roses, Columbus, Ohio
Cranford Rose Garden at the Brooklyn Botanic Garden, New York
Descanso's Rose Garden, La Cañada Flintridge, California
Dr. E. M. Mills Rose Garden, Syracuse, New York
Helen S. Kaman Rose Garden at Elizabeth Park, Hartford, Connecticut
Huntington Library Rose Garden, San Marino, California
International Rose Test Garden, Portland, Oregon
Mable Ringling Rose Garden at The Ringling, Sarasota, Florida
Morcom Rose Garden, Oakland, California
Peggy Rockefeller Rose Garden at the New York Botanical Garden, New York
Tyler Rose Garden Center, Tyler, Texas
San Jose Heritage Rose Garden, San Jose, California

Festivals

Portland Rose Festival, Oregon
Rose Show and Festival, Thomasville, Georgia
Texas Rose Festival, Tyler, Texas
Tournament of Roses, Pasadena, California

REST OF WORLD

Beijing Botanical Garden Rose Garden, Beijing, China
Christchurch Botanic Rose Gardens, Christchurch, New Zealand
David Austin English Rose Garden, Sennan City, Osaka, Japan
Dugald MacKenzie Rose Garden, Palmerston North, New Zealand
Government Rose Garden, Ooty, India
International Rose Garden at Adelaide Botanic Garden, Adelaide, Australia
Keisei Rose Garden, Chiba-ken, Japan
Morwell Centenary Rose Garden, Morwell, Australia
Nanyang Rose Garden, Nanyang, China
National Indo-Africa Friendship Rose Garden, New Delhi, India
National Rose Gardens at Old Parliament House, Canberra, Australia
Nishio City Rose Garden, Nishio City, Japan
Rogers Rose Garden, Hamilton, New Zealand
Rose Garden at Butchart Gardens, Brentwood Bay, Canada
Rose Garden at the Montreal Botanical Garden, Canada
Rosedal de Palermo, Buenos Aires, Argentina
Royal Botanical Gardens Ontario, Burlington, Canada
The City of Sakura Rose Garden, Sakura City, Japan
Shenzhen Renmin Park, Shenzhen, China
Victoria State Rose Garden, Werribee Park, Australia
Zakir Hussain Rose Garden, Chandigarh, India

Festivals

International Rose Garden Festival, Morwell, Australia
Jindai Botanical Garden Rose Festival, Tokyo, Japan
Kelâat M'Gouna Rose Festival, Morocco
Riverland Rose and Garden Festival, Renmark, Australia
Rose Festival, Chandigarh, India
Seoul Rose Festival, South Korea
Spring Rose Festival at Gifu World Rose Garden, Gifu, Japan
Taif Rose Festival, Taif, Saudi Arabia

FURTHER READING

Austin, David. *David Austin's English Roses*. Melton: ACC Art Books, 2012.

Beales, Peter. *Classic Roses*. London: Harvill Press, 1997.

Connolly, Shane. *A Year in Flowers: Inspiration for Everyday Living*. London: Clearview Books, 2022.

de la Haye, Amy. *Ravishing: The Rose in Fashion*. New Haven, CT: Yale University Press, 2020.

Elliott, Brent. *The Rose: The History of the World's Favourite Flower in 40 Roses*. London: Andre Deutsch, 2016.

Fearnley-Whittingstall, Jane. *Rose Gardens: Their History and Design*. London: Chatto & Windus, 1989.

Hall, Tony. *The Kew Gardener's Guide to Growing Roses*. London: Frances Lincoln, 2021.

Horwood, Catherine. *Rose*. London: Reaktion Books, 2018.

Jekyll, Gertrude, and Edward Mawley. *Roses for English Gardens*. London: Country Life, 1902.

Kukielski, Peter E., and Charles Phillips. *Rosa: The Story of the Rose*. New Haven, CT: Yale University Press, 2021.

Lack, H. Walter. *Redouté: Roses*. Cologne: Taschen, 2023.

MacOboy, Stirling. *The Ultimate Rose Book*. New York: H. N. Abrams, 1993.

Marriott, Michael. *RHS Roses: An Inspirational Guide to Choosing and Growing the Best Roses*. London: Dorling Kindersley Publishers, 2022.

Morley, Simon. *By Any Other Name: A Cultural History of the Rose*. London: Oneworld Publications, 2021.

Phaidon Editors. *Flower: Exploring the World in Bloom*. London: Phaidon Press, 2020.

Potter, Jennifer. *The Rose: A True History*. London: Atlantic Books, 2010.

Quest-Ritson, Charles, and Brigid Quest-Ritson. *The Royal Horticultural Society Encyclopedia of Roses*. London: Dorling Kindersley Publishers, 2011.

Taylor, Lindsey. *Art in Flower: Finding Inspiration in Art and Nature*. New York: Monacelli Press, 2023.

Thomas, Christopher. *In Search of Lost Roses*. New York: Summit Books, 1989.

ACKNOWLEDGMENTS

A project of this size requires the commitment and expertise of many people. Special thanks are due to the following for their knowledge, passion, and advice in the selection of the works for inclusion, as well as to Amy de la Haye, Victoria Gaiger, and Kristine Paulus for writing the thematic essays that open the book and Shane Connolly and Michael Marriott for writing the appendices that close it:

Shane Connolly
Floral designer, ambassador for sustainable floristry, and director of Shane Connolly & Co., London

Amy de la Haye
Dress historian, writer, curator, professor at London College of Fashion, and author of *Ravishing: The Rose in Fashion*, London and Sussex, UK

Brent Elliott
Former Librarian, then Historian, Royal Horticultural Society and author of *The Royal Horticultural Society 1804–2004* (2004) and *The Rose* (2016), London

Victoria Gaiger
Co-editor and creative director of *rakesprogress* magazine and *rakesSENSE*

Hazel Gardiner
Botanical artist, floral designer, and broadcaster, London

Marc Jeanson, PhD
Writer and botanist at the National Herbarium at the Muséum National d'Histoire Naturelle, Paris

Michael Marriott
Former head rosarian at David Austin Roses, chairman of the Historic Roses Group, and author of *RHS Roses* (2022)

Kristine Paulus
Collection Development Librarian, LuEsther T. Mertz Library, New York Botanical Garden

Anna Pavord
Gardener and author of *The Tulip* (1999), *The Naming of Names* (2005), and *The Seasonal Gardener* (2022), Dorset, UK

We are particularly grateful to Rosie Pickles for researching and compiling the long list of entries for inclusion. Additional thanks are due to Sarah Bell for her picture research, and to Caitlin Arnell Argles, Theresa Bebbington, Vanessa Bird, Lynne Ciccaglione, Rosanna Fairhead, João Mota, Hope Nicholson, and Elizabeth O'Rourke for their invaluable assistance.

Finally, we would like to thank all the artists, illustrators, photographers, collectors, libraries, institutions, and museums who have given us permission to include their images.

TEXT CREDITS

The publisher is grateful to the following writers for the individual entry texts:

Giovanni Aloi: 55, 173, 179, 182, 205; Matthew Biggs: 11, 42, 74, 139, 181; Shane Connolly: 113; Tim Cooke: 65, 116, 126, 166, 185, 189, 193, 199, 200, 211, 222; Anita Croy: 17, 25, 26, 30, 31, 37, 38, 45, 49, 50, 52, 59, 69, 77, 81, 82, 89, 93, 98, 115, 129, 132, 135, 136, 138, 141, 148, 149, 158, 169, 188, 190, 194, 197, 201, 202, 204, 213, 220, 225, 227, 230, 235, 236; Amy de la Haye: 61, 71, 75, 103, 104, 109, 150, 151, 154, 175, 195, 219, 237; Ellen Mara De Wachter: 28, 86, 99, 102, 128, 145, 159, 171, 186, 216; Brent Elliott: 13, 16, 40, 62, 64, 66, 100, 156, 157, 196, 212, 234; Celia Fisher: 10, 33, 60, 68, 76, 87, 91, 110, 114, 130, 147, 164, 165, 174, 206; Diane Fortenberry: 35, 44, 46, 54, 97, 131, 162, 172, 191, 208, 209; Tom Furness: 118; Victoria Gaiger: 27, 39, 47; Rob Kesseler: 67; Michael Marriott: 48, 51, 53, 72, 80, 84, 88, 90, 105, 120, 210; Marta McDowell: 22, 32, 73, 83, 85, 124, 127, 142, 143, 183, 207, 221, 229; Mairi MacKenzie: 36, 146, 163, 218, 228; Colleen Morris: 70, 108, 125, 140, 160, 167, 180, 192, 217, 231; Kristine Paulus: 24, 29, 34, 79, 106, 112, 117, 119, 122, 152, 184, 203; Rebecca Roke: 133; Gill Saunders: 15, 21, 78, 92, 94, 111, 121, 137, 155, 161, 168, 170, 223, 232; James Smith: 215; David Trigg: 12, 14, 18, 19, 23, 41, 43, 57, 58, 63, 101, 107, 123, 144, 153, 176, 177, 178, 187, 198, 224, 226, 233.

PICTURE CREDITS

Every reasonable effort has been made to identify owners of copyright. Errors and omissions will be corrected in subsequent editions.

© Beverly Allen: 231; Altadena Historical Society: 187; Alte Pinakothek München: 36; Photo: Irini Arakas: 233; © Nobuyoshi Araki/Photo: Phillips Auctioneers Ltd: 186; Archive PL/Alamy Stock Photo: 82; Artelino: 108; © Art Gallery of New South Wales, Image © Art Gallery of New South Wales, 222.1975: 125; © Ashmolean Museum, University of Oxford/Bridgeman Images: 168; From *Flower Fairies of the Summer* by Cicely Mary Barker, © The Estate of Cicely Mary Barker, 1925: 89; Neil Baylis/Alamy Stock Photo: 178; Courtesy Romare Bearden Foundation, Inc./ © 2024 Romare Bearden Foundation/Licensed by VAGA at Artists Rights Society (ARS), NY / DACS, London 2024: 205; Cecil Beaton, *Vogue*, © Condé Nast: 74; Bibliotheek Wageningen Universiteit: 80; Bibliothèque nationale de France: 182, 193, 200, 206; Biodiversity Heritage Library: 73, 100, 112, 127, 136, 180, 196; © Bodleian Libraries, University of Oxford: 174; © The Guy Bourdin Estate 2024, courtesy Louise Alexander Gallery: 17; Bridgeman Images: 137, 234; British Library/Alamy Stock Photo: 165; From the British Library archive/Bridgeman Images: 160, 221; © The Trustees of the British Museum: 60, 79, 128; © Brooklyn Museum/Museum Expedition 1941, Frank L. Babbott Fund/Bridgeman Images: 50; Thom Browne: 227; Courtesy Galerie Buchholz, © VG Bild-Kunst, Bonn/ © DACS 2024, Photo: Timothy Schenck: 58; Jonathan Buckley: 105; Buffalo Bill Center of the West, Cody, Wyoming; McCracken Research Library: 203; Bukowskis: 235; Nils Herrmann, Cartier Collection © Cartier: 199; Courtesy Chisholm Larsson Gallery: 106; Chopard: 39; © Christie's Images/Bridgeman Images: 26, 70, 167, 188; Chronicle/Alamy Stock Photo: 143; © Gillian Condy, South Africa National Biodiversity Institute: 236; Cooper Hewitt, Smithsonian Design Museum, Gift of Isabel Shults: 181; Corbis Historical/Getty Images: 204; © Michael Craig-Martin, courtesy Cristea Roberts Gallery, London: 232; Courtesy Phoebe Cummings: 195; © Salvador Dalí, Fundació Gala-Salvador Dalí, DACS 2024/ © Christie's Images/Bridgeman: 87; Courtesy Amy de la Haye: 175; Patrick Demarchelier/ Trunk Archive: 237; Bouke de Vries: 115; Dinodia Photos/Alamy Stock Photo: 191; © Christian Dior Couture: 189; Xuebing Du: 171; Pari Dukovic/Trunk Archive: 47; © Eli Durst: 211; © The Easton Foundation/VAGA at ARS, NY and DACS, London 2024 / Digital image, The Museum of Modern Art, New York/Scala, Florence: 117; © Awol Erizku: 132; © Cerith Wyn Evans, courtesy Galerie Buchholz: 18; © Estate of Ian Hamilton Finlay: 222; © The Museum at FIT: 49, 150; © Fitzwilliam Museum/Bridgeman Images: 88, 212; © Stephen Flavin/Artists Rights Society, New York an DACS, London 2024, Courtesy David Zwirner, Digital image, The Museum of Modern Art, New York/Scala, Florence: 123; © Florilegius/Bridgeman Images: 210; Courtesy Kate Friend: 216; The J. Paul Getty Museum, Los Angeles, Ms. 101, fol. 63v, 2008.3.63v: 130; T.M. Glass: 63; Simone Gooch and Derek Henderson/ Melanie Roger Gallery, New Zealand, and Michael Reid Gallery, Australia: 11; Courtesy Robin Ketelle and Dale Grahn/Estate of August Grahn: 197; Sourabh Gupta: 43; © Wilhem Hagemann Verlag: 103; Cig Harvey: 145; Rémy Hétreau & Lanvin, Courtesy the Ringling Art Library: 61; © The Hunterian, University of Glasgow: 38; © Imperial War Museum (Art.IWM PST 10996): 166; Italian Lighting Centre: 213; Joseph Bellows Gallery: 66; Josse/Scala, Florence: 111; © Jean Jullien: 23; © Alex Katz/VAGA at ARS, NY and DACS, London 2024 and © Christie's Images, London/Scala, Florence: 20-21; © Ellsworth Kelly Foundation, Courtesy Matthew Marks Gallery, Digital image Whitney Museum of American Art/Licensed by Scala: 15; Nick Knight/Trunk Archive: 31; © Estate of Shiro Kuramata: 133; Courtesy the artist, Ota Fine Arts and Victoria Miro, © YAYOI KUSAMA/Photo Sotheby's: 119; Digital Image Museum Associates/LACMA/ Art Resource NY/ Scala, Florence: 35; Courtesy Des Lawrence, image by Christies: 77; © Leeds Museums and Galleries, UK/Bridgeman Images: 228; Leemage/ Corbis/Getty Images: 41; © Simone Leigh, Courtesy Matthew Marks Gallery: 149; Library of Congress Prints and Photographs Division Washington, DC: 85; © TfL from the London Transport Museum collection: 161; Lyon & Turnbull/Bridgeman Images: 94; © Robert Mapplethorpe Foundation, used by permission: 146; BARBIE® and associated trademarks and trade dress are owned by, and used under license from, Mattel. ©2024 Mattel: 229; © Association Willy Maywald/ ADAGP, Paris and DACS, London 2024 and © Christian Dior Couture: 37; Russ McClintock: 153; © 2000–2024 The Metropolitan Museum of Art. All rights reserved: 29, 30, 54, 72, 96-97, 104, 118; © The Metropolitan Museum of Art/Art Resource/Scala, Florence: 218; © Minneapolis Institute of Art, Bequest of Richard P. Gale: 230; Chris Moore/Catwalking/Getty Images: 219; The Morgan Library: 124; Artwork by Stanley Mouse and Alton Kelley. © 1966, 1984. Rhino Entertainment Company. Used with permission. All rights reserved: 34; München, Bayerische Staatsbibliothek: 156; Macoto Murayama: 67; Musée Marmottan Monet: 169; © 2024 Museum of Fine Arts, Boston. All rights reserved/Bridgeman Images: 46, 95, 192; Digital image, The Museum of Modern Art, New York/ Scala, Florence: 28, 32, 92, 98; Museum of New Zealand Te Papa Tongarewa: 176; Photo: Julianne Nash: 201; The National Gallery/Bequeathed by Mrs M.J. Yates, 1923: 75; Purchased with the aid of The National Lottery Heritage Fund, Art Fund, the Scottish Executive, the Bank of Scotland, the Royal Bank of Scotland, Sir Tom Farmer, the Dunard Fund, Mr and Mrs Kenneth Woodcock (donation made through the American Friends of the National Galleries of Scotland) and private donations, 1999: 91; © National Museums Liverpool/Bridgeman Images: 33; © National Trust Images/Jonathan Buckley: 110; Trustees of the Natural History Museum, London: 48; Naturalis Biodiversity Center: 84; © The Estate of Alice Neel, courtesy The Estate of Alice Neel and David Zwirner: 52; The New York Botanical Garden, C. V. Starr Virtual Herbarium: 93; The New York Botanical Garden, LuEsther T. Mertz Library: 152; New York Public Library: 151; Courtesy NEXT architects: 198; Clive Nichols: 53; © Christoph Niemann, 2024: 225; Courtesy Norton Museum of Art, West Palm Beach, Florida, Gift, from the collection of Juan Rodriguez, 2014.100: 183; Eunike Nugroho: 139; © Georgia O'Keeffe Museum, Santa Fe/Art Resource/ Scala, Florence / © Georgia O'Keeffe Museum / DACS 2024: 147; The Old Design Shop: 121; Old World Auctions: 220; G. Dagli Orti/ © NPL - DeA Picture Library/Bridgeman Images: 162; © Jennifer Packer, courtesy of Sikkema Jenkins & Co., New York, Corvi-Mora, London: 19; Paramount/Kobal/Shutterstock: 101; © Martin Parr/Magnum Photos: 107; Rachel Pedder-Smith: 134-35; © The Irving Penn Foundation: 99; Paris Musées/Petit Palais, Musée des Beaux-Arts de la Ville de Paris: 25; Emilija Petrauskiene: 102; Peter Probst/Alamy Stock Photo: 159; Protolabs: 144; Prudence Cuming Associates: 223; Courtesy The Red Dot Gallery: 138; Retro AdArchives/Alamy Stock Photo: 202; RHS Lindley Collections: 51, 62, 76, 157, 190; RMN-Grand Palais/Jean-Gilles Berizzi/ Dist. Photo Scala, Florence: 173; RMN-Grand Palais/Hervé Lewandowski/RMN-GP/Dist. Foto SCALA, Florence: 114; RMN-Grand Palais/Thierry Olivier/RMN-GP/Dist. Foto Scala, Florence: 68; RMN-Grand Palais/Franck Raux/RMN-GP/Dist. Foto SCALA, Florence: 194; Photo: Shaun Roberts: 141; Gift of Charles Rosenbloom/ Bridgeman Images: 148; © Royal Botanic Gardens, Kew: 217; © Russell-Cotes Art Gallery/Supported by the National Art Collections Fund/Bridgeman Images: 55; Will Ryman Productions/Artist Rights Society (ARS), New York, Photo courtesy of Kasmin, New York / © ARS, NY and DACS, London 2024: 179; Sabancı University Sakıp Sabancı Museum Collection: 172; Natasja Sadi: 177; The Little Prince in the rose garden, from *The Little Prince* (1943), written and illustrated by Antoine de Saint-Exupéry. Reproduced by courtesy of the Saint Exupéry-d'Agay Estate: 83; Courtesy Rosie Sanders/Photo: Paul Biddle: 45; Science & Society Picture Library/Getty Images: 122; Science Photo Library/Alamy Stock Photo: 116; Shirley Sherwood Collection: 42; David Shrigley: 78; Gleb Simonov: 208; Courtesy Shinique Smith and Lower EastSide Printshop, New York: 226; Smithsonian American Art Museum, Gift of Iris and Jay Leonard and Kohler Foundation, Inc: 207; Division of Medicine and Science, National Museum of American History, Smithsonian Institution: 163, 185; Courtesy Smithsonian Libraries and Archives: 22; Smithsonian National Museum of African American History and Culture, Gift of the Black Fashion Museum founded by Lois K. Alexander-Lane: 154; Courtesy Sotheby's: 13, 59; © Sotheby's/Bridgeman Images: 81; Staatsbibliothek zu Berlin, Preussischer Kulturbesitz, Orientabteilung: 44; Städel Museum: 24; The Stapleton Collection/ Bridgeman Images: 64; Luke Stephenson: 14; © 2023 Estate of Alma Thomas/ARS, New York and DACS, London 2024/ © The Metropolitan Museum of Art/Art Resource/Scala, Florence: 224; Courtesy Wolfgang Tillmans, David Zwirner, New York/Hong Kong, Galerie Buchholz, Berlin/Cologne, and Maureen Paley, London: 86; TimeLine Auctions Ltd: 209; © Ben Toms: 158; © Cy Twombly: 214-15; Courtesy UK Parliament, WOA 2593: 164; © Victoria and Albert Museum, London: 71, 126, 170, 184; Walters Art Museum: 65; © Bruce Weber: 27; Wikimedia Commons: 10, 12, 16, 69, 90, 109, 140, 155; Julie Anne Workman: 131.